Fundamentals of

DSL

Technology

Fundamentals of

DSL

Technology

Edited by
Philip Golden
Hervé Dedieu
Krista Jacobsen

Auerbach Publications
Taylor & Francis Group
Boca Raton   New York

Cover designer: Chris Massey

Published in 2006 by
Auerbach Publications
Taylor & Francis Group
6000 Broken Sound Parkway NW, Suite 300
Boca Raton, FL 33487-2742

International Standard Book Number-10: 0-8493-1913-7 (Hardcover)
International Standard Book Number-13: 978-0-8493-1913-6 (Hardcover)
Library of Congress Card Number 2004062330

Library of Congress Cataloging-in-Publication Data

Fundamentals of DSL technology / editors, Philip Golden, Herve Dedieu, Krista Jacobsen.
   p. cm.
  Includes bibliographical references and index.
  ISBN 0-8493-1913-7 (alk. paper)
  1. Digital subscriber lines. I. Golden, Philip. II. Dedieu, Hervé. III. Jacobsen, Krista.

TK5103.78.F86 2005
621.382--dc22   2004062330

Taylor & Francis Group
is the Academic Division of T&F Informa plc.

**Visit the Taylor & Francis Web site at
http://www.taylorandfrancis.com**

**and the Auerbach Publications Web site at
http://www.auerbach-publications.com**

Dedications

Phil Golden dedicates this book to his Mum, Caroline, and Hannah, the three most important

women in his life.

Hervé Dedieu dedicates this book to Claire.

Krista S. Jacobsen dedicates this book to John Cioffi, who arguably started it all.

Acknowledgments

The editorial team is grateful for the participation of so many skilled engineers in the creation of the *Fundamentals of DSL Technology*. The quality of this volume is testimonial to the talent and dedication of its chapter authors, and the editors would like to thank the authors for their outstanding contributions. The high quality of this volume is also due in part to the efforts of the excellent team of reviewers. The editors were fortunate to gather a team of DSL experts whose careful reviews of the material helped to ensure technical accuracy and clarity. The editors would like to thank the reviewers for their role in crafting this volume: Abdelaziz Amraoui, Rodolfo Ceruti, Jim Eyres, Orla Feely, Olivier Grenie, Conor Heneghan, Fred Howett, Ragnar Jonsson, Rob Kirkby, Jae-Chon Lee, Geert Leus, Simon Litsyn, Hannah Massey, Cory Modlin, Tomas Nordström, Vladimir Oksman, Sigurd Schelstraete, Gary Tennyson, Michail Tsatsanis, Jaap van der Beek, Rob van den Brink, and Katie Wilson.

Phil Golden would like to thank both Krista and Hervé for their excellent work. In addition he would like to formally recognize the role that Hervé Dedieu has played as his mentor during his time at LEA. Outside of LEA, both John Cook of BT and Peter Reusens, formerly of Alcatel, have been exceptionally instructive in helping Phil to understand the mysteries of splitters.

Hervé Dedieu would like to thank the talented engineers of LEA with whom he had the privilege to work. He expresses his gratitude to Phil Golden and Guy Nallatamby for their constant support and team spirit. In the success of LEA as a start-up company, the involvement of Phil as the "senior circuit-designer" who represented LEA in various ETSI and ANSI groups was a key point. This book would not have been possible without his work within the different DSL standardization committees and without his talent to gather a team of people who made this book happen.

Krista S. Jacobsen would like to thank Phil and Hervé for inviting her to help create what she believes is the ultimate industry reference on DSL. She would also like to thank her colleagues from Amati and Texas Instruments and the ITU, ETSI, and T1E1.4 DSL standards crowds for creating stimulating (and sometimes frustrating) environments in which to learn about DSL. Finally, she would like to thank Professor John M. Cioffi, who believed in Krista when she didn't believe in herself, and whose teachings and support have opened many doors.

Foreword

DSL's worldwide coronation of broadband Internet service motivates this comprehensive and timely encyclopedia of DSL transmission fundamentals. Rising star, author, and editor Philip Golden joins industry veterans Hervé Dedieu and Krista S. Jacobsen to collect an outstanding set of expert contributors who press the advances of DSL in two volumes — this one and the upcoming *Implementation and Applications of xDSL Technology* — detailing all aspects of DSL modem design. Readers and students of these volumes indeed hold the keys to all aspects of DSL design in their library. One envisions designers and students around the world depending on the enormous information within. This first volume lays a foundation, addressing the basics of DSL.

There have been many texts on DSL in the past few years, including the heavily referenced two-book set by Starr, Sorbara, Cioffi, and Silverman [4] [5], Walter Chen's earliest text [2], John Bingham's final classic [1] before retiring, and Dennis Rauschmayer's pragmatic view [3]. However, none has been as comprehensive as this ambitious collection's realization. The DSL area has been expanding rapidly, making an address by a single author or small group of authors difficult. This text addresses the overall need by combining the strengths of the world's most renowned DSL experts.

A good transmission engineer's first rule is "know your channel," and this text reinforces that notion: a caboodle of Europe's finest DSL engineers adorns an introductory three-chapter examination of DSL copper transmission channels in this first volume. Golden, an internationally acknowledged expert on splitter circuits, joins DSL's premier analog expert, BT engineer John Cook, to overview the basic telephone environment in the first chapter. Fundamentally, strong transmission in DSL depends on a good understanding of the physical-layer twisted pair, provided in Chapter 2 by renowned analog expert Hervé Dedieu. Such good transmission depends also on good noise models, which are provided in the third chapter by another BT expert, Rob Kirkby.

Modulation and equalization expert Ragnar Jonsson of Conexant introduces all line codes in Chapter 4, to set up later chapters that provide greater depth in and understanding of the various transmission methods. Professor Edward Jones of University College Galway, Ireland, relates in Chapter 5 the materialization of the basic objectives in DSL service and compares DSL to other methods of broadband delivery.

Seven succeeding chapters describe the basics of DSL transmission to complete this first volume. The "transmission fest" begins in Chapter 6, which provides an excellent review of single-carrier modulation by highly mobile author Vladimir Oksman of Tadiran, Lucent, Broadcom, and currently Infineon. A professor could not be more proud than to see the fine work in this volume by two former outstanding students, the first of whom is Krista S. Jacobsen, formerly of Texas Instruments, internationally recognized for her exceptional understanding of the discrete multi-tone (DMT) technology. Her Chapter 7 provides a comprehensive treatment of DMT transmission methods that should allow many to understand this simple but high-performing ubiquitous DSL technique, which, following its selection for both ADSL and VDSL, has clearly become the industry favorite. Krista is followed by Broadcom's coding-world superstar Gottfried Ungerboeck, who details his internationally acclaimed trellis codes for DSL in Chapter 8. A second former student, Texas Instruments' frame-format and coding pioneer Cory Modlin, then follows in Chapter 9 with a detailed investigation of the Reed–Solomon codes and interleaving used in almost all DSL systems. Turbo and LDPC codes may find their way into use in DSL's future, and IBM Zurich Research Lab's Evangelos Eleftheriou and Sedat Ölçer are the premier experts on the possibilities, which appear in Chapter 10. Iceland's Ragnar Jonsson returns in Chapter 11 to review basic equalization theory for use in DSL.

Initially noted but under-appreciated, radio-frequency (RF) interference is a major source of performance loss at high speeds in some DSLs. A quartet of Swedish–Bavarian authors — Rickard Nilsson, Thomas Magesacher, Steffen Trautmann, and Tomas Nordström — provide a valuable investigation of RF issues and means for reducing or suppressing RF interference in DSL.

Having completed this first basic volume, excitement builds in anticipation of the ensuing volume that addresses the methods for expansion and growth of DSL's success.

John Cioffi
Stanford, California

References

1. J.A.C. Bingham. *ADSL, VDSL and Multi-Carrier Modulation*. Wiley-Interscience, New York, NY, 2000.
2. W.Y. Chen. *DSL: Simulation Techniques and Standards Development for Digital Subscriber Lines*. Macmillan, New York, 1998.
3. D. Rauschmayer. *ADSL/VDSL Principles: A Practical and Precise Study of Asymmetric Digital Subscriber Lines and Very High Speed Digital Subscriber Lines*. Macmillan Technical Publishing, 1998.
4. T. Starr, J.M. Cioffi, and P.J. Silverman. *Understanding Digital Subscriber Line Technology*. Prentice-Hall, Upper Saddle River, NJ, 1999.
5. T. Starr, M. Sorbara, J.M. Cioffi, and P.J. Silverman. *DSL Advances*. Prentice-Hall, Upper Saddle River, NJ, 2002.

Contents

1   Overview of the POTS Environment — Signals and Circuits 1
     Philip Golden and John Cook

2   The Copper Channel — Loop Characteristics and Models 33
     Hervé Dedieu

3   Noise and Noise Modelling on the Twisted Pair Channel 71
     Rob H. Kirkby

4   The Twisted Pair Channel — Models and Channel Capacity 97
     Ragnar Hlynur Jonsson

5   Introduction to DSL 119
     Edward Jones

6   Fundamentals of Single-Carrier Modulation 143
     Vladimir Oksman

7   Fundamentals of Multi-Carrier Modulation 181
     Krista S. Jacobsen

8   Trellis-Coded Modulation in DSL Systems 211
     Gottfried Ungerboeck

9   Error Control Coding in DSL Systems 233
     Cory S. Modlin

10   Advanced Coding Techniques for Digital Subscriber Lines 271
     Evangelos Eleftheriou and Sedat Ölçer

11   DSL Channel Equalization 299
     Ragnar Hlynur Jonsson

12   Synchronization of DSL Modems 351
     Sverrir Olafsson

13   Radio-Frequency Interference Suppression in DSL 399
     Rickard Nilsson, Thomas Magesacher, Steffen Trautmann,
     and Tomas Nordström

Index 451

1

Overview of the POTS Environment—Signals and Circuits

Philip Golden and John Cook

CONTENTS

1.1 How the Telephony System Is Typically Constructed 2
    1.1.1 Network Structure 3
    1.1.2 Local Exchanges 4
        1.1.2.1 Line Interface Circuits 4
        1.1.2.2 Main Distribution Frames 5
    1.1.3 Cables 5
        1.1.3.1 Dropwires 7
    1.1.4 Network Demarcation Points 7
    1.1.5 Customer Premises Wiring 9
        1.1.5.1 Bus Topology 9
        1.1.5.2 Tree-and-Branch Topology 10
        1.1.5.3 Impedance Presented by Customer Wiring 10
        1.1.5.4 Terminal Equipment Impedance 10
        1.1.5.5 Terminal Equipment State 11
        1.1.5.6 "On-Hook" State 11
        1.1.5.7 "Off-Hook" State 11
1.2 Speech Signals 12
1.3 Hybrid Circuits 14
    1.3.1 Two-Wire Transmission 14
    1.3.2 Two-Wire to Four-Wire Conversion 15
    1.3.3 Conceptual Hybrid Circuit 15
    1.3.4 Choice of Terminating Impedance 16
    1.3.5 Choice of Balance Impedance 16
    1.3.6 Audible Feedback 17
    1.3.7 Gains in the Hybrids 17
1.4 DC Signalling 17
    1.4.1 The Local Exchange Battery 18
    1.4.2 Resistive Feeding from the Local Exchange 18
    1.4.3 Programmable DC Feeding 19
    1.4.4 Resistance of the Local Loop 19
    1.4.5 Resistance of the Terminal Equipment 19
    1.4.6 Additional Uses of DC Signalling 20

1.5   Wetting Current 20
    1.5.1   Corrosion of Cable Joints   20
    1.5.2   Use of Wetting Current 21
    1.5.3   Potential Effect of Wetting Current on DSL 21
1.6   Ringing 21
1.7   Ring Trip 22
1.8   On- or Off-Hook Detection 22
1.9   Dialing 23
    1.9.1   Pulse (Loop Disconnect) Dialing   23
        1.9.1.1   High Voltage Transients Due to Pulse Dialing   24
    1.9.2   Tone Dialing 25
1.10 Subscriber Private Metering 26
1.11 Telephony Speech Coding 26
1.12 Balance about Earth 27
1.13 Testing   27
1.14 Overload 28
1.15 High-Speed Voiceband Modems   29
1.16 CLASS Signalling 29
References   30

ABSTRACT   One of the principal advantages of DSL technology is the use of an existing physical communications infrastructure, namely, the telephone network. This feature, coupled with the fact that the majority of DSL deployments to date co-exist with telephony services on the same line, means that an understanding of the telephony environment is of key importance for the successful design and deployment of DSL technology. This chapter describes the telephony environment, focussing on aspects that have particular relevance to DSL performance.

1.1   How the Telephony System Is Typically Constructed

Although the origin of telephony lies in the year 1876 with Alexander Graham Bell's invention of the telephone, it was in the succeeding few years that the telephony network began to take shape. Telephony was transformed from being a short distance point-to-point service to today's ubiquitous network that enables almost instant communication over large distances. The beginning of this transformation was the development of local exchanges, physical "hubs" that terminate multiple subscriber lines. These are "nodes" of the public switched telephone network (PSTN). In recent years the PSTN has become known as the plain old telephone service (POTS) network. Despite being rather crude, this acronym is now prevalent in technical literature. As well as enabling interconnection between all users of a particular exchange, the development of backbone trunk systems that linked various exchanges enabled first inter-urban, then international, connection.

The original backbone trunks were made up of traditional telephony cables, but nowadays this portion of the telephony network can be made of optical fiber or even satellite links. The part of the network on the other side of the exchange, however, *i.e.*, the subscriber lines leading to each end customer, are to this day for the most part still made up of traditional copper (or aluminium) cables. It is this section of the telephony network, often referred to as the "local loop" or the "last mile," that is of primary interest for DSL technology. A representation of the local loop is given in Figure 1.1.

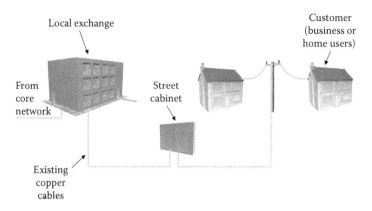

FIGURE 1.1
The local loop.

It should be noted that it is often of interest to minimize the length of the local loop, *i.e.*, effectively decrease the distance between the subscribers and the local exchange. For this reason "remote exchanges" are sometimes used, where the traditional central exchange is connected to a smaller exchange via a high-speed link (which can be either telephony cable or optical fiber). This is generally present in areas where there is significant demographic dispersion in order to minimize the distances between the subscriber and the exchange. In the United States, around 15 percent of telephony subscribers are served via remote multiplexers known as digital loop carriers (DLC). As a general trend, optical fiber is penetrating farther and farther into traditional telephony networks, thus shortening the length of the local loop. Nevertheless, primarily for economic reasons, it is certain that a significant percentage of the telephony network in most countries will remain (at least in the short-to-medium term) as twisted pair cable between the subscriber and the exchange. In particular from the perspective of DSL, one can typically consider the transmission path of most interest as being from the exchange to the subscriber along the local loop and vice versa. Some discussion on the structure of the network shall be given in the text that follows, as well as consideration of each element in the DSL transmission path.

1.1.1 Network Structure

Local exchanges are normally sited according to a transmission standard, typically a national standard that is used to ensure the efficiency of a defined class of telephony connections. Various factors come into play in the formulation of these standards, and for the purposes of this text it is assumed that exchange locations are chosen in order to minimize the length of the local loops. The interested reader is referred to Chapter 7 of [Richards 1973], where a comprehensive discussion of the development of a telephony transmission standard for the United Kingdom is given.

The largest local telephony exchanges are typically found in urban areas, where a significant number of subscribers are located within a relatively small geographical area. It is common for large suburban districts also to have dedicated exchanges; typically these will be of smaller size than those situated in urban areas. In pursuit of ubiquity for the PSTN, hampered by the maximum length of the local loop generally set by signal attenuation, rural customers are typically serviced by even smaller exchanges. The concept of exchange size here is taken to be directly related to the number of subscribers serviced by the exchange.

The original transmission standards were focused on providing adequate telephony service to an existing customer base. More recent network planning must also take into account the fact that the subscriber network will change in form over time; one obvious example of this is the development of large housing estates in previously nonresidential areas. For practical reasons, it is clearly preferable for a residence to be served with a telephone line as soon as it is ready for habitation. This in turn means that network planners have to make some estimations on future developments of the subscriber network, based on a number of different factors. Due to the fact that these estimations are rarely 100 percent accurate, the routing of the deployed telephony cables does not necessarily closely correspond to the optimal network.

1.1.2   Local Exchanges

Local exchanges have gone through repeated technological revolutions since the inception of telephony. In the early years, exchanges relied entirely on manual operators who connected themselves to customers calling attention, asked who they wanted to be connected to, and then provided the appropriate patch cord to make the connection. It was quickly realized that the growth in telecommunications could not sustain this process, and some means of automation would need to be found. The Strowger (q.v.) system was the first of these; it relied on sophisticated electromechanical devices known as selectors for its operation. These devices were operated by pulses sent from a dial attached to the telephone. An appropriate outlet of the selector was selected according to the number of pulses from the dialled digit. Successive ranks of selectors dealt with successive digits of the telephone number in order to select the called customer. In fact, in the United Kingdom some of this technology was still in service until only a decade or so ago, and no doubt it still serves in some places in the world.

Successive generations of electromechanical exchanges tended to reduce the reliance on mechanical functions and increase the reliance on electrical and electronic functions. The end of this line of development was probably exchanges using reed relays for the switching elements and electronics for nearly everything else. At this stage it became possible to make use of tone dialing (DTMF), which dramatically reduced call connection times. Also at this stage, the reliability of the exchanges increased dramatically. There were some more or less abortive attempts to make fully electronic analog telephone exchanges. However, it was not until the era of digitalization in the late 1970s that essentially fully electronic exchanges were successfully in use.

1.1.2.1   Line Interface Circuits

In modern digital exchanges, an electronic line interface circuit is connected to every telephone line. This feeds it with power (typically −50 V, current-limited to 30–40 mA, see Section 1.4), detects signalling (on- or off-hook, etc.), and passes this information to a control system. It also converts the sent and received speech signals to and from 64 kbit/s coded data streams and separates the incoming and outgoing analog speech signals. This is all achieved while maintaining a high level of balance on the wires making up the telephone connection, providing rejection of common-mode interference signals. The circuit must also be able to apply ringing signals (typically 75–100 Vrms at 16–50 Hz) and offer some means of testing itself and the customer's line. The rest of the exchange is then essentially an extended computer for processing and switching the signalling and speech information from these line interface circuits. The techniques used in line interface circuits are described in more detail later on in this chapter, with particular emphasis on the relevance for DSL systems. Most telephony exchanges in service today are still constructed using this technique. There is probably another revolution due in which the exchange switching will be replaced by packetized Internet protocol (IP) data rather than the traditional synchronous transmission

FIGURE 1.2
Subscriber side of main distribution frame (MDF).

systems. As and when this happens, the interfaces may become distributed, with the data passing along the access network lines as bits in part of a DSL data stream.

1.1.2.2 Main Distribution Frames

A single local exchange can service thousands of subscribers. From a practical perspective, this means that cabling can be a challenge. In order to facilitate the distribution of cables in a local exchange, a main distribution frame (MDF) is used. The primary function of the MDF is to allow facile physical access to each of the subscriber lines, in order that the appropriate connections can be made. Network cables are terminated on one side of the MDF, and cables from various exchange[1] equipment are terminated on the other side. The length of cable between the MDF and the terminal equipment can cause issues with broadband services such as DSL; nevertheless, from the perspective of the telephony service, this length of cable does not typically impair operation.[2] Single-pair jumpers are threaded through from the one side to the other to provide appropriate requested services to each customer pair on demand. On the customer side, protection modules are often used to provide both over-voltage and over-current protection.[3] A typical MDF is shown in Figure 1.2.[4] In some newer deployments of ADSL, the MDF can also incorporate the splitter function.

1.1.3 Cables

On the local loop side of the MDF, each cable that radiates out will typically contain from one hundred to five thousand pairs. Farther down the local loop toward the subscriber, each of these cables will typically "separate" into multiple smaller cables that are routed to reflect the geographical distribution of the appropriate subscribers. The wiring junction at which the larger cable divides is known as a flexibility point. The first of these flexibility points is often known as the primary connection point (PCP), and any further flexibility

---

[1] In some unbundled environments some of the terminal equipment may be located in a nearby building.
[2] Although this cable does not typically affect the telephony performance, it may cause other issues, such as with some line testing procedures, as described in Volume 2.
[3] In fact, before the use of integrated protector modules, small open fuses known as "heat coils" were used. To this day, local exchange technicians often refer to the ubiquitous 5-pin protector module as a "heat coil."
[4] Used with permission from "DSL Comes to Munising" by Jonathan Gennick, 30 May 2001 (http://www.oreilly.com/news/dsl_0501.html).

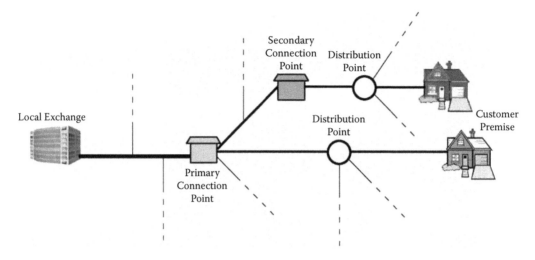

FIGURE 1.3
Flexibility points on typical subscriber lines.

points are known as secondary connection points (SCPs). Any one subscriber line can pass through multiple flexibility points, as shown in Figure 1.3. These provide connection to smaller distribution cables, which may also be made up of cable sections jointed together. A final flexibility point near the customer, the distribution point (DP), allows connection to dropwires to the customers. The DP can be physically either internal or external to a premises, and additionally can be either underground or overhead. At any of these flexibility points, or indeed at any point along the subscriber line, there are potentially cable joints. These are change points used to provide continuation of cable lengths. As previously mentioned, the normal cable used in telephony networks is in the form of two insulated wires twisted uniformly together to form a balanced pair. By twisting four insulated wires together, sometimes using a center string or filament to improve the mechanical uniformity, a quad is formed (see Figure 1.4). The opposite wires of the cross-section are used as the

FIGURE 1.4
Physical structure of a quad.

two elements to form a pair, and hence the overall diameter is smaller than the diameter of two individual pairs twisted together. Consequently, the capacitance and attenuation of the pairs for the same diameter ratio are reduced. For this reason, quads were very much favored in the early days of telephone cables; however, as higher-frequency transmission is used (e.g., DSL), it becomes extremely difficult to control the crosstalk between the pairs within the quad. Hence, quad cables are not used to such a large extent today.

1.1.3.1 Dropwires

Dropwires form the final connection between the DP and the customer when the DP is overhead (*i.e.*, typically at the top of a pole just outside the customer premises). Not all final connections from the DP are overhead; some may be underground, in which case they are generally constructed from cables similar in construction to main access network cables, but with fewer pairs (usually 1, 2, or 5). Even where most of the access network is underground, this last part is still frequently overhead. The upgrade of the distribution network to underground cabling often stops short of replacing overhead dropwires because this final connection is expensive to convert. This is mainly due to the fact that it necessitates an appointment with each and every connected customer. Dropwires are often very different from network cables because they have different requirements placed on them. In particular, they must be suspended above ground, possibly with quite long spans (70 meters or more). This requires a degree of structural strength not required in underground cables. Typically this strength comes from one of three approaches:

- The conductive members may be steel themselves, often copper clad to reduce resistivity and increase connection reliability.
- The conductive members may be made of an alloy, often an alloy of copper and cadmium, which has higher stiffness than pure copper, the members also being typically thicker to further increase strength.
- The dropwire may use conventional copper pairs but have separate steel strength members incorporated into the sheath with the pairs.

Further hazards faced by dropwires are exposure to the elements and close contact with, for example, tree branches and even electrical cables. These hazards typically require the dropwires to have thick protective insulation or sheathing. Copper clad or copper-cadmium dropwires typically are made up into single pair dropwires with thick "figure-of-eight" insulation. Dropwires with copper pairs and separate strength members have a thick overall sheath.

The presence of dropwires in the loop can have some implications for DSL systems, especially over the higher bandwidths, for example, those used by VDSL. This arises because the thick conductor sizes and particularly the thick insulation can result in the dropwires having very different characteristic impedances than conventional 0.5 mm twisted copper pairs. For example, one dropwire in common use in the United Kingdom has a characteristic impedance of 180 Ω at VDSL frequencies, compared to 100 Ω for typical copper pair cable. The mismatch between the two can cause extreme reflections with the apparent impedance of the whole loop fluctuating widely between 55 and 180 Ω.

1.1.4 Network Demarcation Points

In many countries, regulatory regimes have required a demarcation point between the operator part of the network and the customer premises part of the network. In most (but not all) regimes, the network operator is considered to own (and be responsible for) that part of the network on the network side of the demarcation point. The customer is considered to own (and be responsible for) that part of the network on the premises side. This demarcation point is variously named network interface (NI) in the United States and network terminating equipment (NTE) in Europe. The latter term is particularly confusing because it is also used for the customer modem in ISDN service situations. In the telephony case, this "equipment" is substantially empty; it is just a socket. Nevertheless, the term "NTE" is used in both situations, and in Europe the ISDN NTEs are generally operator owned and therefore part of the operator's network anyway.

For residential applications in the United States, the NI is typically located in the network interface device (NID) either at the boundary of the customer premises or mounted on the outside wall of the property. In medium-to-large business applications, there is no NID; however, there is still a network interface. If a NID is used, it houses the primary protector, the NI itself, and a terminal on which the inside wiring is terminated. In this case, the NI is comprised of a plug-and-jack arrangement (allowing the customer to "plug" in an instrument directly, in order to bypass the inside wire altogether). In Europe, the NTE is typically located just inside the customer premises. It consists of some form of socket (depending on the country).

From a regulatory perspective, the demarcation point is very important because it marks the end of the parts of the network the customer is responsible for and with which he can do anything he likes (add to, modify, destroy, connect anything he wishes), and which he must maintain so that if it develops faults, he must repair them at his expense. Equally, it marks the end of the network for the operator so that if provided service is functional at the demarcation point, the operator is fulfilling the contract with the customer. Although the regulatory concept is clear cut, the reality is rather less so, and recent developments in technology have muddied the concept somewhat. The essential problem is that electrical signals flow unhindered across the demarcation boundary, which causes a range of issues, including the following.

- The operator can add ADSL signals to the signals on the loop. The signal then appearing at the NID/NTE is not strictly a telephony signal anymore, because it requires the use of one or more splitters before it can be reliably connected to a telephone. In this case, the demarcation point for telephony needs to be moved to the customer's side of the splitter(s). Strictly, this requires the splitter to be installed, owned, and operated by the network operator, significantly adding to the expense of installation. In the ADSL self-install configuration with multiple filters, one at each extension socket on the customer premises, the situation is particularly confused from a demarcation point of view.[5]

- Home networking signals, such as those defined by the "Home Phoneline Networking Alliance" (HPNA), can be put onto the customer's premises wiring for the purposes of providing a "home LAN" using the premises' phone sockets. However, unless special precautions are taken, these signals can flow through the NID/NTE out onto the public network and may cause interference to network services (notably VDSL). The ownership of this issue is unclear. The International Telecommunications Union (ITU) has defined a filter [ITU-T G.989.3] that can be connected at the NID/NTE to prevent HPNA signals flowing out onto the public network. But there is no way to enforce the connection of these filters by the customer when the HPNA system is installed. Furthermore, to a large extent the correct operation of the HPNA system does not depend on the filter being present. Therefore, customers using HPNA may not be motivated to install the filter.

---

[5] In the United States, a centrally located ADSL splitter is treated as CPE. Both the voice and ADSL signals are considered to flow across the NI.

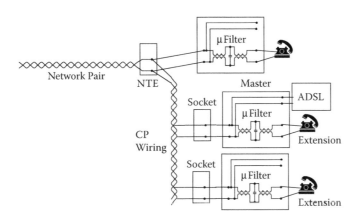

FIGURE 1.5
Distributed filter configuration.

- The customer could connect any signal source to the premise wiring. The signals can then flow out onto the public network, where they may cause radio frequency interference issues with licensed radio services. The ownership of this issue is unclear.

1.1.5   Customer Premises Wiring

The quality of customer premises wiring can best be described as variable, yet it can be of significant importance to the performance of DSL. For example, in the case of the ADSL self-install configuration with multiple filters (as referred to in Section 1.1.4), the ADSL signal can effectively "see" all of the in-house wiring,[6] as in Figure 1.5. In the case of commercial customers, the customer premises wiring may be professionally installed and of high quality; however, in the case of residential deployments, this is often not the case. The vast majority of residential wiring was not installed with high-frequency data services in mind. One of the most obvious issues with customer premises wiring is that of cable balance (see Chapter 3) in the frequencies used for DSL transmission, potentially giving rise to significant emissions. These emissions can in turn interfere both with in-premise transmission systems, and also with the access network on the other side of the demarcation point. A second problem is that in the absence of a master splitter, the impedance of the customer wiring can affect the DSL performance. Although the customer wiring is deregulated, national standards bodies often produce guidance documents such as [Brit. Stand. Inst.]. In general, one of two topologies is used: "bus" or "tree-and-branch." In either case, 24AWG twisted pair cable or flat untwisted pair cable is commonly used. A single installation can, of course, use different types of cable. Results of a customer wiring survey undertaken in the United Kingdom in the late 1990s are given in [Thorne 1998].

1.1.5.1   Bus Topology

Figure 1.6 illustrates the bus topology for customer premises wiring. A possible location of a DSL modem is shown. Typical lengths from the demarcation point to the first phone socket can be in the range of 15 m. The length of cable between each of the phone sockets

---

[6] DSL configurations that use a centralized "master splitter" at the customer premises effectively isolate the DSL service from the premise wiring; however, this configuration generally involves a visit from a qualified technician to the customer premises in order to install the splitter.

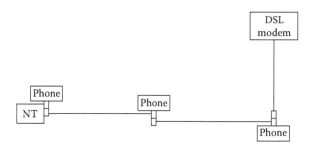

FIGURE 1.6
Bus topology.

is usually around 15 m (this is clearly highly dependent on the premises), and the total length of cable is recommended in [Brit. Stand. Inst.] to total less than 250 m for the bus configuration.

1.1.5.2   *Tree-and-Branch Topology*

The tree-and-branch topology is illustrated in Figure 1.7, again with a possible location of a DSL modem shown. For this topology, the maximum length between the most distant outlet and the demarcation point is recommended by [Brit. Stand. Inst.] not to exceed 50 m, and no more than 100 m of cabling is recommended to be used in total.

1.1.5.3   *Impedance Presented by Customer Wiring*

The impedance presented by the customer wiring is dependent on the actual configuration of the wiring used, the splitter configuration at the customer premises, and indeed the state of the relevant terminal equipment attached to the wiring. The last of these is expanded upon in the next subsection. Some interesting impedance measurements are given in [Thorne 1998].

1.1.5.4   *Terminal Equipment Impedance*

In terms of DSL system design, the impedance of terminal equipment (TE) connected in the telephony network is primarily of interest in the design of splitters. This is due to the fact that one of the primary roles of the splitter function is to isolate the DSL transmission from the telephony TE. In practical terms, these impedances heavily influence the design of the low-pass filter of the splitter. Many of the references in the following paragraphs are valid for European networks. The same principles will generally apply to all networks; however, specifics may vary. Telephony TE is also variously called POTS TE, or voiceband TE. The most common type of POTS TE is, of course, the traditional telephone; however, the general term also refers to voice-grade modems and also most facsimile machines.

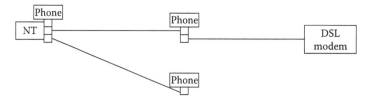

FIGURE 1.7
Tree and branch topology.

1.1.5.5 Terminal Equipment State

Various specification documents describe POTS TE states, with a wide range of terms being used. These can be highly confusing, and the interested reader may wish to consult [ETSI EN.300.001] for an explanation of some less widely used terms.

In very simple terms, the TE can be in one of two states: "on-hook" or "off-hook." This definition is considered satisfactory for documents such as [ANSI T1.421-2002]. In addition, there are transient periods between states.[7]

1.1.5.6 "On-Hook" State

This is the state in which the TE draws insufficient DC current to activate the exchange. It is also known as the "idle state," "offline state," or the "quiescent state" [ETSI TR 101 182]. A further two substates exist within the on-hook state:

- Ringing state: idle state into which a ringing signal is applied [ETSI TR 101 182].
- Idle line signalling state: the state into which a TE, when connected to the network, is placed such that it is capable of receiving or sending speech-band signalling without entering the loop state (see below) [ETSI TR 101 182].

Typically, the steady-state DC resistance of an on-hook TE will be of the order of at least 1 MΩ [ETSI TBR 021], hence limiting the amount of DC current present when the exchange is not "active." An exception to this can occur in certain networks in the case of a TE used to provide display information based on DTMF signalling. In this case, the DC impedance can be in the range of 100 kΩ [ETSI ES 200 778]. The AC impedance[8] of an on-hook TE is typically only defined for the ringing state and the idle line signalling state, and even then only in the frequency band used for this signalling. For the ringing state, the impedance at 25 Hz and 50 Hz is usually greater than 4 kΩ [ETSI TBR 021]. Transient behavior is specified in [ETSI TBR 021]. Two AC impedance conditions for the idle line signalling state are common [ETSI ES 200 778] in European networks:

- An impedance not less than 8 kΩ, but with a phase angle not exceeding +5 degrees over the frequency range 200 Hz to 4 kHz.
- A return loss over the frequency range 1 kHz to 2.5 kHz of not less than 10 dB with respect to a network compromising a resistor of 820 Ω in series with a parallel combination of a 360 Ω resistor and a 180 nF capacitor.

Additionally, an exceptional case based on DTMF signalling is defined in [ETSI ES 200 778], with an AC impedance in the frequency range of 300 Hz to 3.4 kHz greater than 1.8 kΩ and preferably lower than 2.4 kΩ. The AC impedance of on-hook TE at higher frequencies is not typically specified and furthermore can be somewhat variable. Nevertheless, it is of significant importance in the design of DSL splitters. Some valuable measurement results are found in [Thorne 1998], and a sample measurement is given in Figure 1.8.

1.1.5.7 "Off-Hook" State

This is the state in which the TE draws sufficient DC current to activate the exchange, also known as the loop state [ETSI TR 101 182]. A substate of the off-hook state is the "online

---

[7] In documents such as [ETSI TR 101 182], the "loop steady state" is defined as the loop state excluding transients from and to the loop state. A similar definition is given for the "quiescent steady state."

[8] It is critical to note that the different impedance models are defined for specific frequency bands. It is a common mistake to use a given model to calculate an impedance value at a frequency for which the model was not intended.

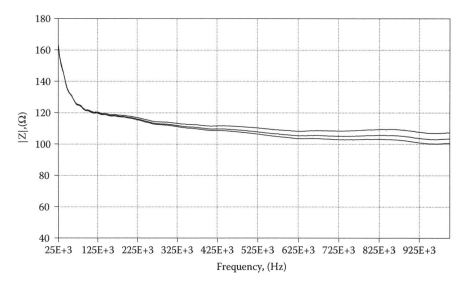

FIGURE 1.8
Impedance comparison of different types of on-hook telephone (see [Thorne 1998]).

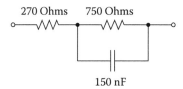

FIGURE 1.9
Complex off-hook voiceband impedance model for terminal equipment.

signalling state," defined as when the TE is capable of receiving frequency shift keying (FSK) data, and the normal transmission functions are suspended. The steady-state DC resistance of an off-hook TE is often approximated by a 400 Ω resistor; however, in practise, electronics in the TE will usually result in an additional voltage drop. A typical specification for the DC behavior of an off-hook TE is given in [ETSI TBR 021]. In terms of AC impedance in the off-hook state, typically only voiceband impedance is specified. For European networks, this is usually specified as a return loss using a reference impedance such as that shown in Figure 1.9. For some other networks, a purely real reference impedance (usually 600 Ω) is used. In this case, there is also an issue of a lack of specification of the TE impedance at higher frequencies.[9] Valuable measurement results are again found in [Thorne 1998], and a sample measurement is given in Figure 1.10. It is not uncommon for network operators to place requirements on their telephone equipment suppliers to ensure adequate rejection of broadcast radio signals (both longitudinal and transverse). The radio suppression often results in some capacitance directly appearing between the line terminals. This can result in a significantly lower line impedance at DSL frequencies than in the voice band, and this in turn affects the requirements for the low-pass section of the DSL splitter.

---

[9] Furthermore, it is clear that designers of legacy TE would not have cared about this parameter.

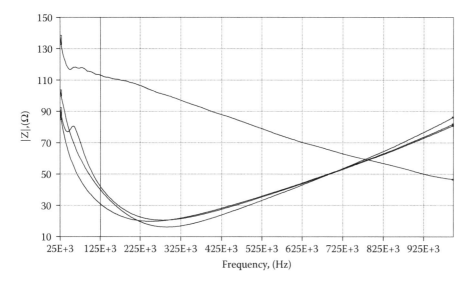

FIGURE 1.10
Impedance comparison of different types of off-hook telephone (see [Thorne 1998]).

1.2   Speech Signals

Telephony speech signals pass bidirectionally on the network pair as differential voltages (see Subsection 1.3.1). The signals for each direction are separated and combined in the telephone and in the local exchange by circuits called hybrids that are described in Section 1.3. The operation of the hybrid relies on knowledge of the impedance of the circuit. Even at telephony frequencies, the pair cable behaves as a transmission line, albeit a rather unusual one. Classic transmission line theory [see Chapter 2] states that the characteristic impedance Z_0 of a transmission line can be derived from a knowledge of the resistance R, inductance L, conductance G, and capacitance C per unit length of the transmission line from the expression:

$$Z_0 = \sqrt{\frac{R + j\omega L}{G + j\omega C}},$$
(1.1)

where ω is the angular frequency in radians/second. It happens that when twisted pair cable is constructed from copper and polyethylene (as it usually is) with the typical dimensions used for telephony cable, at telephony frequencies, L and G are generally negligible. The expression then reduces to:

$$Z_0 = \frac{1 - j}{\sqrt{2}} \sqrt{\frac{1}{\omega}} \sqrt{\frac{R}{C}}.$$
(1.2)

This represents a constant phase angle of −45 degrees and a magnitude that decreases inversely with the square root of frequency. This "complex" characteristic impedance is a fact of life for telephony, and that telephony impedance is often quoted as 600 Ω or 900 Ω resistive is simply a gross approximation. In practise, the two-wire hybrids will have to make some concession to the facts of life for pair cable (see Subsections 1.3.4 and 1.3.5). The complex impedance of pair cable can also have some impact on DSL system design. For example the filters required to separate the telephony signals from the DSL signals.

These filters have to be designed with consideration for the fact that pair cable impedance is nonresistive. The level of the signals on the pair is also of interest. The complex impedance means that care must be taken in the definition of signal levels. The usual practise is to consider the power that would be dissipated in a 600 Ω resistor by the differential voltage present on the pair. A voltage of 775 mVrms is considered 0 dBm, even though the actual circuit power may not be 1 mW. As all transmission of speech signals is done digitally in the core network, and the meaning of these digital signals is standardized by international agreement (so that different telephony networks can interwork), thus the reference signal level is defined in the digital domain, and is known as the "digital milliwatt." When a digital milliwatt in the core network causes (or is caused by) a 0 dBm signal on the access network pair, the relative signal level on the pair is said to be 0 dBr. If it causes (or is caused by) a +6 dBm signal on the pair, the relative signal level on the pair is said to be +6 dBr. The signal levels on the pair may not be the same in each direction. For example, in the United Kingdom, on medium-to-long loops the relative signal level from the telephone is designed to be +1 dBr as received at the exchange, and the relative signal level to the telephone is −6 dBr sent by the exchange. This is achieved by incorporating loss pads of 1 dB and 6 dB into the transmit and receive arms of the local exchange hybrid. Many local exchanges providing constant current feed (see Subsection 1.4.3) have some form of "regulation" (a kind of automatic gain compensation), meaning that as the loop length changes, the relative signal levels also change. Again considering short loops in the United Kingdom, an additional 3 dB loss is inserted into both the transmit and receive loss pads of the local exchange to change the relative signal levels at the exchange to +4 dBr from the telephone and −9 dBr to the telephone. This regulation is operated by the change in loop DC voltage detected at the exchange, and it is necessary to prevent telephone conversations between subscribers on short loops from being too loud.[10]

1.3 Hybrid Circuits

1.3.1 Two-Wire Transmission

As with any electrical circuit, telephony transmission requires both a forward and a return path. The first telephony transmission circuits used a single conducting wire with the ground acting as the return path. However, it was soon discovered that the efficiency of transmission could be greatly increased when a second conductor was used as the return path instead of ground. This development was quite innovative, as almost everyone at the time (including power companies) was using ground as a return path for their circuits, which in turn created significant interference. This two-wire configuration means that telephony signals are always transmitted as differential mode signals (see Chapter 2). Conversely, many of the interfering signals (see Chapter 3) appear on telephony lines as common mode signals (see Chapter 2). It was later discovered that twisting the pair of wires together made the circuit much less susceptible to electrical interference, such as crosstalk from other nearby pairs (see Chapter 3). Hence the local loop is almost ubiquitously made up of twisted pair circuits with copper as the conductor.[11]

---

[10] It should be noted that telephone sets provide regulation, *i.e.*, they change their sending and receiving sensitivities according to the loop current, but this facility is effectively disabled when they are fed from a constant current exchange.

[11] There are exceptions to this. In particular, the last part of the local loop before the customer premises may not be a twisted pair, and also aluminium has been used instead of copper in some existing telephony cabling.

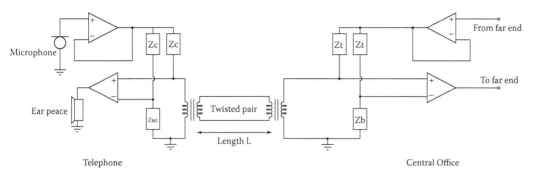

FIGURE 1.11
Equivalent circuit of access network telephony transmission (from [Cook 1995]).

1.3.2 Two-Wire to Four-Wire Conversion

At either end of the local loop, there is a conversion from two-wire to four-wire transmission, as illustrated in Figure 1.11. A four-wire circuit can always provide better transmission quality than a bidirectional two-wire circuit because each direction can be optimized, and the effects of impedance mismatches are more controllable [Reeve 1999]. The trunk circuit "behind" the local exchange is four-wire partially because the digital amplifiers and repeaters used for long distance telephony are inherently unidirectional. The reason four-wire transmission is not generally used for the entire loop is predominantly economic, as twice the amount of cabling would be required to provide service to the same customer base. An explicit example of the conversion from two wires to four is evident in the traditional telephone, in which there is an interface between a two-wire circuit (telephone line) to a four-wire circuit (the earpiece and microphone being physically separate). The circuit providing the interface between two-wire and four-wire transmission at both ends of the local loop is called a hybrid. Early hybrid implementations would have used purely passive components (typically with two interconnected transformers); however, modern hybrids will almost always contain active electronics. A representation of a hybrid implementation is shown in Figure 1.11.

1.3.3 Conceptual Hybrid Circuit

The basic hybrid circuit can be considered as a bridge. One of the ports is connected to the twisted pair; the opposite port (as in Figure 1.11) receives a signal to be transmitted over the twisted pair. Signal detection is (at least conceptually) performed by differential amplifiers that sense the bridge output. The bridge is characterized by two impedances: a terminating impedance and a balance impedance. The terminating impedance is the impedance presented by the bridge to the twisted pair, whereas the balance impedance is that impedance the twisted pair must present to the bridge in order to keep it correctly balanced. A third impedance is needed to complete the bridge, and in Figure 1.11 this is chosen to be equal to the terminating impedance. However, no extra degree of freedom results from an alternative choice of impedance [Cook 1995].[12] One can consider the case of a signal V_{in} arriving from the microphone in Figure 1.11, where Z_{SO} is assumed to be exactly equal to the input impedance of the line. Here it is evident that there will be zero potential difference at the input of the differential amplifier leading to the earpiece; *i.e.*,

---

[12] Any effect of varying this third impedance could also be achieved just by suitably varying Z_{SO}.

there is no feedback, and the signal $V_{in}\frac{Z_{SO}}{Z_{SO}+Z_C}$ appears at the transformer. In the case where there is a difference between Z_{SO} and the input impedance of the line (this will always be the case in reality), there will be a nonzero voltage at the input of the differential amplifier and, hence, some feedback of the transmitted signal to the receive path. The magnitude and phase of this (sidetone) signal will clearly be dependent on the relationship between Z_{SO} and the input impedance of the line. Alternatively, a signal arriving at the transformer from the twisted pair will see the impedance Z_C and hence be detected by the differential amplifier and heard at the earpiece.

1.3.4   Choice of Terminating Impedance

Section 1.2 noted that the characteristic impedance of a twisted pair cable has a significant phase angle in the voiceband, and hence the magnitude of its impedance varies with frequency. Terminating the line in its own characteristic impedance would enable the input impedance of the loop to be known or easily estimated regardless of loop length, which would in turn simplify the design of circuits interfacing with the loop. Using the line's characteristic impedance as a termination can also significantly reduce unwanted audible feedback (see Subsection 1.3.6). In practise, it is impossible to exactly match the characteristic impedance of the loop due to the variation of line characteristics in any real network. For maximum power transfer, however, one would terminate the loop in the complex conjugate of its characteristic impedance [Pozar 1997]. This is not always advisable, however, as there can be significant issues of audible feedback due to impedance mismatches. (The reflection coefficient of an ideal transmission line is zero when it is terminated in its characteristic impedance rather than the conjugate of this.) Furthermore, the difficulties in measuring and specifying complex impedances and, in particular, the difficulties of measuring transmission in the presence of complex impedances, has meant that some administrations, including the United States, have retained nominally resistive terminations for both the central office and terminal equipment terminating impedances. These are not usually equal, partially due to the fact that the need to squeeze many more pairs into ducts near the central office results in a finer wire being used for the cables close to the central office than those farther out in the network. Most modern European networks have been designed and implemented with complex impedances.

1.3.5   Choice of Balance Impedance

In the case of a hybrid used at a local exchange, the balance impedance is usually set as a compromise of the impedance shown by the average user loop (an example of a practical balance impedance is 1100 Ω in parallel with a 33 nF capacitor), so as to give the minimal amount of feedback. This allows better separation of the transmit and return paths. Due to the fact that the input impedance of telephony lines at voice frequencies will always have a nonzero phase angle, this balance impedance will in general be complex. In practical terms, the input impedance of the loop can vary significantly between cables, and hence many hybrids have a selection of balance impedances available for matching. The balance impedance value is chosen either when the circuit is designed (it can be assigned by prescription), or it can be found by testing the loop and adjusting the balance impedance for maximum return loss [Reeve 1999].[13] For the hybrid circuit in a telephone set, however,

---

[13] Especially in the United States, the vast majority of local exchange switches will have at least two balance networks (one for "loaded" and one for "nonloaded" loops) available. Some line cards have three or more different balance networks available.

the unavoidable slight unbalancing resulting from the practical implementation of the hybrid is welcome, because a small amount of the transmitted signal is fed to the earpiece of the telephone, this signal being the dominant component of the sidetone (see below). In fact, sidetone is necessary so that the speaker can hear his or her own voice, in order to determine how loudly to speak. It is immediately apparent that the amount of sidetone must be controlled or else the person will speak at an inappropriate level. If the sidetone is too quiet, the talker assumes the telephone is dead, whereas if it is too loud, it can disturb the speaking process.

1.3.6 Audible Feedback

Sidetone is one example of audible feedback (when the speaker hears his or her own voice), and is predominantly caused by the imbalance in the speaker's telephone set. Imbalance of the hybrid in the local exchange to which the receiver is connected can produce another type of feedback known as echo. In addition, any impedance mismatches in the local loop (either between speaker and local exchange or local exchange and receiver) can also cause audible feedback. Speech that is fed back to the talker will be perceived as either sidetone or echo depending on the time delay from speaking to hearing the feedback (*i.e.*, the round trip delay). If the delay is very short, less than approximately 5 ms, then the feedback will be indistinguishable from the original utterance (sidetone). As the delay is increased and the feedback becomes distinguishable from the original utterance, it is classified as echo. At lower delays, less than approximately 20 ms, this produces a "barrel" effect (as if the talker were speaking into a barrel); at higher delays a distinct echo can be perceived. It should be noted that echo cancellation techniques may be used in the digital circuitry between central offices to reduce echo. These echo cancellers use adaptive filters to minimize any portion of the returned signal that resembles the incoming speech signal. They need only a short amount of time to adapt to the properties of the connection (predominantly the line characteristics of the two local loops involved), and a careful listener can hear the quick disappearance of an echo at the very beginning of a call.

1.3.7 Gains in the Hybrids

The gains of the amplifiers and the efficiency of the transducers in the telephone can be taken together as acoustic-to-electrical transmission sensitivity and vice versa. The resulting acoustic-to-trunk (digital) signal and trunk-to-acoustic signal gains then depend on the gains of the driving and differential amplifiers in the transmission bridges. These gains need to be set so that the range of normal acoustic speech levels make appropriate use of the dynamic range of the analog-to-digital (A/D) and digital-to-analog (D/A) converters in the local exchange [Cook 1995]. Within this overall constraint, it is possible to apportion more or less gain to the amplifiers in the telephone or the central office. Having more gain in the telephone makes any imbalance in the telephone's bridge more significant, causing increased sidetone, whereas having more gain in the local exchange makes imbalance in the local exchange bridge more significant, causing increased echo [Cook 1995]. A network transmission plan (see Subsection 1.1.1) is needed to specify how losses in a network are partitioned so as to ensure that levels of echo, sidetone, and overall loudness are within acceptable limits for the vast majority of connections. Each country has historically had a different loss plan (and impedance strategy), which means that the relative importance of sidetone versus echo will tend to be different. For example, the network in the United Kingdom is more sensitive to sidetone than the North American network, which in turn is more sensitive to echo [Cook 1995].

1.4   DC Signalling

Although actual voice transmission usually takes place at frequencies greater than 200 Hz, DC signalling still plays an important role in telephony. In particular, the fundamental distinction between a "live" POTS connection (*i.e.*, either a call is taking place or it is about to take place), and a "dormant" line is the presence of significant DC line current in the former case. This DC line current is sourced from a battery in the local exchange. For the purposes of this chapter, it is assumed that loop start lines [Fike 1983] are being considered. For telephone lines connecting local exchanges to smaller private exchanges, an alternative form of feeding known as ground start may be used. Details of this are found in [Fike 1983].

1.4.1   The Local Exchange Battery

The common battery used in local exchanges is a 24-cell lead-acid battery, which when fully charged has an open circuit voltage of 48 V.[14] This battery provides power to a variety of equipment in the local exchange, including the POTS subscriber line interface circuits (SLICs), *i.e.*, the physical cards that provide the interface between the local loop and the POTS switch. Conceptually, this supply can be thought of as having two functions in terms of the SLIC operation: (a) provision of power to the active electronics in the voice transmission circuitry, and (b) provision of line current via the DC feed circuit in the SLIC. For practical reasons, multiple lines will use a single local exchange battery. The actual minimum line feed voltage needed for each line depends on that line's DC resistance. For this reason, each line is usually interfaced to the battery by a DC-to-DC converter, enabling a degree of power saving. In some older SLICs, there may be a direct resistive physical connection between the DC feed circuit on the card and the local exchange battery; however, this interfaces the line via large inductive components (the windings of a relay) that effectively provide shunting of common-mode interference to ground. This method of producing line current via large magnetic components is known as constant voltage (or resistive) feeding, and is common on older line cards. Modern SLICs usually also have the capacity to provide constant current feeding, which provides further advantages in terms of power consumption. It should be noted that the actual DC feeding method used is usually proprietary to the SLIC in question. The type of DC feeding used can be extremely important in the case of equipment connected to the telephone line whose electrical performance depends on DC signalling (for example, some distributed filters used in DSL).

1.4.2   Resistive Feeding from the Local Exchange

The purely resistive feeding method is very common in older analog exchanges. Here, a given DC voltage is applied to the line through source resistances (typically the series resistance of the physical relay coil), irrespective of the actual loop length. This approach can result in unnecessary power dissipation on short lines. The voltage applied to the line will nominally be seen as a negative DC voltage applied to one wire in the pair, with the other grounded (for example, 0 V and −48 V). However, in reality, the voltage could be offset slightly from this (for example, −5 V and −53 V). Positive voltages are avoided to try to minimize the effect of electrochemical reactions on wet telephone wires. In the case where a wet wire was at a positive potential with respect to the ground, electrolysis could cause corrosion due to metal ions going from the wire to the ground.

---

[14] The battery is float charged at a slightly higher voltage (52.1 V) to maintain a full charge [Reeve 1999].

1.4.3 Programmable DC Feeding

In many cases of newer digital exchanges, the SLIC can produce a programmable DC loop current that is independent of the resistance seen at the line port of the SLIC, up to a certain threshold resistance value. In the case when the resistance seen at the line port of the SLIC achieves or exceeds this threshold value, the voltage at the line port of the SLIC will typically be maintained at this value for higher resistances. This threshold will typically be set at the maximum possible value to maintain the functionality of the SLIC while conserving power. A certain minimum drop (*i.e.*, difference between the battery voltage and the line voltage) will be necessary in order to maintain the internal minimum biasing voltages of the SLIC; not respecting this threshold could potentially result in distortion of the voice signal from the SLIC line driver amplifiers.

1.4.4 Resistance of the Local Loop

DC resistance is one of the two parameters[15] commonly used for loop design, the intent of which are to provide satisfactory transmission and signalling while taking into account the economics of the situation [Reeve 1999]. The elementary physical properties of a conductor mean that the DC resistance will ideally be proportional to the line length (for a given radius), and inversely proportional to the square of the radius (for a given line length). A typical rule for resistive design (in North America) is that the conductor loop resistance be limited to 1300 Ω, whereas in many European countries, the longest loops may have significantly less DC resistance than this.[16]

1.4.5 Resistance of the Terminal Equipment

The DC behavior of the terminal equipment is of key importance in terms of POTS service. In particular, it is necessary that very little DC current flows in the case where no POTS transmission is in progress (such as when the phone is on-hook). In addition, it is necessary that the voltage drop of the terminal equipment is bounded during the normal operating phase of the call (when the phone is off-hook), to ensure that there is a good current flow for signalling to the exchange.

It is assumed for the moment that only one piece of POTS terminal equipment is connected to the line at the customer end; however, these arguments can easily be extended to the case of multiple terminal equipment in parallel. An on-hook (or quiescent state) POTS terminal equipment will typically have a DC resistance of at least 1 MΩ, whereas in the off-hook state, the voltage drop should not generally exceed between 9 and 14.5 V, depending on loop current (see Clause 4.4.1 of [ETSI TBR 021]).

When the terminal equipment transitions from the on-hook to the off-hook state (*e.g.*, when somebody lifts the handset of the telephone), the transition is signalled to the local exchange equipment by a flow of DC current. Hence, the DC resistance of the POTS terminal equipment must be below a certain maximum value in order for the off-hook (or loop) state to be recognized by the local exchange, which then responds appropriately. In practice, POTS terminal equipment will usually contain active electronics that cause its DC characteristic to be nonlinear, and thus the electrical requirements on POTS terminal equipment will usually specify a maximum allowable DC voltage drop over a given DC current range.

In order to operate reliably, terminal equipment generally needs to have a minimum amount of DC current passing through its terminals. This property effectively sets the reach

---

[15] The other is insertion loss at a given frequency, typically 1 kHz or 1.6 kHz.

[16] For certain European countries, however, the longest lines may have significantly more DC resistance than 1300 Ω (for example, many Scandinavian countries).

of a POTS service. In the case of older telephones that used carbon granule transmitters as their microphones, this current was roughly 23 mA [Fike 1983]. However, most existing terminal equipment can operate at loop currents much lower than this [Bingel 2000]. Indeed, in modern transmission the limiting component specification is the minimum current required to operate the line relay in the central office (*i.e.*, the component used to recognise that the phone has gone off-hook). Under these conditions, the on-hook current drawn by the electronics inside the telephone set must be well below the minimum current required by the line relay, otherwise, the line relay would energize and incorrectly indicate an off-hook condition. The steady-state online DC requirements for terminal equipment are given in Clause 4.7.1 of [ETSI TBR 021].

1.4.6   Additional Uses of DC Signalling

As well as indicating whether a POTS line is "live" or quiescent, DC signalling also has other common uses in POTS networks (see Clause 14 of [ETSI ES 201 970]). For instance, a reversal in the polarity of the voltage applied to the line is commonly used in many networks in order to signal various events to the terminal equipment. In addition, techniques such as register recall signalling [ETSI ES 201 729], loop disconnect dialing [ETSI ES 201 187], K-break [ETSI ES 201 970], and calling line identification [ETSI EN 300 659-1] can all use DC signalling.

1.5   Wetting Current

In the case of a DSL service that does not have an underlying low frequency service,[17] as is the case for many symmetric DSL deployments, one can have operation on lines with the absence of any DC signals (commonly known as "dry loops"). It is highly likely that such lines are very susceptible to a physical phenomenon that can have an adverse effect on the performance of the DSL in question, namely, the corrosion of cable joints.

1.5.1   Corrosion of Cable Joints

A typical subscriber loop can be expected to traverse around two dozen joints (splices) connecting each cable segment together. The method used to join the wires can vary, but generally all are of some form of metallic contact. Corrosion or oxidation is harmful to the electrical performance of metallic contacts and is vastly accelerated by ingress of moisture and other contaminants through the protective housing of the cable joining sections [Beaumont 2001]. Oxidation penetrates into the metallic surfaces and between the points of contact, getting deeper as time goes on. The resulting metal oxide has a semiconducting property, which increases the resistance of the contact. Over time this can lead to a service-affecting fault [Beaumont 2001]. In particular, the balance of the cable is very susceptible to changes of contact resistance. Failure to address this problem on a large scale can have an adverse impact on network crosstalk levels and possibly void the assumptions made in access network frequency plans. In order to protect against this, network operators use petroleum jelly-filled cables, jelly-filled splicing connectors, and high-performance

---

[17] This low frequency service would typically be POTS or integrated services digital network (ISDN), both of which use DC signalling.

joint housings. These prevent moisture and contaminant ingress [Beaumont 2001]. Unfortunately, the majority of access networks contain both old and new cables, and not every enclosure is hermetically sealed. In addition, maintenance causes disturbance to joint enclosures, and the resulting damage can go unnoticed.

1.5.2 Use of Wetting Current

In order to mitigate these corrosive effects, wetting current can be used. Wetting current (also known as sealing current in the United States) is a method whereby a small (6–20 mA) electrical current is passed through transmission loops in order to reverse the effects of high resistance contact faults owing to corrosion of wire joints [Beaumont 2001].[18] Varying theories exist as to the mechanism behind the advantageous effect of wetting current, and the interested reader is referred to [Holm 1967], [Bennett 1998], and [Schubert 1991]. Although not accepted by all, the practical benefits of wetting current are advocated by many.

1.5.3 Potential Effect of Wetting Current on DSL

The most obvious effect of wetting current on DSL is the need to add a DC current-providing feature at the line interface at the local exchange DSL equipment. Furthermore, the DSL equipment at the customer end will need to be able to sink whatever DC current is determined necessary. It should be noted that wetting current does not need to be continuously applied; its beneficial effects can result when it is applied over relatively short periods [Beaumont 2001]. Indeed, most existing wetting current generators for older digital services have a "flash" feature that initially pulses the loop with a high current value to clear splice oxidation [Reeve 1999]. The application of pulses rather than continuous current has clear advantages with regard to power consumption. Traditionally, levels of wetting current have been set between 6 mA and 20 mA. These current levels are predicted to limit the contact resistance of copper contacts to around 30 Ω and 10 Ω, respectively [Beaumont 2001]. While it is applied, however, it may cause harmful noise in the transmission frequency spectrum, and changes in the contact resistance may alter the properties of the transmission path at a rate such that the DSL transmission is adversely affected [Beaumont 2001]. Hence, there is a compromise between the positive effect on the physical properties of the cable and the negative effect on the performance of the DSL. These issues are currently under study in DSL standardization committees, and the use of wetting current is currently an option in such symmetric DSL standards as [ETSI TS 101 524].

1.6 Ringing

When an incoming call is received on a telephony connection, the local exchange has to cause the customer's telephone(s) to ring. This is done by sending a high-level low-frequency signal differentially on the pair. The specifics of this signal vary from country to country; the frequency can be from 16 to 50 Hz, although 20 to 30 Hz is usual, and the amplitude is from 50 Vrms to as much as 120 Vrms. This signal actually provides the AC power to operate the ringing in the customer premises. Originally, it would have operated an electromechanical actuator such as a bell to generate the ringing sound. The actuator would have been designed to be mechanically resonant with the applied ringing signal frequency.

---

[18] Although the etymology would suggest that "whetting" is a more appropriate spelling, the term "wetting" is more widely used in the industry.

In a modern phone, the AC-coupled ringing signal is rectified in a semiconductor circuit to generate DC power to operate an electronic/piezo-electric tone sounder.

As explained earlier, the local exchange line interface converts and provides all the signals needed on the telephony circuit, and a modern exchange is generally a largely electronic circuit. Sometimes, even the ringing signal application will be done electronically, either by directly modulating the line feed current or by switching in a separate source with an electronic switch, such as a high-voltage FET or triac. However, in many circuits where higher ring voltages are required, this function is still done with a separate ringing generator applied via an electromechanical relay. The requirements for ring switching of low on-hook resistance, very high off-hook isolation, and low cost are still best met by such a device even today, and such relays have now been miniaturized dramatically. As mentioned earlier, ringing is essentially a differential signal, *i.e.*, between the wires of the pair, although it may not be balanced because all the AC signal may actually be sent on just one of the wires of the pair, with the other wire essentially acting simply as a return path. During ringing, line feed power ($\approx$50 Vdc) is generally maintained in addition to the AC ringing waveform, although in some networks it is commonly reversed in polarity with respect to line feed during the nonringing phase of the call. This latter feature is sometimes used by more complex CPE to detect ringing, as it is easier and quicker to detect than the low-frequency ringing signal itself. The use of relays in ringing application to switch in an independent ringing supply has implications for DSL system design. Even modern electromechanical relays exhibit contact bounce, and so the transients generated during ringing application can be large in amplitude and (by electronics standards) extensive in time. Although in some telephone exchange interfaces the operation of the ring relay may be synchronized to the phase of the ringing voltage, in many it is not. In such a case, there can be a 100 V transient from the change from -50 to $+50$ V line feed plus the sudden application of ringing AC voltage at the peak of its cycle (for example, 106 V peak for the United Kingdom's 75 Vrms 25 Hz ringing waveform), giving a total transient step of over 200 V. Such large transients have to be very well suppressed if they are not to cause havoc with data transmission by DSL, and particularly ADSL, on the same pair. These transients are most extreme at the exchange end of the loop, where the relays reside. And it should be remembered that a large number of transients can be involved, because it is common practice to generate the ring cadence (the pattern of ring bursts) by using the ring relay in the line interface.

1.7   Ring Trip

When the phone is answered by the customer, it effectively puts a near-short (an impedance of, at most, a few hundred Ohms) across the pair. This starts to draw current from the DC component of the ringing signal. It is the increase of this DC current flow that is detected in the local exchange and used to cease the application of the ringing signal and then connect the call. The process is known as ring-trip, and it has implications for DSL, as it is another source of transients. The transients are similar in magnitude to those caused by ring switching in the exchange, because most of the ring voltage and line feed voltage are also present at the customer end of the loop (low frequencies are not attenuated much on the loop), and this voltage is suddenly shorted out by the phone going off-hook. Again the shorting is typically done by a mechanical switch (*i.e.*, the hook switches off the phone), so contact bounce is still an issue. The number of such transients is much lower, though, as they only occur once when the call is actually answered. The other key difference is that the source of the transient is at the customer end of the loop, rather than at the local exchange.

1.8 On- or Off-Hook Detection

The customer also causes current to be drawn from the exchange line feed voltage when the phone goes off-hook at the beginning of a new call. It is this current flow that is detected in the local exchange and results in the application of dial tone and the connection of a subsystem to detect the customer's dialled digits. Similarly, when the phone goes on-hook at the end of the call, the lack of current flow signals to the local exchange indicates that the call has ended. Again transients are generated in these processes, but they are clearly smaller than those caused during ring-trip.

1.9 Dialing

Dialing is the process by which the party initiating a POTS connection attempts to generate the signals (to be sent to the local exchange) that identify the party at the other end of the desired connection. These signals may be in the form of dial pulses (sequential on-hook or off-hook of the terminal equipment causing current pulses) or tones. The local exchange indicates that it is ready to receive these signals by sending a "dial tone" signal to the terminal equipment. Upon receipt of the appropriate pulses or tones, the local exchange equipment interprets the signals to appropriately set up a connection. In older telephone sets, a mechanical technique known as pulse dialing (or loop disconnect dialing) is often used, whereas newer telephones tend to use tone dialing using dual-tone multi-frequency (DTMF) signalling. One of the key features of DSL is that it can be deployed as an overlay service; *i.e.*, it can "peacefully" co-exist on lines with legacy services. In many administrations, replacement of telephony terminal equipment is at the user's discretion and expense. This at least partially explains why many older telephone sets still exist in networks used for DSL, and hence why pulse dialing is relevant to DSL.

1.9.1 Pulse (Loop Disconnect) Dialing

Once a dial tone has been received by the calling party, a rotary dial such as that shown in Figure 1.12 can provide called party address information to the central office in the

FIGURE 1.12
Rotary dial telephone.

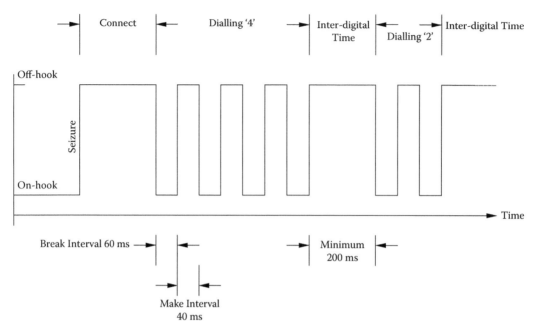

FIGURE 1.13
Dial pulse sequences with a pulse rate of 10 PPS for the digits "4,20."

form of dial pulses, *i.e.*, short duration breaks in the loop current caused by the opening
and closing of the loop. A conventional telephone set will have ten equally spaced finger-
holes, as shown in Figure 1.12. The mechanical operation of pulse dialing is described in
[Fike 1983]. The nominal pulse rate is ten pulses per second (*i.e.*, 10 Hz), but a typical rotary
dial instrument can vary from eight to twelve PPS. The make/break (off-hook/on-hook)
ratio applied to the loop is nominally 40/60; *i.e.*, the loop is closed 40 percent of the time
and opened 60 percent. A sample dial pulse pattern is shown in Figure 1.13. Detection of
the dial pulses by modern local exchange equipment is primarily based on time-domain
signal processing (knowledge of the pulse rate, make/break ratio); however, an additional
consideration is the distortion introduced by the copper pair.

1.9.1.1   *High Voltage Transients Due to Pulse Dialing*

Voltage spikes are produced each time the dial pulsing contacts break the circuit, *i.e.*, inter-
rupt the flow of loop current. These spikes can have a large amplitude, in part due to the
inductive DC feed used in many local exchanges.

Conventional telephones have an inherent protection against these voltage spikes, which
can cause the bell of the ringing circuit in the terminal equipment (see Section 1.6) to sound.
The consequent "ringing" is usually very soft, and hence is referred to as "tinkle" (or "bell
tap"). In order to prevent this effect, an "anti-tinkle" circuit such as that shown in Figure 1.14
is used. Basically when the dial is rotated, the ringing circuit is shunted with a resistor *R*
in order to prevent bell tinkle. The ringing capacitor *C* in Figure 1.14[19] serves as a spark
quencher to suppress arcing at the dial pulsing contacts. Additionally, the speech circuit is

---

[19] The primary function of this capacitor is to prevent direct current from passing through the ringing circuit.

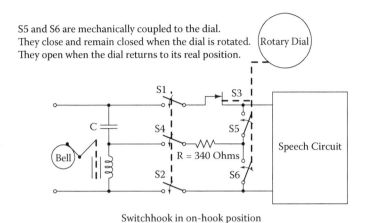

S5 and S6 are mechanically coupled to the dial.
They close and remain closed when the dial is rotated.
They open when the dial returns to its real position.

Switchhook in on-hook position

FIGURE 1.14
Anti-tinkle circuit.

also shorted by switches 1 and 2 to prevent loud clicks in the receiver during pulse dialing. In terms of DSL, the previously described transient voltages can have significant energy in the frequency band used for DSL transmission and can thus cause significant interference with DSL performance if not adequately filtered out [ETSI TR 102 139]. This is one of the functions of the splitter.

1.9.2 Tone Dialing

Most newer telephones use tone dialing, not least because it is much faster than pulse dialing. Tone dialing involves a dial pad with push-buttons and almost always uses DTMF signalling. The most common form of DTMF dialing uses a dial pad similar to that shown in Figure 1.15. Each button (and hence each touch-tone digit) is represented by a unique combination of two single-frequency tones. The frequencies are arranged in a matrix, as shown in Figure 1.15. As the button is pressed for a specific digit, the appropriate combination of tones is generated, corresponding to the horizontal and vertical position of the push-button. (For example, pushing "1" would generate tones at 697 Hz and 1209 Hz.) At the central office, the received tones are filtered and detected in order to determine the required digit. For European networks, DTMF equipment and transmission are specified by [ETSI ES 201 235-1],

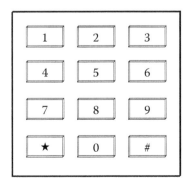

FIGURE 1.15
DTMF dial pad.

whereas for North America both [ANSI T1.401-2000] and [ANSI/EIA/TIA-470-A-1987] contain DTMF specifications. DTMF signalling takes place at a frequency range well below those used in DSL and hence does not typically cause interference with DSL transmission.

1.10   Subscriber Private Metering

Many customers have a desire to keep track of their expenditure on telecommunications; this is of course still important today but even more so in years gone by when these services were much more expensive than they are today. The early automated technology for charging for services consisted of an electromechanical pulse counter (much like the odometer in a motor vehicle) at the local exchange, with one counter for each customer. Pulses were supplied to this counter by the exchange mechanism at a rate that depended on the type of call being made. Billing was based on the pulse count, so that each pulse would be valued at a known amount. A common method of providing the customer with visibility of his expenditure was to relay these pulses to a secondary counter in the customer premises. This kind of service is known as subscriber private metering or SPM. To avoid the need for extra pairs to transmit the pulses to the customer premises, ways were found of combining the necessary signals on the same pair used for the telephony. The most common method was to add a burst of a high-frequency tone to the differential signals on the customer's pair. These tone bursts were then detected by a special toneburst-detecting meter in the customer premises. The tone frequencies were chosen to be beyond the normal responsiveness of typical telephony equipment (but not necessarily beyond the responsiveness of the customer's ears!) at either 12 or 16 kHz, and at a level that ensured their transmission to the customer despite the high frequency, typically 2 Vrms. An alternative method of transmitting the pulses is to impose a large low-frequency common-mode signal on the customer's loop, typically 20 Vrms of 50 Hz. Again, this signal is detected in the customer's premises with a specially designed meter. In this case, the customer is spared hearing the bursts by the good balance of the cable, so that little or no pulse is experienced in the differential mode used to transmit speech.

The proliferation of various telecommunications services and the desire to be able to reward customers with special discounts and billing strategies has resulted in much more complex billing systems in modern networks. Usually the actual service usage information is recorded by the exchange's control system, and this information is passed on electronically to an offline computer-based billing system that is able to provide detailed itemized billing. The advent of these billing strategies has made it practically impossible to maintain a meaningful SPM system in many modern networks. At the same time, the itemized billing given to customers has greatly decreased the need for such a service. As a result, SPM as a service today is typically in decline or even withdrawn. However, it is still found in minority applications such as for some call-box functions. The presence of SPM can have implications for DSL deployment, especially for ADSL and its splitter filters. Where ADSL may be needed on the same line as SPM, the splitter filters must be capable of passing the SPM signals. This can place extra constraints on the splitter filter design.

1.11   Telephony Speech Coding

The analog speech signals for each direction of transmission are coded and decoded in a voice codec, which is part of the line interface circuit in the local exchange. The most common coding process used by the majority of trunk networks is essentially a simple A/D and D/A conversion with an 8 kHz sampling rate. However, the quantization levels are approximately logarithmically spaced rather than by the conventional linear spacing, which has the effect of achieving an adequate signal-to-noise ratio over a wide range of signal levels. At high signal levels, the coding is approximately equivalent to 5-bit linear quantization, giving about a 35 dB signal-to-noise ratio and, at lower signal levels, the effective quantization is equivalent to 12 or 13 bit linear quantization, maintaining the same signal-to-noise ratio. There are 256 quantization levels in all, so they can be coded into 8 bit sample words, giving an overall bit rate of just 64 kbit/s. The converters are preceded by anti-aliasing filters with approximately 3.4 kHz of passband width and 30 dB of alias band suppression. Two slightly different nonlinear quantization schemes are used worldwide, one known as A-law in Europe and another known as μ-law in the United States; both are described in [ITU-T G.712] and, fortunately for international communications, translation between the two schemes is reasonably simple and has no discernible effect on speech quality.

1.12 Balance about Earth

Mention has been made in passing of the fact that the use of twisted pairs for telephony transmission enables the important differential transmission signals to be sent without interference of many sorts that generally affect both wires of the twisted pair equally and so exist in the common-mode. This property of twisted pairs is absolutely vital to the functionality of telephony networks, as fundamentally it not only enables transmission to occur free of external interference, but also facilitates the packing of many twisted pairs into a single cable without their interfering significantly with one another. In other words, crosstalk between pairs is greatly reduced. More is said about the residual levels of crosstalk and the interference that does remain in Chapter 3. In passing, it is worth mentioning that the differential and common-mode modes of propagation on the twisted pair are sometimes given other names. Specifically the differential mode may also be called the transverse mode or the metallic mode, and the common-mode is sometimes called the longitudinal mode.

The degree of balance at voice frequencies is quite prodigious. It can be measured as the level of differential mode signal that is detected on a pair when the pair is exposed to a perfectly balanced common-mode signal (the longitudinal conversion loss or LCL). At voice frequencies, for a cable in good condition, LCL invariably measures better than 60 dB, and 70 or 80 dB is not unusual. Balance degrades with increasing frequency at roughly 10 dB per decade so that it is usually better than 50 dB at ISDN frequencies (40 kHz), better than 40 dB at ADSL frequencies (1 MHz), and better than 30 dB at VDSL frequencies (up to 12 MHz). At the higher frequencies the balance takes on another significant role. It obviously continues to help with the fight against external interference, usually radio transmissions at these frequencies, but also helps to reduce the tendency for intended differential mode signals (such as high-speed DSL signals) from leaking out of the cable and radiating as interference to nearby radio receivers. More is said about this in Chapter 13.

1.13 Testing

In Section 1.1 it was stated that testing is a vital function for access network maintenance and has to be supported by the hardware such as line interface circuits and other systems. The essential requirement is to determine if there is a fault and, if so, approximately where it is so at least the right kind of engineer can be despatched to fix it. The whole subject of testing and maintenance processes is complex and warrants further treatment. It is covered in depth in the chapter on line testing in Volume 2.

1.14   Overload

Access network copper cables, by their very nature, may be stretched over miles of all sorts of urban and rural landscape and come into proximity with all sorts of hazards on their way. A common hazard is lightning strikes. There is no practicable defence against a direct lightning strike on the cable, as lightning will simply vaporize the copper within the cable! However, more commonly, lightning merely strikes close to cables and induces enormous surges in them. Although these induced surges may reach 5 kV or so, they do not generally damage the cable because they are quite short lived, but they do travel along the cables and enter exchange equipment and customer premises equipment (CPE).

If equipment is to be reliable, it must be able to cope with these large surges without damage. A two-stage approach is generally taken in the exchange to protect equipment against overload. The first stage involves very high-powered surge arresters fitted at the connecting blocks on the MDF. These arresters (or primary protection circuits) have traditionally been gas discharge tubes, although in recent years semiconductor devices have become available with ratings sufficient to replace the gas discharge tubes and have the advantage of lower cost and better precision. The primary protection absorbs most of the energy of the largest surges, but arresters are to some extent limited in what they can do, because they need to provide protection while still allowing quite large signals out of the exchange, such as ringing signals and line feed current. Line feed current can reach 180 V or more on some systems, depending on local safety regulations. It follows that surges getting past the MDF arresters must be dealt with by the equipment connected behind it, which contains further protectors of lower rating carefully placed so that they can better protect more vulnerable circuit parts, such as line interface circuits (behind the ring relays) and DSL transceiver chips. A similar approach can be used at the customer end. Primary protection can be placed in the NTE or NID and secondary protection in the CPE itself. There is a special circumstance here, as unlike at the exchange, neither may have access to a ground connection into which to discharge common-mode transients. For this reason, it is common to deal with common-mode transients at this end of the network simply with reinforced insulation.

Another overload hazard that must be considered is that of contact with main power supplies. Fortunately, the prevalence of this kind of overload is quite low, but it has to be taken seriously because of the risk that it could cause fires, either in the exchange or in the customer's premises. The comparatively low resistivity of the copper pairs means that they do not themselves melt and fuse until the current is quite high, around 6 Amps or so. However, currents much less than this flowing in terminal equipment have the potential to cause hazardous fires or smoke and fumes. The larger currents can be dealt with simply by fuses in series with the line. Simple fuses that require manual repair often suffice because the overload conditions are so rare, although self-resetting fuses are also available and have become economical. More difficult to deal with are the lower current levels that may be insufficient to cause fuse rupture but are still sufficient to cause excess dissipation and are thus a fire hazard in the terminal equipment. These hazards must also be considered, as the consequences (exchange destruction, loss of life) are so great. Thermal fusing can be effective in preventing this hazard.

1.15   High-Speed Voiceband Modems

As described in Section 1.11, the speech codec provides a speech channel with approximately 35 dB signal-to-noise ratio and 3.4 kHz bandwidth. Application of Shannon's equation for limiting information throughput with a channel of this description, and allowing for a near state-of-the-art 9 dB implementation gap, yields an expected information capacity of about 30 kbits/s. This compares well with the actual maximum throughput of the V.34bis voiceband modem standard of 33.6 kbit/s. However, the underlying data channel is making use of a total data rate of 64 kbit/s, and somehow V.90 and V.92 modems manage to get access to a higher proportion of this capacity, at the much hyped rate of 56 kbit/s, $\frac{7}{8}$ of the underlying data channel rate.

This remarkable achievement makes use of the knowledge of the nonuniform sampling levels in the voiceband codec. By carefully shaping its transmit signals, these devices make use of the fact that the signal-to-noise ratio is preserved at low signal levels by expanding the "constellation" of allowable and distinguishable symbols that can be sent. The result is that V.90 signals have a particularly wide dynamic range and bandwidth (although they are still constrained to the 4 kHz voiceband).

Again, this can have some impact on DSL system design. ADSL splitters must be transparent and linear when exposed to V.90 signals, and the leakage of noise from DSL systems must be kept very low if these modems are to function at full capacity in the voiceband sharing the pair with an ADSL service.

Increasingly, DSL systems are being used to transport voice circuits using technologies such as VoIP. When this is done, there is often motivation to use more modern voiceband codecs which transport voice adequately on a narrower underlying data channel (28.8 kbit/s or lower), allowing the transport of more voice signals in a given DSL capacity limit. Although it may seem perverse to try to operate a V.90 modem on these channels, it doesn't necessarily prevent user disappointment at the resulting data rate achieved.

1.16   CLASS Signalling

CLASS signalling is used to support calling line identification (CLI) [ETSI EN 300 659-1]. This is a system that enables the customer receiving a phone call to know what number is calling before the call is answered, and so to vet or be prepared for that call. CLASS signalling can also support other functions including Advice of Charge and Message Waiting Indication.

The signalling protocol is essentially an elaboration of the ringing system. At the beginning of the ringing phase of the call, the onset of the actual ringing pattern (or cadence) is delayed for a few hundred milliseconds and, during this interlude, an FSK modem data burst is sent to the line. This data burst in principle could contain any information about the incoming call but generally contains the calling line number, if known.

To save power in the CLI receiving device (so that it does not have to be prepared for the data burst at all times), the FSK modem data burst is preceded by a terminal equipment alerting signal (TAS). The sequence is:

1. TAS
2. (pause)

3. Data burst (V.21 voiceband data)

4. (pause)

5. Onset of normal ringing pattern

The TAS may consist of a polarity reversal, a dual-tone burst, a short burst of ringing voltage, or a combination of these. The CLI device faces some challenges in its design because the 100 V transient represented by a line feed reversal does not really die away completely before the data burst starts. Also, the data burst occurs at a time when there is no line feed present on the line (*i.e.*, when the line is said to be "dry," as described in Subsection 1.5.2).

This feature of telephony can affect DSL delivery. For various reasons described in the chapter in Volume 2 concerning splitters, the customer end splitter function for ADSL may not be fully functional in the dry state, causing noise on or distortion of the CLI signal. These problems have to be mitigated if CLI is to remain functional in the presence of ADSL.

References

[ANSI T1.401-2000] ANSI T1.401-2000, *Network to Customer Installation Interfaces.*

[ANSI T1.421-2002] ANSI T1.421-2002, *In-Line Filter for Use with Voiceband Terminal Equipment Operating on the Same Wire Pair with High Frequency (up to 12 MHz) Devices.*

[ANSI/EIA/TIA-470-A-1987] ANSI/EIA/TIA-470-A-1987, *Telephone Instruments with Loop Signaling.*

[Beaumont 2001] S. Beaumont and J. MacDonald, *BT Requirements for SDSL Wetting Current*, ETSI TM6 contribution TD10 Sophia Antipolis, February 2001.

[Bennett 1998] B.W. Bennett, *The effect of current on stationary contact behaviour*, Proceedings of the 34th IEEE Holm conference on electrical contacts, 1998.

[Bingel 2000] T. Bingel, *Two Parallel Off-Hook Telephone Current Feed is Satisfied by Contribution 239R1 V-I Template*, ANSI Contribution T1E1.4/200-354, November 2000.

[Brit. Stand. Inst.] British Standards Institution DISC PD1002, *A Guide to Cabling in Private Telecommunications Systems*, 1997.

[Cook 1995] J. Cook and P. Sheppard, *ADSL and VADSL Splitter Design and Telephony Performance*, IEEE Journal on Selected Areas in Communications, Volume 13, No. 9, December 1995.

[ETSI EN.300.001] ETSI EN 300 001 v1.1.5 (1998-10), *Attachments to the Public Switched Telephone Network (PSTN); General technical requirements for equipment connected to an analogue subscriber interface in the PSTN.*

[ETSI EN 300 659-1] ETSI EN 300 659-1 v1.3.1 (2001-01), *Access and Terminals (AT); Analogue access to the Public Switched Telephone Network (PSTN); Subscriber line protocol over the local loop for display (and related) services.*

[ETSI ES 200 778] ETSI ES 200 778 v1.2.2 (2002-11), *Access and Terminals (AT); Analogue access to the Public Switched Telephone Network (PSTN);Protocol over the local loop for display and related services; Terminal equipment requirements.*

[ETSI ES 201 970] ETSI ES 201 970 v1.1.1 (2002-08), *Access and Terminals (AT); Public Switched Telephone Network (PSTN); Harmonized specification of physical and electrical characteristics at a 2-wire analogue presented Network Termination Point (NTP).*

[ETSI ES 201 729] ETSI ES 201 729 v1.1.1 (2000-02), *Public Switched Telephone Network (PSTN); 2-wire analogue voice band switched interfaces; Timed break recall (register recall); Specific requirements for terminals.*

[ETSI ES 201 187] ETSI ES 201 187 V1.1.1 (1999-03), *2-wire analogue voice band interfaces; Loop Disconnect (LD) dialing specific requirements.*

[ETSI ES 201 235-1] ETSI ES 201 235-1 v1.1.1 (2000-09), *Specification of Dual Tone Multi-Frequency (DTMF) Transmitters and Receivers.*

[ETSI TBR 021] ETSI TBR 021 ed.1 (1998-01), *Terminal Equipment (TE); Attachment requirements for pan-European approval for connection to the analogue Public Switched Telephone Networks (PSTNs) of TE (excluding TE supporting the voice telephony service) in which network addressing, if provided, is by means of Dual Tone Multi Frequency (DTMF) signalling.*

[ETSI TR 101 182] ETSI TR 101 182v1.1.1 (1998-08), *Analogue Terminals and Access (ATA); Definitions, abbreviations and symbols.*

[ETSI TR 102 139] ETSI TR 102 139 v1.1.1 (2000-06), *Compatibility of POTS terminal equipment with xDSL systems.*

[ETSI TS 101 524] ETSI TS 101 524 v1.1.3 (2001-11), *Transmission and Multiplexing (TM); Access transmission system on metallic access cables; Symmetrical single pair high bitrate Digital Subscriber Line (SDSL).*

[Fike 1983] J.L. Fike, and G.E. Friend *Understanding Telephone Electronics*, Ft. Worth: Texas Instruments Learning Center, 1983.

[Holm 1967] R. Holm, *Electric Contacts*, 4th edition, New York: Springer-Verlag, 1967.

[ITU-T G.712] ITU-T Recommendation G.712, *Transmission performance characteristics of pulse code modulation channels*, November 2001.

[ITU-T G.989.3] ITU-T Recommendation G.989.3, *Phoneline networking transceivers—Isolation function*, March 2003.

[Pozar 1997] D.M. Pozar, *Microwave Engineering*, New York: Wiley, 2nd edition, August 1997.

[Reeve 1999] W.D. Reeve, *Subscriber Loop Signaling and Transmission Handbook : Analog*, New York: Wiley-IEEE Press, October 1999.

[Richards 1973] D.L. Richards, *Telecommunication by Speech: The Transmission Performance of Telephone Networks*. London: Butterworth, 1973.

[Schubert 1991] R. Schubert, *Sealing current and regeneration of copper junctions*, IEEE Transactions on Components, Hybrids and Manufacturing Technology, No. 14, pp 214-217, 1991.

[Thorne 1998] D. Thorne et al., *A Proposal for a Revised Customer Network Model*, ADSL forum contribution 98–061, June 1998.

2

The Copper Channel—Loop Characteristics and Models

Hervé Dedieu

CONTENTS

2.1 Historical Use of the Copper Network—Digital vs. Analog 34
2.2 Physical Characteristics of Copper Lines—The Shielded Twisted Pair 35
    2.2.1 Differential Mode Signalling 35
    2.2.2 Insulation, Core Assembly, and Shielding of Twisted Pairs 37
    2.2.3 The Advantages of Twisting 38
    2.2.4 Wire Diameter and Gauge 38
    2.2.5 Bridged Taps 39
    2.2.6 Load Coils 39
    2.2.7 Radio Filters 40
    2.2.8 Shared Line Multiplexers 40
2.3 Electrical Characteristics of Twisted Pairs 41
    2.3.1 Primary Parameters of Lines 41
    2.3.2 Line Terminated by Its Characteristic Impedance Z_0 42
    2.3.3 Characteristic Impedance in POTS Band 43
    2.3.4 Characteristic Impedance at DSL Frequencies 44
    2.3.5 Transmission of Signals Across the Line 44
    2.3.6 Power Transmission Across Lines 46
    2.3.7 ABCD Line Modelling 48
    2.3.8 Transfer Function and Insertion Loss Associated
        with DSL Lines 50
    2.3.9 Scattering Parameters 51
    2.3.10 Impedance and Admittance Modelling Parameters 55
2.4 Generic Models of DSL Cables 56
    2.4.1 The British Telecom Models 0 and 1 (*RLCG* Modelling) 56
        2.4.1.1 Empirical Model for Resistance 56
        2.4.1.2 Empirical Model for Inductance 57
        2.4.1.3 Appropriate Model for Conductance 57
        2.4.1.4 Empirical Model for Capacitance 57
    2.4.2 The British Telecom Model #0 Parameters 57
    2.4.3 The British Telecom Model #1 Parameters 57
    2.4.4 Examples of Cable Modelling Using the BT #0 Model 58
    2.4.5 The KPN Models #0 and #1 (*RLCG* Modelling) 59

2.4.6   The KPN Model #0 59
2.4.7   The KPN Model #1 60
2.4.8   Examples of Cable Modelling Using the KPN #1 Model 60
2.4.9   Other Models of Interest 63
2.5   Loop Configuration 63
2.5.1   Examples of Loop Configurations 63
2.5.2   Insertion Loss of Some North American Reference Loops 69
2.5.3   Insertion Loss of Some European Reference Loops 69
References 69

ABSTRACT   This chapter first describes the physical characteristics of DSL copper lines. The differential signalling principle combined with the pair twisting principle are shown to provide immunity to external perturbations. Some specific elements deployed in copper lines, such as bridged taps and load coils, are then briefly described. The main electrical characteristics of copper lines are reviewed; the concepts of primary and secondary parameters and characteristic impedance are described in detail. Some very useful models of line parametrization are reviewed; the so-called $ABCD$ parameters as well as the scattering parameters (of great importance in test and measurements) are explained. The generic DSL models that are commonly used for simulation purpose are then presented with an emphasis on the British Telecom models and the KPN models. Some loop configurations that have been proposed by standards bodies are then listed with their main transfer function characteristics.

2.1   Historical Use of the Copper Network—Digital vs. Analog

As explained in the last chapter, the origin of the telephony can be traced back to Alexander Graham Bell's invention[1] of the telephone in 1876 [Bell 1876]. Nevertheless, many of the copper lines that were to be used as the medium for these early telephony networks had been installed. Indeed, the utilization of metallic lines for commercial communications had been demonstrated when Samuel Morse constructed the first commercial telegraph line between Baltimore and Washington in 1844. This was a digital communication system with a communication rate of about 20 symbols per minute. Indeed, Bell's invention some 32 years later was initially presented as an "Improvement in Telegraphy" [Bell 1876], implying the potential replacement of a slow digital communication system by a more efficient analog one. Upon forming the Bell Telephone Company with Sanders and Hubbard in 1877, Bell had to compete against the Western Union Company, which had more than 250,000 miles of telegraph wire already deployed. Around this time, Western Union began to deploy analog telephony across their already established "digital" network. This copper network has grown as described in Chapter 1 to become the worldwide telephony network that exists today. Analog technology was almost exclusively used from the late 1800s to the 1960s. Around this time, digital systems were introduced in between central exchanges (see Section 1.1.1 for a description of the network structure) in order to address issues of noise for transmissions over long distances. Since then, digital technology

---

[1] There is some controversy over who actually invented the telephone. Others such as Antonia Meucci (1850), Philippe Reiss (1860), and Elisha Gray (1876) certainly had similar ideas.

has penetrated further and further into the copper network, and the deployment of integrated services digital network (ISDN) in 1985 meant that an end-to-end digital service was now available on a large scale [Star 1999]. The success of DSL has furthered the penetration of digital technology in the copper network, and the current trend suggests a circular progression back to an all-digital network.

2.2 Physical Characteristics of Copper Lines—The Shielded Twisted Pair

As mentioned in Section 1.3.1, the initial approach to using copper lines for telephony involved a single conducting wire with the ground acting as the return path. The quality of early telephony was often extremely poor; however, a drastic improvement came with the concepts expressed in Alexander Graham Bell's patent [Bell 1881, patent US244,426]. These can be listed succinctly as follows:

- Use of differential mode instead of common mode signalling.
- Use of shielded pairs.
- Use of twisted pairs instead of flat pairs.

The benefits of each of these concepts is described in Sections 2.2.1, 2.2.2, and 2.2.3, respectively. This is followed in Section 2.2.4 by a description of the significance of wire diameter.

2.2.1 Differential Mode Signalling

As explained in Section 1.3.2, transmission of differential signals involves the use of a second conductor to make up the return path instead of ground. In order to understand the concept of differential mode signalling, it is useful to consider a model of a simple transmission line. The global model is of a pair of wires that is fed with a voltage source of amplitude V_{ac} (with internal impedance Z_g), and is terminated with an impedance Z_{load}. In accordance with usual terminology, one wire of the twisted pair is called ring, and the other is called tip.[2] Each wire of the pair is assumed to exhibit uniform impedance per unit length. Furthermore, each section of wire located at a given length x from the generator is assumed to be connected to the ground via the same parasitic capacitance $C_p(x)$ (not necessarily uniformly distributed along the line). This simple model is shown in Figure 2.1, which shows both the global line model as well as a model of an infinitely small section. The section is made of a series inductance $L dx$, a series resistance $R dx$, a shunt resistance of admittance $G dx$, and a shunt capacitance $C dx$. As shown in Figure 2.1, these elements form a dissipative low-pass filter. If a is a point located at distance x from the generator on the tip wire, and b is located at a distance x from the generator on the ring wire, it is observed that due to line symmetry and balanced connection of the two wires with respect to ground, one has the equalities

$$i_a = i_b \qquad i'_a = i'_b \qquad i''_a = i''_b. \tag{2.1}$$

Elementary application of Kirchhoff's law for currents gives

$$i_{ap} = -i_{bp} \tag{2.2}$$

---

[2] These names, tip and ring, are derived from the electrical contacts of old telephone plugs. These were physically similar to modern audio plugs. In Europe they are sometimes referred to as the "A wire" and the "B wire."

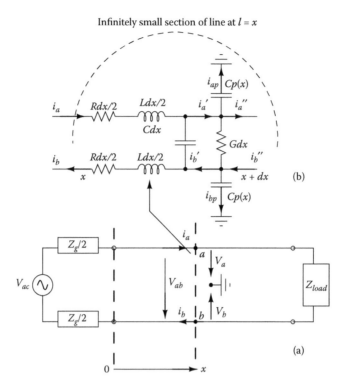

FIGURE 2.1
Line signalling in differential mode: (a) Infinitely small portion of line. (b) Full line.

and, therefore,

$$V_a = -V_b. \tag{2.3}$$

In addition, no current flows to ground because of the symmetry of voltages (the sum of currents flowing to ground $i_{ap} + i_{bp}$ is identically zero). When the line is perfectly symmetric and well balanced with respect to ground, the voltage V_a between tip and ground at $l = x$ is equal in absolute value and opposed in phase to the voltage V_b at $l = x$ between ring and ground. With respect to Figure 2.1(a), the voltage V_{ab} can be defined as a differential mode signal[3] applied between the two wires. This means that at the receiver end, the voltage information can be processed by determining the difference between the two voltages across tip and ring. In the case of Figure 2.1(b), one can consider the line model to remain unchanged; however, now both the generator and terminating impedance are connected to ground. It is clear that the voltage on the line will be such that the parasitic capacitors will be charged with the same amplitude and phase. This voltage is said to be a common mode signal, because it is equally seen by both wires. The common mode signal is responsible for currents travelling in the same direction along the x axis of both tip and ring wires, and hence these currents are sometimes referred to as longitudinal signals. In general, common mode signals in telephony are perturbations occurring due to coupling of physically adjacent lines (electrostatic and magnetic coupling), lightning storms, radio frequency interferences, etc. Assuming that there are both differential mode and common mode signals on the line, then, in theory, if the line were well balanced and if the information at the receiver end were differentially processed, the common mode signals would be completely removed

---

[3] Differential mode signals are sometimes called metallic signals.

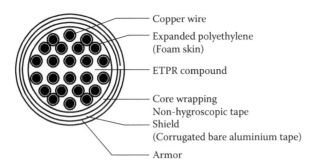

- Copper wire
- Expanded polyethylene (Foam skin)
- ETPR compound
- Core wrapping Non-hygroscopic tape
- Shield (Corrugated bare aluminium tape)
- Armor

FIGURE 2.2
View of a cable section.

at the receiver. In practice a perfectly differential receiver is impossible to implement, and hence common mode rejection chokes are used to enhance practical rejection of common mode signals.

2.2.2 Insulation, Core Assembly, and Shielding of Twisted Pairs

The second of Bell's three innovations detailed in his 1881 patent is the concept of shielding the copper pair. This involves surrounding the pair with a metallic shield in order to isolate it from outside sources of electromagnetic interference (see Chapter 13). As explained in Section 1.1.3, the copper pairs used in practical networks will almost always be grouped into a cable. Hence, it is apparent that some form of insulation is also needed, in order to prevent the pairs touching each other (or indeed anything else). The following section gives physical details of a typical cable.

Figure 2.2 shows a view of a telephone cable section made of several pairs of copper wires. All modern cables use conductors with a polyolefin (plastic) insulation. This kind of insulation is described under the generic term of PIC[4] (polyolefin insulated cables). Modern cable insulation can be classified as solid (polyethylene or polypropylene) or expanded foam (expanded polyethylene, polyvinyl chloride (PVC)). Expanded foam cables are nowadays preferred for local loop wiring because they are slightly less expensive than their solid insulated counterparts. Many copper pairs in modern networks are insulated with expanded polyethylene (foam skin) surrounded with an external layer of solid high-density polyethylene of distinctive color. The latter can be used to identify the pair in the cable. The core is filled with an ETPR[5] compound in order to avoid interstices (spacing) between pairs. A nonhygroscopic tape (*i.e.*, one that does not absorb or propagate moisture) with an overlap usually surrounds the cable core. A corrugated bare aluminium tape is then applied longitudinally in order to form the shield. The shield is grounded at the cable ends in order to reduce interference from external sources. To increase corrosion resistance, the aluminum is bonded to a thin plastic. Finally, an armor made of steel tape protects the cable. This armor is usually covered with a black jacket of low-density polyethylene. In cable with 25 pairs or fewer, the pairs are usually laid forming a cylindrical shape. For larger cables, units of 25 pairs or super-units of 50 or 100 pairs are assembled to form the final core, each group having a color-coded unit binder. For a more detailed description of cable construction and assembly, the reader is referred to [Hugh 1997].

---

[4] Older cables use a paper (pulp) type insulation.
[5] Extended-thermoplastic rubber, a material having excellent dielectric properties and resistant to oxidative breakdown.

2.2.3   The Advantages of Twisting

Bell's third innovation was the concept of the twisted pair. A simple explanation for the benefit of twisting the wires can be found by inspection of the model of Figure 2.1. One can see that no current flows toward ground if the line is both homogeneous and well balanced with respect to earth. An assumption made was that the parasitic capacitors C_p connecting tip and ring to ground were perfectly matched. If this is not the case, the line becomes unbalanced, currents flow to ground and cancellation of common mode signals is no longer possible due to the unbalanced action of interferers on both wires. Therefore, it is of primary importance to ensure that both tip wire and ring wire remain at equal distance from any disturbance forcing electrostatic and magnetic interferences to be equal on tip and ring. The simplest way to achieve this symmetry (*i.e.*, the minimization of capacitive unbalance) is to twist the line. The benefits of twisting are further discussed in Sections 2.4 and 2.5.

2.2.4   Wire Diameter and Gauge

Wire diameter is an important characteristic when assessing the ability of a loop to act as a medium for information transmission. It is well known that the DC resistance of a twisted pair varies in proportion with the inverse of the squared value of the diameter. Unfortunately for DSL, however, the radio-frequency (RF) resistance varies inversely with the diameter (the skin effect puts current only in its circumference). Thus the thickened wires do not give the same improvement to DSL signals as to voice transmission. Even more important for DSL are the inductance and the capacitance per unit of length in the model described in Figure 2.1. These have a complex non-linear dependence on the wire diameter, as well as other parameters such as the type of insulation, the number of twists per units of length, and the frequency. Nonetheless, as a general rule one can consider that the ability of a channel (of given length) to carry information (*i.e.*, the channel capacity) will increase with diameter. In North America, twisted pairs are characterized by the American wire gauge designation (AWG), which is indicative of wire diameter. Typical twisted pair gauges are #19, #22, #24, and #26. In most markets outside of North America, wires are classified according to their approximate diameter in millimeters. For example, 0.4 mm (comparable to 26 gauge) and 0.5 mm (comparable to 24 gauge[6]) are the most common, although in many developing countries heavy gauges of 0.6 mm to 0.9 mm can be found in newly urbanized areas. Table 2.1 indicates the correspondence between AWG, metric size diameter, and resistance in ohms per kilometer. Loop resistance is an important parameter for POTS equipment, and networks were engineered to a maximum limit for loop resistance. Hence, thinner wires were used for shorter loops (cheaper in cable, cheaper in duct space) and thicker were used for longer loops. For manageability, long pairs tend to share cables with the thinner pairs for some distance, so the build of a long pair typically starts thin from the central office (CO) and gets progressively thicker.

It is common to deploy 0.4 mm twisted pair (or AWG26) along the first 3 km from the CO to some primary or secondary connection point. Beyond this, successively heavier gauge can be used in order to avoid an excessive loop resistance. If possible, the overall DC resistance is maintained below 1300 Ω. This practice is sometimes referred to as the 1300 Ω resistance design rule. Along its path from central office to customer, a loop can consist of several sections having different diameters. Given that copper wire is generally stored in the form of 150 m spools, a loop of 3 km can exhibit about 20 splices. Signal

---

[6] In many European networks, the majority of the local loop will be made of 0.4 mm wire. The final few hundred meters to the customer premises will often be 0.5 mm wire. In certain urban regions, thinner wire may be used in the cables out of the central office due to space constraints.

TABLE 2.1

Common Twisted Pair Gauges Characteristics

AWG	Wire Diameter (mm)	Loop Resistance (Ω/km), (20 degrees C)
19	0.9	55.4
22	0.63	110.9
24	0.5	175.2
26	0.4	281.4

reflection can occur at splices, which correspond to change of diameter due to impedance discontinuity.

2.2.5 Bridged Taps

In certain networks, it can be quite common for a given twisted pair line to have another section of twisted pair connected at some point along its length, the final end of this unused twisted pair being open circuited. This is known as a bridged tap: a section of wire pair connected to a loop on one end and not terminated at the other end. Figure 2.3 shows a loop of length 2 km that connects a CO and a CPE with a bridged tap of about 500 m at the loop midpoint. According to [Starr 1999], approximately 80 percent of the loops in the United States have bridged taps, and sometimes several bridged taps exist along the same loop. The main reason for the existence of bridged taps is the wish on the part of the network owner to discontinue service to an existing customer and to route rapidly to a new customer in the same neighborhood. Although bridged taps should not have any discernible effect on POTS performance, this is not true for DSL transmission. The bridged taps are open stubs and will resonate in the DSL signal band, resulting in spectral nulls and phase distortion.

2.2.6 Load Coils

The technique of loading a loop was invented in 1900 by Michael Pupin [Pupin 1900]. In essence, it involves placing a series of physical inductors called load coils at equally spaced intervals along the loop. A typical value of 88 mH placed at 1.8 km intervals has been used in the past for long line deployment. The benefit of Pupin's technique is to increase the overall inductance per unit length of the twisted pair. This results in lower loss characteristics in the POTS band (POTS transmission is improved) at the expense of greater attenuation at frequencies above the POTS band. Pupin's method is easily understood after reading Section 2.3, in which it is shown that a current or voltage source generated at one end of the

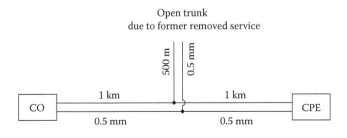

FIGURE 2.3
Example of a bridged tap.

twisted pair is damped at l meters of its source by a factor $e^{(-\gamma l)}$, where γ is the propagation constant given in Equation 2.9. This propagation constant in the POTS band depends on the cable resistance per unit of length R and cable capacitance C per unit of length. (It can be considered for frequencies lower than 10 kHz, as is the case in the POTS band, that the inductance per unit of length $L \approx 0$, and the conductance per unit of length $G \approx 0$.) Applying the general Equation 2.9 in the POTS band, one finds for the propagation constant

$$\gamma \approx \sqrt{\frac{RC\omega}{2}} + j\sqrt{\frac{RC\omega}{2}}. \tag{2.4}$$

In his patent [Pupin 1900], Pupin imagines a cable between New York and Boston (approx. 250 miles) having a capacitance of 50 nF per mile and a resistance of only 20 Ω per mile. By virtue of Equation 2.4, this cable would have a propagation constant of real part $\alpha = 0.0974$ per mile at 3 kHz. This corresponds to a damping factor $e^{-250 \times 0.094} = 2.7e^{-11}$ between New York and Boston, making practical communication impossible. Pupin notes that the situation would be more favorable for communication if the same cable were to have inductance per mile of 50 mH. In this case, by applying Equation 2.9 at 3 kHz, one finds that $\alpha = 0.01$, and therefore a current initiated in Boston with amplitude I will induce in New York a current with amplitude $Ie^{-2.5} = I \times 8$ percent. The trick invented by Pupin is therefore to implicitly change the uniformly distributed inductance of the line by placing a discrete inductance of several mH each mile. Although it is impractical, this cable can be approximated by lumped loading. Pupin's modified cable exhibits minimized attenuation in the POTS band and no phase distortion. One unfortunate consequence of this is that DSL transmission cannot take place on loaded loops, because the rejection of frequencies above tens of kHz is magnified by the presence of these discrete coils. According to [Starr 1999], 10 to 15 percent of loops in the United States have load coils; however, they are rarely found on loops shorter than 5.5 km. If any loading coils exist on lines intended for DSL, they must be located and removed in order for DSL to be successfully deployed.

2.2.7 Radio Filters

Somewhat similar to load coils, radio filters are series inductive components that have been deliberately introduced. The purpose was to eliminate noise arising from nearby radio and radar transmitters, where low-frequency amplitude modulation of the radio carrier could be heard as noise in the telephone device. Whereas a load coil has an inductance of between 20 and 90 mH, radio filters are typically 2 or 3 mH. This gives a cut-off frequency that is very much higher than voiceband, allowing use of ISDN service but not so high that DSL services would work. Detecting radio filters is altogether much more difficult than load coils for three reasons:

- They are of much smaller value.
- Load coils are placed at intervals within the access network, whereas radio filters are more likely to be placed very close to or in the subscriber premises.
- Radio filters may be incorporated in the telephone device, which of course would have no effect on DSL transmission.

2.2.8 Shared Line Multiplexers

A shared line multiplexer is a common telephone network operator device that allows two or more customers to share the same physical network pair. Such devices are used when there are more customers than serviceable pairs at a final distribution point. Two or more line cards in the exchange are connected to a near-end multiplexer unit, and at the far end, two or more drop wires provide telephony services. For obvious reasons, DSL cannot operate on the same line as a shared line multiplexer. Unfortunately, the presence of such devices is not always recorded in cable record information for individual subscribers.

2.3   Electrical Characteristics of Twisted Pairs

2.3.1   Primary Parameters of Lines

One can consider an infinitely small portion of uniform line such as depicted in Figure 2.4. This is the unbalanced version of the circuit shown in Figure 2.4, which can be regarded as a two-port acting as a low-pass filter made up of the concatenation of a series impedance and a shunt impedance. Without loss of generality, the series impedance is made of a resistance Rdx and an inductance Ldx, where R is the resistance per unit length and L is the inductance per unit length. The shunt impedance is made up of a parallel resistance (of admittance Gdx) and a capacitance Cdx, where G is the conductance per unit length and C is the capacitance per unit length. The parameters R, L, G, and C are the so-called primary parameters. Elementary circuit theory shows that

$$dV = -(R + jL\omega)I\,dx \quad dI = -(G + jC\omega)V\,dx. \tag{2.5}$$

Hence,

$$\frac{dV}{dx} = -(R + jL\omega)I \quad \frac{dI}{dx} = -(G + jC\omega)V. \tag{2.6}$$

Therefore,

$$\frac{d^2V}{dx^2} = -(R + jL\omega)\,\frac{dI}{dx} = (R + jL\omega)(G + jC\omega)V, \tag{2.7}$$

and

$$\frac{d^2I}{dx^2} = -(G + jC\omega)\,\frac{dV}{dx} = (R + jL\omega)(G + jC\omega)I. \tag{2.8}$$

If γ is defined such that

$$\gamma = \alpha + j\beta = \sqrt{(R + jL\omega)(G + jC\omega)}, \tag{2.9}$$

Equations 2.7 and 2.8 can be written as

$$\frac{d^2V}{dx^2} = \gamma^2 V \quad \frac{d^2I}{dx^2} = \gamma^2 I, \tag{2.10}$$

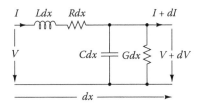

FIGURE 2.4
Line section of length dx with primary parameters of R, L, G, and C per unit of length.

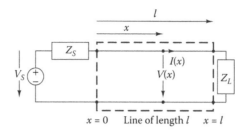

FIGURE 2.5

Line section of length l loaded with Z_L and powered by a voltage generator V_S of source impedance Z_S.

which are the so-called Telegrapher's equations. Now consider a line of length l that is closed by a load impedance Z_L and powered by a voltage source generator V_S with internal impedance Z_S, as described in Figure 2.5.

The solution to Equation 2.10 for a line of length l at a point located a distance x from the source is the sum of two voltage or current waves travelling in opposite directions:

$$V(x) = V_0^+ e^{-\gamma x} + V_0^- e^{\gamma x}, \tag{2.11}$$

$$I(x) = I_0^+ e^{-\gamma x} + I_0^- e^{\gamma x}. \tag{2.12}$$

The voltage or current wave travelling with decay $e^{-\gamma x}$ is the voltage or current incident wave, and the voltage or current wave with decay $e^{\gamma x}$ is referred to as the reflected wave. The boundary conditions allowing the determination of V_0^+, V_0^-, I_0^+, and I_0^- are

$$V(l) = Z_L I(l) \quad V(0) = V_S + Z_S I(0). \tag{2.13}$$

The frequency-dependent variable γ is the propagation constant of the line. Its real part α is called the attenuation constant, and the imaginary part β is the phase constant.

2.3.2   Line Terminated by Its Characteristic Impedance Z_0

The characteristic impedance[7] Z_0 of a line is defined as the load impedance $Z_L = Z_0$ that causes the impedance $V(x)/I(x)$ at any location x of the line to be equal to Z_0. By inserting Equation 2.6 into Equation 2.11, one finds that

$$I(x) = V_0^+ \sqrt{\frac{G + jC\omega}{R + jL\omega}} e^{-\gamma x} - V_0^- \sqrt{\frac{G + jC\omega}{R + jL\omega}} e^{\gamma x}. \tag{2.14}$$

The impedance Z_0 is defined as

$$Z_0 = \sqrt{\frac{R + jL\omega}{G + jC\omega}}, \tag{2.15}$$

and the line is terminated with $Z_L = Z_0$ such as depicted in Figure 2.6.

The boundary condition $V(l) = Z_L I(l) = Z_0 I(l)$ together with Equations 2.11, 2.14, and 2.15 applied at $x = l$ show that

$$V(l) = V_0^+ e^{-\gamma l} + V_0^- e^{\gamma l} \tag{2.16}$$

---

[7] Another definition commonly accepted is that the characteristic impedance of a line is the input impedance of an infinitely long section of the line, irrespective of its terminating impedance. See Subsection 2.3.5 for a proof.

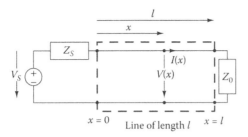

FIGURE 2.6
Line section of length l loaded with its characteristic impedance.

and

$$I(l) = \frac{V_0^+}{Z_0}e^{-\gamma l} - \frac{V_0^-}{Z_0}e^{\gamma l} = \frac{V(l)}{Z_0}.$$ (2.17)

Combining Equations 2.17 and 2.16 shows that

$$V_0^- = 0 \quad \text{when} \quad Z_L = Z_0.$$ (2.18)

Therefore, the impedance $V(x)/I(x)$ at any location x of the line is such that

$$\frac{V(x)}{I(x)} = Z_0 \quad \text{when} \quad Z_L = Z_0.$$ (2.19)

Equation 2.18 shows that when the line is terminated in its characteristic impedance, the reflected wave is identically zero. Exploiting the boundary condition $V(0) = V_S + Z_S I(0)$ and the fact that the line input impedance is Z_0, it can be deduced that

$$V(x) = V_S\frac{Z_0}{Z_S + Z_0}e^{-\gamma x} \quad I(x) = V_S\frac{1}{Z_S + Z_0}e^{-\gamma x} \quad \text{when} \quad Z_L = Z_0.$$ (2.20)

2.3.3 Characteristic Impedance in POTS Band

In the POTS band, common twisted pairs deployed in telephony have an inductance per km in the range of several hundreds of micro-Henry (400 to 800 µH). In the same band, the resistance per km is between 50 and 500 Ω depending on the diameter used. For instance, a polyethylene insulated twisted pair of 0.40 mm diameter per conductor has a resistance that is approximately a constant of 280 Ω in the POTS band (see Figure 2.18). The inductance per kilometer of this cable in the POTS band is approximately a constant of 580 µH (see Figure 2.19). Therefore, $L\omega$ is lower than 5 percent of R over the entire POTS band. As G can be neglected with respect to $C\omega$, at low frequencies, the characteristic impedance in Equation 2.15 can be approximated as the following "complex" impedance

$$Z_0 \approx \sqrt{\frac{R}{jC\omega}} \approx \frac{1-j}{\sqrt{2}}\sqrt{\frac{R}{C\omega}}.$$ (2.21)

As already noted in Section 1.2, this represents a constant phase angle −45 degrees and a magnitude that decreases inversely with the square root of frequency.

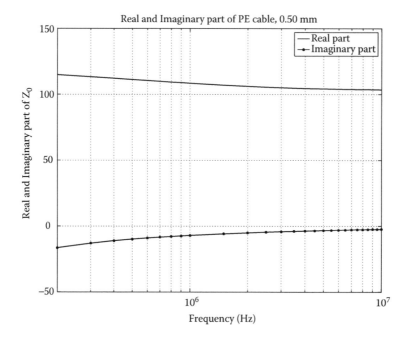

FIGURE 2.7
Real and imaginary part of Z_0 for a polyethylene insulated cable of diameter 0.50 mm.

2.3.4 Characteristic Impedance at DSL Frequencies

At high frequencies, $L\omega$ in Equation 2.15 becomes dominant over R, and therefore one can use the approximation given in Equation 2.22:

$$Z_0 \approx \sqrt{\frac{L}{C}}. \tag{2.22}$$

For instance, above 200 kHz, polyethylene insulated twisted pair with 0.50 mm wire has the R and L primary parameters given in Figure 2.18 and Figure 2.19, respectively, and exhibits the characteristic impedance behavior shown in Figure 2.7. One can see that the real part of the impedance converges to about 100 Ω. The imaginary part is converging toward 0.

2.3.5 Transmission of Signals Across the Line

In the most general case, when the line is not connected to a load equal to the characteristic impedance Z_0 as described in Figure 2.5, Equations 2.11 and 2.12 or Equations 2.14 and 2.13 hold. Therefore, from Equations 2.11 and 2.14, one obtains the following.

$$\begin{bmatrix} V(x) \\ I(x) \end{bmatrix} = \begin{bmatrix} V_0^+ e^{-\gamma x} + V_0^- e^{\gamma x} \\ \dfrac{V_0^+}{Z_0} e^{-\gamma x} - \dfrac{V_0^-}{Z_0} e^{\gamma x} \end{bmatrix}. \tag{2.23}$$

Writing Equation 2.23 at $x = l$ relates V_0^+ and V_0^- to $V(l)$ and $I(l)$ as follows.

$$V_0^+ = \frac{1}{2}(V(l) + Z_0 I(l))e^{+\gamma l} \qquad V_0^- = \frac{1}{2}(V(l) - Z_0 I(l))e^{-\gamma l}. \tag{2.24}$$

The system of equations in 2.23 evaluated at $x = 0$ shows that

$$V(0) = V_0^+ + V_0^- \qquad I(0) = \frac{V_0^+}{Z_0} - \frac{V_0^-}{Z_0}. \tag{2.25}$$

By combining Equations 2.24 and 2.25, it is straightforward to find the relationship between currents and voltage at both side of the line; *i.e.*,

$$\begin{bmatrix} V(0) \\ I(0) \end{bmatrix} = \begin{bmatrix} \cosh(\gamma l) & Z_0 \sinh(\gamma l) \\ \dfrac{1}{Z_0} \sinh(\gamma l) & \cosh(\gamma l) \end{bmatrix} \begin{bmatrix} V(l) \\ I(l) \end{bmatrix}. \tag{2.26}$$

The relationship in Equation 2.26 is of fundamental importance when simulating lines as well as concatenations of lines. Defining Z_i as the input impedance of the line and calculating the ratio of both equations in the equations of 2.26 yields

$$Z_i = \frac{V(0)}{I(0)} = Z_0 \frac{Z_L \cosh(\gamma l) + Z_0 \sinh(\gamma l)}{Z_L \sinh(\gamma l) + Z_0 \cosh(\gamma l)} = Z_0 \frac{Z_L + Z_0 \tanh(\gamma l)}{Z_0 + Z_L \tanh(\gamma l)}. \tag{2.27}$$

It should be noted that for a very long line $\tanh(\gamma l) \longrightarrow 1$, and the input impedance of Equation 2.27 reduces to $Z_i \longrightarrow Z_0$, irrespective of the value of Z_L. According to this property, the characteristic impedance of a uniform line is often defined as the input impedance of an infinitely long section of this line, irrespective of its loading impedance Z_L. When the line is terminated with $Z_L = Z_0$, it is clear from Equation 2.27 that $Z_i = Z_0$ is in accordance with Equation 2.19. When $Z_L = 0$, the input impedance of the short-circuited line reduces to $Z_i = Z_{sc}$ given by

$$Z_{sc} = Z_0 \tanh(\gamma l). \tag{2.28}$$

When $Z_L \to \infty$, the input impedance corresponding to this open circuit reduces to $Z_i = Z_{oc}$ given by

$$Z_{oc} = Z_0 \coth(\gamma l). \tag{2.29}$$

It follows from Equations 2.28 and 2.29 that the characteristic impedance can be deduced from Z_{sc} and Z_{oc}; *i.e.*,

$$Z_0 = \sqrt{Z_{sc} Z_{oc}}. \tag{2.30}$$

In a similar fashion the propagation constant can be deduced from Z_{sc} and Z_{oc} according to

$$\gamma = \frac{1}{l} \tanh^{-1} \sqrt{\frac{Z_{sc}}{Z_{oc}}}. \tag{2.31}$$

By inserting the relationships in Equation 2.24 into Equation 2.23 and taking into account the boundary condition $V(l) = Z_L I(l)$, one finds the following.

$$V(x) = \frac{1}{2} [V(l) + Z_0 I(l)] \left[e^{-\gamma(x-l)} + \frac{Z_L - Z_0}{Z_L + Z_0} e^{+\gamma(x-l)} \right] \tag{2.32}$$

and

$$I(x) = \frac{1}{2} \left[\frac{V(l) + Z_0 I(l)}{Z_0} \right] \left[e^{-\gamma(x-l)} - \frac{Z_L - Z_0}{Z_L + Z_0} e^{+\gamma(x-l)} \right]. \tag{2.33}$$

Equations 2.32 and 2.33 can be rewritten as functions of initial conditions $V(0)$ and $I(0)$; *i.e.*,

$$V(x) = \frac{V(0)}{e^{+\gamma l} + \frac{Z_L - Z_0}{Z_L + Z_0} e^{-\gamma l}} \left[e^{-\gamma(x-l)} + \frac{Z_L - Z_0}{Z_L + Z_0} e^{+\gamma(x-l)} \right] \tag{2.34}$$

and

$$I(x) = \frac{I(0)}{e^{+\gamma l} - \frac{Z_L - Z_0}{Z_L + Z_0} e^{-\gamma l}} \left[e^{-\gamma(x-l)} - \frac{Z_L - Z_0}{Z_L + Z_0} e^{+\gamma(x-l)} \right]. \tag{2.35}$$

As $V(0) = V_S \frac{Z_i}{Z_S + Z_i}$ and $I(0) = \frac{V_S}{Z_S + Z_i}$, the initial conditions can be derived using Equation 2.27 as

$$V(0) = \frac{Z_0 [Z_L + Z_0 \tanh(\gamma l)]}{Z_S [Z_0 + Z_L \tanh(\gamma l)] + Z_0 [Z_L + Z_0 \tanh(\gamma l)]} V_S \tag{2.36}$$

and

$$I(0) = \frac{Z_0 + Z_L \tanh(\gamma l)}{Z_S [Z_0 + Z_L \tanh(\gamma l)] + Z_0 [Z_L + Z_0 \tanh(\gamma l)]} V_S. \tag{2.37}$$

It should be noted that for a given source voltage V_S, a given source impedance Z_S, and a known termination Z_L, the voltage and current across the line as described by Equations 2.34 and 2.35 are functions only of characteristic impedance Z_0 and propagation constant γ. For this reason, Z_0 and γ are the so-called secondary parameters[8] of the line. The primary parameters, *i.e.*, the R, L, G, and C of Figure 2.4, can be deduced from the secondary parameters. From Equations 2.9 and 2.15, it is straightforward to find

$$R(\omega) = Re(\gamma(\omega)Z_0(\omega)) \quad G(\omega) = Re\left(\frac{\gamma(\omega)}{Z_0(\omega)} \right) \tag{2.38}$$

and

$$L(\omega) = \frac{1}{\omega} Im(\gamma(\omega)Z_0(\omega)) \quad C(\omega) = \frac{1}{\omega} Im\left(\frac{\gamma(\omega)}{Z_0(\omega)} \right). \tag{2.39}$$

It should be noted that at any distance x of the line, the ratio of amplitude of the reflected wave $e^{+\gamma(x-l)}$ with respect to the incident wave $e^{-\gamma(x-l)}$ is given by

$$\rho = \frac{Z_L - Z_0}{Z_L + Z_0}. \tag{2.40}$$

The frequency-dependent variable ρ is the reflection coefficient of the line. When $Z_L = Z_0$, the reflection coefficient is zero, and therefore the reflected wave is identically null along the line, and Equations 2.34 and 2.35 are equivalent to Equation 2.20.

2.3.6 Power Transmission Across Lines

As is evident from Equation 2.40, when the line is terminated by its characteristic impedance, there is no reflection of signals at the termination. Considering a DSL transmission system with full duplex communication across the line, if the generators at both sides were to send signals through source impedances of exactly Z_0, this would prevent reflection of signals at both ports. As mentioned in Section 1.3.4, this situation does not exactly correspond to the best possible exchange of power between the source and the load. Considering a generator of source impedance $Z_S = Z_0$, the maximum power available is achievable when the load on which this generator is connected is Z_0^*. Therefore,

$$P_{max.\ av.} = Re(V(0)I^*(0)) = Re\left(\frac{Z_0^*}{Z_0 + Z_0^*} V_S \frac{V_S^*}{Z_0 + Z_0^*} \right) = \frac{|V_S|^2}{4\,Re(Z_0)}. \tag{2.41}$$

---

[8] Z_0 and γ are functions of ω as described by Equations 2.15 and 2.9.

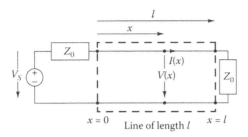

FIGURE 2.8

Line section of length l loaded with Z_0 and powered with a voltage generator V_S of source impedance Z_0.

For a line connected to its characteristic impedance such as that of Figure 2.8 with $Z_S = Z_0$, the power transferred to the load is found using Equation 2.20:

$$P_{deliv.} = Re(V(l)I^*(l)) = Re\left(\frac{V_S}{2}e^{-\gamma l}\frac{V_S^*}{2Z_0^*}e^{-\gamma^*l}\right) = \frac{|V_S|^2 Re(Z_0)}{4|Z_0|^2}e^{-2\alpha l}. \qquad (2.42)$$

The ratio of power delivered to the load to maximum power available from the voltage generator is the square value of the transmission coefficient, which is defined as the scattering parameter s_{21} in classical circuit theory; *i.e.*,

$$|s_{21}|^2 = \frac{P_{deliv.}}{P_{max.\ av.}} = \frac{(Re(Z_0))^2}{|Z_0|^2}e^{-2\alpha l}. \qquad (2.43)$$

At frequencies such as typically used in DSL transmission, the primary parameter R becomes negligible with respect to $L\omega$, and primary parameter G becomes negligible with respect to $C\omega$. Hence, the characteristic impedance becomes real and tends towards $\sqrt{\frac{L}{C}}$. In this situation, if the terminations on both ports are equal to Z_0, reflection of signals is prevented, and optimum transfer of power is guaranteed as $|s_{21}|^2 \to e^{-2\alpha l}$.

For a general line of input impedance Z_i, the reflected power at the source port with a voltage generator V_s of internal impedance Z_0 is

$$P_{reflec.} = \frac{|V_S|^2}{4\,Re(Z_0)} - \frac{Re(Z_i)|V_S|^2}{(Z_0 + Z_i)(Z_0^* + Z_i^*)} = \frac{|V_S|^2}{4\,Re(Z_0)}\left|\frac{Z_i - Z_0}{Z_i + Z_0}\right|^2. \qquad (2.44)$$

The ratio of the reflected power with respect to the maximum power available from a source generator of impedance Z_0 is therefore

$$\frac{P_{reflec.}}{P_{max.\ av.}} = \left|\frac{Z_i - Z_0}{Z_i + Z_0}\right|^2. \qquad (2.45)$$

The return loss of the transmission line is the inverse ratio of reflected power to maximum power available from the load. This ratio is usually given in decibels. From Equation 2.45,

$$RL = 10\log_{10}\left(\left|\frac{Z_i + Z_0}{Z_i - Z_0}\right|^2\right). \qquad (2.46)$$

When $Z_i \to Z_0$, the return loss becomes infinite and the reflected power tends to zero. When $Z_S = Z_0$, by using Equation 2.27, Equation 2.46 can be expressed in terms of $\gamma, l, Z_0, Z_L,$ and l, giving

$$RL = 10\log_{10}\left(\left|\frac{Z_L + Z_0}{Z_L - Z_0}\right|^2\left|\frac{1 + \tanh(\gamma l)}{1 - \tanh(\gamma l)}\right|^2\right). \qquad (2.47)$$

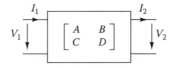

FIGURE 2.9
Uniform line of length l modelled as a two-port with $ABCD$ parameters.

Equation 2.47 shows that the return loss will become infinite if $Z_L = Z_0$ (this case corresponds to $Z_i = Z_0$), or when $\tanh(\gamma l) \to 1$ (case of a long line, which induces $Z_i \to Z_0$).

2.3.7   ABCD Line Modelling

The relationship in Equation 2.26 shows that a uniform portion of line of length l can be modelled as a two-port network, where the voltage V_1 and current I_1 of the left-hand port are linked to the voltage V_2 and current I_2 of the right-hand side port through the so-called $ABCD$ model shown in Figure 2.9.

$$\begin{bmatrix} V_1 \\ I_1 \end{bmatrix} = \begin{bmatrix} A & B \\ C & D \end{bmatrix} \begin{bmatrix} V_2 \\ I_2 \end{bmatrix},$$
(2.48)

where

$$A = \frac{V_1}{V_2}\bigg|_{I_2=0} = \cosh(\gamma l) \qquad B = \frac{V_1}{I_2}\bigg|_{V_2=0} = Z_0 \, \sinh(\gamma l),$$
(2.49)

and

$$C = \frac{I_1}{V_2}\bigg|_{I_2=0} = \frac{1}{Z_0} \sinh(\gamma l) \qquad D = \frac{I_1}{I_2}\bigg|_{V_2=0} = \cosh(\gamma l).$$
(2.50)

The values of A, B, C, D are dependent only on the secondary parameters; *i.e.*, they are independent of the termination impedance. $ABCD$ models are particularly useful when one wants to model lines consisting of several concatenations of sublines having different electrical characteristics (*i.e.*, different secondary parameters). For instance, one can consider a line that is formed by the concatenation of two uniform sublines of length l_1 and l_2 with parameters A_1, B_1, C_1, D_1, and A_2, B_2, C_2, D_2, respectively. This situation is depicted in Figure 2.10, and the $ABCD$ matrices are shown in Equation 2.51.

$$\begin{bmatrix} V_1 \\ I_1 \end{bmatrix} = \begin{bmatrix} A_1 & B_1 \\ C_1 & D_1 \end{bmatrix} \begin{bmatrix} V_2 \\ I_2 \end{bmatrix} \qquad \begin{bmatrix} V_2 \\ I_2 \end{bmatrix} = \begin{bmatrix} A_2 & B_2 \\ C_2 & D_2 \end{bmatrix} \begin{bmatrix} V_3 \\ I_3 \end{bmatrix}.$$
(2.51)

The product of the two $ABCD$ matrices models the concatenation of the two sublines. From Equation 2.51,

$$\begin{bmatrix} V_1 \\ I_1 \end{bmatrix} = \begin{bmatrix} A_1 & B_1 \\ C_1 & D_1 \end{bmatrix} \begin{bmatrix} A_2 & B_2 \\ C_2 & D_2 \end{bmatrix} \begin{bmatrix} V_3 \\ I_3 \end{bmatrix}.$$
(2.52)

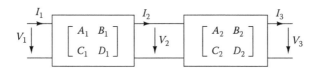

FIGURE 2.10
Concatenation of two uniform sublines.

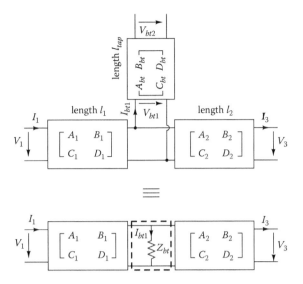

FIGURE 2.11
Concatenation of two uniform sublines with a bridged tap in between.

This result can be extended to a concatenation of n uniform sublines having n sets of secondary parameters; *i.e.*,

$$\begin{bmatrix} V_1 \\ I_1 \end{bmatrix} = \prod_{i=1}^{n} \begin{bmatrix} A_i & B_i \\ C_i & D_i \end{bmatrix} \begin{bmatrix} V_{n+1} \\ I_{n+1} \end{bmatrix}. \tag{2.53}$$

When a line contains a bridged tap, $ABCD$ modelling can also be used. One can consider that two sections of uniform lines of length l_1 and l_2 are interconnected with a bridged tap in between. The unused section of twisted pair (the bridged tap) with length l_{tap} is equivalent to an impedance Z_{bt} at the interconnection of the two sections of lines, as illustrated in Figure 2.11.

$$\begin{bmatrix} V_{bt1} \\ I_{bt1} \end{bmatrix} = \begin{bmatrix} A_{bt} & B_{bt} \\ C_{bt} & D_{bt} \end{bmatrix} \begin{bmatrix} V_{bt2} \\ 0 \end{bmatrix}. \tag{2.54}$$

From Equation 2.54, it is clear that

$$Z_{bt} = \frac{A_{bt}}{C_{bt}} = Z_{0bt} \coth(\gamma_{bt} \, l_{tap}), \tag{2.55}$$

where Z_{0bt} is the characteristic impedance of the bridged tap section, and γ_{bt} is its propagation constant. The equivalent $ABCD$ model of a single impedance Z that forms a two-port as shown in Figure 2.12 is given in the same figure. The overall $ABCD$ model of the concatenation of sublines involving a bridged tap as shown in Figure 2.11 is given by

$$\begin{bmatrix} V_1 \\ I_1 \end{bmatrix} = \begin{bmatrix} A_1 & B_1 \\ C_1 & D_1 \end{bmatrix} \begin{bmatrix} 1 & 0 \\ \dfrac{C_{bt}}{A_{bt}} & 1 \end{bmatrix} \begin{bmatrix} A_2 & B_2 \\ C_2 & D_2 \end{bmatrix} \begin{bmatrix} V_3 \\ I_3 \end{bmatrix}, \tag{2.56}$$

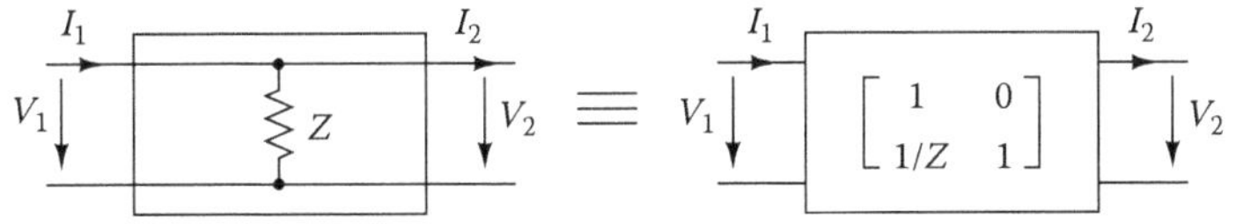

FIGURE 2.12
ABCD modelling of an impedance in parallel.

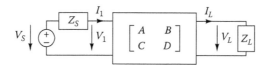

FIGURE 2.13
Two-port with *ABCD* model loaded with termination impedances Z_S and Z_L.

and therefore in this specific case,

$$\begin{bmatrix} A & B \\ C & D \end{bmatrix} = \begin{bmatrix} A_1 & B_1 \\ C_1 & D_1 \end{bmatrix} \begin{bmatrix} 1 & 0 \\ \dfrac{\tanh(\gamma_{bt}l_{tap})}{Z_{0bt}} & 1 \end{bmatrix} \begin{bmatrix} A_2 & B_2 \\ C_2 & D_2 \end{bmatrix}. \tag{2.57}$$

2.3.8 Transfer Function and Insertion Loss Associated with DSL Lines

One can consider a DSL line as a two-port network with *ABCD* parameters loaded with termination impedances Z_S and Z_L (see Figure 2.13). It can be of interest to express the loop transfer function between V_L and V_S, i.e.,

$$H = \frac{V_L}{V_S}, \tag{2.58}$$

as a function of A, B, C, D, Z_S, and Z_L. From the Equations 2.59,

$$\begin{bmatrix} V_1 \\ I_1 \end{bmatrix} = \begin{bmatrix} A & B \\ C & D \end{bmatrix} \begin{bmatrix} V_L \\ \dfrac{V_L}{Z_L} \end{bmatrix}, \quad V_S = Z_S I_1 + V_1, \tag{2.59}$$

the transfer function is obtained as

$$H = \frac{Z_L}{AZ_L + B + Z_S(CZ_L + D)}. \tag{2.60}$$

From a measurement perspective, it can be difficult to access V_S due to internal loading of the generator V_S, and hence, for practical reasons, telecommunication engineers prefer to use the concept of insertion loss instead of the transfer function. The principle of insertion loss can be understood from Figure 2.14. Setting the switches of Figure 2.14 to position 1, 1′, the two-port network is effectively absent, and Z_L and Z_S are directly connected. Setting the switches to 2, 2′, the two-port network is now inserted. The ratio of power across Z_L

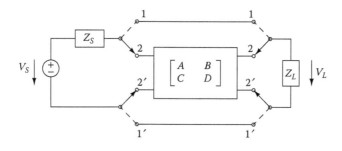

FIGURE 2.14
Insertion loss measurement principle.

before and after insertion of the two-port is called the insertion loss:

$$IL = 10 \log 10 \frac{P_{before\ ins.}}{P_{after\ ins.}}. \tag{2.61}$$

If V_{L1} is the voltage across Z_L before insertion of the two-port network (line), and V_{L2} is the voltage after insertion of the line,

$$IL = 10 \log 10 \frac{Re\left(\frac{|V_{L1}|^2}{Z_L}\right)}{Re\left(\frac{|V_{L2}|^2}{Z_L}\right)} = 20 \log 10 \left|\frac{V_{L1}}{V_{L2}}\right|. \tag{2.62}$$

Equation 2.63 follows by expressing Equation 2.62 in terms of A, B, C, D, V_S, Z_S, and Z_L.

$$IL = -20 \log 10 \left|\frac{Z_L + Z_S}{AZ_L + B + Z_S(CZ_L + D)}\right|. \tag{2.63}$$

Using Equation 2.46, one can also express the return loss in terms of A, B, C, D, V_S, Z_S, and Z_L; i.e.,

$$RL = 20 \log 10 \left|\frac{AZ_L + B + Z_0(CZ_L + D)}{AZ_L + B - Z_0(CZ_L + D)}\right|. \tag{2.64}$$

2.3.9   Scattering Parameters

ABCD parameter modelling is one modelling solution out of many. The *ABCD* method is particularly suited when a cable transfer function made of a concatenation of several pieces of homogeneous lines is to be computed. The *ABCD* parameters are also very intuitive because they deal with voltages and currents across two-port networks. As soon as measurements are involved, it is less appropriate to deal directly with *ABCD* parameters, because it is difficult to measure them at high frequency with reliable precision. At high frequencies, engineers prefer to measure transmitted and reflected powers instead of dealing with *ABCD* parameters, which are subject to a considerable range of magnitude and imprecision due to sensitivity of measurement. The modelling technique that has been introduced by circuit theorists in order to deal with transmitted and reflected power across two-port networks is the scattering parameter formalism. Less intuitive than the *ABCD* parameter modelling, the scattering parameter formalism deals with travelling waves, which are linear combinations of the voltages and the current in networks. The scattering parameters (so-called *S* parameters) are particularly suitable for problems of power transfer of networks designed to be terminated by resistive loads, as is the case in DSL at high frequencies. They involve reflection and transmission coefficients having finite range, which are measurable with high precision by modern network analyzers. This section presents the main results linked with the *S* parameters with a short introduction for the reader who is encountering the scattering matrix for the first time.

The two-port of Figure 2.15 is considered, which is powered with a generator of resistance R_G and loaded with a resistive load R_L. A normalization impedance R_n is defined, the role of which will become clear later. The incident waves at port 1 and port 2 are (respectively) arbitrarily defined as ξ_1 and ξ_2 such that

$$\xi_1 = \frac{V_1}{\sqrt{R_n}} + i_1 \sqrt{R_n}, \tag{2.65}$$

$$\xi_2 = \frac{V_2}{\sqrt{R_n}} + i_2 \sqrt{R_n}. \tag{2.66}$$

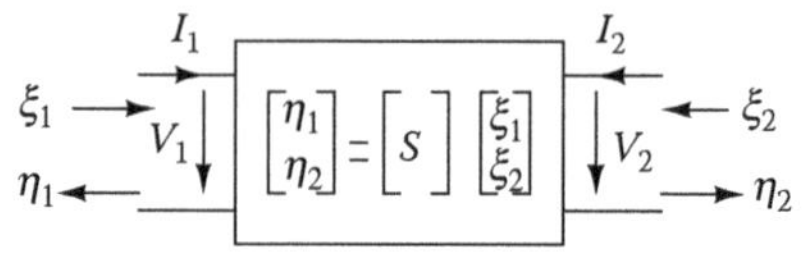

FIGURE 2.15
Two-port network.

The reflected waves at port 1 and port 2 are (respectively) arbitrarily defined as η_1 and η_2 such that

$$\eta_1 = \frac{V_1}{\sqrt{R_n}} - i_1 \sqrt{R_n}, \qquad (2.67)$$

and

$$\eta_2 = \frac{V_2}{\sqrt{R_n}} - i_2 \sqrt{R_n}. \qquad (2.68)$$

The scattering matrix S is defined as the matrix providing the relationship between the reflected and incident waves; *i.e.*,

$$\begin{bmatrix} \eta_1 \\ \eta_2 \end{bmatrix} = S \begin{bmatrix} \xi_1 \\ \xi_2 \end{bmatrix} = \begin{bmatrix} S_{11} & S_{12} \\ S_{21} & S_{22} \end{bmatrix} \begin{bmatrix} \xi_1 \\ \xi_2 \end{bmatrix}. \qquad (2.69)$$

By using elementary circuit theory, one can find the relationship between S and the $ABCD$ parameters; *i.e.*,

$$S = \frac{1}{A + \frac{B}{R_n} + C R_n + D} \begin{bmatrix} A + \frac{B}{R_n} - C R_n - D & 2(AD - BC) \\ 2 & -A + \frac{B}{R_n} - C R_n + D \end{bmatrix}. \qquad (2.70)$$

By using the relationships in Equations 2.49 and 2.50, one can deduce from Equation 2.70 the expression of the scattering matrix in terms of γ and Z_0 for a homogeneous line of length l; *i.e.*,

$$S = \frac{1}{2 + \left(\frac{Z_0}{R_n} + \frac{R_n}{Z_0}\right) \tanh(\gamma l)} \begin{bmatrix} \left(\frac{Z_0}{R_n} - \frac{R_n}{Z_0}\right) \tanh(\gamma l) & \frac{2}{\cosh(\gamma l)} \\ \frac{2}{\cosh(\gamma l)} & \left(\frac{Z_0}{R_n} - \frac{R_n}{Z_0}\right) \tanh(\gamma l) \end{bmatrix}. \qquad (2.71)$$

Checking the power flow, one can now justify the definitions of incident and reflected waves. It is straightforward to show that the power flow into port 1 is

$$\Re(V_1 i_1^*) = \frac{[\xi_1 + \eta_1]}{2\sqrt{R_n}} \sqrt{R_n} \left[\frac{\xi_1 + \eta_1}{2}\right]^* = \frac{|\xi_1|^2 - |\eta_1|^2}{4}. \qquad (2.72)$$

In a similar fashion, one can find the power flow into port 2; *i.e.*,

$$\Re(V_2 i_2^*) = \frac{[\xi_2 + \eta_2]}{2\sqrt{R_n}} \sqrt{R_n} \left[\frac{\xi_2 + \eta_2}{2}\right]^* = \frac{|\xi_2|^2 - |\eta_2|^2}{4}. \qquad (2.73)$$

Equations 2.72 and 2.73 show that the power entering a port is the difference between the power induced by the incident wave and the power induced by the reflected wave. The two-port network of Figure 2.16 is now considered, in which the loads at both ports are such that $R_G = R_L = R_n$. The power flow across the two-port network is of interest. In the DSL

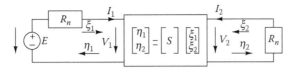

FIGURE 2.16
Two-port network loaded with R_n.

transmission band this resistance is the usual load impedance which is in a range between 100 Ω and 135 Ω. In this specific case, one obtains

$$\xi_1 = \frac{V_1}{\sqrt{R_n}} + \sqrt{R_n}i_1 = \frac{V_1 + R_n i_1}{\sqrt{R_n}} = \frac{E}{\sqrt{R_n}}, \tag{2.74}$$

$$\xi_2 = \frac{V_2}{\sqrt{R_n}} + \sqrt{R_n}i_2 = \frac{-R_n i_2 + R_n i_2}{\sqrt{R_n}} = 0, \tag{2.75}$$

$$\eta_1 = S_{11}\xi_1 \qquad \eta_2 = S_{21}\xi_1. \tag{2.76}$$

The active power transmitted to the load of port 2 is

$$\Re(-V_2 i_2^*) = \frac{|\eta_2|^2}{4}. \tag{2.77}$$

In a similar fashion, the active power reflected at port 1 is

$$\Re(-V_1 i_1^*) = \frac{|\eta_1|^2}{4}. \tag{2.78}$$

Recalling that the maximum power available from the generator is $\frac{|E|^2}{4R_n}$, and using Equations 2.74, 2.76, and 2.77, one finds for the power delivered to the load

$$P_{deliv.} = |S_{21}|^2 P_{max\ av.}. \tag{2.79}$$

Similarly, power reflected at port 1 is obtained by using Equations 2.74, 2.76, and (2.78); *i.e.*,

$$P_{reflect.} = |S_{11}|^2 P_{max\ av.}. \tag{2.80}$$

For a passive device such as a cable, S_{21} and S_{11} are such that $|S_{21}(j\omega)| <= 1$ and $|S_{22}(j\omega)| <= 1$. The coefficient $S_{21}(j\omega)$ is defined as the transmission coefficient (from port 1 to port 2), whereas $S_{11}(j\omega)$ is the reflection coefficient (of port 1). Similar definitions hold for S_{12} and S_{22} by exchanging port 1 for port 2 and vice versa. Observe that the return loss already defined as the inverse ratio of reflected power to maximum power available from load is simply given as

$$RL = -10\log 10|S_{22}|^2. \tag{2.81}$$

In a similar way, if one wants to express the insertion loss using the measurement principle shown in Figure 2.14 with $Z_S = Z_L = R_n$, one finds that the power before cable insertion is $\frac{|E|^2}{4R_n}$, and the power after cable insertion is given by Equation 2.79. Therefore, one obtains the simple formula

$$IL = -10\log 10|S_{21}|^2. \tag{2.82}$$

For a homogeneous cable of length l, with characteristic impedance Z_0, propagation constant γ, and loaded on both sides with a resistance R_n, Equations 2.71 and 2.81 show that

$$RL = -10\log 10 \left\| \frac{\left(\frac{Z_0}{R_n} - \frac{R_n}{Z_0} \right)}{2 + \left(\frac{Z_0}{R_n} + \frac{R_n}{Z_0} \right)\tanh(\gamma l)} \right\|^2. \tag{2.83}$$

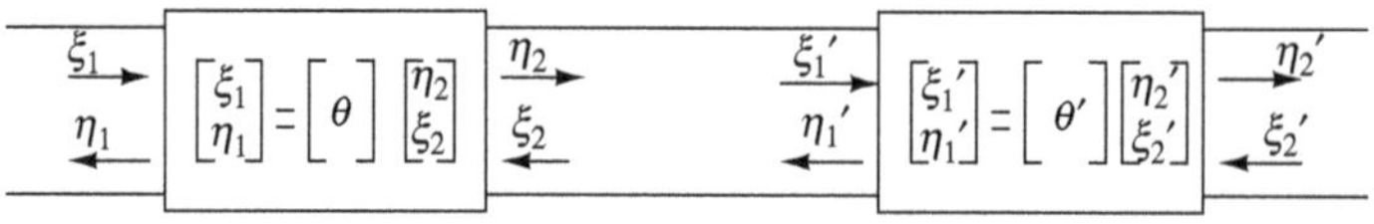

FIGURE 2.17
Transfer matrix principle for a cascade of two-ports.

For the same cable, the expressions in Equations 2.71 and 2.82 show that

$$IL = -10 \log 10 \left\| \frac{2}{2\cosh(\gamma l) + \left(\frac{Z_0}{R_n} + \frac{R_n}{Z_0}\right)\sinh(\gamma l)} \right\|^2 . \tag{2.84}$$

Although the scattering formalism is very practical in terms of measurements and brings simple expressions of return loss and insertion loss, it turns out to be very impractical when one has to compute the effect of a cascade of several homogeneous cables. In other words, if two two-ports are cascaded, the overall scattering matrix of this cascade is not the product of the individual matrices. Therefore, instead of working directly with the scattering matrix, engineers prefer to work with the wave transfer matrix, the manipulation of which is easier when cascading two-ports. The wave transfer matrix θ is such that

$$\begin{bmatrix} \xi_1 \\ \eta_1 \end{bmatrix} = \theta \begin{bmatrix} \eta_2 \\ \xi_2 \end{bmatrix} = \begin{bmatrix} \theta_{11} & \theta_{12} \\ \theta_{21} & \theta_{22} \end{bmatrix} \begin{bmatrix} \eta_2 \\ \xi_2 \end{bmatrix} . \tag{2.85}$$

When cascading two two-port networks, the waves at the interface port are such that the reflecting wave of one port is the incident wave of the other and vice versa. The arrangement of waves in Equation 2.85 is such that the wave transfer matrix of a cascade of two-port networks is the product of the individual transfer matrices. This is illustrated in Figure 2.17. If one defines θ as the transfer matrix of the two-port network on the left and θ' as the transfer matrix of the two-port network on the right, one finds

$$\begin{bmatrix} \xi_1 \\ \eta_1 \end{bmatrix} = \theta \begin{bmatrix} \eta_2 \\ \xi_2 \end{bmatrix} \qquad \begin{bmatrix} \eta_2 \\ \xi_2 \end{bmatrix} = \begin{bmatrix} \xi_1' \\ \eta_1' \end{bmatrix} = \theta' \begin{bmatrix} \eta_2' \\ \xi_2' \end{bmatrix} , \tag{2.86}$$

and it follows from Equation 2.86 that

$$\begin{bmatrix} \xi_1 \\ \eta_1 \end{bmatrix} = \theta\theta' \begin{bmatrix} \eta_2' \\ \xi_2' \end{bmatrix} . \tag{2.87}$$

Elementary circuit theory shows that the relationship between the wave transfer matrix and the $ABCD$ parameters is given by

$$\theta = \frac{1}{2} \begin{bmatrix} A + \dfrac{B}{R_n} + C R_n + D & A - \dfrac{B}{R_n} + C R_n - D \\ A + \dfrac{B}{R_n} - C R_n - D & A - \dfrac{B}{R_n} - C R_n + D \end{bmatrix} . \tag{2.88}$$

By substituting Equations 2.49 and 2.50 into Equation 2.88, one finds that for a homogeneous portion of cable of length l of characteristic impedance Z_0 and propagation constant γ terminated in R_n, the wave transfer matrix θ is

$$\theta = \frac{1}{2} \begin{bmatrix} 2\cosh(\gamma l) + \sinh(\gamma l)\left(\dfrac{Z_0}{R_n} + \dfrac{R_n}{Z_0}\right) & -\sinh(\gamma l)\left(\dfrac{Z_0}{R_n} - \dfrac{R_n}{Z_0}\right) \\ \sinh(\gamma l)\left(\dfrac{Z_0}{R_n} - \dfrac{R_n}{Z_0}\right) & 2\cosh(\gamma l) - \sinh(\gamma l)\left(\dfrac{Z_0}{R_n} + \dfrac{R_n}{Z_0}\right) \end{bmatrix} . \tag{2.89}$$

By working with wave transfer matrices when cascading several pieces of homogeneous cables, one can obtain the overall transfer matrix of the overall network and compute the corresponding scattering matrix by using the following correspondence between the wave transfer matrix and scattering matrix.

$$S = \begin{bmatrix} \dfrac{\theta_{21}}{\theta_{11}} & \dfrac{\theta_{11}\theta_{22} - \theta_{12}\theta_{21}}{\theta_{11}} \\ \dfrac{1}{\theta_{11}} & -\dfrac{\theta_{12}}{\theta_{11}} \end{bmatrix}. \tag{2.90}$$

2.3.10 Impedance and Admittance Modelling Parameters

Perhaps of less practical importance are the classical impedance and admittance parameter models; they are included here for completeness. Considering the two-port network of Figure 2.15, the Z-impedance parameter model is defined as the Z matrix such that

$$\begin{bmatrix} V_1 \\ V_2 \end{bmatrix} = Z \begin{bmatrix} i_1 \\ i_2 \end{bmatrix} = \begin{bmatrix} Z_{11} & Z_{11} \\ Z_{21} & Z_{22} \end{bmatrix} \begin{bmatrix} i_1 \\ i_2 \end{bmatrix}. \tag{2.91}$$

Elementary algebra shows that the Z-parameter model is related to the *ABCD*-parameter model under the relationship in Equation 2.92:

$$Z = \begin{bmatrix} \dfrac{A}{C} & \dfrac{AD - BC}{C} \\ \dfrac{1}{C} & \dfrac{D}{C} \end{bmatrix}. \tag{2.92}$$

Inserting the relations in Equations 2.49 and 2.50 in Equation 2.92, one finds that a homogenous cable of length l, with characteristic impedance Z_0 and propagation constant γ, obeys the following Z-parameter model.

$$Z = \begin{bmatrix} \dfrac{Z_0}{\tanh(\gamma l)} & \dfrac{Z_0}{\sinh(\gamma l)} \\ \dfrac{Z_0}{\sinh(\gamma l)} & \dfrac{Z_0}{\tanh(\gamma l)} \end{bmatrix}. \tag{2.93}$$

Considering the two-port network of Figure 2.15, the Y(admittance)-parameter model is defined as

$$\begin{bmatrix} i_1 \\ i_2 \end{bmatrix} = Y \begin{bmatrix} V_1 \\ V_2 \end{bmatrix} = \begin{bmatrix} Y_{11} & Y_{11} \\ Y_{21} & Y_{22} \end{bmatrix} \begin{bmatrix} V_1 \\ V_2 \end{bmatrix}. \tag{2.94}$$

Derivation of the relationship between the Y-model and *ABCD*-model shows that

$$Y = \begin{bmatrix} \dfrac{D}{B} & \dfrac{BC - AD}{B} \\ \dfrac{1}{B} & \dfrac{A}{B} \end{bmatrix}. \tag{2.95}$$

Inserting the relations in Equations 2.49 and 2.50 into Equation 2.95, one finds that a homogeneous cable of length l, with characteristic impedance Z_0 and propagation constant

γ, obeys the following Y-parameter model.

$$Y = \begin{bmatrix} \dfrac{1}{Z_0 \tanh(\gamma l)} & \dfrac{-1}{Z_0 \sin(\gamma l)} \\ \dfrac{-1}{Z_0 \sinh(\gamma l)} & \dfrac{1}{Z_0 \tanh(\gamma l)} \end{bmatrix}. \tag{2.96}$$

2.4 Generic Models of DSL Cables

In addition to huge databases of measurements that have been made available through co-operative studies encouraged by standardization bodies, models of cables have been derived in order to accurately describe the behavior of the primary parameters. Utilizing these models avoids referencing large tables of measurements when computing the two-port characteristics of any DSL line. Although physical principles have inspired the models, many are rather empirical. Ease of use and compactness have been the principal requirements. This section follows part of the presentation in [Van den Brink 1998]. The report introduces the principle of measuring large sections of homogeneous cables (rather than short ones as recommended previously) and performing the full two-port extraction of the four primary parameters on these long sections. This approach resulted in an improvement in accuracy, as cable sections of several hundred meters can now be characterized as full two-port networks (magnitude and phase) with a phase accuracy corresponding to 1 cm of cable length uncertainty. Two classes of model parameters are proposed: (a) model parameters focusing on modelling the primary parameters, $RLCG$, and (b) model parameters focusing on the modelling of secondary parameters. The first class of models was proposed by British Telecom [Cook 1996] [Lawrence 1996] and KPN [Van den Brink 1997a] [Van den Brink 1997b] [Van den Brink 1998]. These models have proven to be especially useful to describe cable behavior over a range of frequencies from DC to tens of MHz with good precision. The second class of models was proposed by Swisscom [Pythoud 1998] and Deutche Telekom [Pollakowski 1996]. In this chapter, the presentation is restricted to the first class of model; for details on the second class of model, the interested reader is referred to the ETSI report [Van den Brink 1998].

2.4.1 The British Telecom Models 0 and 1 (*RLCG* Modelling)

Used all over the world now, these two empirical models were first proposed by John Cook of British Telecom [Cook 1996] [Lawrence 1996].

2.4.1.1 Empirical Model for Resistance

As frequency increases, the current flow in a wire becomes less uniform across the cross section and tends to concentrate close to the wire surface; this behavior is known as the skin effect. As the skin effect accounts for the current flow at high frequencies, the resistance of the wire increases drastically. It is well known that once the skin effect becomes dominant, it increases in proportion to $\sqrt{f}$. At a range of low frequencies below where the skin effect is dominant, the wire resistance is close to the DC resistance. This has suggested an empirical model of the form

$$R(f) = \sqrt[4]{\left(R_{oc}^4 + a_c f^2\right)}. \tag{2.97}$$

To further complicate matters, some types of aerial drop-wire are bimetallic, where a copper outer conductor is mechanically reinforced by a steel inner core. As steel is a material with a large relative permeability, the skin effect dominates more at lower frequencies than in the copper conductor. It has been found that a good empirical model can be obtained by extending the model Equation 2.97 to include a hypothetical separate conductor with the same model as above but different parameters, suggesting the overall model:

$$R(f) = \frac{1}{\frac{1}{\sqrt[4]{(R_{oc}^4 + a_c f^2)}} + \frac{1}{\sqrt[4]{(R_{os}^4 + a_s f^2)}}}, \tag{2.98}$$

where R_{oc} is the DC resistance due to copper and R_{os} is the DC resistance due to steel. The separate skin effects for copper and steel are accounted for by a_c and a_s.

2.4.1.2 Empirical Model for Inductance

At low frequencies where the skin effect is not dominant, the parameter L exhibits a constant inductance L_0. At high frequencies, when the skin effect is dominant, the L parameter tends toward a constant inductance L_∞. The inductance model has been empirically modelled through

$$L(f) = \frac{L_0 + L_\infty \left(\frac{f}{f_m}\right)^b}{1 + \left(\frac{f}{f_m}\right)^b}, \tag{2.99}$$

where b and f_m are parameters that control the transition between L_0 and L_∞ across the frequency axis.

2.4.1.3 Appropriate Model for Conductance

A suitable model for cable conductance has been found to be

$$G(f) = g_0 \, f^{g_e}, \tag{2.100}$$

where g_0 and g_e control the behavior of an exponentially increasing dielectric loss.

2.4.1.4 Empirical Model for Capacitance

A suitable model for capacitance has been found to be

$$C(f) = C_\infty + C_0 f^{-c_e}. \tag{2.101}$$

For good dielectrics, C_0 can be considered to be negligible, and the capacitance model is C_∞. Poorer dielectrics such as PVC may need the complete model given by Equation 2.101.

2.4.2 The British Telecom Model #0 Parameters

The so-called BT model #0 is described by the 11 parameters R_{oc}, a_c, L_0, L_∞, f_m, g_0, N_{ge}, C_∞, C_0, N_{ce}, and N_b. These parameters allow the modelling of primary parameters $RLCG$ through Equations 2.97, 2.99, 2.100, and 2.101.

2.4.3 The British Telecom Model #1 Parameters

The so-called BT model #1 is described by the 13 parameters R_{oc}, a_c, R_{os}, a_s, L_0, L_∞, f_m, g_0, N_{ge}, C_∞, C_0, N_{ce}, and N_b. These parameters allow the modelling of primary parameters $RLCG$ through Equations 2.98, 2.99, 2.100, and 2.101. Note that when R_{os} or a_s goes to infinity, the BT model #1 reduces to the BT model #0.

TABLE 2.2

BT #0 Modelling Parameters for European Cables of Several Sections as Described by ETSI in the G.996.1 Recommendation

Cable Section	0.32 mm	0.40 mm	0.5 mm	0.63 mm	0.90 mm
r_{0c} (Ω/km)	409	280	179.2	113	55.1
a_c (Ω^4/km^4 Hz^2)	0.3822	0.0969	0.0561	0.0257	0.0094
L_0 (μH/km)	607	587.3	674.6	699.4	750.9
L_∞ (μH/km)	500	426	532.7	477.2	520.5
b	5.269	1.385	1.195	1.0956	0.9604
f_m (Hz)	609000	745900	664700	265800	123800
C_∞ (nF/km)	40	50	50	45	40
C_0 (nF/km)	0	0	0	0	0
c_e	1	1	1	1	1
g_0 (S/km)	0	0	0	0	0
g_e	1	1	1	1	1

2.4.4   Examples of Cable Modelling Using the BT #0 Model

Table 2.2 shows the modelling parameters of cables of different sections according to the BT model #0 previously described. The parameters were picked from the ITU-T document [G996.1 2003], which specifies test procedures for DSL transceivers. Applying Equations 2.97, 2.99, 2.100, and 2.101 by directly using the parameters in Table 2.2 will provide R (in Ω/km), L (in μH/km), C (in nF/km), and G (in S/km). The corresponding R and L parameter behaviors with respect to frequency are shown respectively in Figure 2.18 and Figure 2.19. The C parameter is, in this modelling case, constant and equal to C_∞. The G parameter has been considered negligible with respect to $C\omega$.

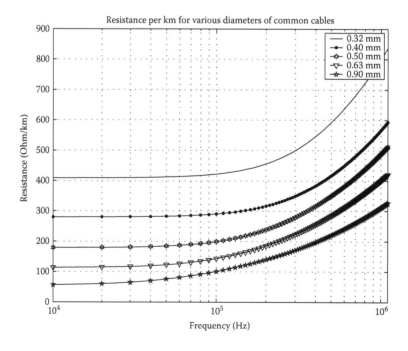

FIGURE 2.18
Resistance per km for various diameters of common cables, BT model #0.

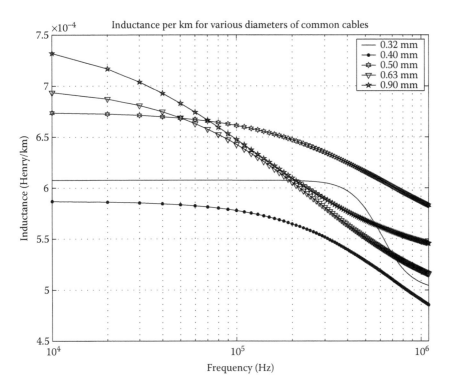

FIGURE 2.19
Inductance per km for various diameters of common cables, BT model #0.

2.4.5 The KPN Models #0 and #1 (*RLCG* Modelling)

Introduced by Rob van den Brink [Van den Brink 1997a] [Van den Brink 1997b], the empirical models KPN #0 and #1 of the Royal PTT Netherlands are focussed on modelling *RLCG* parameters. These models attempt to describe the skin effect in a manner that closely relates to the underlying physics, and are valid from DC to tens of MHz.

2.4.6 The KPN Model #0

This model needs only four line constants. The *RLCG* parameters are defined as follows.

$$R(f) = \Re \left[j2\pi f \frac{Z_{0\infty}}{c} + R_{ss00} \left[\frac{1}{4} + \chi \coth \left(\frac{4}{3}\xi \right) \right] \right], \tag{2.102}$$

$$L(f) = \Re \left(\frac{1}{j2\pi f} \left[j2\pi f \frac{Z_{0\infty}}{c} + R_{ss00} \left[\frac{1}{4} + \chi \coth \left(\frac{4}{3}\chi \right) \right] \right] \right), \tag{2.103}$$

$$G(f) = 2\pi f \frac{\tan(\phi)}{c Z_{0\infty}}, \tag{2.104}$$

$$C(f) = \frac{1}{c Z_{0\infty}}, \tag{2.105}$$

where χ in Equations 2.102 and 2.103 is such that

$$\chi = \chi(f) = (1 + j)\sqrt{f \frac{\mu_0}{R_{ss00}}}, \tag{2.106}$$

and $\Re$ denotes the real part function in Equations 2.102 and 2.103. The *RLCG* model of Equations 2.102, 2.103, 2.104, and 2.105 is made of four parameters $Z_{0\infty}$, R_{ss00}, ϕ, and c (c is the propagation speed lower than c_0, the speed of light). In this model, $\mu_0 = 4\pi 10^{-7}$ H/m is the permeability of a vacuum.

2.4.7   The KPN Model #1

For fine tuning purposes, seven empirical parameters have been added to the KPN model #0. The model [Van den Brink 1998] obeys the following equations.

$$R(f) = \Re\left[j2\pi f \frac{Z_{0\infty}}{c} + R_{ss00}\left[1 + K_l K_f \left(\chi \coth\left(\frac{4}{3}\chi\right) - \frac{3}{4}\right)\right]\right], \tag{2.107}$$

$$L(f) = \Re\left(\frac{1}{j2\pi f}\left[j2\pi f \frac{Z_{0\infty}}{c} + R_{ss00}\left[1 + K_l K_f \left(\chi \coth\left(\frac{4}{3}\chi\right) - \frac{3}{4}\right)\right]\right]\right), \tag{2.108}$$

$$G(f) = \frac{\tan(\phi)}{c\,Z_{0\infty}}(2\pi f)^M, \tag{2.109}$$

$$C(f) = \frac{1}{c\,Z_{0\infty}}\left(1 + \frac{(K_c - 1)}{1 + \left(\frac{f}{f_{c0}}\right)^N}\right), \tag{2.110}$$

where χ in Equations 2.107 and 2.108 is such that

$$\chi = \chi(f) = (1+j)\sqrt{f\frac{\mu_0}{R_{ss00}}\frac{1}{K_n K_f}} \tag{2.111}$$

and $\Re$ denotes the real part function in Equations 2.107 and 2.108. There are in total eleven parameters, which are $Z_{0\infty}$, R_{ss00}, ϕ, c, K_l, K_f, K_c, f_{c0}, K_n, M, and N. In many cases, it has been observed that fixing $K_n = 1$ and $M = 1$ and extracting the nine other parameters provides good matching between model and measurements.

2.4.8   Examples of Cable Modelling Using the KPN #1 Model

Some example loop models that are valid for the KPN #1 model have been borrowed from [Van den Brink 1998]. The majority of the Dutch distribution cable is primarily composed of 0.5 mm underground cables; occasionally, 0.8 mm wires are used to reach long distances. Four types of cables have been sampled from [Van den Brink 1998]: so-called KPN_L1 (0.5 mm), KPN_L2 (0.5 mm), KPN_L3 (0.8 mm), and KPN_L4 (0.5 mm). The eleven parameters corresponding to these four types of cables are given in Table 2.3. The corresponding primary parameters *RLCG* are given in Figures 2.20 to 2.23.

TABLE 2.3

Parameters for Dutch Cables Using the KPN #1 Model (from [Van den Brink 1998])

Cable	$Z_{0\infty}$	$\frac{c}{c_0}$	R_{ss00}	$2\pi \tan(\phi)$	K_f	K_l	K_n	K_c	N	f_{c0}	M
L1	136.651	0.79766	0.168145	0.13115	0.72	1.2	1	1.08258	0.7	4521710	1
L2	136.047	0.798958	0.168145	0.169998	0.7	1.1	1	1.08201	1	1862950	1
L3	137.527	0.850608	0.065682	0.114526	1	1	1	1.06967	1	559844	1
L4	137.005	0.787661	0.168145	0.153522	0.9	1	1	1.07478	1	557458	1

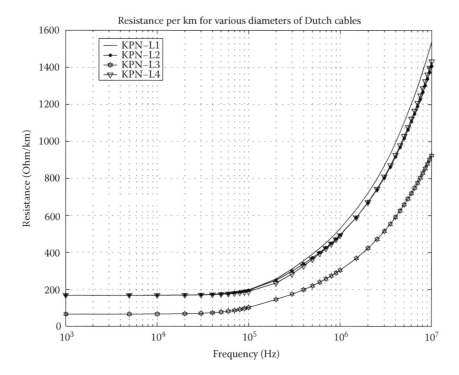

FIGURE 2.20
Resistance per km for various diameters of Dutch cables, KPN model #1.

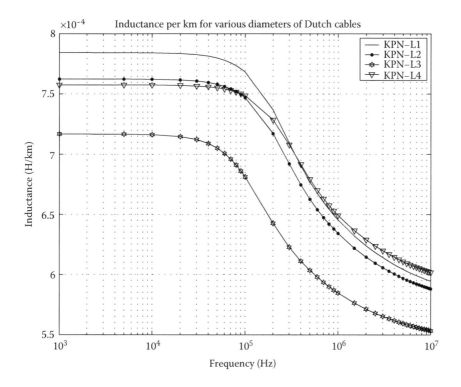

FIGURE 2.21
Inductance per km for various diameters of Dutch cables, KPN model #1.

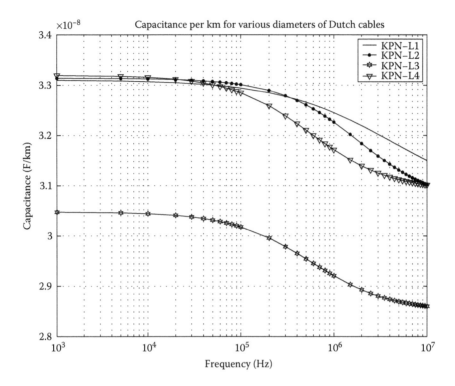

FIGURE 2.22
Capacitance per km for various diameters of Dutch cables, KPN model #1.

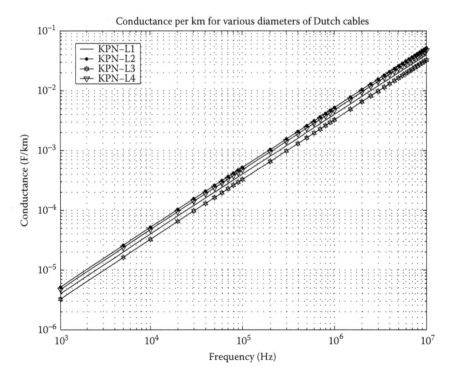

FIGURE 2.23
Conductance per km for various diameters of Dutch cables.

2.4.9 Other Models of Interest

In the previous models, an intrinsic limitation is that the series impedance $R(\omega) + jL(\omega)\omega$ as well as the parallel conductance $G(\omega) + jC(\omega)\omega$ have both real and imaginary parts that do not fulfill the Hilbert relationships. It can be useful to note that a function $F(s) = R(s) + jX(s)$ for $s = \sigma + j\omega$ fulfills the Hilbert conditions if

$$R(\omega) = \frac{2}{\pi} \int_0^\infty \frac{\omega' X(\omega') - \omega X(\omega)}{\omega'^2 - \omega^2} d\omega' \tag{2.112}$$

and

$$X(\omega) = \frac{2}{\pi} \int_0^\infty \frac{\omega' R(\omega') - \omega R(\omega)}{\omega'^2 - \omega^2} d\omega'. \tag{2.113}$$

When the series impedance $R(\omega) + jL(\omega)\omega$ and the parallel conductance $G(\omega) + jC(\omega)\omega$ have real and imaginary components that are not related according to Equations 2.112 and 2.113, the corresponding cable model does not exhibit a real impulse response. Therefore, time domain simulation is meaningless when using the previous BT and KPN models. In an effort to produce models accurate in both the frequency and time domains, the empirical MAR#1 and MAR#2 models [Musson 1998] have been introduced. These two models have series impedance $R(\omega) + jL(\omega)\omega$ and parallel conductance $G(\omega) + jC(\omega)\omega$, which fulfill the Hilbert conditions.

A very interesting nonempirical model of cable has been developed by Jason Jon Yoho in his doctoral thesis [Yoho 2001]. This physical model includes the skin effect, the influence of the permittivity of the insulating material, and the twisting effects. This physical model has been shown to accurately model the primary parameters between 10 kHz and tens of MHz.

2.5 Loop Configuration

2.5.1 Examples of Loop Configurations

Sections 2.2.4 and 2.3.1 give a good indication of why a large variety of loop configurations can exist in a given network. It is of value to consider the more representative loop configurations that can be encountered in practice. In a considerable effort to study, qualify, and improve the performance of DSL transmission, various standardization bodies have

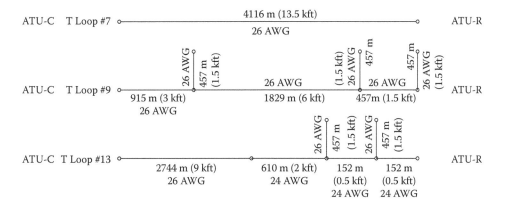

FIGURE 2.24
North American test lines of G.996.1 recommendation (ANSI loops).

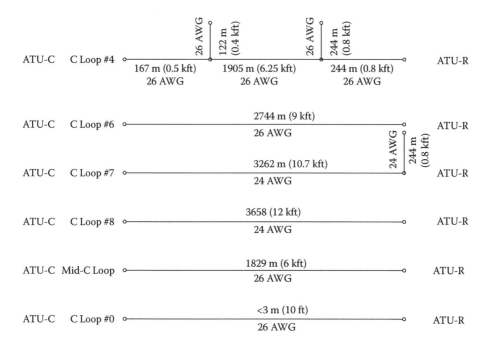

FIGURE 2.25
North American test lines of G.996.1 recommendation (CSA loops).

defined a number of loop configurations reflecting the scenarios of loop architectures expected to be found in real deployments. Due to differences in cable technology, network constraints, use of bridged taps, etc., the North American topologies differ from those used in Europe. Figures 2.24 to 2.27 show some loop topologies that have been adopted for North America and Europe. These are taken from the International Telecommunication

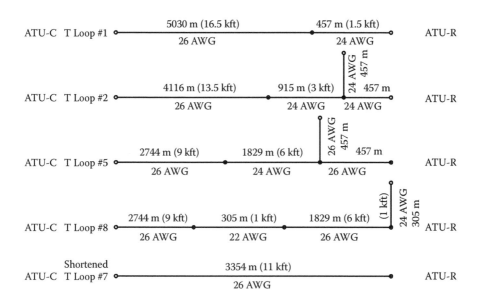

FIGURE 2.26
North American test lines of G.996.1 recommendation (ANSI loops).

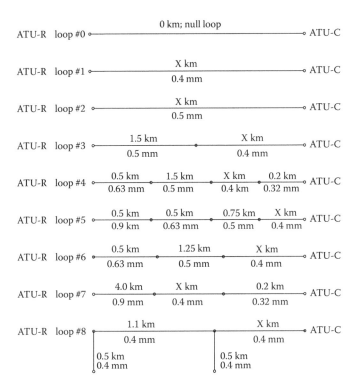

FIGURE 2.27
ETSI test lines of G.996.1 recommendation (ETSI loops).

Union Standardization G.996.1 recommendation, which details test procedures for digital subscriber line transceivers. Depending on the type of DSL transmission to be deployed (for example, VDSL, ADSL, HDSL, etc.), other configurations of test loops have been considered. The interested reader is referred to the relevant standards documents. Figures 2.24 to 2.26 describe the North American test lines included in the G.996.1 recommendation. Figure 2.27 corresponds to the European test loops approved by the European Telecommunications Standards Institute (ETSI). The reader will note that the lengths of the European loops are not fully specified and, instead, the variable X is used to designate some variable portion of loop length. The nominal length for the variable segment of each European loop is given in Table 2.4. Associated with this average length is an insertion loss value for the

TABLE 2.4

Examples of Loop insertion Loss for European Testing

Loop #	Nominal Length X for Insertion Loss 36 dB at 300 kHz	Nominal Length X for Insertion Loss 51 dB at 300 kHz	Nominal Length X for Insertion Loss 61 dB at 300 kHz
1	2.55 km	3.60 km	4.30 km
2	3.40 km	4.80 km	5.70 km
3	1.40 km	2.45 km	3.15 km
4	0.90 km	1.90 km	2.65 km
5	1.45 km	2.50 km	3.20 km
6	1.30 km	2.35 km	3.05 km
7	0.60 km	1.60 km	2.20 km
8	0.75 km	1.80 km	2.50 km

entire loop. For repeatability of measurement results (measurements are performed using analog line simulators), it is required that this insertion loss is achieved at a prescribed frequency. The line simulator consists of a concatenation of elementary analog circuits that emulate a small section of line (for example, 10 m sections). Therefore the engineer responsible for the measurements can add or remove a small section of simulated line in order to reduce the impairment due to inexact line simulation.

2.5.2 Insertion Loss of Some North American Reference Loops

The insertion loss associated with the North American loops of the G996.1 Recommendation has been computed and is shown in Figures 2.28 to 2.31. The primary parameters were computed using the BT model #1, with its 11 parameters taken from Table 2.5. The insertion loss is computed with 100 Ω source and termination impedances.

2.5.3 Insertion Loss of Some European Reference Loops

The insertion loss associated with the European loops of the G996.1 Recommendation has been computed and is shown in Figures 2.32 and 2.33. The primary parameters were computed using the BT model #1, with its 11 parameters taken from Table 2.2. The insertion loss is computed with 100 Ω source and termination impedances.

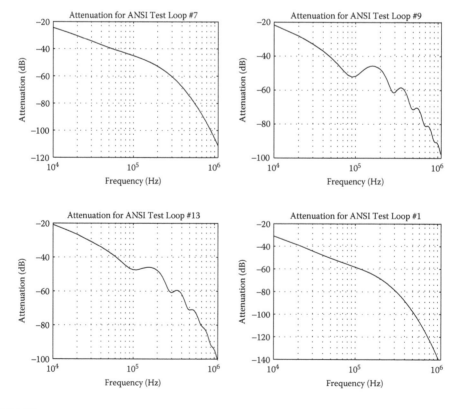

FIGURE 2.28
G.996.1 recommendation (ANSI loops).

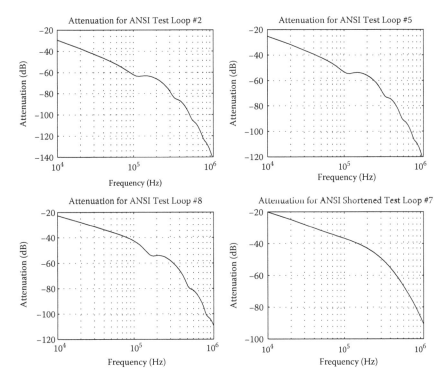

FIGURE 2.29
G.996.1 recommendation (ANSI loops).

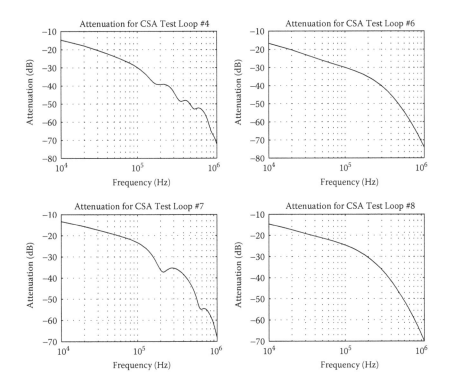

FIGURE 2.30
G.996.1 recommendation (CSA loops).

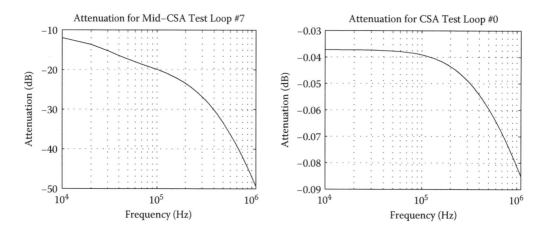

FIGURE 2.31
G.996.1 recommendation (CSA loops).

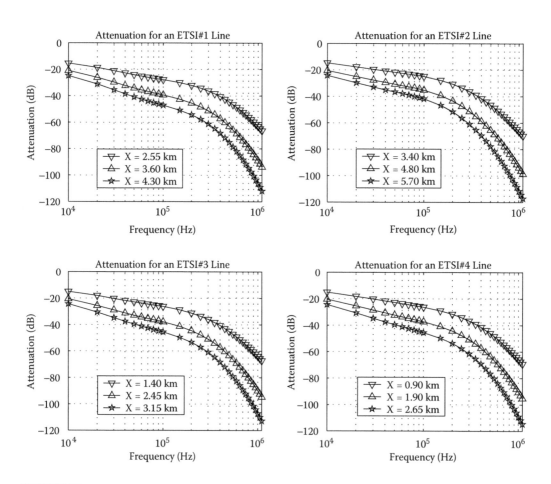

FIGURE 2.32
G.996.1 recommendation (ETSI loops).

TABLE 2.5

24-Gauge and 26-Gauge AWG Twisted Pair *RLCG* Modelling Parameters

Parameter	#24 Gauge	#26 Gauge
r_{0c} (Ω/km)	174.559	286.176
a_c (Ω^4/km^4 Hz^2)	0.05307	0.147697
L_0 (H/km)	617.296 10^{-6}	675.369 10^{-6}
L_∞ (H/km)	478.971 10^{-6}	488.952 10^{-6}
b	1.15298	0.929
f_m (Hz)	553760	806339
C_∞ (F/km)	51.56 10^{-9}	51.56 10^{-9}
C_0 (F/km)	0	0
c_e	0	0
g_0 (S/km)	0.235 10^{-12}	4.3 10^{-8}
g_e	1.38	0.70

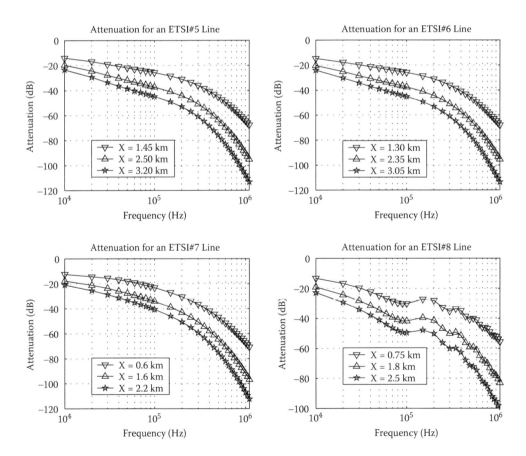

FIGURE 2.33

G.996.1 recommendation (ETSI loops).

References

[Bell 1876] A.G. Bell. *Improvement in Telegraphy.* US174,465 Letters Patent, dated March 7, 1876, application filed February 14, 1876.

[Bell 1881] A.G. Bell *Telephone-Circuit.* US244,426 Letters Patent, dated July 19, 1881, application filed June 4, 1881.

[Carlin 1956] H.J. Carlin. *The Scattering Matrix in Network Theory.* IRE Transactions on Circuit Theory, June 1956, pp. 88–97.

[Cook 1996] J.W. Cook. *Parametric Modeling of Twisted Pair Cables for VDSL.* ETSI contribution TD22, Vienna, Austria, March 1996.

[G996.1 2003] ITU-T G996.1 *Test Procedures for Digital Subscriber Line (DSL) Transceivers.*, 2003.

[Hughes 1997] H. Hughes. *Telecommunications Cables.* Wiley, Chichester, 1997.

[Lawrence 1996] V. B. Lawrence et al. *Broadband Access to the Home on Copper.* Bell Labs Technical Journal, Summer 1996, pp. 100–114.

[Musson 1998] *Maximum Likelihood Estimation of the Primary Parameters of Twisted Pair Cables.* ETSI-TM6 contribution TD8, [981t08a0], Madrid, Spain, 26–30 Jan. 1998.

[Pollakowski 1996] M. Pollakowski. *DTAG Cables Transmission Characteristics.* ETSI-TM6 Contribution TD40, Vienna, Austria, 18–22 March, 1996.

[Pupin 1900] M.I. Pupin. *Art of Reducing Attenuation of Electrical Waves and Apparatus Therefor.* US652,230 Letters Patent, dated June 19, 1900, application filed Dec. 14, 1899.

[Pythoud 1998] F. Pythoud. *Model of Swiss Access Network Cables.* ETSI-TM6 Contribution TD48, Madrid, Spain, 26–30 January, 1998.

[Rauschmayer 1999] D.J. Rauschmayer. *ADSL/VDSL Principles: A Practical and Precise Study of Asymmetric Digital Subscriber Lines and Very High Speed Digital Sunscriber Lines.* Macmillan, Indianapolis, IN, 1999.

[Starr 1999] T. Starr, J.M. Cioffi, and P.J. Silverman. *Understanding Digital Subscriber Line Technology.* Prentice Hall PTR, Upper Saddle River, NJ, 1999.

[Van den Brink 1998] R.F.M. van den Brink. *Cable Reference Models for Simulating Metallic Access Networks.* ETSI STC TM6 permanent document, June 1998.

[Van den Brink 1997a] R.F.M. van den Brink. *Measurements and Models on Dutch Cables.* ETSI-TM6 Contribution TD15, [971t15r1], Revision 1, Tel Aviv, Israel, 10–14 March, 1997.

[Van den Brink 1997b] R.F.M. van den Brink. *A Round Robin Test on Cable Measurements.* ETSI-TM6 Contribution TD16, [971t16r1], Revision 1, Tel Aviv, Israel, 10–14 March, 1997.

[Yoho 2001] J.J. Yoho. *Physically-Based Realizable Modeling and Network Synthesis of Subscriber Loops Utilized in DSL Technology.* Doctor of Philosophy Dissertation of the Virginia Polytechnic Institute and State University, October 2001, Blacksburg, VA.

3

$\blacksquare$

Noise and Noise Modelling on the Twisted Pair Channel

Rob H. Kirkby

CONTENTS

3.1   Crosstalk 72
   3.1.1   Empirical Models 74
      3.1.1.1   3CXT 78
      3.1.1.2   Cable Filling and the Meaning of N 78
      3.1.1.3   Regional Differences 79
      3.1.1.4   Verification 79
   3.1.2   Summary of Crosstalk PSD Formulae 80
3.2   Impulsive Noise 81
3.3   Noise from Faults 82
3.4   Engineering Measures 82
   3.4.1   Screening (Does Not Work) 82
   3.4.2   Enforced Continence 83
   3.4.3   Deployment Discipline 83
   3.4.4   Band Duplexing 83
   3.4.5   Interleaving 84
3.5   Modelling 84
   3.5.1   Use of Noise Models 84
   3.5.2   Production 85
      3.5.2.1   Network Model 85
      3.5.2.2   Sources 85
      3.5.2.3   Standards 85
      3.5.2.4   The Experts 86
      3.5.2.5   Measurement 86
      3.5.2.6   Coupling 86
      3.5.2.7   Summation 86
   3.5.3   Impulsive Noise 86
3.6   Mathematical Modelling 87
   3.6.1   Worst-Case Length for FEXT Coupling 87
   3.6.2   Length Dependence of NEXT 88
   3.6.3   NEXT and FEXT as Scattering 89
      3.6.3.1   NEXT 93
      3.6.3.2   FEXT 94
References 95

ABSTRACT   In the DSL channel, the dominant type of noise is crosstalk from other DSL systems; it typically sets the maximum data rate one may expect. Impulsive noise is also significant, being the cause of most errors in the delivered payload. This chapter discusses crosstalk in detail and practical noise modelling in general.

3.1   Crosstalk

Crosstalk is the leakage into one channel of signal power from another channel. For DSL, this means coupling between pairs in the same cable. The coupling mechanism is a consequence of the cable's construction. It increases both with cable length and frequency. It is worst between adjacent pairs. For any two pairs in a cable section, the coupling function does not usually change appreciably over time, and it is symmetrical in that the same coupling function is observed between two ends when measured in either direction. Having said that, the coupling functions appear random: any one coupling function shows dramatic nulls in arbitrary positions between peaks that follow a trend, but not particularly closely. The coupling functions between different pairs of wire-pairs are unrelated. As an example, see Figure 3.1, which shows the measured NEXT coupling function between a pair and its three strongest coupling neighbors, in a sample of 10-pair 0.5 mm cable 149 meters long. All the pair ends were terminated (in 100 Ω). The models discussed later are plotted as "trends" for comparison. The spacing between the dips in each curve is often 525 kHz, suggesting that some of this dynamic behavior is resonance effects between the ends of the cable section. Indeed, in real lines with bridged taps, the resonance nulls are prominent. However, in Figure 3.1, the nulls don't line up and are not regular in any of the curves, so resonance is not a full explanation. The complicated and unpredictable forms of the individual coupling functions mean that most practical work uses simplified models of the trends. However, an installed system that uses multiple pairs could theoretically measure the couplings relating

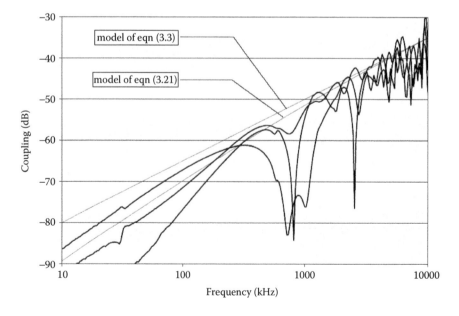

FIGURE 3.1
Typical NEXT coupling functions.

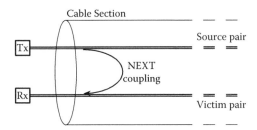

FIGURE 3.2
NEXT coupling.

to its own pairs and equalize out coupling from itself. Such systems are being proposed [Ginis 2003].

Crosstalk coupling follows different trends, depending on network configuration: it all has the same physical cause, but different configurations produce different mixes of the crosstalk leakage and cable attenuation.

Near-end crosstalk (NEXT) is the coupling between transmitters and receivers at the same end of a cable, as shown in Figure 3.2. NEXT typically imposes the limit to DSL system performance when the co-located transmitters and receivers use the same bandwidth. Considering crosstalk between like systems provides the "self-NEXT" bounds on symmetric technologies (for example, ISDN, HDSL, and SHDSL). Between unlike systems, the consideration of who disturbs whom is the subject of spectrum management.

Far-end crosstalk (FEXT) is the coupling between transmitters and receivers at the opposite ends of a cable, as shown in Figure 3.3. The self-FEXT bound shows higher capacity than the self-NEXT bound, and it is important to note that capacity diminishes more slowly with distance. However, FEXT is negligible[1] when NEXT is also present,[2] and so the higher capacity is only available when NEXT is managed away. An example is ADSL deployed from the exchange. Considering the part of the downstream band not shared with the upstream band, crosstalk from all technologies must be controlled for ADSL to have its best performance. Where the transmitters are not co-located, for example in VDSL's upstream channels, there are extra complications [Kirkby 1995].

NEXT and FEXT are exhibited by cables with two ends. A real cable network can have a more complicated structure, with branches and with access points along its length. This gives rise to some extra variants of crosstalk which are of importance in special cases: secondary NEXT and third-circuit crosstalk.

Secondary NEXT is NEXT where the noise sources are separated from the receiver under study, joining the cable some distance away and then transmitting away from the receiver, as illustrated in Figure 3.4. The coupling function is the simple combination of ordinary NEXT and attenuation. Ordinary NEXT would be more powerful than this, so secondary NEXT is only significant when circumstances force the spatial separation of the interfering transmitters from the victim receiver. One example is where the victim is the ADSL downstream channel, and the interferer is a symmetrical system. ADSL typically has a longer reach[3] than the high-rate symmetrical systems, making secondary NEXT significant in standardization work. Also, some administrations limit the deployment of SDSL in order to protect ADSL, making secondary NEXT significant in spectral management work.

---

[1] For a qualitative comparison see Section 3.6.
[2] NEXT coupling is always present. It is common usage to use "crosstalk" for both the coupling and the power coupled.
[3] Reach here means viable deployment distance.

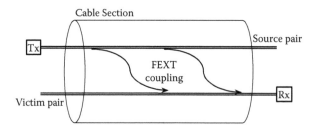

FIGURE 3.3
FEXT coupling.

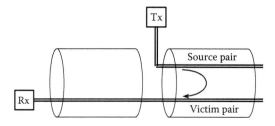

FIGURE 3.4
Secondary NEXT coupling.

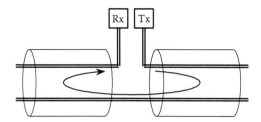

FIGURE 3.5
3CXT coupling.

Third-circuit crosstalk (3CXT) is crosstalk where the transmitter and receiver are co-located at an interior point of the cable network, their signals are flowing in the same direction, and there are other lines in the cable that are unbroken at this connection point. Figure 3.5 illustrates the mechanism that results in 3CXT. This configuration is typical for interference between repeaters of asymmetric systems.[4] The reader is cautioned that although the diagram may suggest that 3CXT is NEXT squared, real 3CXT is more powerful than this.[5]

3.1.1 Empirical Models

The presently used models for crosstalk in common use are empirically based. The original study [Chen 1993] was by Bell Labs, in which a sample of American 50-pair cable was

---

[4] For symmetric systems' repeaters, ordinary NEXT will dominate; note the interference is between distinct repeaters, as a modern system will be able to cancel its own couplings.
[5] It also has a different power law in frequency.

measured extensively, and the trends abstracted as simple power laws. The power coupling trends were found to be:

- NEXT has a $f^{3/2}$ power law in frequency and is substantially independent of cable length and cable attenuation.
- FEXT has a f^2 power law in frequency and is proportional to cable length; it also suffers the same attenuation as signals in the cable.

The sum of couplings was also found to have trends with respect to cable fill. Of particular practical interest is the sum of couplings into one pair from N others, which was found to approximate $\left(\frac{N}{49}\right)^{0.6}$.

Crosstalk from one pair into another pair can be characterized by a transfer function. Such transfer functions describe the relationship between a signal transmitted into one end of the disturbing pair and the received NEXT or FEXT on the disturbed pair. If the disturber has a power spectral density (PSD) $S_d(f)$ then the crosstalk PSD due to the disturber is

$$S_{x-talk}(f) = S_d(f)\,|H_{x-talk}(f)|^2, \tag{3.1}$$

where $|H_{x-talk}(f)|^2$ is the crosstalk power transfer function. Such transfer functions are heavily dependent upon the characteristics and structure of the individual cables, and are very different from pair to pair and from binder to binder. Because of the variability and unpredictability of the crosstalk transfer functions, crosstalk is usually analyzed using worst-case crosstalk transfer functions. The worst-case transfer functions are constructed based on cable measurements (such as those in [Chen 1993]) and computer simulations such that most real crosstalk functions introduce less crosstalk than the worst-case approximation. It is very common to use "1% worst case crosstalk models," which are constructed such that only one percent of all crosstalk transfer functions should be worse than the "1% worst case crosstalk model" predicts. Figure 3.6 shows examples of individual NEXT transfer functions and a

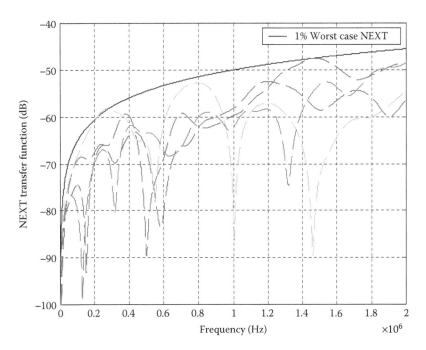

FIGURE 3.6
NEXT transfer functions for individual cases and a "1% worst case" transfer function.

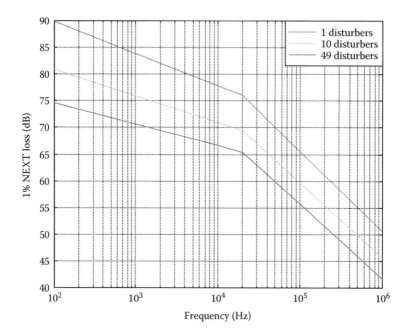

FIGURE 3.7
Unger NEXT model.

one percent worst case transfer function. This figure illustrates that the actual crosstalk transfer functions introduce less crosstalk than the worst-case transfer functions. It also illustrates that the actual crosstalk transfer functions have the same general trend as the worst case transfer functions, but unlike the models which are smoothly increasing with frequency, the actual crosstalk transfer functions vary significantly with frequency.

There are three worst-case crosstalk coupling models in wide use in the DSL industry. One is the so-called Unger NEXT model [ANSI T1.413-2001], which is illustrated in Figure 3.7. Another is the "simplified NEXT model," which is a simplification of the Unger model [ANSI T1.413-2001]. The simplified NEXT model is given by

$$|H_{NEXT}(f, N)|^2 = N^{0.6}0.8536 \cdot 10^{-14}(f/1\,\mathrm{Hz})^{1.5}, \tag{3.2}$$

where N is the number of disturbing pairs, and f is the frequency. The third model is the European Telecommunications Standards Institute (ETSI) crosstalk model, which is also similar to the Unger model but is slightly more sophisticated than the simplified NEXT model [ETSI TS 101 388] [ETSI TS 101 524]. The ETSI crosstalk model is given by

$$|H_{NEXT}(f, L, N)|^2 = N^{0.6}10^{-5}(f/1\,\mathrm{MHz})^{1.5}(1 - |s_{T0}(f, L)|^4), \tag{3.3}$$

where $s_{T0}(f, L)$ is ETSI's notation for the transmission function of the disturbed loop, which for simplicity can also be denoted as $H_{channel}(f, L)$. These three models are compared in Figure 3.8. The Unger and simplified NEXT coupling models are usually used to model NEXT for North American applications, but the ETSI crosstalk coupling model is usually used for European applications.

Similar to worst-case NEXT coupling, worst-case FEXT coupling is given by

$$|H_{FEXT}(f, L, N)|^2 = N^{0.6}K_{FEXT}f^2L\,|H_{channel}(f, L)|^2, \tag{3.4}$$

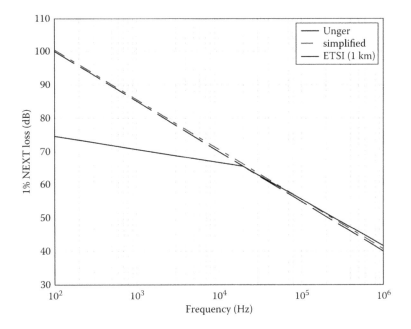

FIGURE 3.8
Comparison of the Unger NEXT model (the solid line), the simplified NEXT model, and the ETSI crosstalk model (assuming 1 km loop).

where L is the loop length and K_{FEXT} is a constant. For North American cables, the FEXT constant is typically considered to be

$$K_{FEXT} = 8 \cdot 10^{-20} \tag{3.5}$$

when L is measured in feet. For European cables, the FEXT constant is typically considered to be

$$K_{FEXT} = 10^{-4.5} \frac{1}{\text{MHz}^2 \text{km}} = 10^{-19.5}. \tag{3.6}$$

In all of these transfer functions, the crosstalk experienced increases with the number of disturbing pairs N. The mean power coupling must be proportional to N, but the models are interested in a near worst-case bound. Therefore, the models provide a less than linear reduction of the bound as the cable fill is reduced. Empirically, a law of the form $(N/49)^{0.6}$ was found, and this has become the accepted wisdom.

This assumption works well as long as all the disturbers are of the same type. But if the disturbers are of different type (*i.e.*, they have different PSDs), then the disturbers have to be combined into a single overall crosstalk PSD that takes into account that some pairs cause worse crosstalk than others. One way to combine more than one PSD for crosstalk computations is to use the so-called FSAN approach.[6] According to the FSAN approach, the NEXT PSD is given by

$$S_{NEXT}(f) = |H_{NEXT}(f, 1)|^2 \left(\sum_k N_k S_k^{1/0.6}(f) \right)^{0.6}, \tag{3.7}$$

[6] "Full-service access network" (FSAN) is a headline under which delegates of the major telcos sometimes cooperate to provide a common view to standards bodies.

where N_k is the number of disturbers of type k and $S_k(f)$ is the PSD for disturber of type k, and $|H_{NEXT}(f)|^2$ is the NEXT power transfer function for a single pair (*i.e.*, $N = 1$). The FEXT PSD is given by

$$S_{FEXT}(f) = |H_{FEXT}(f, 1)|^2 \left(\sum_k N_k \left(S_k(f) |H(f, L_k)|^2 \right)^{1/0.6} \right)^{0.6}, \tag{3.8}$$

where L_k is the length of loops from disturber k, and $|H_{FEXT}(f)|^2$ is the FEXT power transfer function for a single pair. An alternative approach that is sometimes used to calculate the combined crosstalk from more than one type of disturbers is given by

$$S_{NEXT}(f) = \left| H_{NEXT} \left(f, \sum_k N_k \right) \right|^2 \frac{\sum_k N_k S_k(f)}{\sum_k N_k}. \tag{3.9}$$

This approach is mostly used in North America for historical reasons, but it is gradually being replaced by the more rational FSAN method. Aspects of the FSAN method that make it plausible include:

- If two sources of crosstalk are in fact identical but have been labelled differently, then the method gets the same sum as if they had been labelled the same.
- If one source is dominant, then the sum is equal to the dominant source alone.

3.1.1.1 3CXT

Empirical work suggests that 3CXT has a $f^{5/2}$ power law in frequency and is substantially independent of both cable length and cable attenuation.

For 3CXT,

$$|H_{3CXT}(f, N)|^2 = K_3 \left(\frac{f}{f_0} \right)^{5/2} \left(\frac{N}{49} \right)^{0.6} \tag{3.10}$$

with $K_3 = 1.033 \times 10^{-9}$ [dimensionless] and $f_0 = 100\,\text{kHz}$. In dB form, Equation 3.10 becomes

$$\left| H_{3CXT_{dB}}(f, N) \right|^2 = -89.8588 + 25 \log 10 \left(\frac{f}{f_0} \right) + 6 \log 10 \left(\frac{N}{49} \right). \tag{3.11}$$

3.1.1.2 Cable Filling and the Meaning of N

The basic empirical data underlying the crosstalk coupling models was all gathered using 50-pair cables. In those measurements, N was the number of other pairs in use, which was a number between 0 and 49. The $N^{0.6}$ term in each model accounts for the fact that not all pairs inducing crosstalk disturb the victim pair equally. For cables of other sizes, different authorities extrapolate the term differently. First, it has become common to assume that most of the coupling is between pairs in the same binder group,[7] and to either neglect coupling between pairs in different binder groups or to discount it by 10 dB. This becomes significant when the local spectrum management regime attempts to impose fill control on a per-binder basis. The author takes the view, common in Europe, that the quantity $(N/49)$ represents the probability that a given neighboring pair is live, so for a large full cable would evaluate

---

[7] The binder group is called a "unit" in British English. Note that older cables may be of layered construction (*i.e.*, not made of binder groups), in which case the cable is effectively one binder group.

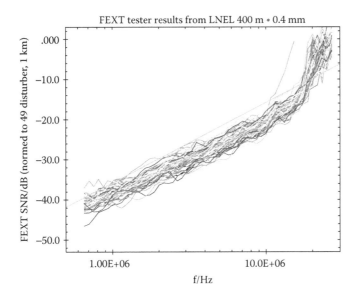

FIGURE 3.9
FEXT SNR measurements.

($N/49$) as 1 irrespective of cable size. Hence the denominator should actually be one less than the number of pairs. However, for smaller cables the author keeps the denominator equal to 49 to represent that the victim pair has fewer actual neighbors, as even when the cable is full, it receives less actual crosstalk. A common need is for a worst-case[8] bound on the noise environment to be anticipated, in the absence of knowledge of cable configuration details. In this case, the interpretation of ($N/49$) as cable fill is extremely convenient. Pessimistic evaluations can assume the cable is full and simply remove the $N^{0.6}$ term from the coupling expressions.

3.1.1.3 Regional Differences

British Telecom found by measurement that their cables typically have FEXT coupling about 5 dB worse than the standard model. This is believed due to the longer twist pitch in cables in the United Kingdom, compared with the United States. The American spectrum management standard [ATIS 2001] uses a length-dependent variation on the standard model for NEXT, as do some ETSI documents [ETSI ADSL 2002]. (See Subsection 3.6.3 below.)

3.1.1.4 Verification

A telco may wish to measure its own cables to verify that the models are valid for them. In an example published by BT [Cook 1999], measurements were made on a sample of cable 400 m long and containing one hundred 0.4 mm copper pairs. All the 9900 individual pair to pair FEXT couplings were measured as functions of frequency. The pairs' insertion losses were also measured,[9] and an average insertion loss calculated. Then, for each pair, the power sum of the couplings into it was calculated and divided into the average insertion loss, simulating that signals of a common spectrum are transmitted over the pairs. The quantity calculated for a pair is then "the signal-to-noise ratio" (SNR). Figure 3.9, from that

---

[8] One percent worst case, that is.
[9] Insertion loss and FEXT are together the 100×100 end-to-end coupling set for a 100-pair cable.

study, shows the SNR functions obtained and the SNR that would be predicted using the corresponding FSAN model:

$$SNR(f) = \frac{|H_{FEXT}(f, L, N)|^2}{|H(f)|^2}.$$

(3.12)

With $N = 49$ (100 percent cable fill) and $L = 0.4$ km, $SNR(f)$ is such that

$$SNR(f) = 0.4 \cdot K_{FEXT} \left(\frac{f}{f_0}\right)^2,$$

(3.13)

where $f_0 = 100$ kHz.

Inspection of Figure 3.9 shows the difficulties of abstracting simple models from empirical data.[10] Nonetheless, the data shows the expected f^2 trend. It was adjudged that the FSAN line is about 5 dB above the 99th percentile of the cluster of results. In passing, it should be noted that this cluster of results is spread by a consistent 15 dB, which may be of interest if consideration of something less extreme than the 99th percentile becomes desirable.

3.1.2 Summary of Crosstalk PSD Formulae

For European cables, the following formulae apply:

$$S_{NEXT}(f, L, N) = K_{NEXT} \cdot \left(\frac{f}{100 \text{ kHz}}\right)^{3/2} \cdot \left(\frac{N}{49}\right)^{0.6} \cdot \left(1 - |H_{channel}(f, L)|^4\right) \cdot S_d(f), \quad (3.14)$$

where $K_{NEXT} = 3.266 \times 10^{-6}$ [dimensionless], and $S_d(f)$ is the PSD of transmissions on lines causing NEXT.

$$S_{FEXT}(f, L, N) = K_{FEXT} \cdot \left(\frac{f}{100 \text{ kHz}}\right)^2 \cdot \left(\frac{L}{1 \text{ km}}\right) \cdot \left(\frac{N}{49}\right)^{0.6} \cdot \left(1 - |H_{channel}(f, L)|^2\right) \cdot S_d(f),$$

(3.15)

where $K_{FEXT} = 3.266 \times 10^{-6}$ [dimensionless], and $S_d(f)$ is the PSD of transmissions on lines causing FEXT.

$$S_{3CXT}(f, N) = K_{3CXT} \cdot \left(\frac{f}{100 \text{ kHz}}\right)^{5/2} \cdot \left(\frac{N}{49}\right)^{0.6} \cdot S_d(f),$$

(3.16)

where $K_{3CXT} = 1.033 \times 10^{-9}$ [dimensionless], and $S_d(f)$ is the PSD of transmissions on lines causing 3CXT.

For American cables, the following formulae can be used to calculate the crosstalk PSDs.

$$S_{NEXT}(f, L, N) = K_{NEXT} \cdot \left(\frac{f}{100 \text{ kHz}}\right)^{3/2} \cdot \left(\frac{N}{49}\right)^{0.6} \cdot \left(1 - |H_{channel}(f, L)|^4\right) \cdot S_d(f), \quad (3.17)$$

where $K_{NEXT} = 2.789 \times 10^{-6}$ [dimensionless], and $S_d(f)$ is the PSD of transmissions on lines causing NEXT.

$$S_{FEXT}(f, L, N) = K_{FEXT} \cdot \left(\frac{f}{100 \text{ kHz}}\right)^2 \cdot \left(\frac{L}{1 \text{ kft}}\right) \cdot \left(\frac{N}{49}\right)^{0.6} \cdot \left(1 - |H_{channel}(f, L)|^2\right) \cdot S_d(f),$$

(3.18)

---

[10] Compared to Figure 3.1, these curves are rather tame; they are averages based on power addition, so a null would only appear on a victim pair's curve if the contributors' coupling function nulls all lined up.

where $K_{FEXT} = 8.264 \times 10^{-6}$ [dimensionless], and $S_d(f)$ is the PSD of transmissions on lines causing FEXT.

$$S_{3CXT}(f, N) = K_{3CXT} \cdot \left(\frac{f}{100 \text{ kHz}}\right)^{5/2} \cdot \left(\frac{N}{49}\right)^{0.6} \cdot S_d(f), \tag{3.19}$$

where $K_{3CXT} = 1.033 \times 10^{-9}$ [dimensionless], and $S_d(f)$ is the PSD of transmissions on lines causing 3CXT.

3.2 Impulsive Noise

Impulsive noise is noise characterized by short pulses with significant energy. Its occurrence correlates strongly with human activity and is typically due to switching transients. The steady-state noise discussed in the previous section, crosstalk and so forth, is called "stationary" noise when discussed in contrast to impulsive noise. Impulsive noise was first investigated in the early sixties by Mandelbrot [Berger 1963][Mandelbrot 1965], who was studying errors induced into voiceband modems over dial-up links. In the early days of DSL, impulsive noise was mostly caused by electromechanical activity in telephone exchanges, including both switch activity [Henkel 1994] and application of ringing cadences [Cook 1993]. Since electronic switching has become the norm, these sources have vanished, and the remaining source appears to be electrical power switching near the customer's premises. Demonstrated sources include railway traction control, arc welding, and switching of domestic fluorescent lights. Figure 3.10 shows an example impulse captured during a survey. The pulses vary widely in form, duration, amplitude, and spacing; there is no typical shape. There is no bound on duration or energy of impulse noise, although the longer and more energetic pulses are rarer. There are no commonly accepted models of impulsive noise, though there have been several attempts at such models, usually based on observations

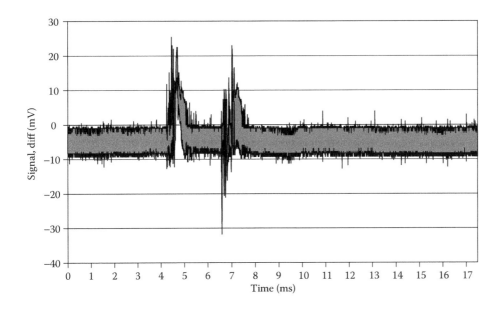

FIGURE 3.10
A noise impulse.

gathered from real networks. Mandelbrot [Mandelbrot 1965] proposed an interarrival time model, although it was so intractable (for example, predicting zero mean error rate) as to be of theoretical interest only.[11] Bond [Bond 1987] also proposed an interarrival time model.

Cook [Cook 1993] concluded that the salient feature of an impulse was its peak voltage V_b as received at the quantizer inside a receiver, and reported power laws relating this to the DSL bandwidth f and the period of observation[12] τ: V_b is proportional to $f^{3/4}$ and to $\tau^{1/3}$. He also proposed a "symbolic" test pulse based on these observations, which has since been incorporated into standards (such as G.test), although it is not really a representative stimulus for DSL with long equalizers. More lifelike models have been proposed by Kessler et al. [Henkel 1999], McLaughlin et al. [McLaughlin 1999], and Kirkby et al. [Kirkby 2001], but they are also more complicated, and none has received acceptance in the standards community.

3.3   Noise from Faults

This chapter primarily addresses the noise to be expected on a normal fault-free line, however, certain faults can increase the noise on a line. One example is jointing errors ("rectified split pairs"), in which the wires of two pairs are misconnected at two joints, so for some distance each wire of one pair is twisted with a wire of the other pair. This grossly increases crosstalk coupling between the affected pairs. Broken joints spoil the symmetry of the pair, giving it poor balance. This typically enhances RFI ingress. Corroded joints introduce a rectifying action. This nonlinearity will intermodulate the line signal with itself, producing a noise component that cannot be equalized out, even though it is due to the system's own signal. Corroded wet joints can also generate noise that is believed to be due to the randomness of electrolysis. This noise can easily disable a DSL system. These are among the faults discussed in more detail in the chapter on planning rules in Volume 2 of this series.

3.4   Engineering Measures

3.4.1   Screening (Does Not Work)

From time to time, the suggestion is raised that cables with individually screened pairs should be used to reduce crosstalk. Although this can be dismissed on economic grounds, at least for DSL,[13] there are also technical objections.

- Screening increases the attenuation of wanted signals [Hughes 1997].
- Screening makes the cables heavier, larger, and harder to bend.
- Effective screening requires great care with earthing.
- Screening does eliminate capacitive coupling, but not magnetic coupling at DSL frequencies.

To summarize, screening attenuates legitimate signals, is expensive, and does not eliminate crosstalk. There are other changes to construction that would be beneficial for DSL use,

---

[11] Mandelbrot took it further and invented fractals.
[12] The period in which V_b is observed to find one peak value.
[13] If network operators could afford to re-cable the local access network, they would do it in glass!

for example, the tighter twisting now used in CAT5 cables. However, the crosstalk properties of existing cables are not a mistake by the manufacturers: these cables were designed for voice use and are fit for purpose. Control of voice crosstalk was once a major design problem that was effectively solved in the 1930s by twisting neighboring pairs (or quads) with different pitches. This solution is, in reality, so good that DSL is viable even though it uses the cables at frequencies two or three orders of magnitude above the cables' design frequencies.

3.4.2 Enforced Continence

In a classical telecommunications channel (*e.g.*, the deep space channel), one way of improving the SNR at the receiver is to transmit more power; a system designer will use the most powerful transmitter that is affordable. However, the access network channel is dominated by crosstalk, so increasing the transmit power would increase the noise environment correspondingly, with no gain in SNR.[14]

Superficially, one might expect a newcomer to see the status quo as an opportunity to be exploited. In the past (the days of monopoly telcos), this did not happen because the existing environment was investment by the same people as the newcomer: big loss + small gain = no action. In the brave new world of local loop unbundling (LLU), in which incumbent network operators must make lines available for lease by competitive access carriers, even an aggressive newcomer realizes the "gain" only exists until he becomes an investor and is as vulnerable to crosstalk as the rest of the potential victims. The result is that anyone who wants to do real business wants his or her neighbors' behavior constrained. The local access network being crosstalk limited leads directly to the need for spectral management, which is discussed in detail in Volume 2 of this series.

3.4.3 Deployment Discipline

In many LLU regimes, a system is permitted to be connected on lines on which it will work and on some lines on which it will not.[15] Then it is the operator's problem to choose those lines on which they will offer service. This decision will be conditioned by, among other things, the expectation of the noise environment. It should be noted that this expectation can be different for different operators in the same network: they may, for example, take different views on how widespread DSL will be in five years time. This is discussed in more detail in Volume 2 of this series.

3.4.4 Band Duplexing

Because NEXT is a worse impairment than FEXT, a given frequency band (when used in one direction) will have higher capacity if all the neighbors use the same frequency band in that one direction only. At low frequencies, however, the most efficient use of telephone pairs is to send signals bidirectionally, whereas at high frequencies, the best overall capacity is achieved if separate bands are dedicated to each direction. In a virgin network, one might imagine engineering the regime change to best advantage, but in a real ("heritage") network, this band dedication can only happen for bands not already in use. Thus in practise, unidirectional working is attractive for the bandwidth above the present bidirectional

---

[14] Decreasing the transmit power is also possible and is attractive until other sources of noise become significant. In practise, the standard transmit powers are chosen so crosstalk will dominate comfortably, but not beyond that.
[15] A major exception is the United States, where there are performance guarantees; there the discipline is on the guarantor.

systems. The usual form of duplexing is frequency division, dedicating different nonover-lapping bands to each direction. This is the case for VDSL and for the higher frequencies in the downstream channel of ADSL. Time division duplexing ("ping-pong") is also possible, where nonoverlapping periods of time are dedicated to each direction. This is used in Japan, originally for ISDN and now for ADSL too.

3.4.5   Interleaving

ADSL has two countermeasures against impulsive noise: interleaving with a forward error correction code, and, for DMT ADSL, a judicious choice of symbol length.[16]

Interleaving works against burst errors by breaking the data stream into blocks, each of which is protected using a Reed–Solomon code. The blocks are then further broken up into small pieces that are shuffled by an interleaving process. Following this procedure, the pieces of any one of the original blocks are widely separated. The intention is to ensure a noise burst will only damage a correctable number of the pieces of each block. At the receiver, the pieces are reassembled into blocks, each of which is then corrected by the code. The detailed schemes used are discussed further in Chapter 9.

3.5   Modelling

This section discusses how to implement a noise model. Eventually a noise model becomes a specification, perhaps a formula for a power spectral density (PSD) in a simulation, or perhaps a table of numbers for an arbitrary waveform generator on a laboratory bench. This section is concerned with how and why the model is constructed.

3.5.1   Use of Noise Models

Noise models must be implemented for simulation work and for laboratory testing. In the former case, the "model" is a means of generating numbers in a computer program. In the latter case the "model" is a means of generating electrical waveforms in physical equipment. Simulation provides predictions of system performance and indirectly supports a variety of engineering and managerial processes, including:

- Optimizing the design of DSL modems (typically by manufacturers)
- Deciding whether performance is sufficiently promising to warrant continued development (manufacturers, standards bodies)
- Determining the performance to test for in a standard (standards bodies)
- Optimizing the DSL population in a network (part of spectrum management)
- Determining the service limits for the modem (operators)

Laboratory testing is used to verify the conformance of real equipment to standards and to calibrate subsequent simulations. Despite the obvious differences in form of the models for simulation and laboratory use, the logic of construction is similar,[17] and the differences will be left as an exercise for the reader.

---

[16] Other DSL technologies, such as SDSL, do not include impulse suppression because of strict latency requirements.

[17] At least one laboratory generates its noise via its simulation suite.

3.5.2 Production

To predict the performance of DSL, its environment must be represented in some detail: perhaps a particular configuration of cables with a particular population of systems running in it, and afflicted with particular sources of noise. The noise model will represent the noise as experienced at the receiver of interest. Therefore, different noise models will result from different assumptions about:

- The physical size of the network (for example, different countries, whether in town or rural).
- The cables used (varies by country).
- Where the receiver is (for example, in the local exchange, in a cabinet, in a house, in an office).
- The neighboring systems (varies by locale, with time, and with simulator's optimism[18]).

In keeping with industry practice, this chapter shall pursue models in the frequency domain, eventually producing a PSD for the noise as experienced at a modem receiver. The development shall start with the noise sources and their spectra, then take into consideration the coupling between these sources and the receiver, and finally determine how to combine the noise components.

3.5.2.1 Network Model

Each simulator will have to take a view on what the network reality is. Much guidance can be obtained from published materials, especially the normative tests in standards (for example, [ETSI ADSL 2002]), where the operators involved try to ensure the required performance is tested in an environment representative of their network. However, other aspects, such as how full a network will be in five years' time, are guesswork; companies take commercial risks based on these guesses and usually regard the details as proprietary. To obtain an initial assessment of a proposed DSL system, one common assumption is that the cable is a simple single section with two ends, 100 percent filled with instances of the modem under consideration. The initial objective of the simulations is to find the longest cable in which the modem still works. This is called "self-limited reach."

An assumption popular with naive proponents of a new system is that a single instance of the system operates in cables without any other system present. This unsurprisingly predicts operation over great distances. Although such an analysis might tell of the quality of the engineering in the modem, and is certainly easy to check in a laboratory, it says nothing about operational performance in a real network. Hence, such a simulation is valueless to a network operator, who is only concerned with providing services that can be guaranteed under even worst-case noise conditions.

3.5.2.2 Sources

Because crosstalk is the dominant impairment for DSL, the most important sources are the other DSL systems in the network.

3.5.2.3 Standards

Standards define spectral masks, so a standards-conformant modem has an upper bound on the transmitted PSD. In recent standards, these masks have become fairly tight in the

---

[18] Different operators in the same unbundled network may take different commercial views of risk.

signal bandwidth. Out of the transmitted band, the modems typically cut off faster than the mask. Older standards specified more generous masks, so the true signal cannot really be inferred from the mask. (Really ancient standards, such as G703 30 channel PCM, only included time domain pulse masks; as the systems did not include spectral scrambling they cannot be said to have a "spectrum," because the spectrum changes depending on the traffic being carried.)

3.5.2.4 *The Experts*

In order to get consistent results between the participants of international debates, the experts have published nominal spectra ("templates") for most systems of interest. These are derived from measurements and or detailed understanding of the systems' modulations and are generally the best models available [ETSI Spect. Manag. 2002].

3.5.2.5 *Measurement*

If a new system is to be included in a simulation (or the effects of a fault are to be studied), it is necessary for the simulator to determine the spectrum for himself, typically by measurement. It is also common to include other forms of noise, for example, a -140 dBm/Hz "background" noise almost by default,[19] and sometimes radio-frequency (RF) tones to represent RF interference (RFI). (See Chapter 13 for details of RFI.)

3.5.2.6 *Coupling*

Coupling is by a combination of crosstalk and line attenuation. Crosstalk models are discussed above in Section 3.1, and line attenuation is discussed in Section 2.3.5. In general, there will be a coupling path from every transmitter in the network to every receiver.[20] The noise sources other than crosstalk are usually provided in the form in which they are expected at the receiver, so they are not modified by coupling.

3.5.2.7 *Summation*

At first sight one would expect that, because the various noise sources are independent physical processes, the power of their sum would be the sum of their powers (and at each frequency independently, so the PSD of their sum is the sum of their PSDs). However, this conflicts with the FSAN models of crosstalk, and crosstalk contributions should be added by the FSAN sum method.

3.5.3 Impulsive Noise

At present, it is not possible to predict the effect of impulsive noise on system performance either by simulation or by laboratory tests, because it has no adequate statistical characterization. Statistics that have been gathered by field tests (for example, in [Cook 1993]) simply show so much variation that predictions cannot be made with confidence.

As a result, mainstream modelling of performance considers stationary noise only. Impulsive noise is "modelled" in system design usually by a consensus of experts that a level

---

[19] A value apparently set by agreement of the experts. Johnson noise is an insignificant part of this value. Johnson noise, the classical noise due to thermal movement of the electrons in a conductor, is -174 dBm/Hz at 20 degrees C.
[20] It is possible to save computer time by omitting insignificant terms, although these days it is usually more efficient to implement all terms rather than spend human effort in deciding what to leave out.

of protection is appropriate[21] and in testing to verify that the protection built in is, in fact, operating properly. A typical standards conformance test[22] is to exercise the modem in a typical stationary noise environment, to stress it realistically but so it should work properly. Next, the modems are disturbed with bursts of white noise of a specified duration, where each burst has an amplitude sufficient to ruin the modem's data carriage but not to damage the electronics, and where the bursts are sufficiently separated that they are independent events to the modem. To pass, a modem must deliver undamaged payload.

3.6 Mathematical Modelling

This section uses mathematical modelling to explore some of the phenomena discussed in the main body. The results here are speculative (insufficient peer review) but may be of interest for their physical insight and for their prediction of details.

3.6.1 Worst-Case Length for FEXT Coupling

Inspection of the FEXT coupling model—or indeed thinking about the physical causes of FEXT—leads one to expect FEXT at any given frequency to be worst for some particular length of cable: any shorter, and the reduction in length will reduce the FEXT; any longer and the propagation loss will reduce the FEXT. The FSAN formulae are eminently suitable for calculus. Denoting the FEXT coupling as $FEXT(f, L)$ (for notational simplicity), it is possible to seek the value of L at which $\frac{\partial FEXT(f,L)}{\partial L}$ is zero. First, note that $|H(f)|^2$ may be written in terms of the secondary propagation parameters $\gamma = \alpha + j\beta$, i.e., $|H(f)|^2 = e^{-2\alpha l}$, and that α is itself a function of frequency. At DSL frequencies, α is dominated by the skin effect, so it is approximately proportional to root f. Therefore, α can be modelled at $\alpha = k\sqrt{f}$, where for a definite cable type k is some constant. Hence,

$$|H(f)|^2 = e^{-2kL\sqrt{f}} \tag{3.20}$$

and

$$FEXT(f, L, N) = K_{FEXT} \left(\frac{N}{49}\right)^{0.6} \left(\frac{f}{f_0}\right)^2 \left(\frac{L}{L_0}\right) e^{-2kL\sqrt{f}}. \tag{3.21}$$

Taking the partial derivative,

$$\frac{\partial FEXT(f, L)}{\partial L} = \frac{K_{FEXT}}{L_0} \left(\frac{N}{49}\right)^{0.6} \left(\frac{f}{f_0}\right)^2 e^{-2kL\sqrt{f}} - 2k\, K_{FEXT} \sqrt{f} \left(\frac{N}{49}\right)^{0.6}$$
$$\times \left(\frac{f}{f_0}\right)^2 \left(\frac{L}{L_0}\right) e^{-2kL\sqrt{f}}. \tag{3.22}$$

Equation 3.22 equals zero at

$$L_{max} = \frac{1}{2k\sqrt{f}}. \tag{3.23}$$

---

[21] This usually means "affordable:" the cost in transistors and in delay being balanced against the unquantified fear of harm from this source.
[22] There is a philosophical conflict here between the needs of operators and the needs of standards bodies. Typically, the operators want to know how performance will be for real, whereas standards bodies need conformance tests real modems can pass. Impulsive noise is believed to cause some uncorrectable errors, and the operators want to know how many, but conformance tests would consider any to be failures.

Note that at this value, the cable loss is

$$|H(f)|^2 = e^{-1}, \tag{3.24}$$

which represents a loss of about 4.3 dB.[23] From Equations 3.21, 3.23, and 3.24, one can find the maximum value for FEXT coupling:

$$FEXT(f, N)_{\max} = \frac{K_{FEXT}}{ke\sqrt{f}} \left(\frac{N}{49}\right)^{0.6} \left(\frac{f}{f_0}\right)^2, \tag{3.25}$$

which can be rewritten

$$FEXT(f, N)_{\max} = \frac{K_{FEXT}}{ke\sqrt{f_0}} \left(\frac{N}{49}\right)^{0.6} \left(\frac{f}{f_0}\right)^{3/2}. \tag{3.26}$$

Interestingly, because the first term is now a constant, this is the same form as NEXT coupling. Denoting NEXT coupling as $NEXT(f, N)$,

$$NEXT(f, N) = K_{NEXT} \left(\frac{N}{49}\right)^{0.6} \left(\frac{f}{f_0}\right)^{3/2}, \tag{3.27}$$

which invites a numerical comparison. To get an example value for k, consider 0.5 mm copper cable, which has about 10 dB insertion loss per km at 300 kHz, suggesting

$$k_{0.5 \text{ mm Cu}} = \frac{\log(0.1)}{-2\sqrt{300000}} = 9.1 \times 10^{-4} [\text{km}^{-1}\text{Hz}^{-1/2}]. \tag{3.28}$$

Noting that because in this case $K_{NEXT} = K_{FEXT}$, the ratio of worst-case FEXT coupling to NEXT coupling can be written as

$$\frac{FEXT(f, N)_{\max}}{NEXT(f, N)} = \frac{1}{2ek\sqrt{f_0}} = 0.276. \tag{3.29}$$

So FEXT coupling at its worst length is still less than NEXT coupling but is of the same order. At its worst, FEXT is about 5.5 dB less for this gauge of cable. For 0.4 mm cabling, it is about 4 dB less.

Therefore, it can be concluded that:

- FEXT at a given frequency has a critical length at which it couples most power.
- The FSAN model predicts this critical length is that which attenuates the signal by e.
- At this maximal coupling, FEXT is less than NEXT but comparable with it.

3.6.2   Length Dependence of NEXT

As discussed above, the FSAN NEXT model does not vary with length of the common cabling. However, common sense suggests there should be a relationship: in the limit, a zero-length piece of cable should have zero coupling. By ignoring length, the FSAN NEXT

---

[23] Exactly half a neper, for those who find logs base e more natural.

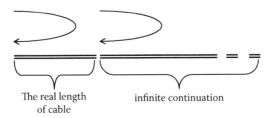

The real length
of cable

infinite continuation

FIGURE 3.11
NEXT coupling to infinity.

model effectively approximates the length of the common cabling as infinite. This is a good approximation for most lines. This model is denoted as $NEXT(\infty)$. More recent discussion in standards bodies has used a more elaborate model for NEXT, as illustrated in Figure 3.11. This model accounts for the length-dependence of NEXT [ETSI Spect. Manag. 2002]:

$$NEXT(L) = NEXT(\infty)(1 - |H(L)|^4). \tag{3.30}$$

$H(.)$ is the scalar transfer function of the line's channel, so $|H(L)|^2$ is the attenuation of signals along the length L of the cable, and $|H(L)|^4$ is the attenuation of signals along the length of the cable and back. This length-dependent model is quite credible, if one assumes every part of a cable has the same propensity for coupling NEXT: one considers an infinite continuation of the cable. Clearly one gets the same coupling at the real end of interest as one would at the start of the infinite continuation, as both are the coupling for infinite length. At the end of interest, this is the sum of the coupling of the real length of cable and the attenuated coupling of the infinite continuation. The attenuation is two passes along the real length, that is, $|H|^4$. Subtracting the continuation results in the formula above. The same reasoning is plausible for 3CXT. This argument gives an intuitive explanation of why neglecting length has been an acceptable approximation in the DSL industry: the cable has to be really short for NEXT to be much different from its value on an infinitely long loop. For example, to lose half the coupling, the cable shortens from an insertion loss of infinity to 1.5 dB. Therefore, at ADSL frequencies, approximately 150 m of cable will get half the coupling of an infinite length of cable. Cables of practical interest are typically of the order of one to a few kilometers.

3.6.3   NEXT and FEXT as Scattering

In Chapter 2, the physics of the twisted pair channel was described and the transmission model derived. It was noted that if the wires of a pair are equally coupled to all the other wires in the cable, then by symmetry there would be no leakage into or out of the differential mode. The chapter also pointed out that the randomized twisting of pairs approximates this symmetry rather closely. Whatever crosstalk there is can be considered as shortcomings of the symmetry. Following Hughes [Hughes 1997], one may model crosstalk by assuming the asymmetries are small, random, and independent. Then their effects will add per the central limit theorem, and scalar coupling will be Gaussian.[24]

Here, the couplings in which asymmetries are most likely to be significant are assumed to be the electrostatic and magnetic linkages between neighboring pairs. Two pairs are

---

[24] Hughes then uses measurement to show how small the perturbations are in real cables.

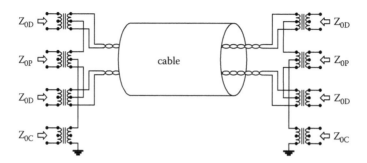

FIGURE 3.12
The circuits used for modelling crosstalk in a cable.

considered, starting with the effect of single elements of unbalance. The derivation is in terms of waves and the scattering of waves (where later the scattered waves are added vectorially). The scattering will produce waves in the differential modes of the pairs, and leaked waves to the common modes;[25] to support the development, the common modes will be assumed also to have characteristic impedances. To illustrate the modes, the two pairs are considered as normal differential circuits with a notional balanced phantom circuit between them, and a notional common mode ground return phantom circuit, as illustrated in Figure 3.12, which shows the cable section, the circuits (the transformer coupling serves to define the circuits), and the circuits' nominal characteristic impedances. The two Z_{0D} differential mode circuits are serviceable, their virtues including a consistent character-istic impedance along their length. The phantoms would in practice not have consistent characteristic impedances,[26] but for the purposes of this analysis, the absence of consis-tency in the characteristic impedances will not matter. Their impedances need only exist locally to support the egress of waves, and their values can safely vary from place to place. Inside the cable, the conductors are mutually affected by each others' changes in voltage (capacitive coupling) and current (inductive coupling). However, while symmetry is main-tained these effects all cancel exactly, and to each circuit the various real capacitive and inductive linkages just appear as the mode's intrinsic capacitance and inductance per unit length (the C and L of the usual RLCG parameters of the telegrapher's equation). The effects of a small deviation in a short element of the cable can be considered. Figure 3.13 shows a small capacitive perturbation between two of the conductors[27] in the element. There is, of course, capacitance between all of the conductors; the deviation is just the difference between reality and the ideal symmetrical configuration. Also shown is a signal in the top pair's differential mode, as a wave travelling from left to right. In the symmetrical case, it would pass directly through; in this perturbed case some of it will scatter. Qualitatively, the signal stimulates a current in the capacitor, and this current generates four equal waves radiating away from it in the normal pairs' modes (see Figure 3.14), and two equal waves radiating from it in the balanced phantom's modes. There is no coupling into the common phantom. The incident wave also continues as normal, so that which is transmitted is the

---

[25] What to do with the common modes is a recurring problem in modelling. Paul Clayton [Clayton 1994] considers the full multidimensional telegrapher's equations, handling the conductors separately and with channels defined as boundary conditions. Other authors consider a hierarchy of phantom circuits.

[26] In the past, a shortage of lines has led to practical use of phantoms, but they have always been of low quality even for voice.

[27] The choice is of one conductor from one pair and one conductor from the other. By symmetry, effects of equal magnitude are obtained by the other choices. Throughout this analysis, the sign convention that positive is "up" has been used in the figure; this is, of course, arbitrary, and different choices just invert some of the waves.

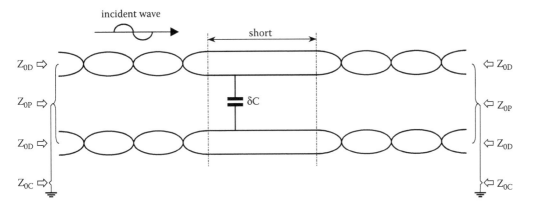

FIGURE 3.13
Capacitive perturbation.

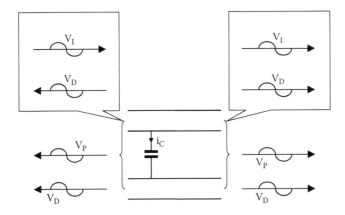

FIGURE 3.14
Scattering δC.

sum of the incident wave and an extra wave. Quantitatively if the incoming wave has scalar voltage v_I, then the capacitor current i_C is

$$i_c = \frac{-j\omega\delta C}{2} \left(\frac{1}{1 + j\omega\delta C \left(\frac{Z_{0D}}{4} + \frac{Z_{0P}}{2} \right)} \right) v_I \qquad (3.31)$$

and the scattered waves have amplitudes:

$$\text{differential modes} \qquad v_D = \frac{Z_{0D}}{4} i_C, \qquad (3.32)$$

$$\text{balanced phantom} \qquad v_P = \frac{Z_{0P}}{2} i_C, \qquad (3.33)$$

$$\text{common phantom} \qquad v_C = 0. \qquad (3.34)$$

The transmitted wave is

$$v_0 = v_I + v_D. \qquad (3.35)$$

Figure 3.15 shows a mutual induction perturbation between two of the conductors in the element. Qualitatively, the signal current stimulates a voltage across the inductance,

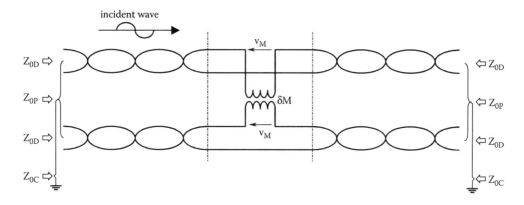

FIGURE 3.15
Inductive perturbation.

and this voltage generates a pair of complementary waves[28] radiating away from it in each differential pair's mode (see Figure 3.16), and a complementary pair of waves in the common phantom. There is no coupling into the balanced phantom. The incident wave also continues as normal, so that which is transmitted is the sum of the incident wave and an extra wave. Quantitatively, if the incoming wave has scalar voltage v_I, then the inductance voltage v_M is

$$v_M = \frac{j\omega\delta M}{Z_{0D}} \left(\frac{1}{1 + j\omega\delta M \left(\frac{1}{Z_{0D}} + \frac{1}{8Z_{0DC}} \right)} \right) v_I \qquad (3.36)$$

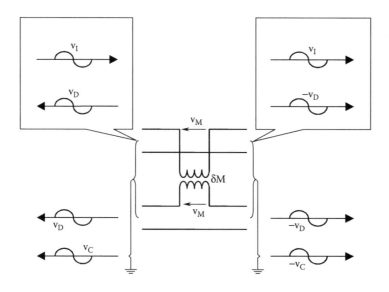

FIGURE 3.16
Scattering δM.

---

[28] Equal in magnitude, opposite in sign.

and the scattered waves have amplitudes:

$$\text{differential modes} \quad v_D = \frac{v_M}{2}, \tag{3.37}$$

$$\text{balanced phantom} \quad v_P = 0, \tag{3.38}$$

$$\text{common phantom} \quad v_C = \frac{v_M}{4}. \tag{3.39}$$

The transmitted wave is

$$v_0 = v_I - v_D. \tag{3.40}$$

In both capacitive and inductive coupling, therefore:

- NEXT, FEXT, and echo are generated simultaneously, and are equal at point of generation.
- All the waves are proportional to v_I.[29]
- For small imbalances, all the couplings are approximately proportional to the frequency and to the imbalance (δC or δM). (A good approximation except when considering v_0; when there is any scattering at all, $|v_0|^2 < |v_I|^2$ by conservation of energy, and so the higher-order terms are significant here.)
- At higher frequencies, the Z_0 become nearly resistive and the scattered waves will be approximately $\pm\frac{\pi}{2}$ out of phase with the incident wave.

Interestingly, capacitive imbalance seems to not have coupling to the common mode, so common mode currents from external noise sources should be unable to enter the differential modes by this route. External noise ingress, RFI, and so forth must be entering through inductive imbalances only.

Because capacitive and inductive coupling elements have similar effects on crosstalk, they may be treated together. For any given pair of adjacent pairs (and for a study of either FEXT or NEXT), suppose a short cable element of length δ has a random coupling variate δX such that the direct coupling from the first pair to the second is

$$\frac{v_D}{v_I} = j\omega\delta X. \tag{3.41}$$

It is assumed that δX is real (only strictly true when Z_{0D} is real) and has variance proportional to δ (reasonable if δ is long enough for separate lengths to be independent[30]). Furthermore, it is assumed that δX has variance $\delta\sigma_x^2$, which is equivalent to declaring the pairs to have coupling of variance σ_x^2 per unit length.

3.6.3.1 NEXT

Of interest is the total NEXT from the first pair to the second. Along the length of the cable, each perturbation δX contributes a direct NEXT scatter component, and, at the near end of the second pair, their sum emerges with each component attenuated and delayed by the length of the first pair to the perturbation and again by the length of the second pair from the perturbation. Contributions indirectly coupled by more than one act of scattering may be neglected,[31] although the perturbations are each small, as the indirect contributions

---

[29] No surprises here: this circuitry is all linear. That is why analysis may consider one incident wave in isolation without loss of generality.

[30] This approximation allows integration; physically, at very short lengths one is inside a single physical cause of coupling and so as δ tends to 0 the lengths cease to be independent. For analysis, this effect can be neglected.

[31] For the same reason, it was safe to neglect the common modes' transmission properties.

will be of the order of δ^2 or smaller. At any one frequency, the coupling will be a complex number formed as the sum

$$NEXT_{coupling} = \sum_0^\infty j\omega\delta\, X_s \exp(-2\gamma s), \qquad (3.42)$$

where s is distance along the cable, stepping in increments of δ, and $\gamma = \alpha + j\beta$ is the propagation coefficient of the pairs.

In this sum of phasors, the succeeding contributions rotate with distance (as well as get smaller). Supposing that the (many) perturbations δX_s are independently random and of comparable size, the conditions for a two-dimensional variation of the central limit theorem are satisfied: a random complex variate of variance[32] $\frac{\omega^2\sigma_x^2}{4\alpha}$, whose real and imaginary parts are *i.i.d* Gaussian[33] with zero mean and variance $\frac{\omega^2\sigma_x^2}{8\alpha}$, is expected.

The δX_s do not change with time,[34] so the crosstalk between a given pair of pairs is stable. They also do not change with frequency; however, the phasor rotations (and attenuations) do, so a substantial change of frequency results in a new sum. Above 100 kHz the loss of cables is dominated by the skin effect, so the loss in dB is proportional to $\sqrt{f}$. Hence, as a function of frequency, α is proportional to $\sqrt{f}$. For some particular cable type, assume $\alpha = k\sqrt{f}$. Hence, this analysis predicts that scalar NEXT coupling between given adjacent pairs at a given frequency is a random complex value with variance $\frac{\pi^2 f^{3/2}\sigma_x^2}{k}$. In noise modelling, the power coupling properties are of primary interest. The square of the magnitude of the distribution is a negative exponential distribution whose mean is the variance above. Hence, this analysis has predicted the form of the FSAN NEXT model, with some additional information: the power coupling is random and has the negative exponential distribution.

The FSAN NEXT model predicts the same coupling for all cables, which seems unlikely according to the above, as k definitely varies between cables, and σ_x^2 seems to be related to the physical processes of manufacture. For the record, one would expect less NEXT from lossier cables. The analysis above also suggests that there is backscatter in telephone cables, and its law has the same form as NEXT, with a bigger constant of proportionality, because it is comparable with all the NEXT couplings to a pair added together. In practice, echo for DSL systems seems dominated by mismatches, particularly at the terminations, so presumably this backscatter is lost in the noise. Incidentally, if the sum above is taken over a finite length of cable (Σ_0^L), one gets the length-dependent formula of Section 3.6.3.

3.6.3.2   *FEXT*

One can also consider the total FEXT from the first pair to the second. Along the length of the cable, each perturbation δX contributes a direct FEXT scatter component, and at the far end of the second pair their sum emerges. Each component has travelled the length of the cable and it is reasonable to assume their attenuations are equal. It is tempting to approximate their delays as equal too, but looking ahead this would lead to the false conclusion that the sum of the scattering terms is frequency-independent, because the δX_s are frequency-independent. If this were the case then different pairs of pairs would have different coupling functions, but the coupling function for any one pair of pairs would be smooth. However, lab measurements show that each coupling function varies randomly with frequency. Thus, in this analysis it must be assumed there is some variation in delay between the two pairs along their length. Each scatter component can be modelled as having random phase in the

---

[32] Which is obtained by integration, taking $\delta \to 0$ in the sum above.

[33] This distribution is radially symmetric, so has phase uniformly distributed over 0 to 2π; the magnitude is independent of phase and has the "Rayleigh" distribution.

[34] Actually, they change as the rest of the cable properties change: with temperature, physical disturbance of the cable, and age.

sum. Again, contributions coupled by more than one act of scattering are neglected. At any one frequency, the coupling will be a complex number formed as the sum

$$FEXT_{coupling} = \sum_0^L j\omega\delta X_s \exp(-\gamma L) = j\omega\exp(-\alpha L)\sum_0^L \delta X_s \exp(-j\beta L), \qquad (3.43)$$

where s is distance along the cable, stepping in increments of δ, $\gamma = \alpha + j\beta$ is the propagation coefficient of the pairs, and α, which relates to attenuation, may be taken as equal for all paths, whereas β which relates to delay may not.

In this sum of phasors, the succeeding contributions rotate arbitrarily with distance. Supposing that the perturbations δX_s are independently random, of comparable size, and many, the conditions for the two-dimensional variation of the central limit theorem are again satisfied: the result should be a random complex variate of variance $\omega^2 \exp(-2\alpha L)L\sigma_x^2 = \omega^2|H(f)|^2 L\sigma_x^2$, whose real and imaginary parts are i.i.d. Gaussian with zero mean and variance $\frac{\omega^2|H(f)|^2 L\sigma_x^2}{2}$.

As for NEXT, the crosstalk coupling between a given pair of pairs is stable with time, but a substantial change of frequency results in an independent random value. Again, only the power coupling properties are of interest, and the resulting analysis yields a negative exponential distribution whose mean is the complex variate's variance. Hence, the analysis has predicted the form of the FSAN NEXT model, with some additional information: the power coupling is random and has the negative exponential distribution. The reservation above about k in the NEXT model does not emerge from this FEXT analysis, suggesting the FSAN FEXT model adequately represents the cable properties in the $|H(f)|^2$ term.

References

[ANSI T1.413-2001] ANSI T1.413-2001, *Spectrum Management for Loop Transmission Systems, American National Standard*, 2001.

[ATIS 2001] ATIS T1E1.4. *Spectrum Management For Loop Transmission Systems*. T1.417-2001 may be downloaded from https://www.atis.org/atis/docstore/index.asp.

[Berger 1963] J.M. Berger and B. Mandelbrot. *A New Model for Error Clustering in Telephone Circuits*. IBM Journal, July 1963.

[Bond 1987] D.J. Bond. *A theoretical study of burst noise*. BT Technology Journal, Vol. 5, No 4, October 1987.

[Chen 1993] W.Y. Chen and D.L. Waring. *DMT ADSL Performance Simulation for CSA*. ANSI T1E1.4/93-166, August 1993.

[Clayton 1994] R.P. Clayton. *Analysis of Multiconductor Transmission Lines*. Wiley, 1994.

[Cook 1993] J.W. Cook. *Wideband impulsive noise survey of the access network*. BT. Technol. Journal, Vol. 11, No. 3, July 1993.

[Cook 1999] J.W. Cook, R.H. Kirkby, M.G. Booth, K.T. Foster, D.E.A. Clarke, and G. Young. *The Noise and Crosstalk Environment for ADSL and VDSL Systems*. IEEE Communication Magazine, pp. 73–78, May 1999.

[ETSI ADSL 2002] *Transmission and Multiplexing (TM); Access Transmission Systems on Metallic Access Cables; Asymmetric Digital Subscriber Line (ADSL)—European Specific Requirements*. ETSI TS 101 388, V1.3.1, May 2002.

[ETSI Spect. Manag. 2002] ETSI TM6. *Transmission and Multiplexing (TM); Spectral management on metallic access networks; Part 2: Technical methods for performance evaluations*. TM6(01)20, latest version is Dec 02; to eventually be issued as TR 101 830 2.

[ETSI TS 101 388] ETSI TS 101 388 V1.3.1 (2002–05), *Asymmetric Digital Subscriber Line (ADSL)—European specific requirements*, ETSI Technical Specification, 2002.

[ETSI TS 101 524] ETSI TS 101 524 V1.2.1 (2003–03), *Symmetric single pair high bitrate Digital Subscriber Line (SDSL)*, ETSI Technical Specification, 2003.

[Ginis 2003] G. Ginis and J. Cioffi. *Vectored Transmission for Digital Subscriber Line System.* Submitted to IEEE JSAC special issue on twisted pair transmission may be downloaded from http://www-isl.stanford.edu/cioffi/dsm/vectorpap/vector.ps.

[Hagelbarger 1959] D.W. Hagelbarger. *Recurrent Codes: Easily Mechanized, Burst-Correcting, Binary Codes.* BSTJ, pp. 969–985, July 1959.

[Henkel 1994] W. Henkel and T. Kessler. *A Wideband Impulsive Noise Survey in the German Telephone Network: Statistical Description and Modeling.* Archiv fur Elekronik und Ubertragungstechnik, Vol. 48, No. 6, 1994.

[Henkel 1999] W. Henkel and T. Kessler. *An Impulse-Noise Model—a Proposal for SDSL.* ETSI TM6, May 1999, Grenoble meeting, Technical document 45.

[Hughes 1997] H. Hughes. *Telecommunications Cables.* Wiley, Chichester, 1997.

[Kirkby 1995] R. Kirkby (idea due to John Cook, of BT Laboratories). *FEXT Is Not Reciprocal.* ANSI T1E1.4/95-141, Orlando meeting, November 1995.

[Kirkby 2001] R. Kirkby. *Text for "Realistic Impulsive Noise Model"* ETSI TM6 Feb. 2001, Sophia Antipolis meeting, Technical document 20.

[Mandelbrot 1965] B. Mandelbrot. *Self-Similar Error Clusters in Communications Systems and the Concept of Conditional Stationarity.* IEEE Trans. Communication Technology, March 1965.

[McLaughlin 1999] S. McLaughlin et al. *Statistics of Impulse Noise.* ETSI TM6, Edinburgh meeting, Sept 1999, Technical documents 18–21.

4

The Twisted Pair Channel—Models and Channel Capacity

Ragnar Hlynur Jonsson

CONTENTS

4.1   Introduction 97
4.2   Information in Digital Communication Systems 98
4.3   Entropy and Information 99
4.4   Transmission Rate and Channel Capacity 101
4.5   Channel Capacity in Presence of Additive Gaussian Noise 102
4.6   Theoretical Rate Computations for PAM, QAM, CAP,
        and DMT Systems 105
        4.6.1   PAM Error Probability 105
        4.6.2   QAM/CAP Error Probability 108
        4.6.3   Ideal DFE Data Rate Calculations 111
        4.6.4   Ideal DMT Data Rate Calculations 114
4.7   Examples of Data Rate Calculations 115
References 117

ABSTRACT   This chapter addresses how channel models can be used to calculate the channel capacity and estimate achievable data rates for DSL systems. The basic concepts of information theory are introduced including entropy, information metrics, transmission rate, and channel capacity. The channel capacity in the presence of additive Gaussian noise is addressed and error probability of PAM, QAM, CAP, and DMT systems is discussed.

4.1   Introduction

In 1948, Claude E. Shannon laid the foundations of information theory in his landmark paper "A Mathematical Theory of Communication" [Shannon 1948]. This paper provided mathematical methods to compute the performance of communications systems given a mathematical model of the system. The previous chapters have provided theoretical models of the DSL loop characteristics. By using these models and the principles introduced by Shannon, it is possible to estimate the theoretical (optimal) performance of DSL systems.

Shannon's model of a general communication system is shown in Figure 4.1. The information source generates a message that could, for example, be a sequence of symbols. For the purpose of modelling the communication system, the information source can usually be viewed as a random signal generator of some sort. The transmitter operates on the message

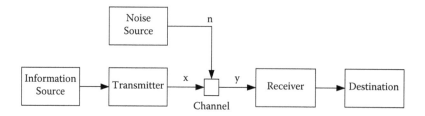

FIGURE 4.1
Shannon's model of a general communication system.

to generate a signal suitable for transmission over the channel. The channel is the medium used to carry the transmitted signal to the receiver. The channel can distort the signal (see Chapter 2) and introduce noise (see Chapter 3). The receiver attempts to reconstruct the original message and passes it on to its destination. In this model, the difference between distortion and noise is that the distortion is a deterministic operation whereas noise involves random perturbations of the signal. Ideally, the distortion can be compensated for by applying some form of an inverse distortion operator, but because of its random nature, noise will cause some loss of information about the transmitted signal (but not necessarily the transmitted message). In DSL systems, the channel distortion will be substantially equivalent to a linear infinite impulse response filter. The frequency response for the loop transfer function is addressed in Section 4.7, and the impulse response for an equivalent filter can be derived from the frequency response by using the inverse Fourier transform. The channel distortion may limit the useful bandwidth of the received signal but can otherwise be compensated for by an inverse distortion operation (equalization). Shannon's theory links the channel capacity (the maximum achievable rate over the channel) with the level of noise over the channel and the channel bandwidth.

4.2   Information in Digital Communication Systems

One of the key questions addressed by Shannon in his 1948 paper [Shannon 1948] was how to measure information. To answer the question, it makes sense to list the desired characteristics of a good information metric. The information conveyed by the occurrence of an event can be linked to the probability of occurrence of the event. Let $P(E)$ be the event probability and $I(E)$ the associated measure of information satisfying the following criteria.

1. If an event is certain to occur, then the occurrence of the event does not convey any information (*i.e.*, if $P(E) = 1$ then $I(E) = 0$).

2. The occurrence of an unlikely event conveys more information than the occurrence of a likely event (*i.e.*, if $P(E_1) > P(E_2)$, then $I(E_1) < I(E_2)$).

3. The information conveyed by the occurrence of the two independent events should be the sum of the information conveyed by the occurrence of each event taken independently (*i.e.*, if $P(E_1, E_2) = P(E_1)P(E_2)$, then $I(E_1, E_2) = I(E_1) + I(E_2)$).

The information metric

$$I(E) = -K \log(P(E)), \tag{4.1}$$

where K is a positive constant, satisfies all three criteria. Criterion 1 is clearly satisfied because

$$-K \log(1) = 0. \tag{4.2}$$

Criterion 2 is also satisfied, because log(.) is a monotonically increasing function such that

$$a > b \Rightarrow \log(a) > \log(b) \Rightarrow -K \log(a) < -K \log(b). \qquad (4.3)$$

Criterion 3 is satisfied because

$$-K \log(P(E_1, E_2)) = -K \log(P(E_1) \times P(E_2)) = -K \log(P(E_1)) - K \log(P(E_2)). \qquad (4.4)$$

Furthermore, because the probability of an event occurring is always in the range from zero to one, criteria 1 and 2 imply that the information metric is never negative.

The choice of the constant K is actually equivalent to choosing a base for the logarithmic function, because

$$-K \log(P(E)) = \log_a(P(E)), \qquad (4.5)$$

where

$$K = \frac{1}{\log(a)}. \qquad (4.6)$$

For practical reasons, binary signals are widely used in communication systems, and therefore it is common to measure information in bits. To measure the information in bits, the information metric uses a logarithm with base 2 and $K = 1$. Therefore, the information metric in bits is given by

$$I(E) = -\log_2(P(E)) \text{ [bits]}. \qquad (4.7)$$

It is possible to use a logarithm with another base in the information calculations, but then the unit is no longer bits. For example, if logarithm with base 256 is used, then the information is measured in octets (8 bit bytes).

4.3 Entropy and Information

It is assumed that an information source produces symbols that take on the M values of $E_1, E_2, \ldots, E_M$ with probability of occurrence $p_1, p_2, \ldots, p_M$. Then the entropy of the information source (*e.g.*, a random signal generator) is defined as

$$H = -\sum_{m=1}^{M} p_m \log_2(p_m) \text{ [bits]}, \qquad (4.8)$$

where p_m is the probability of the information source producing symbol E_m.

From Equations 4.7 and 4.8, it is obvious that entropy is the average value of the information metric or the expected value of the information metric. Thus entropy has the same units as information and can also be expressed in bits. Entropy is a measure of the uncertainty associated with a signal or an information source, but it can also be viewed as a measurement of the information content in a signal or the average information generated by an information source.

Because p_m is always in the range from zero to one, it is trivial to show that the entropy is always greater than or equal to zero. A random variable, **a**, that can take M discrete values all with equal probability has entropy

$$H = -\sum_{m=0}^{M-1} \frac{1}{M} \log_2 \left(\frac{1}{M} \right) = -\log_2 \left(\frac{1}{M} \right) = \log_2(M) \text{ [bits]}. \qquad (4.9)$$

For two discrete random variables, **a** and **b**, (*e.g.*, samples of random signals) with joint probability $p_{m,n} = P(\mathbf{a} = a_m, \mathbf{b} = b_n)$, the joint entropy is given by

$$H(\mathbf{a}, \mathbf{b}) = -\sum_m \sum_n p_{m,n} \log_2(p_{m,n}) \text{ [bits]} . \tag{4.10}$$

By observing that $p_n = \sum_m p_{m,n}$ and $p_m = \sum_n p_{m,n}$, it is easy to show that

$$H(\mathbf{a}, \mathbf{b}) = -\sum_m \sum_n p_{m,n} \log_2(p_{m,n}) \leq -\sum_m p_m \log_2(p_m) - \sum_n p_n \log_2(p_n) = H(\mathbf{a}) + H(\mathbf{b}), \tag{4.11}$$

with equality if and only if $p_{m,n} = p_m p_n$, *i.e.*, if **a** and **b** are independent. The conditional entropy of **a** given $\mathbf{b} = b_n$ is given by

$$H(\mathbf{a}|\mathbf{b} = b_n) = -\sum_m p_{m|n} \log_2(p_{m|n}), \tag{4.12}$$

where $p_{m|n}$ is the conditional probability of $\mathbf{a} = a_m$ knowing that $\mathbf{b} = b_m$, *i.e.*,

$$p_{m|n} = \frac{p_{m,n}}{\sum_m p_{m,n}} = \frac{p_{m,n}}{p_n}. \tag{4.13}$$

In turn, the conditional entropy of **a**, assuming that **b** is given, can be expressed by

$$H(\mathbf{a}|\mathbf{b}) = -\sum_n p_n H(\mathbf{a}|b_n) = -\sum_m \sum_n p_{m,n} \log_2(p_{m|n}). \tag{4.14}$$

By substituting Equation 4.13 into Equation 4.14, it is trivial to show that

$$H(\mathbf{a}|\mathbf{b}) = -\sum_m \sum_n p_{m,n} \log_2(p_{m,n}) + \sum_n p_n \log_2(p_n) = H(\mathbf{a}, \mathbf{b}) - H(\mathbf{b}). \tag{4.15}$$

Therefore,

$$H(\mathbf{a}, \mathbf{b}) = H(\mathbf{b}) + H(\mathbf{a}|\mathbf{b}). \tag{4.16}$$

If the random variables **a** and **b** are not independent, then

$$H(\mathbf{a}) > H(\mathbf{a}|\mathbf{b}), \tag{4.17}$$

which implies that knowledge about **b** provides some knowledge about **a** (and vice versa). The information that knowledge about **b** provides about **a** can be measured by

$$I(\mathbf{a}, \mathbf{b}) = H(\mathbf{a}) - H(\mathbf{a}|\mathbf{b}). \tag{4.18}$$

The function $I(\mathbf{a}, \mathbf{b})$ is known as the mutual information of **a** and **b**. By using Equation 4.16, the mutual information can be written as

$$\begin{aligned}
I(\mathbf{a}, \mathbf{b}) &= H(\mathbf{a}) - H(\mathbf{a}|\mathbf{b}) \\
&= H(\mathbf{b}) - H(\mathbf{b}|\mathbf{a}) \\
&= H(\mathbf{a}) + H(\mathbf{b}) - H(\mathbf{a}, \mathbf{b}).
\end{aligned} \tag{4.19}$$

The mutual information is an important concept for computing transmission rate and channel capacity, as discussed in Section 4.4 below.

For random signal generators with continuous probability distributions, the entropy is given by

$$H = -\int_{-\infty}^{\infty} f(x) \log_2(f(x))dx,$$ (4.20)

where $f(x)$ is the probability density function for the random signal generator. For example, a random variable $\mathbf{a}$ having Gaussian distribution with variance σ^2 has entropy

$$H = -\int_{-\infty}^{\infty} \frac{1}{\sigma\sqrt{2\pi}} \exp\left(-\frac{(x-m)^2}{2\sigma^2}\right) \log_2\left(\frac{1}{\sigma\sqrt{2\pi}} \exp\left(-\frac{(x-m)^2}{2\sigma^2}\right)\right) dx = \log_2\left(\sigma\sqrt{2\pi}\right) + \frac{1}{2}.$$ (4.21)

For two random variables, $\mathbf{a}$ and $\mathbf{b}$, with continuous distributions, the joint entropy is given by

$$H(\mathbf{a}, \mathbf{b}) = -\iint f_{\mathbf{a},\mathbf{b}}(x, y) \log_2(f_{\mathbf{a},\mathbf{b}}(x, y))dx\,dy,$$ (4.22)

where $f_{\mathbf{a},\mathbf{b}}(x, y)$ is the joint probability density function. In the same manner as for the discrete distributions, the conditional entropy for random variables with continuous distributions is given by

$$H(\mathbf{a}|y) = -\int f_{\mathbf{a}|\mathbf{b}}(x, y) \log_2(f_{\mathbf{a}|\mathbf{b}}(x|y))dx,$$ (4.23)

and

$$H(\mathbf{a}|\mathbf{b}) = -\iint f_{\mathbf{a},\mathbf{b}}(x, y) \log_2(f_{\mathbf{a}|\mathbf{b}}(x|y))dx\,dy,$$ (4.24)

where

$$f_{\mathbf{a}|\mathbf{b}}(x|y) = \frac{f_{\mathbf{a},\mathbf{b}}(x, y)}{\int f_{\mathbf{a},\mathbf{b}}(x, y)dx} = \frac{f_{\mathbf{a},\mathbf{b}}(x, y)}{f_{\mathbf{b}}(y)}.$$ (4.25)

The entropy for continuous distributions shares most of the properties of the entropy for discrete distributions, except that the entropy for continuous distributions can be negative. This is because, unlike the discrete distribution entropy, the continuous distribution entropy is not an absolute measure of information, but rather a measure of information relative to some coordinate system (see [Shannon 1948]). However, the difference of two continuous distribution entropy values that share the same coordinate system is an absolute measure of information. Because of this relative nature of the entropy for continuous distributions, the continuous distribution entropy is sometimes referred to as differential entropy. For a more detailed and a more rigorous discussion about entropy, see [Papoulis 2002].

4.4 Transmission Rate and Channel Capacity

As discussed above, entropy is a measure of information content of a random signal. When a signal is sent over a noisy channel, some of the information content is "lost" due to transmission errors induced by the noise. When a signal $\mathbf{x}$ is sent over a noisy channel, and a signal $\mathbf{y}$ is received ($\mathbf{x}$ and $\mathbf{y}$ could be discrete or continuous signals), then the information content of the transmitted signal is given by $H(\mathbf{x})$, and the loss of information due to channel noise is given by $H(\mathbf{x}|\mathbf{y})$ (*i.e.*, the uncertainty about $\mathbf{x}$ given $\mathbf{y}$). The transmission rate of a communication system can be defined as the rate of information that is actually delivered over the communication channel. Ideally, the transmitted rate would be delivered completely error free over the channel, but due to the random nature of the channel noise, error events cannot be completely excluded. However, error events can be made to have arbitrarily low probability. Therefore, the transmission rate R of a communication system is defined as the rate of information delivered over the communication channel, with arbitrarily low error probability [Shannon 1948].

The transmission rate is then given as the transmitted information rate minus the information rate lost due to noise. That is,

$$R = H(\mathbf{x}) - H(\mathbf{x}|\mathbf{y}) = I(\mathbf{x}, \mathbf{y}). \tag{4.26}$$

The transmission rate clearly increases with increasing entropy of the transmitted signal, $H(\mathbf{x})$, but decreases with increasing information loss due to channel errors, $H(\mathbf{x}|\mathbf{y})$. It is clear from Equation 4.19 that it is possible to rewrite the transmission rate as

$$R = H(\mathbf{y}) - H(\mathbf{y}|\mathbf{x}). \tag{4.27}$$

This implies that the transmission rate is equal to the entropy of the received signal minus the entropy due to the channel noise. Yet another formulation for the transmission rate is

$$R = H(\mathbf{x}) + H(\mathbf{y}) - H(\mathbf{x}, \mathbf{y}). \tag{4.28}$$

This can be interpreted as the transmission rate being the rate of the information that is common in the transmitted and received signal. The transmission rate representations in Equations 4.26, 4.27, and 4.28 are all equivalent, but each gives a different insight into the concept of transmission rate.

The transmission rate clearly depends on both the characteristics of the transmitted signal (the transmitter) and the channel characteristics. In other words, the transmission rate is a metric of the overall system, including both the transmitter and the channel (assuming an ideal receiver). But what is the highest transmission rate that can be achieved over a given channel? The answer is obviously the transmission rate for a (hypothetical) transmitter that has the highest transmission rates for the given channel. That maximum transmission rate over a given channel is referred to as the channel capacity and is defined as

$$C = \max_{P(\mathbf{x})} \{H(\mathbf{x}) - H(\mathbf{x}|\mathbf{y})\}. \tag{4.29}$$

The channel capacity is the highest information rate (data rate) that can be achieved over a given channel with arbitrarily low error probability.[1]

If there are no constraints on the transmit signal $\mathbf{x}$, then it is possible to increase infinitely the entropy, $H(\mathbf{x})$, of the signal. Therefore, the channel capacity of virtually any channel can be made infinite if there are no constraints on the transmitted signal. Real systems, however, always have some constraints on the transmitted signal, and channel capacity with unconstrained transmit signals is seldom of interest. It is important to keep in mind that channel capacity is not only defined in terms of the channel characteristics, but also in terms of the constraints on the transmitted signal.

4.5   Channel Capacity in Presence of Additive Gaussian Noise

Many communications channels, including DSL loops, can be modelled (with reasonable accuracy) as having only additive noise, where the noise is not correlated with the

---

[1] As discussed above, due to the random nature of noise, transmission errors cannot be completely eliminated. Nevertheless, the probability of errors can be made arbitrarily small.

transmitted signal. For such channels, the received signal (ignoring channel distortion) is given by

$$\mathbf{y} = \mathbf{x} + \mathbf{n}, \tag{4.30}$$

where $\mathbf{n}$ is the additive noise. Then the conditional probability of receiving $\mathbf{y}$ given $\mathbf{x}$ is

$$P_{\mathbf{x}|\mathbf{y}}(y|x) = P_{\mathbf{n}}(y - x), \tag{4.31}$$

where $P_{\mathbf{n}}(\mathbf{n})$ is the probability distribution of the noise. For this case, the conditional entropy is

$$H(\mathbf{y}|\mathbf{x}) = H(\mathbf{n}), \tag{4.32}$$

the transmission rate becomes

$$R = H(\mathbf{y}) - H(\mathbf{n}), \tag{4.33}$$

and the channel capacity becomes

$$C = \max_{P(\mathbf{x})}\{H(\mathbf{y}) - H(\mathbf{n})\}. \tag{4.34}$$

Because $\mathbf{n}$ is independent of $\mathbf{x}$, the channel capacity is the transmission rate for the $P(\mathbf{x})$ that maximizes the entropy, $H(\mathbf{y})$, of the received signal.

For a channel with additive white Gaussian noise (AWGN), the noise entropy is

$$H(\mathbf{n}) = \log_2\left(\sqrt{2\pi}\, S_n\right) + \frac{1}{2}, \tag{4.35}$$

where S_n is the average power of the noise $\mathbf{n}$ [Shannon 1948]. For a band-limited channel with bandwidth ω, the noise entropy per second is

$$H'(\mathbf{n}) = \left(\log_2\left(\sqrt{2\pi}\, S_n\right) + \frac{1}{2}\right)\omega. \tag{4.36}$$

If the average transmit power is limited to S_p, then the power of the received signal is limited to $S_p + S_n$. In this case, the optimum transmit signal is a Gaussian signal, and the received signal also becomes a Gaussian signal [Shannon 1948]. The entropy of the received signal then becomes

$$H(\mathbf{y}) = \log_2\left(\sqrt{2\pi}\,(S_p + S_n)\right) + \frac{1}{2}, \tag{4.37}$$

and for a band-limited channel with bandwidth ω, the received signal entropy per second is

$$H'(\mathbf{y}) = \left(\log_2\left(\sqrt{2\pi}\,(S_p + S_n)\right) + \frac{1}{2}\right)\omega. \tag{4.38}$$

Therefore, for a channel with AWGN, the channel capacity is

$$C = \max_{P(\mathbf{x})}\{H'(\mathbf{y}) - H'(\mathbf{n})\} = \left(\log_2\left(\sqrt{2\pi}\,(S_p + S_n)\right) + \frac{1}{2}\right)\omega - \left(\log_2\left(\sqrt{2\pi}\, S_n\right) + \frac{1}{2}\right)\omega$$

$$= \omega\log_2\left(\frac{S_p + S_n}{S_n}\right). \tag{4.39}$$

In terms of signal-to-noise ratio (SNR), this becomes

$$C = \omega\log_2\left(\frac{S_p + S_n}{S_n}\right) = \omega\log_2\left(1 + \frac{S_p}{S_n}\right) = \omega\log_2(1 + SNR). \tag{4.40}$$

If the channel noise is not white but colored Gaussian noise, then the channel capacity can be derived by breaking the frequency band into very narrow sub-bands, each with bandwidth Δf. Then the noise is practically white in each sub-band, and Equation 4.40 applies. By taking the limit as $\Delta f \rightarrow 0$, the channel capacity formula for additive Gaussian noise becomes

$$C = \int_0^\omega \log_2\left(1 + \frac{S_\mathbf{p}(f)}{S_\mathbf{n}(f)}\right) df. \tag{4.41}$$

This channel capacity formula gives the "Shannon limit" for systems in the presence of additive Gaussian noise [Shannon 1948]. This formula is very useful for computing the maximum achievable performance on a channel, but as the following sections will demonstrate, it is also useful for computing the performance of real systems with known deficiency relative to the Shannon limit.

Figure 4.2 shows the channel capacity of a twisted-pair wire in the presence of additive white Gaussian noise. The transmit power level was limited to 16 dBm (40 mW), and the channel capacity was computed for different noise levels. As can be seen from the figure, it should be possible to transmit about 1 Gbps over a 1 km length loop if the only noise on the loop is thermal noise at -174 dBm/Hz (4×10^{-21} W/Hz). This is obviously far more than can be achieved in real DSL systems, because the noise levels are much higher, especially when crosstalk is present. For the more realistic noise level of -80 dBm/Hz (10^{-11} W/Hz) the channel capacity is down to about 10 Mbit/s over a 1 km loop, which is of the same order of magnitude as could be achieved for real DSL systems.

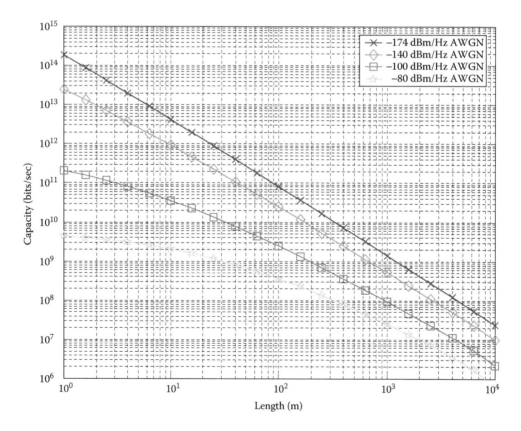

FIGURE 4.2

Channel capacity for twisted-pair wires in the presence of AWGN, assuming 16 dBm transmit power.

4.6 Theoretical Rate Computations for PAM, QAM, CAP, and DMT Systems

The most commonly used modulation methods in DSL systems today are pulse amplitude modulation (PAM), quadrature amplitude modulation (QAM), carrierless amplitude phase modulation (CAP), and discrete multi-tone (DMT). All these systems are based on amplitude modulation of one form or another. It is possible to derive the theoretically achievable data rates for DSL systems by computing the bit error rate as a function of the signal-to-noise ratio for each modulation scheme.

4.6.1 PAM Error Probability

Basic PAM modulation (see Section 6.2.2) uses M discrete amplitude levels (constellation points) to transmit $\log_2(M)$ bits on each symbol. Normally the amplitude levels are uniformly spaced so there is the same distance between all points, but in some communication systems (such as V.90 and V.92 voiceband modems), the constellation points are not uniformly spaced.

Figure 4.3 illustrates a typical PAM system with four uniformly spaced constellation points. In the 4PAM (2B1Q) mapping, two bits are mapped in four-level symbols. The figure also illustrates how additive Gaussian noise might be distributed around the constellation points. Due to the noise, the amplitude of the received signal is not confined to the

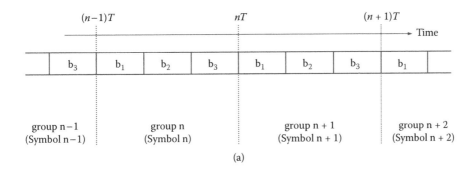

(a)

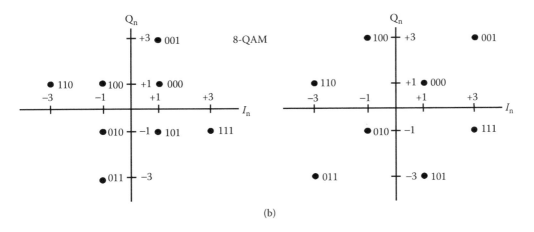

(b)

FIGURE 4.3
4-PAM constellation mapping and noise distribution.

transmitted amplitude levels, but is distributed around them. The receiver estimates which symbol was sent by choosing the constellation point closest to the received signal amplitude. If the amplitude of the noise is more than half the distance d between constellation points, then the receiver will make a detection error (a transmission error occurs). Therefore, for an M-PAM system with uniformly distributed constellation points, the probability of error is

$$P_e(\sigma) = \frac{2(M-1)}{M} Q\left(\frac{d}{2\sigma^2}\right),$$ (4.42)

where

$$Q(x) = \frac{1}{\sqrt{2\pi}} \int_x^\infty \exp(-\frac{y^2}{2})dy$$ (4.43)

is the probability of the value of the Gaussian noise exceeding the value x.

Given the relationship between the signal energy and d, the previous equation can be easily translated in terms of SNR; *i.e.*,

$$P_e(SNR) = \frac{2(M-1)}{M} Q\left(\sqrt{\frac{3SNR}{M^2-1}}\right),$$ (4.44)

where SNR is the signal-to-noise ratio of the received signal. The factor 2 in Equation 4.44 is because both positive and negative noise values can cause errors, so both tails of the Gaussian distribution must be considered. The reason for the $\frac{M-1}{M}$ factor is that for the outermost constellation points, only one tail of the Gaussian distribution will cause errors (refer to the explanation in Figure 4.3). The factor $\frac{2(M-1)}{M}$ can be viewed as the average number of closest (nearest) neighbors. All points have two closest neighbors, except the two end points, which only have one closest neighbor. If the amplitude of the PAM signal is wrapped onto itself as is done by a modulo operator in Tomlinson–Harashima precoding (see Chapter 11), then the highest and lowest points have the same number of closest neighbors as all the other points. In this case, the average number of closest neighbors becomes 2 instead of $\frac{2(M-1)}{M}$.

Figure 4.4 shows the error probability for M-PAM systems with M in the range from 2 to 64. It is interesting to note that as the constellation size is doubled (*i.e.*, one more bit is carried by each symbol), the PAM system needs an SNR increase of approximately 6 dB to maintain the same error probability. This can be explained by observing that the argument of the $Q(\cdot)$ function in Equation 4.44 has the term $\frac{SNR}{(M^2-1)}$. Therefore, to maintain a constant value of $Q(\cdot)$ as M increases, the SNR value must be adjusted to maintain a constant value for $\frac{SNR}{(M^2-1)}$. For reasonably large M, doubling of M implies that the SNR must increase approximately fourfold (*i.e.*, by approximately 6 dB) to maintain the same error probability. For small M, the SNR must increase by slightly more than a factor of four (6 dB) because of the scaling due to the average number of closest neighbors and due to the -1 term in $\frac{SNR}{(M^2-1)}$.

The idea of maintaining a constant value of the argument of the $Q(\cdot)$ function gives rise to the concept of normalized SNR [Forney 1998]. The normalized SNR can be defined as

$$SNR_{norm} = \frac{SNR}{M^2-1}.$$ (4.45)

For M-PAM systems, the number of bits per symbol is $R = \log_2(M)$. Therefore, the normalized SNR value can be expressed in terms of bits per symbol as

$$SNR_{norm} = \frac{SNR}{2^{2R}-1}.$$ (4.46)

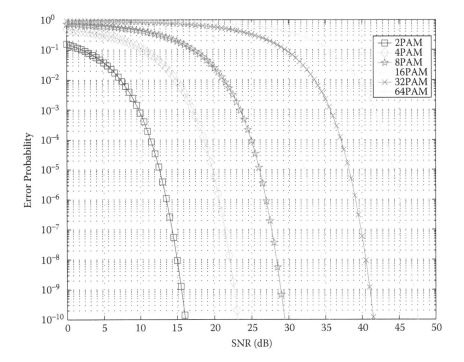

FIGURE 4.4
Error probability as a function of SNR for M-PAM signals.

The error probability of M-PAM can now be expressed in terms of SNR_{norm} as

$$P_e(SNR_{norm}) = \frac{2(M-1)}{M} Q\left(\sqrt{3SNR_{norm}}\right). \tag{4.47}$$

Apart from the term $\frac{2(M-1)}{M}$, due to the average number of neighbors, the error probability is independent of M. For large M, the average number of neighbors is almost constant, and the error probability is therefore virtually constant for a fixed value of SNR_{norm}.

Figure 4.5 shows the error probability for M-PAM as a function of normalized SNR. The curves for all M-PAM systems fall between the $M = 2$ and the $M \to \infty$ curves, and the error probability for the $M = 2$ case is exactly half the error probability for the $M \to \infty$ case. As M increases, the error probability rapidly approaches the $M \to \infty$ error probability.

The Shannon limit is for AWGN and is given in Equation 4.40. The effective bandwidth w of baseband signals is half the symbol rate, or $\frac{1}{2T}$. Therefore, for a baseband signal, the Shannon limit for the maximum number of bits per symbol in the presence of AWGN is given by

$$C = \omega T \log_2(1 + SNR) = \frac{1}{2}\log_2(1 + SNR). \tag{4.48}$$

This can be rewritten as

$$SNR = 2^{2C} - 1. \tag{4.49}$$

This means that the minimum SNR that can achieve R bits per symbol, with arbitrarily few errors, is

$$SNR_{Shannon}(R) = 2^{2R} - 1. \tag{4.50}$$

This gives an alternative interpretation of the normalized SNR, because this implies that

$$SNR_{norm} = \frac{SNR}{SNR_{Shannon}(R)}. \tag{4.51}$$

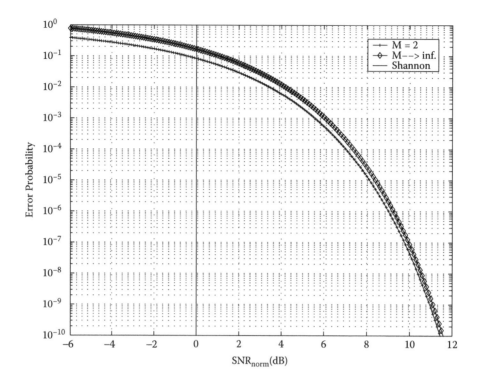

FIGURE 4.5
Error probability as a function of normalized SNR for M-PAM signals.

Thus, the normalized SNR is the SNR normalized by the Shannon limit for the given rate. This implies that the 0 dB line in Figure 4.5 is the Shannon limit, and the normalized SNR value required to achieve a given error probability is the Shannon gap for M-PAM at that error rate, which can be denoted as $\Gamma(P_e)$. For example, the Shannon gap for M-PAM at error probability of 10^{-7} is approximately 9.95 dB.

   Another way to view the Shannon gap is to think of it as the degradation in performance of real systems relative to the theoretically optimal Shannon performance. This degradation can be represented as a reduction of effective SNR, and the achievable rate for an M-PAM system is then given by

$$R = \omega \log_2\left(1 + \frac{SNR}{\Gamma(P_e)}\right),\qquad(4.52)$$

where $\Gamma(P_e)$ is a constant reflecting the Shannon gap. For a bit error rate of 10^{-7}, this constant is approximately $\Gamma(P_e) = 9.89$ (*i.e.*, 9.95 dB).

4.6.2   QAM/CAP Error Probability

Unlike (baseband) PAM systems, which use one-dimensional constellations, QAM and CAP systems use two-dimensional constellations (see Section 6.2). Each QAM/CAP symbol can be viewed as a complex number that takes on a value corresponding to one of the constellation points. The x-axis of the constellation is the "in-phase" component (denoted I) of the constellation, and the y-axis is the "quadrature" (denoted Q) component. Figure 4.6 and Figure 4.7 show examples of QAM/CAP constellations. The constellation in Figure 4.6 is a square 16-point constellation that can represent 4 bits per symbol. The constellation in Figure 4.7 is a 32-point cross constellation that can represent 5 bits per symbol. Both constellations are arranged on a uniformly spaced grid.

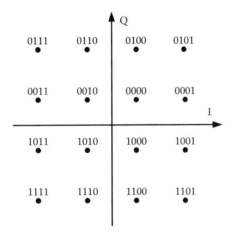

FIGURE 4.6
16-QAM square constellation.

The error probability for uniformly spaced QAM/CAP constellations, in the presence of additive Gaussian noise, can be calculated using the same basic approach as was used for PAM. The error probability can be calculated based on the average number of closest neighbors multiplied by the probability of the noise being large enough to cause an error. If either the I or the Q component of the noise becomes larger than $\frac{d}{2}$, then a transmission error will occur. Therefore, for a square $M \times M$-QAM constellation, the bit error probability is the same as for M-PAM and is given by Equation 4.44, even though M has a slightly different meaning in the $M \times M$-QAM case than in the M-PAM case. However, if both the I and the Q components of the noise are larger then $\frac{d}{2}$, then the wrong detection becomes the

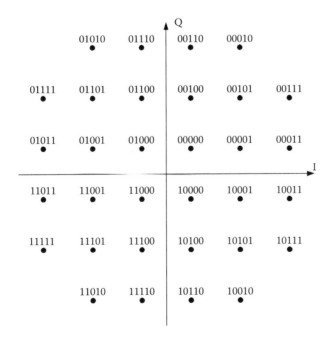

FIGURE 4.7
32-QAM cross constellation.

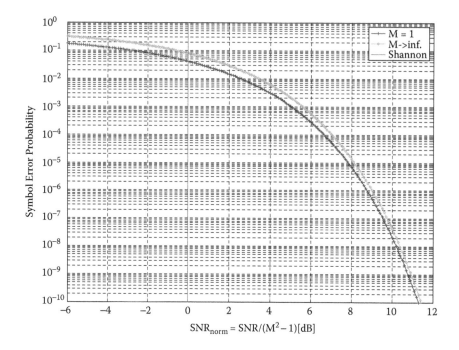

FIGURE 4.8

Symbol error probability for square $M \times M$ QAM/CAP constellations in the presence of additive Gaussian noise.

diagonal neighbor (and beyond), but still only one symbol error occurs (*i.e.*, there is only one symbol error, but two bit errors). Therefore, the probability of an error is the average number of closest neighbors multiplied by the probability of either the I or the Q noise component exceeding $\frac{d}{2}$ minus the average number of second closest neighbors multiplied by the probability of both noise components exceeding $\frac{d}{2}$. For a square $M \times M$ constellation, the average number of closest neighbors is $\frac{4(M-1)}{M}$ and the average number of second closest neighbors is $\frac{4(M-1)^2}{M^2}$, so the symbol error probability becomes

$$P_e(SNR_{norm}) = 4\frac{(M-1)}{M}Q\left(\sqrt{3SNR_{norm}}\right) - 4\left(\frac{(M-1)}{M}Q\left(\sqrt{3SNR_{norm}}\right)\right)^2. \tag{4.53}$$

The normalized SNR is again defined according to Equation 4.45, even though in this context M has a slightly different meaning.

Figure 4.8 shows the symbol error probability for an $M \times M$ QAM/CAP constellation as a function of normalized SNR. As before, the gap between the $M = 2$ and $M \to \infty$ curves is due to the effect of the average number of closest neighbors. All square, uniformly spaced QAM/CAP constellations with four or more points will fall between these two curves.

The effective bandwidth ω of passband signals is the symbol rate, $\frac{1}{T}$. Therefore, for a passband signal, the Shannon limit for the maximum number of bits per symbol in the presence of AWGN is given by

$$C = \omega T \log_2(1 + SNR) = \log_2(1 + SNR). \tag{4.54}$$

This can be rewritten as

$$SNR = 2^C - 1. \tag{4.55}$$

This means that for passband signals, the minimum SNR that can achieve R bits per symbol (with arbitrarily few errors) is

$$SNR_{Shannon}(R) = 2^R - 1. \tag{4.56}$$

As for the baseband case, this implies that

$$SNR_{norm} = \frac{SNR}{SNR_{Shannon}(R)}.$$ (4.57)

That is, for square QAM/CAP constellations, the normalized SNR is the SNR normalized by the Shannon limit for the given rate. This implies that the 0 dB line in Figure 4.8 is the Shannon limit, and the normalized SNR value required to achieve a given error probability is the Shannon gap for M-PAM at that error rate. For example, the Shannon gap for $M \times M$-QAM at symbol error probability of 10^{-7} is approximately 9.75 dB.

It is worth noting that the Shannon gap for a given symbol error probability is slightly less for QAM and CAP modulation than it is for PAM modulation (see Subsection 4.6.1 above). This is because error events that cause false detection of both the I and the Q component on the same symbol are only counted as one symbol error event, even though they cause two bit errors. However, for a given bit error probability (as opposed to symbol error probability), the Shannon gap is the same for square QAM and CAP constellations as for PAM constellations, and the Shannon gap for $M \times M$-QAM at bit error probability of 10^{-7} is approximately 9.95 dB. Just as for PAM systems, the Shannon gap for QAM and CAP systems can be viewed as the degradation in performance of real systems relative to the theoretically optimal Shannon performance. The achievable rate for a QAM and CAP system is then given by

$$R = \omega \log_2 \left(1 + \frac{SNR}{\Gamma(P_e)} \right),$$ (4.58)

where $\Gamma(P_e)$ is again a constant reflecting the Shannon gap. Just as for PAM systems, this constant is approximately $\Gamma(P_e) = 9.89$ (*i.e.*, 9.95 dB) for a bit error rate of 10^{-7}.

4.6.3   Ideal DFE Data Rate Calculations

When signals are transmitted over a DSL loop, the signals are distorted by the loop transfer function and corrupted by additive crosstalk and background noise (see Section 4.7). In ideal receivers, the distortion due to the loop transfer function can be compensated for by an ideal receiver, but due to the inherently unpredictable nature of noise, the additive noise cannot be completely removed. If the noise is colored, then the noise is correlated with time shifts of itself. This implies that for colored noise, it is possible to reduce the noise power by removing the correlated (predictable component) from the noise, making it white (noncorrelated). The process of compensating for the channel distortion and minimizing the errors due to noise is referred to as equalization (see Chapter 11). An ideal equalizer both removes the channel distortion and minimizes the error rate due to noise.

For additive Gaussian noise, an ideal minimum mean square error decision feedback equalizer (MMSE-DFE) is close to being the ideal equalizer [Lee 1994]. It is possible to use the DFE structure for equalization in PAM, QAM, and CAP systems, and the achievable performance of these systems can be evaluated in terms of ideal DFE performance. It is not practical, however, to use the DFE structure with DMT signals, so the DMT performance calculations will be discussed separately in Subsection 4.6.4 below.

It can be demonstrated (see (10.84) in [Lee 1994]) that the SNR for an ideal MMSE-DFE is

$$SNR_{eq} = \frac{\exp\left(\frac{1}{\omega} \int_{\omega} \log_e(S_y(f))df \right)}{\exp\left(\frac{1}{\omega} \int_{\omega} \log_e(S_n(f))df \right)},$$ (4.59)

where $S_y(f)$ is the PSD of the received signal, $\mathbf{y}$, and $S_n(f)$ is the noise PSD. Because the noise and the signal are not correlated, the received PSD is given by

$$S_y(f) = S_p(f) + S_n(f), \tag{4.60}$$

where

$$S_p(f) = S_x(f)\,|H(f)|^2 \tag{4.61}$$

is the component of the received PSD that is due to the transmit signal. Therefore, the equalizer output SNR can be rewritten as

$$
\begin{aligned}
SNR_{eq} &= \exp\left(\frac{1}{\omega}\int_{\omega} \log_e(S_y(f))df - \frac{1}{\omega}\int_{\omega} \log_e(S_n(f))df\right) \\
&= \exp\left(\frac{1}{\omega}\int_{\omega} \log_e\left(\frac{S_p(f) + S_n(f)}{S_n(f)}\right)df\right) \\
&= \exp\left(\frac{1}{\omega}\int_{\omega} \log_e\left(1 + \frac{S_p(f)}{S_n(f)}\right)df\right).
\end{aligned}
\tag{4.62}
$$

Strictly speaking, the equalized error is not Gaussian unless the transmit symbols have a Gaussian distribution, but for cases of interest the noise is almost Gaussian. In this chapter, the equalized error will be treated as if it were Gaussian.

As discussed in Subsections 4.6.1 and 4.6.2, the achievable data rate for PAM, QAM, and CAP systems is

$$R = \omega \log_2\left(1 + \frac{SNR}{\Gamma}\right), \tag{4.63}$$

where ω is the signal bandwidth, and Γ is the factor accounting for the Shannon gap and, if present, coding gain and noise margin. Therefore, the data rate for an ideal MMSE-DFE system is

$$R = \omega \log_2\left(1 + \frac{SNR_{eq}}{\Gamma}\right) = \omega \log_2\left(1 + \frac{1}{\Gamma}\exp\left(\frac{1}{\omega}\int_{\omega} \log_2(1 + SNR_{Rx}(f))\,df\right)\right), \tag{4.64}$$

where $SNR_{Rx}(f)$ is the received SNR.

DSL systems are conventionally operated with some margin to ensure reliable operation and to avoid unacceptable error rates in the event of minor increase in crosstalk noise. Typically, the noise margin is at least 5 dB or 6 dB. The SNR margin can be accounted for in the Γ coefficient in Equation 4.64 according to

$$\Gamma = \Gamma(P_e) \cdot \gamma_m. \tag{4.65}$$

The data rate for ideal MMSE-DFE operating with margin γ_m is then

$$R = \omega \log_2\left(1 + \frac{SNR_{eq}}{\Gamma(P_e) \cdot \gamma_m}\right). \tag{4.66}$$

This can be rewritten as

$$\Gamma(P_e) \cdot \left(2^{R/W} - 1\right) = \frac{SNR_{eq}}{\gamma_m}. \tag{4.67}$$

TABLE 4.1

SNR Required for PAM Constellations and Square QAM Constellations
to Achieve Desired Bit Error Rate (BER)

	Required SNR [dB]								
BER	10^{-1}	10^{-2}	10^{-3}	10^{-4}	10^{-5}	10^{-6}	10^{-7}	10^{-8}	10^{-9}
2PAM 2×2-QAM	2.15	7.33	9.80	11.41	12.60	13.54	14.32	14.98	15.56
4PAM 4×4-QAM	10.52	14.86	17.12	18.63	19.77	20.68	21.43	22.08	22.64
8PAM 8×8-QAM	17.19	21.28	23.47	24.95	26.07	26.96	27.71	28.35	28.91
16PAM 16×16-QAM	23.45	27.44	29.59	31.06	32.17	33.06	33.80	34.44	35.00
32PAM 32×32-QAM	29.57	33.51	35.65	37.11	38.22	39.11	39.85	40.48	41.04
64PAM 64×64-QAM	35.63	39.55	41.68	43.14	44.25	45.13	45.88	46.51	47.07
128PAM 128×128-QAM	41.68	45.58	47.71	49.17	50.27	51.16	51.90	52.53	53.09
256PAM 256×256-QAM	47.71	51.61	53.74	55.19	56.29	57.18	57.92	58.56	59.11

This in turn can be rewritten in dB as

$$SNR_{margin} = 10 \log_{10}\left(\frac{1}{\gamma_m}\right) = SNR_{eq} \text{ [dB]} - SNR_{req}, \tag{4.68}$$

where SNR_{req} is the required minimum margin given by

$$SNR_{req} = 10 \log_{10}\left(\Gamma(P_e) \cdot \left(2^{R/W} - 1\right)\right). \tag{4.69}$$

The Shannon gap, $\Gamma(P_e)$, depends on the error rate and can be derived from Equations 4.44 and 4.53 or read from the plots in Figure 4.5 for PAM as well as for QAM/CAP. The factor $\frac{R}{W}$ is translated into bits per symbol for QAM and CAP and two times bits per symbol for PAM. Required SNR values are tabulated in Table 4.1.

Combining Equations 4.64 and 4.68 gives the operating margin for PAM systems as

$$SNR_{margin} = 2T \int_0^{1/2T} 10 \log_{10}(1 + SNR(f))\, df - SNR_{req-PAM}, \tag{4.70}$$

where $SNR(f)$ is the frequency-dependent SNR of the receiver input, and $SNR_{req-PAM}$ can be obtained from Table 4.1.

The operating margin for QAM and CAP systems is

$$SNR_{margin} = T \int_{f_c-1/2T}^{f_c+1/2T} 10 \log_{10}(1 + SNR(f))\, df - SNR_{req-QAM}, \tag{4.71}$$

where again $SNR(f)$ is the frequency-dependent SNR of the receiver input and $SNR_{req-QAM}$ can again be obtained from Table 4.1. Fractionally spaced equalizers (see Chapter 11) can make use of frequency folding (aliasing) to enhance the received SNR. For PAM systems, the effective received SNR for fractionally spaced equalizers becomes (see [ANSI T1.413-2001])

$$SNR_{PAM-fold}(f) = \sum_{n=0}^{N_{os}-1} SNR_{rx}\left(\frac{n}{2T} + (-1)^n f\right), \tag{4.72}$$

where N_{os} is the oversampling ratio. For QAM and CAP systems, the effective received SNR for fractionally spaced equalizers becomes (see [ANSI T1.413-2001])

$$SNR_{QAM-fold}(f) = \sum_{n=0}^{N_{os}-1} SNR_{rx}\left(f_c + n/T + (-1)^n(f - f_c)\right). \tag{4.73}$$

When computing ideal DFE performance for fractionally spaced equalizers, the input SNRs in Equations 4.70 and 4.71 are replaced by the folded SNRs in Equations 4.72 and 4.73, respectively.

The formulae in Equations 4.70 and 4.71 are widely used to compute performance of PAM, QAM, and CAP systems. For example, they are part of the alternative "Method B" to demonstrate spectral compatibility according to Annex A of the North American spectrum management document T1.417 [ANSI T1.413-2001]. The PAM performance formula in Equation 4.70 was also used as the basis for setting performance requirements for SHDSL in ITU Recommendation G.991.2 [ITU-T G.991.2] and for ETSI SDSL in [ETSI TS 101 524].

In the discussion above, it was assumed that no coding was used. By using trellis coding (see Chapter 8) and other similar techniques, it is possible to improve performance compared to non-coded systems. Such improvement in performance due to coding is referred to as coding gain. The coding gain reduces the gap of the system from Shannon capacity, and in performance calculations it can be accounted for by factoring it into the Γ coefficient in Equation 4.64 according to

$$\Gamma = \frac{\Gamma(P_e) \cdot \gamma_m}{\gamma_c}, \tag{4.74}$$

where γ_c denotes the coding gain. In Equations 4.70 and 4.71, the coding gain can be accounted for by subtracting the coding gain (in dB) from the required margin. Coding gain due to trellis coding is typically in the range from 3 dB to about 5 dB, but slightly higher coding gain is possible. Forward error correction (FEC) of the received data, such as with Reed–Solomon codes (see Section 9.2.5), can also provide coding gain, but FEC at the received data level usually implies that some overhead needs to be subtracted from the net data rate to account for FEC overhead.

4.6.4   Ideal DMT Data Rate Calculations

In DMT systems (see Chapter 7), a discrete Fourier transform (DFT) is used to combine many narrowband QAM signals to form one broadband signal. For each narrowband channel (each subchannel or tone), the signal and noise are almost white within the band. Therefore, the data rate for each subchannel can be approximated by

$$C_k = \frac{1}{T} \log_2 \left(1 + \frac{SNR_k}{\Gamma} \right), \tag{4.75}$$

where $\frac{1}{T}$ is the data symbol rate of each subchannel, SNR_k is the "average" SNR in the subchannel, and Γ is the overall gap from Shannon capacity for QAM at the desired error rate. If the DMT system operates with a noise margin or if coding is used, then the margin and coding gain are incorporated into the Γ factor according to Equation 4.74.

The aggregate data rate using all the narrowband subchannels is the sum of the data rates of the individual subchannels:

$$C = \sum_{k=K_0}^{K_1} C_k = \frac{1}{T} \sum_{k=K_0}^{K_1} \log_2 \left(1 + \frac{SNR_k}{\Gamma} \right). \tag{4.76}$$

The SNR_k values should be computed as the ratio of the average signal power divided by the average noise power. However, based on the assumption of almost flat signal and noise spectrum within each channel, the SNR for each channel can be approximated by

$$SNR_k = \frac{S_p(f_k)}{S_n(f_k)}, \tag{4.77}$$

where f_k is the sub-carrier (center) frequency for each subchannel. The combination of Equations 4.76 and 4.77 provides a good first approximation of an ideal DMT system. This approximation, however, does not take into account several issues related to real DMT implementations, including minimum and maximum constellation size, DFT windowing effect, and various overhead issues. These issues are discussed in more detail in Chapters 7 and 13.

The formula in Equation 4.76, or a variant of that formula, is widely used to compute performance of DMT systems. For example, they are part of the alternative "Method B" to demonstrate spectral compatibility according to Annex A of the North American spectrum management document [ANSI T1.413-2001]. Variation of this formula was also used as the basis for setting performance requirements for ETSI ADSL in [ETSI TS 101 388].

4.7   Examples of Data Rate Calculations

The discussion above provides all the components needed to do basic capacity calculations for DSL systems. The characteristics of twisted-pair wires as a communication medium are addressed in Chapters 2 and 3. The frequency-dependent attenuation of the transmitted signal can be computed based on the loop models given in Chapter 2. Given an *ABCD* matrix for any loop, its insertion loss can be computed (as in Equation 2.63) according to

$$H_{iloss}(f) = \frac{Z_S(f) + Z_L(f)}{Z_S(f)\,(C(f)Z_L(f) + D(f)) + A(f)Z_L(f) + B(f)}, \tag{4.78}$$

where $A(f)$, $B(f)$, $C(f)$, and $D(f)$ are the frequency elements of the *ABCD* matrix (see Section 2.3.7), and $Z_S(f)$ and $Z_L(f)$ are the source and termination impedances at either end of the loop. The PSD of the received signal is then given by

$$S_p(f) = S_x(f)\,|H_{iloss}(f)|^2, \tag{4.79}$$

where $S_x(f)$ is the PSD of the transmitted signal.

As discussed in detail in Chapter 3, crosstalk is the primary noise source in DSL systems. For the purpose of estimating the performance of a DSL system, the crosstalk PSD needs to be computed, in most cases by using one of the models presented in Section 3.6. Both the crosstalk noise and the background noise are modelled as additive Gaussian noise.

Subsection 4.6.3 describes basic data rate calculations for PAM, QAM, and CAP systems in the presence of additive Gaussian noise (for example, crosstalk and background noise). Subsection 4.6.4 describes the corresponding basic data rate calculations for DMT systems. By combining all the above information, it is possible to do basic data rate calculations for DSL systems.

The following example demonstrates basic data rate calculations for PAM, QAM, CAP, and DMT systems. In this example, the channel is a loop consisting of a 1 km long 0.4 mm wire with worst-case crosstalk noise according to the ETSI ADSL FB disturber model (see [ETSI TS 101 388]). The data rates are calculated for four hypothetical DSL systems that are identical except for the line code used. All four hypothetical DSL systems use the same transmit PSD with flat -40 dBm/Hz transmit power from 4 kHz up to 1.1 MHz. The loop insertion loss is computed according to Equation 4.78 and using the PE04 wire parameters from Table A.1 of [ETSI TS 101 388]. The received signal PSD is computed according to Equation 4.79 and is plotted in Figure 4.9. The crosstalk PSD profile is taken from Tables 14 and 15 of [ETSI TS 101 388], and the NEXT and FEXT noise are computed as described in Chapter 3. The received NEXT and FEXT noise PSDs are also shown in Figure 4.9. The

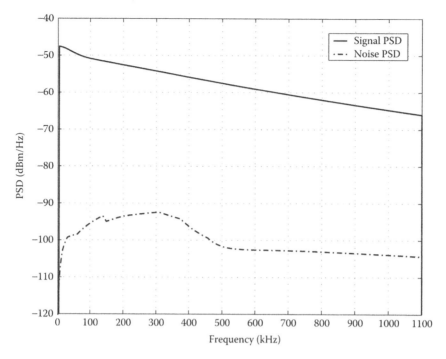

FIGURE 4.9
Signal and noise PSD for example rate calculations.

signal and (combined) noise profiles in Figure 4.9 are used to calculate SNR values to use in the data rate calculations.

The PAM data rate can be calculated according to Equation 4.64 using the SNR derived from the signal and noise profiles in Figure 4.9. Because the signal bandwidth in this example is 1.1 MHz, it is appropriate to use a symbol rate of 2.2 MHz. The QAM and CAP data rates are computed according to Equation 4.40 assuming carrier frequency $f_c = 550$ kHz and symbol rate 1.1 MHz. The DMT data rate is computed according to Equation 4.76 assuming 255 carriers with spacing of $\Delta f = 4.3125$ kHz and a symbol rate of 4 kHz.

Table 4.2 gives the calculated data rate for all the systems. For comparison, the table also contains the Shannon channel capacity for the loop. Comparison of the data rates for the different systems shows that they all have very similar performance. This is not surprising, considering that the data rate formulas for all these systems are very similar in nature. The PAM modulation and QAM/CAP modulation have exactly the same (theoretical) performance. When bit error rate is considered, PAM and QAM/CAP (square constellations) have the same Shannon gap and therefore the same performance. The slightly lower data rate for

TABLE 4.2

Example Calculation of Transmission Rates

Modulation	Γ [dB]	f_s[kHz]	f_c[kHz]	Calculated Rate [kbps]
Shannon	0	N/A	N/A	15260
PAM	9.95	2200	N/A	11635
QAM/CAP	9.95	1100	550	11635
DMT	9.95	4 ($\times$255)	k$\times$4.3125	10827

DMT is because the calculations account for the 7.8 percent overhead due to cyclic-prefix (see Section 7.3) by assuming a symbol rate of 4 kHz but sub-carrier spacing of 4.3125 kHz. Practical PAM and QAM/CAP systems require some excess bandwidth (see Chapter 6), which hasn't been included in the calculations. The excess bandwidth would reduce the data rates from the values given here.

When all the different implementation issues are considered, there can be slightly larger variations in the calculated performance, but even then there is relatively little difference in performance among the PAM, QAM, CAP, and DMT line codes. The main differences among the four line codes are various practical implementation issues, where each line code has its strengths and its weaknesses.

References

[ANSI T1.413-2001] ANSI T1.413-2001, *Spectrum Management for Loop Transmission Systems, American National Standard*, 2001.

[ETSI TS 101 388] ETSI TS 101 388 V1.3.1 (2002–05), *Asymmetric Digital Subscriber Line (ADSL)—European specific requirements*, ETSI Technical Specification, 2002.

[ETSI TS 101 524] ETSI TS 101 524 V1.2.1 (2003–03), *Symmetric Single Pair High Bitrate Digital Subscriber Line (SDSL)*, ETSI Technical Specification, 2003.

[Forney 1998] G.D. Forney Jr. and G. Ungerböeck, *Modulation and coding for linear Gaussian channels*, IEEE Transactions on Information Theory, Vol. 44, Issue 6, pp. 2384–2415, Oct. 1998.

[ITU-T G.991.2] ITU-T G.991.2 (02/2001), Single-pair high-speed digital subscriber line (SHDSL) transceivers, ITU-T Recommendation, 2001.

[Lee 1994] E.A. Lee and D.G. Messerschmitt, *Digital Communication*, 2nd ed., Boston: Kluwer, 1994.

[Papoulis 2002] A. Papoulis and S.U. Pillai, *Probability, Random Variables and Stochastic Processes*, 4th ed., New York: McGraw-Hill, 2002.

[Richards 1973] D.L. Richards, *Telecommunication by Speech: The Transmission Performance of Telephone Networks*. London: Butterworth, 1973.

[Shannon 1948] C.E. Shannon, *A mathematical theory of communication*, Bell Syst. Tech. J., Vol. 27, pp. 379–423 and 623–656, July and Oct. 1948.

[Shannon 1949] C.E. Shannon, *Communication in the presence of noise*, Proc. IRE, Vol. 37, pp. 10–21, 1949.

[Shannon 1959] C.E. Shannon, *Probability of error for optimal codes in a Gaussian channel*, Bell Syst. Tech. J., Vol. 38, pp. 611–656, May 1959.

5

Introduction to DSL

Edward Jones

CONTENTS

5.1 Introduction 120
5.2 History 120
5.3 Alternative Broadband Access Technologies 122
    5.3.1 Fiber 122
    5.3.2 Wireless 123
    5.3.3 Cable Modem 123
    5.3.4 Power Line Communications 125
    5.3.5 Digital Subscriber Lines 125
5.4 Overview of DSL Technology 125
    5.4.1 Introduction 125
    5.4.2 Performance Requirements of DSL Systems 127
    5.4.3 Basic Rate ISDN (BRI) 127
    5.4.4 HDSL 128
    5.4.5 HDSL2 and HDSL4 129
    5.4.6 SDSL 130
    5.4.7 G.shdsl 130
    5.4.8 ADSL 131
    5.4.9 Splitterless ADSL 133
    5.4.10 ADSL2, ADSL2plus 133
    5.4.11 VDSL 134
    5.4.12 Related Topics 134
        5.4.12.1 Spectrum Management 134
        5.4.12.2 Deployment and Testing 135
        5.4.12.3 End-to-End Architectures 135
        5.4.12.4 Ethernet in the First Mile (EFM) 135
5.5 Representative DSL Transceivers 136
    5.5.1 Symmetric DSLs 136
        5.5.1.1 Scrambler and Descrambler 136
        5.5.1.2 Trellis-Coded Modulation (Encoder and Decoder) 136
        5.5.1.3 Equalizer and Precoder 137
        5.5.1.4 Transmit Filter 137
        5.5.1.5 Transmit Analog Front End (AFE) 137
        5.5.1.6 Receive AFE 138
        5.5.1.7 Echo Canceller 138
        5.5.1.8 Timing Recovery 138
        5.5.1.9 Additional Functions 138

    5.5.2   Passband Single-Carrier Systems 138
            5.5.2.1   Error Correction and Interleaving 139
            5.5.2.2   Modulation and Demodulation 139
            5.5.2.3   Carrier Recovery 139
            5.5.2.4   Additional Points 140
    5.5.3   Multi-Carrier Systems 140
    5.5.4   Closing Remarks 140
5.6   Summary 141
References 141

ABSTRACT   This chapter introduces some of the technologies that are used in DSL systems to try to realize the information-transmission potential of twisted-pair lines, and also briefly describes the different varieties of DSL that are in use or are under development. The motivation for the development of DSL services and technology is considered; in particular, alternative access technologies are reviewed, following which the different varieties of DSL are discussed in order to set the context for later chapters.

5.1   Introduction

Previous chapters of this book have discussed the fundamentals of the copper access network: its architecture, characteristics, and in particular how the transmission channel affects a signal. In addition, Chapter 3 described in detail how the signal is degraded by noise and interference from various sources (crosstalk, impulse noise, etc.). Some fundamental limits on the data-carrying capacity of the copper access network were presented in Chapter 4. The present chapter introduces some of the technologies that are used in DSL systems and briefly describes the varieties of DSL that are in use or are under development. In addition, simple block diagrams for DSL transceivers are presented, highlighting the various subsystems required and how these are used to compensate for the various sources of signal distortion described in previous chapters. Many of these subsystems are described in considerable detail in later chapters.

The focus of this chapter is very much on physical layer aspects of DSL transmission systems; later chapters will describe end-to-end system architectures that utilize DSL transmission systems, and will outline how the DSL physical layer interacts with high layers to realize useful services. Furthermore, the intention here is not to go into exhaustive detail on the different DSL standards; this is covered in Volume 2 of this series (and in many other books and papers already published, for example [Starr 1999][Starr 2003] and references therein). Rather, the intention is to give the reader a basic overview of DSL, and to set the context for the chapters that follow.

5.2   History

This section is intended to briefly summarize the various transmission technologies that have been deployed on the telephone network. For a detailed description of this network, the reader is referred to Section 1.1. The public switched telephone network (PSTN) was originally designed to carry voice signals, and the bandwidth of these signals was limited to the frequency range from approximately 200 Hz to 3.4 kHz (with some variations, depending on location). Although a voice signal is analog by its nature, digitization of many of the links in the telephone network is very common now (as explained in Chapter 1), particularly in the trunk network between telephone company central offices (COs), which is heavily based on optical fiber and microwave links. For the most part, a voice signal travels in analog form from the originating user to the local CO across a copper twisted pair (the local loop), where it is digitized by a codec ("coder/decoder"), following which it is transmitted over the trunk network to the CO serving the user at the other end. Here it is converted from digital form back to analog by another codec, before being transmitted across this user's local loop to the receiving telephone.

The 1950s saw the introduction of voiceband modems for the purpose of transmitting data across the PSTN. Early modems (for example, the Bell 103) transmitted at low bit rates (300 bits per second (bit/s)) using frequency shift keying (FSK) modulation. Modem technology quickly developed to provide higher bit rates and also enabled full-duplex transmission. For example, the CCITT (now ITU-T) V.22 standard provided for communication at 1200 bit/s, and the later V.22bis recommendation extended this to 2400 bit/s. Subsequent developments led to V.32 (9600 bit/s), V.32bis (14, 400 bit/s), and later V.34, which uses very sophisticated signal processing techniques to achieve bit rates up to 33.6 kbit/s, with various fall-back options. In the late 1990s, pulse coded modulation (PCM) modems were developed and standardized as ITU-T Recommendation V.90. This recommendation provides for up to 56 kbit/s in the downstream direction (from the CO to the user), where an all-digital path is assumed to exist between the data source and the CO serving the user. This is a reasonable assumption in practice, because many information sources, such as Internet service providers (ISPs), have direct digital connections to the PSTN. In V.90, the upstream direction of transmission uses V.34 modulation, limiting upstream bandwidth to 33.6 kbit/s. A good overview of the technologies used in voiceband modems may be found in [Forney 1984] [Forney 1996] [Ayanoglu 1998]. Figure 5.1(a) shows a block diagram of a typical voiceband modem communication link, and Figure 5.1(b) shows a link using a V.90 modem.

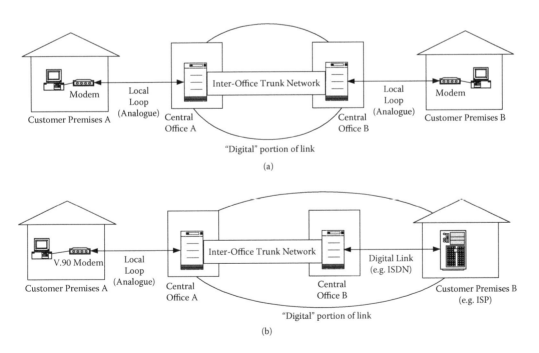

FIGURE 5.1
(a) Conceptual block diagram of a typical voiceband modem link (pre-V.90); (b) block diagram of a V.90 modem link, indicating the all-digital link between (for example) an ISP and the serving central office.

Voiceband modems continue to push the limits of the technology; however, there is a limit to what is achievable within the existing PSTN framework with its limited bandwidth (although, given the remarkable advances in voiceband modem technology over the years, it may not be entirely clear where this limit lies). At the same time, users (both residential and commercial) continue to demand ever-increasing bit rates for many different applications, so the local loop of the PSTN as it stands has essentially become a bottleneck. Residential users demand faster transmission rates for Internet access, and the multitude of applications it enables (Web browsing, e-mail, online shopping and gaming, and many other applications). Typically, the traffic pattern for residential users is asymmetric, in the sense that applications like Web browsing generally demand higher bit rates in the downstream direction than in the upstream direction. Business users, particularly of the small office-home office (SOHO) variety, also need faster access for remote office connectivity, LAN-extension, file sharing, video-conferencing, etc. In this case, the bit rate requirements tend to be more symmetric, because remotely located business customers often tend to transmit as much as they receive.

These user requirements for higher-speed local access have driven the need for transmission systems capable of providing transmission speeds of hundreds of kilobits, or even megabits, per second. The next subsection briefly introduces some of the access mechanisms that may be used to achieve this, including DSL.

5.3   Alternative Broadband Access Technologies

5.3.1   Fiber

It is generally accepted that the ultimate goal in local access is the provision of fiber-optic transmission to every user, so-called Fiber-to-the-home (FTTH) (with this term implicitly encompassing Fiber-to-the-business as well). A common architecture for deploying fiber-based communications is the passive optical network (PON), which has a single transceiver in the CO serving multiple customers, with splitters and couplers to distribute the service among the different subscribers.[1]

Although FTTH would satisfy the bandwidth requirements of even the most demanding user, this scenario is unlikely to be achieved for some considerable time to come. The reason is, quite simply, the cost involved in the installation of an FTTH network (particularly labor and other nonequipment costs), costs that would be extremely difficult for the service provider to recover in a reasonable time frame.

However, as was mentioned in Section 1.1, although the goal of FTTH for all users is still some way off, some progress toward this end is being made. For example, it is common for fiber to be deployed to serve new offices and residential buildings and developments, and to replace an existing telephony plant that has reached the end of its useful life. This encompasses both fiber deployments directly to the home, as well as to intermediate points in the distribution network, for example, at the end of a residential street. This type of architecture is often referred to as Fiber-to-the-curb or Fiber-to-the-cabinet. As a half-way point to full FTTH, these installations are capable of serving many subscribers with a single fiber, and effectively reducing the distance over which subscribers need to be served by means of other access technologies (including DSL, as will be described later). Some further information on optical fiber access may be found in [Cramer 2002] and [IEEE Com. Mag. Dec. 2001].

---

[1] An alternative is point-to-point, whereby each subscriber has a dedicated optical transceiver at the CO.

An interesting variation on this theme is the development of systems for optical wireless (or free-space optical) transmission for local access; see, for example, [IEEE Com. Mag. March 2003].

5.3.2 Wireless

Wireless remote access comes in a number of different variations [Boelcskei 2001] [IEEE Com. Mag. 2002] [IEEE Com. Mag. Sept. 2001], which are sometimes generically referred to as wireless local loop (WLL). Wireless may seem like the obvious choice for a (fixed-position) local access technology, because it does not require the installation of a transmission medium. This can be particularly important in developing countries, where the level of installed communications infrastructure significantly lags behind that in developed countries. However, there are a number of issues that have hampered the deployment of wireless local access. For example, the available radio spectrum is becoming increasingly congested, forcing broadband wireless access systems to move to higher frequencies, where line-of-sight (LOS) operation may become necessary. This applies, for example, with the local multi-point distribution system (LMDS) and similar systems operating between 20 and 40 GHz. Furthermore, there are still challenges and costs associated with deploying the necessary infrastructure where it is required, for example, planning issues associated with location of base stations, as well as the challenge of developing user-friendly customer premises equipment (CPE).

Systems operating at lower frequencies, where non-LOS transmission is more reliable, have also been deployed (*e.g.*, microwave multi-point distribution system (MMDS) in the region of 2–4 GHz), though greater bandwidth efficiency may be required to increase bit rates. On the other hand, fading and multi-path propagation make it more difficult to use higher-order modulation to achieve the necessary spectral efficiency. At the same time, the fact that the transmitter and receiver are in fixed locations means that directional (and multiple) antennae may be used to increase performance.

A related area of standards development is the so-called wireless metropolitan area network (wireless MAN), currently under the auspices of IEEE Working Group 802.16 [IEEE 802.16].

5.3.3 Cable Modem

For many years, coaxial cable has been used to distribute television services to subscribers. It was quickly realized that this same medium could also be used to carry broadband data and even voice, and so represented another means of broadband access. However, much of the cable network is unidirectional, in the sense that it was originally intended for broadcast applications (*i.e.*, cable TV delivery). In recent times however, large portions of the cable infrastructure have been made bidirectional, in order to allow for two-way transmission. A typical architecture consists of fiber-optic cable carrying signals between the cable head-end and fiber nodes in the network, from which existing coaxial cable is used to cover the "last mile" to the subscribers' premises. This architecture is generally referred to as hybrid fiber-coax (HFC) and is illustrated in Figure 5.2. A special cable modem is used to terminate the connection in the subscriber premises. The need for interoperability between cable modem equipment from different vendors resulted in the development of the Data over Cable Service Interface Specifications (DOCSIS).

Quadrature amplitude modulation (QAM) (see Chapter 6) is commonly used for cable modem transmission in the downstream direction. Using 64-QAM, a single 6 MHz analog channel originally used for cable TV transmission is capable of carrying around 30 Mbit/s (allowing for roll-off of the signal spectrum and also allowing for a guard band between

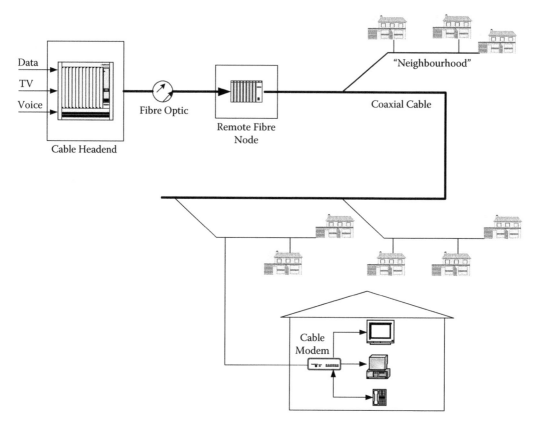

FIGURE 5.2
Simplified representation of Hybrid Fiber Coax architecture for broadband access.

channels). Increasing the number of QAM levels to 256 would increase the data-carrying capacity of each channel to around 40 Mbit/s. In the downstream direction, the signal (consisting of a combination of data, TV, and perhaps voice) is converted into an optical signal and carried by optical fiber to a "fiber node" in the distribution network. At this point it is converted to an electrical signal, and distributed to subscribers using the existing coaxial cable. The cable modem separates the composite received signal into its various components (for example, Internet data, voice, and TV) and distributes them to their respective destinations (PC, telephone, TV set). In the upstream direction, lower-order modulation, for example, quaternary phase shift keying (QPSK), is typically used. This has lower spectral efficiency than 64- or 256-QAM, but it is more robust and better able to deal with the harsh conditions to which the upstream signal is subjected. In any case, the asymmetric bandwidth requirements of applications such as Internet access mean that a lower bit rate can be tolerated in the upstream direction.

One of the problems associated with HFC for broadband data access is the cost associated with converting the existing cable network from unidirectional to bidirectional operation, in particular, provision of bidirectional amplifiers and related equipment. A further problem is the fact that HFC is a shared medium; that is, all of the available bandwidth is shared among all of the subscribers served by a particular fiber node. As more subscribers join the network, the bandwidth available to any single subscriber decreases.

Further details on cable access and HFC technology may be found in [IEEE Com. Mag. June 2001] .

5.3.4 Power Line Communications

A recent development in the broadband access field is the use of the electric power supply network for the transmission of broadband data. One of the major motivations for this approach is (as explained in the abstract to Chapter 1) the ubiquitous nature of this network. In addition to using the electricity supply network for access, there is also the possibility of using existing in-home electric wiring as a form of local area network (LAN) in the home. However, a number of technical issues still need to be fully addressed, including the design of systems able to perform well in the very harsh environment of this network, regulatory issues, and issues relating to safety. An overview of initial developments in this area may be found in [IEEE Com. Mag. May 2003].

5.3.5 Digital Subscriber Lines

Although there are several other media that can be used to provide broadband access to residential and business subscribers, none of them has the ubiquity (or the level of maturity of development) of the telephone network. Telephony service is provided to almost every business and residential subscriber in most of the world, with several hundred million twisted-pair telephone lines installed globally to date. Furthermore, as discussed in Chapter 4, the data-carrying capacity of telephone twisted pairs greatly exceeds what is currently achievable with voiceband modem technology.

The next section of this chapter briefly describes the different varieties of DSL technology, and the following section introduces some of the details of the physical layers of these systems. Chapters 6 and 7 provide additional detail on the physical layers used in DSL.

5.4 Overview of DSL Technology

5.4.1 Introduction

The range of DSL technologies is quite broad, and this breadth can be somewhat confusing to the uninitiated. This section briefly describes the different types of DSL technology that have been developed or are currently under development. Much of this development has taken place in various regional and global standards committees, for example, ANSI committee T1E1.4 (Digital Subscriber Loop Access), ETSI Working Group TM6 (Transmission and Multiplexing), and ITU-T Study Group 15/Question 4, as well as in-industry forums such as the DSL Forum. The work of the various standards committees will be described in more detail in Volume 2 of this series.

In simple terms, DSL technologies can be subdivided into two broad classes:

- **Symmetric**. Within this class, the data rate transmitted in both directions (downstream and upstream) is the same. This is a typical requirement of business customers.
- **Asymmetric**. In this case, there is asymmetry between the data rates in the downstream and upstream directions, with the downstream data rate typically higher than the upstream (usually appropriate for applications such as Web browsing).

This division is quite crude however, and, to confuse matters, some of the various technologies are capable of both asymmetric and symmetric operation. To further complicate things, many DSL systems are capable of multi-rate operation, which adds a further dimension of variability.

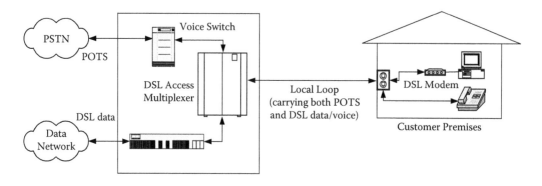

FIGURE 5.3
Block diagram of "generic" DSL reference model. It should be noted that DSL is an "overlay" on the existing switched telephone network.

An additional point to note is that symmetric DSLs generally use baseband modulation such as pulse amplitude modulation (PAM) (see Section 6.2.2), where the bandwidth of the transmitted signal extends all the way down to 0 Hz (notwithstanding the effect of any coupling transformers or other filtering), whereas the asymmetric technologies generally use passband modulation, which avoids the lowest frequencies that would be used by voiceband services such as analog telephony (see Chapters 6 and 7 for further information on digital modulation techniques). This is generally because the residential users who would typically make use of asymmetric DSLs still need to be able to make use of "lifeline" POTS, even when the DSL service is unavailable (for example, due to a power failure in the customer premises). Provision of lifeline POTS service is generally less of an issue for business users, who might typically carry all of their business voice traffic on the DSL link anyway.

A block diagram of a typical DSL configuration is shown in Figure 5.3. Note that the term "digital subscriber line" generally refers to the analog local loop between each customer premises and its local central office, and a DSL modem is required at each end of the loop. Furthermore, the DSL service can be regarded as being provided by means of an "overlay" network that is not part of the normal switched telephone network. This means that the service provider CO needs to be able to separate the DSL service from the POTS service, with the voice service being sent onward by means of the ordinary trunk network, whereas the data carried by the DSL may be sent to a data network that is separate from the switched voice network. The CO will generally provide DSL service to the user premises using a DSL access multiplexer (DSLAM), which is described in Volume 2 in this series. The DSLAM usually contains many DSL modems serving multiple customers.

A block diagram of a typical DSL configuration is shown in Figure 5.3: the key point to note here is that in essence, a "digital subscriber line" exists on a *single* local loop between the customer premises and the central office, unlike the voiceband modem case where, technically, the modem link includes *two* local loops (plus the network elements in between). Furthermore, the DSL service can be regarded as being provided by means of an "overlay" network that is not part of the normal switched telephone network. This means that the service provider central office needs to be able to separate the DSL service from the POTS service, with the voice service being sent onward by means of the ordinary trunk network, whereas the data carried by the DSL may be sent to a data network that is separate from the switched voice network. The CO will generally provide DSL service to the user premises using a DSL access multiplexer, which is described in Volume 2 in this series.

Generally speaking, before they can be used for the transmission of user data, DSL transceivers must go through an activation phase, whereby various receiver (and transmitter) elements must be initialized. In particular, the receiver functional blocks (equalizer, timing recovery, etc.) must be adapted so that reliable communication can take place under the particular loop and noise conditions at the time. The details of these functional blocks are covered in upcoming chapters.

In some cases, when DSL systems have a number of possible configurations to choose from (for example, multiple bit rates), the activation phase is also used to allow the transceivers on either end of the line to agree on what configuration they will use, through a session of "handshaking."

In the interest of simplicity, the description given here will follow a broadly "historical" approach, with reference made to the above classifications where appropriate. This section will merely introduce the various DSL technologies, and the following section will introduce functional block diagrams of the physical layers of different categories of systems. Subsequent chapters discuss the different blocks in some detail. Volume 2 includes details and features of specific standards for DSL (some of which are referred to in passing in this chapter).

5.4.2   Performance Requirements of DSL Systems

In a practical sense, all varieties of DSL have a specified environment in which they are expected to operate reliably. This specified environment includes the types of loops over which the service is expected to operate, as well as a definition of the expected noise environment (including impulse noise and crosstalk). DSL technologies that have been developed in recent times are also expected to be spectrally compatible with services already in use in the loop plant, in the sense that the presence of the new DSL will not unduly degrade the performance of existing services.

Chapters 6 and 7 provide detailed descriptions of the modulation techniques most commonly used in DSL systems; in particular, the signal-to-noise ratio (SNR) required for these modulation techniques to operate with some specific probability of error is discussed in detail. These SNR requirements impose fundamental performance limits that can be achieved in practical DSL systems.

Basic DSL performance requirements are usually specified in terms of acceptable bit error ratio (BER) with specified noise margin while operating in certain conditions. The BER usually used in DSL development is 10^{-7}, and the noise margin is usually either 5 or 6 dB. The noise margin specification means that the system is expected to operate at an actual BER no greater than the specified BER when the noise is increased by a level equal to the noise margin. Typically, the specified test conditions mirror anticipated worst-case conditions. The inclusion of the noise margin means that DSL systems generally operate in normal conditions with BER much less than 10^{-7}, and it also allows for reliable operation when the noise conditions are worse than normal (*e.g.*, due to the presence of unexpected sources of noise).

The next few subsections largely deal with symmetric varieties of DSL, and subsequent sections deal with the asymmetric variations.[2]

5.4.3   Basic Rate ISDN (BRI)

Basic rate integrated services digital network (ISDN) is regarded by many as the "original" DSL [Starr 1999]. ISDN was intended to provide a global digital network for the integrated transmission of voice and data signals. The focus in ISDN was on transmission of voice signals, and low-speed data signals. Basic rate ISDN (BRI) is capable of transmitting up to

---

[2] As noted previously, some of the "asymmetric" technologies can also be used for transmission of symmetric bit rates.

160 kbit/s, symmetrically, over distances of approximately 5.5 km (18,000 ft) on a single span. Operation over longer distances is possible with the use of repeaters. The transmission bit rate is divided into two "B" channels, each carrying 64 kbit/s, and one "D" (Data) channel carrying 16 kbit/s. The remaining 16 kbit/s are used for framing and control. The majority of installed ISDN uses 2-binary, 1-quaternary (2B1Q) PAM (see Section 4.5.1) at a symbol rate of 80 kHz. Bidirectional transmission in the same bandwidth is achieved by the use of echo cancellation.

Further details on ISDN may be found in [Starr 1999], [ANSI T1.601 1992], [ETS 300 403], [Stokesberry 1993], and [IEEE Com. Mag. Aug. 1992].

5.4.4   HDSL

High-bitrate digital subscriber line (HDSL) is the term that is usually applied to the provision of symmetric T1 (1.544 Mbit/s) or E1 (2.048 Mbit/s) rates over one, two, or three copper twisted pairs. Development and deployment of HDSL technology started in the late 1980s and early 1990s, and it is now widely used throughout the developed world, especially in North America. One of the motivations for the development of this technology arose from the increased usage of T1 and E1 transmission to customer premises; the traditional technology used for this purpose (based on alternate mark inversion (AMI) and high-density bipolar 3 (HDB3) modulation) was problematic from the point of view of plant engineering (requiring loop qualification) and crosstalk generation. HDSL alleviated many of these problems. Most deployed HDSL uses similar technology to BRI, *i.e.*, 2B1Q modulation, plus echo cancellation. However, European specifications also include provision for the use of single-carrier modulation, in particular, carrierless amplitude/phase (CAP) modulation (see Section 6.2.3).

The most commonly deployed variant of HDSL uses two twisted pairs, whereby half of the transmitted data is sent (in both directions) on each. For example, transmission of 1.544 Mbit/s is accomplished by transmitting half of the data (784 kbit/s including overhead) over each twisted pair. Furthermore, the use of "one pair" of two-pair HDSL to provide fractional-rate T1 or E1 bit rates (*i.e.*, half of the full rate) is quite common. This type of service would be used, for example, to serve "small" business customer sites where the volume of traffic does not justify the cost of a full T1 or E1 link. In Europe, HDSL is also specified for operation over a single pair carrying 2.320 Mbit/s; two pairs, each carrying 1.168 Mbit/s; and three pairs, each carrying 784 kbit/s. More information on HDSL as used in North America may be found in [ANSI T1 1994], and details on the European specification may be found in [TS 101 135 2000].

An interesting extension of HDSL technology was the development of DSL systems that transmitted data at rates of $n \times 8$ kbit/s, where n is an integer. These systems were based on both 2B1Q and also CAP technology and, though official recommendations were never actually produced, this technology has been widely deployed, largely by competitive local exchange carriers. These systems are often referred to as symmetric digital subscriber line (SDSL), which is not to be confused with ETSI's SDSL specification (described in Section 5.4.6).

Two-pair HDSL can operate over a single span of up to 3.7 km (12,000 ft) of 0.5 mm wire; however, its range can be greatly extended with the use of repeaters. The use of more than one pair helps to ensure longer reach, because the bandwidth used on each pair is less than would be used with a single pair, and hence the attenuation suffered by the signal per km of reach is also less. Lower bandwidth also helps to facilitate spectral compatibility with existing systems. However, the use of more than one pair per customer means that fewer customers can be served with a given number of twisted pairs; single-pair systems have the advantage that more customers can be served. The late 1990s saw the development of a high-performance replacement for HDSL, which would use a single pair, but would be spectrally compatible with existing systems. This technology is described in the next subsection.

5.4.5 HDSL2 and HDSL4

As noted above, there was a strong requirement for a technology to replace "HDSL-like" systems, utilizing a single twisted copper pair and providing high performance (adequate reach) while retaining spectral compatibility with other services. This led to the development within T1E1.4 of "second generation HDSL," or so-called "HDSL2." Like HDSL, the technology for HDSL2 is based on echo-cancelled PAM, but HDSL2 incorporates many innovations that were not present in HDSL. Among these is the use of error-correcting codes to enhance performance. HDSL2 uses powerful trellis-coded modulation (TCM), which is discussed in more detail in Chapter 8. The use of TCM can provide several dB of extra performance (in the form of coding gain) to the system. This was found to be particularly important for HDSL2, given the ambitious performance targets (essentially doing the same thing as HDSL, but with only one twisted pair). The HDSL2 standard is flexible enough to allow vendors to choose the parameters of the system (including a programmable convolutional code) to give just the required amount of coding gain, and hence trade-off complexity in the TCM decoder against performance. However, the standard includes an example of a code that provides up to 5 dB of gain. HDSL2 uses coded 16-PAM modulation with three information bits and one redundant bit per symbol, resulting in a symbol rate of 517.33 kHz for transmission of T1 rates (including HDSL2 overhead). Because of the presence of TCM, HDSL2 requires the use of precoding [Tomlinson 1971] [Harashima 1972], whereby some of the equalization normally carried out by a traditional equalizer in the receiver (see Chapters 6 and 11) is instead carried out by the transmitter (*i.e.*, the signal is pre-equalized before transmission).

A further innovation is the use of asymmetric spectra for the signals in the upstream and downstream directions. Unlike 2B1Q HDSL, where the shape of the transmit signal spectrum is the same in both directions of transmission, the HDSL2 transmit signal has two different (overlapping) shapes for the two directions of transmission. (Note that this type of asymmetry should not be confused with asymmetry in the transmitted bit rate.) Among the reasons for this were the need to reduce the effect of self-crosstalk (and thus ensure that the performance requirements would be met), and also to enhance spectral compatibility.

The general performance requirements for HDSL2 are outlined in [ANSI T1 2000]; in essence, the service is expected to operate in the presence of a number of different types of disturber, over a particular set of test loops conforming to carrier serving area (CSA) design rules (in simple terms, up to 2.7 km (9,000 ft) of 0.4 mm wire, or 3.7 km (12,000 ft) of 0.5 mm wire). Operation over longer loops is possible through the use of repeaters; however, the use of repeaters has implications for spectral compatibility with other services.

To allow for the provision of T1 service over loops beyond carrier service area (CSA) limits while retaining spectral compatibility, T1E1.4 developed a two-pair version of HDSL2, which is commonly referred to as "HDSL4."[3] As with two-pair HDSL, half of the transmitted data is carried on each of the two pairs. The fact that the bit rate on each pair is lower means that less bandwidth is required, which in turn means that the signal on each pair suffers less attenuation and hence longer reach can be achieved (around 3.4 km of 0.4 mm wire). HDSL4 uses largely the same technology as HDSL2, *i.e.*, coded 16-PAM modulation with

---

[3] There is an ironic closing of the circle in this development, in the sense that HDSL2 was a single-pair version of the multi-pair HDSL, whereas HDSL4 is a multi-pair version of HDSL2.

asymmetric upstream and downstream spectra. However, the spectra for HDSL4 are quite different from those for HDSL2.

5.4.6 SDSL

In Section 5.4.4, reference was made to "SDSL" based on 2B1Q technology, essentially a single-pair version of HDSL capable of operation at a number of different bit rates. In the late 1990s, ETSI TM6 started work on a symmetric multi-rate technology that supports mainly business customers with bit rates of $n \times 64$ kbit/s, up to a maximum of 2.304 Mbit/s ($n = 36$), plus overhead. The bit rate provided is a function of the loop length over which service is provided: the shorter the loop, the greater the bit rate that can be delivered. This technology is referred to as symmetrical single-pair high-bitrate digital subscriber line, using the same acronym "SDSL."

In some ways, ETSI SDSL is similar to 2B1Q SDSL referred to earlier (symmetric bit rate, multi-rate operation). However, ETSI SDSL also has much in common with HDSL2 and HDSL4 in that it makes use of TCM and transmits three information bits per symbol, as opposed to two bits per symbol with 2B1Q. Also, although the transmit spectra in ETSI SDSL are symmetric, some provision is made for asymmetric spectra at the highest bit rates (2.048 and 2.304 Mbit/s) in order to enhance performance and increase spectral compatibility. Further details on ETSI SDSL may be found in [TS 101 524 2001].

5.4.7 G.shdsl

Thus far, the discussion has covered a number of symmetric bit rate technologies: basic rate ISDN, HDSL, 2B1Q SDSL, HDSL2, HDSL4, and "ETSI SDSL." Development of these individual technologies has generally taken place under the auspices of regional standards bodies such as T1E1.4 and ETSI TM6. However, the ITU-T has also been quite active in DSL standards development through its Study Group 15/Question 4 working group. For example, ITU-T has published Recommendation G.991.1, covering HDSL technology. In addition, it has developed recommendations for symmetric bit rate, multi-rate DSL technology that draws heavily on the developments in the regional standards bodies. In ITU-T parlance, this is referred to as single-pair high-speed digital subscriber line, with the acronym SHDSL; the relevant recommendation is G.991.2, also commonly known as G.shdsl [ITU-T G.991.2 2001].

G.shdsl defines operation at payload bit rates from 192 kbit/s up to 2.304 Mbit/s, in increments of 8 kbit/s, over a single wire pair (though there is an optional two-pair mode that can be used for greater reach). It includes many of the features of HDSL2/4 and ETSI SDSL, including symmetric bit rates, multi-rate operation, and the use of 16-level trellis-coded (TC) PAM. Many of the operational elements of G.shdsl are region-specific, and the ITU-T recommendation includes a number of annexes that contain details specific to a particular geographical region. For example, Annex A contains information that is specific to North America, and Annex B contains details of how G.shdsl systems would be deployed in Europe. The technical content of these annexes has largely originated in the respective regional standards bodies (*i.e.*, ANSI T1E1.4 and ETSI TM6), so in a sense, G.shdsl encompasses a number of the DSL technologies discussed above, in particular, HDSL2, HDSL4, and SDSL.

G.shdsl systems have a two-phase start-up sequence: pre-activation and core activation. The purpose of the pre-activation phase is to allow the transceivers on either end of the line to exchange information about their capabilities and to agree upon the best configuration given the loop and noise conditions. This is necessary because the G.shdsl recommendation covers many possible configurations, both mandatory and optional, so it is important that

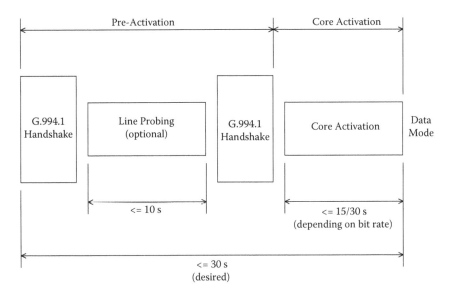

FIGURE 5.4
Timeline of G.shdsl activation sequence.

the transceiver on one end of the loop know the capabilities of the other one. Once this exchange has taken place, the core activation takes place, whereby the receiver (and transmitter) functional blocks are adapted so that reliable communication at the agreed-upon bit rate can take place. The pre-activation sequence also includes an optional line probe during which each transceiver may examine the loop and noise conditions and determine (for example, through SNR measurements) the bit rates it is capable of supporting under these conditions. Figure 5.4 shows the timeline of the entire G.shdsl activation sequence; the handshake portions are where the two transceivers exchange information about their respective capabilities, and the optional line probe portion is where particular signals are transmitted from one transceiver to the other, in order that information about the operating conditions may be obtained. ITU-T Recommendation G.994.1 [ITU-T G.994.1 2003], commonly known as "G.hs," describes the manner in which the handshaking takes place.

Since the initial publication of the G.shdsl recommendation, further developments have taken place in the various standards committees that have resulted in enhancements to the basic specification. For example [ITU-T PF-R15 2003]:

- Two-pair operation has been extended to multi-pair operation.
- Optional provision has been made to allow for transmission of bit rates up to 5.696 Mbit/s.
- The capability to carry out a shorter-duration "warm-start" has been added.
- Support for transport of packet-mode data has been included.

5.4.8   ADSL

The discussion thus far has concentrated mainly on the varieties of DSL that transmit symmetric bit rates (both single-rate like HDSL2 and multi-rate like G.shdsl). For the most part, such systems are of most benefit to business customers. Of perhaps greater demographic importance are the DSL technologies that support asymmetric bit rates, which more closely match the requirements of most residential customers (for Web browsing, etc.).

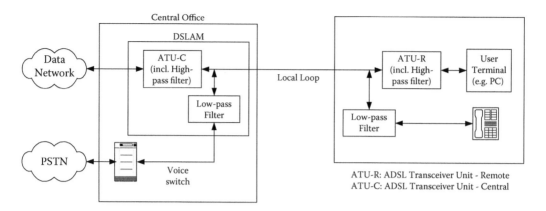

FIGURE 5.5
Block diagram of ADSL reference model.

The most widely deployed form of this technology is the original definition of asymmetric digital subscriber line (ADSL), which is capable of providing data rates of up to 8 Mbit/s downstream (*i.e.*, toward the consumer) and up to 896 kbit/s in the upstream direction. Transmission uses one pair of wires. The original motivation for the development of this technology was video on demand; however, the commercial motivation for ADSL quickly changed to high-speed Internet access in the mid-1990s.

As noted above, most of the asymmetric DSLs make use of modulation that avoids the lowest few kHz of the available spectrum on the loop. This is to ensure that lifeline POTS can still function, whether or not the DSL service is operational. The spectrum of the ADSL signal starts around 25 kHz. To separate the two services (ADSL and POTS), a pair of low-pass and high-pass "splitter" filters is required (see the splitter chapter in Volume 2 for more details on splitters and related technology). A block diagram of a typical ADSL reference model is shown in Figure 5.5. A variation of this basic configuration permits the operation of an ADSL system above ISDN, which has a substantially higher bandwidth than POTS (80 or 120 kHz, depending on the modulation used for ISDN). This is of particular interest in some European markets where ISDN is widely deployed.

As with symmetric DSL, ADSL technology has been standardized by the various regional and global standardization bodies; both DMT and single-carrier technologies were originally proposed for ADSL, though the standards are based on DMT modulation. The North American recommendation can be found in [ANSI T1 1998], and the European recommendation developed by ETSI TM6 is in [TS 101 388 2002]. Furthermore, ITU-T has also created Recommendation G.992.1 ("G.dmt") [ITU-T G.991.1 1999] to cover this type of "full-rate" ADSL, incorporating many of the features of the regional standards.

Like the more advanced symmetric DSLs, ADSL makes use of techniques such as error-correcting codes (see Chapter 9), as well as various techniques for choosing the optimum bit rate for the line conditions (in much the same way as G.shdsl uses line probing). For example, ADSL uses variable constellation sizes and sophisticated bit allocation algorithms to determine where best to distribute energy in the usable bandwidth (see Chapter 7).

Because the upstream rate in ADSL is much lower than the downstream rate, it follows that the bandwidth required for upstream transmission is much less than that required for downstream transmission. Upstream transmission generally uses the frequency range from around 25 kHz to around 138 kHz for transmission. Downstream transmission may occur in either of two bands, depending on the mode of operation of the ADSL transceiver. Full overlap between the downstream and upstream bands may be utilized, in which case downstream transmission uses the band from 25 kHz to around 1.104 MHz.

This results in more available bandwidth for downstream transmission, which may increase transmission rate. Sophisticated echo cancellation techniques are needed in this case, and crosstalk into the upstream channel is increased. A much more commonly used mode of operation is frequency division duplexing (FDD), where the downstream and upstream bands do not overlap. In this case, downstream transmission starts at around 138 kHz.

5.4.9 Splitterless ADSL

Besides the "full rate" ADSL described in the previous subsection, a simpler variant that does not require a splitter has also been developed. One of the motivations for this was to enable easier installation of ADSL at the customer's premises, in particular, to avoid installation of the splitter and (possibly) new premise's wiring. However, removal of the filters means that interference from the POTS service can leak into the ADSL transmission and vice versa. Therefore, it is necessary to use a so-called "in-line" low-pass filter in series with each telephone in the user premises to reduce these effects; these filters are available in modular form for easy installation by the customer. This type of installation of ADSL is referred to as "splitterless," even though in-line filters are required. The ITU Recommendation G.992.2 ("G.lite") was written to specify splitterless ADSL, and it places restrictions on the data rate to ensure reliable performance: the downstream data rate is limited to 1.5 Mbit/s. However, experience has shown that "splitterless" full-rate ADSL is achievable using in-line filters, and this installation mode is now common for ADSL [ITU-T G.992.3 2002]. As a consequence, G.lite never really gained traction in the marketplace.

5.4.10 ADSL2, ADSL2plus

Since the development of the original ADSL specification, a number of enhancements have been made in order to increase performance (in terms of higher bit rate, increased reach, and better management and diagnostic control). As before, many of the developments have been driven by the work of regional standards bodies, and these enhancements to the original G.992.1 have been captured by ITU-T as G.992.3, also known as "ADSL2" [ITU-T G.992.3 2002]. The enhancements (many of which are optional) include:

- The addition of a single-bit constellation for more robust performance over longer loops (thus enabling operation over loops that were previously unusable), and the inclusion of mandatory trellis coding (previously optional).
- The inclusion of "seamless" rate adaptation to enable almost continuous changes to the bit rate, and the distribution of bits across the used bandwidth (online "bit swapping").
- Changes to the error-correction coding (including greater flexibility).
- Inclusion of an optional "all digital" mode that allows the use of the POTS band for transmission of additional data by the ADSL modem.
- Several features to combat interference (including RFI).
- Improved initialization procedures, and an optional "fast" initialization mode (around 3 s).
- Flexibility in the amount of ADSL overhead that is used (allowing more bits/s for user data), and more comprehensive diagnostic features.
- Supports for bonding several ADSL channels together, as well as support for transport of packet-based services such as Ethernet.

An additional specification based on ADSL2, called "ADSL2plus," has also been written. The main additions in ADSL2plus are:

- Extension of the upper limit of the downstream bandwidth from the original 1.1 MHz to 2.2 MHz. This results in higher downstream bit rates on short-to-medium length loops.
- Spectral shaping of the downstream transmit PSD, to allow greater flexibility in configuration for particular conditions (for example, to meet regional requirements and to improve the spectral compatibility of CO and remote deployments).

5.4.11   VDSL

Very-high bit rate DSL (VDSL) is currently the highest-speed DSL variant, which provides tens of Mbit/s to users in order to extend the performance of existing applications in Internet access, video-conferencing, provision of digital video, telemedicine, and distance learning. In a sense, VDSL is an extension of existing ADSL technology; however, provision of higher bit rates can only be carried out over shorter loops. In fact, the deployment architecture for VDSL is quite similar to the hybrid fiber-coax network described earlier; *i.e.*, fiber-optic transmission is used to connect the central office with a remote optical network unit (ONU), which may be located, for example, at the end of a street. The remaining (short) distance between the ONU and the customer premises is covered using VDSL transmission over the usual twisted copper pair. This architecture could be referred to as hybrid fiber–copper.

VDSL can support both symmetric and asymmetric bit rates. In particular, first-generation VDSL (known as VDSL1) can support 13 or 26 Mbit/s symmetrically, whereas asymmetric transmission can provide up to 52 Mbit/s downstream with 6.4 Mbit/s upstream. The highest downstream rates can only be achieved over short loops. To achieve high bit rates, VDSL uses up to 12 MHz of bandwidth (as opposed to the 1.1 or 2.2 MHz used by ADSL and ADSL2plus, respectively). In one sense, VDSL may be suitable for both broad categories of application, *i.e.*, "residential" applications (largely asymmetric-rate) and "business" (largely symmetric-rate). Like ADSL, VDSL is also capable of operating in the presence of existing POTS or ISDN transmission, using the same "splitter" concept. Although theoretically VDSL could operate using either FDD (no overlap between downstream and upstream transmission) or echo-cancellation (some overlap between downstream and upstream), practical echo-cancelled operation would be extremely difficult to achieve due to the high bandwidths, so VDSL development has largely concentrated on the use of FDD.

From the perspective of standards development, both regional standards bodies and the ITU-T have active projects in this area. In particular, these standards bodies have defined a number of PSD masks for use with VDSL, including the use of multiple disjoint frequency bands for upstream and downstream transmission. Both DMT and QAM/CAP technologies are used in VDSL1, whereas second-generation VDSL (VDSL2) specifies only DMT for the physical layer. Further details on VDSL standards may be found in Volume 2 of this series, and additional information on VDSL technology may be found in [Cioffi 1999] and [IEEE Com. Mag. May 2000].

5.4.12   Related Topics

5.4.12.1   Spectrum Management

Previous chapters have covered the impact of crosstalk on the performance of a DSL system. Clearly, one type of DSL may interfere with another type. Apart from the fundamental performance requirements of DSL systems, an additional requirement is that new types of DSL systems should be spectrally compatible with existing (legacy) systems; that is, they should not cause any undue degradation in the performance of these legacy systems [Starr 2003]. Typically, limits are placed on the PSDs of the transmit signals of new services in order to ensure spectral compatibility. In addition, rules or guidelines exist as to how different types of DSLs may be deployed, particularly in nonhomogeneous situations. For example, where different DSLs with very different PSDs are present in the same binder cable, care must be taken to ensure that all services have acceptable performance. These guidelines constitute what is generally called spectrum management. For example, [ANSI T1 2001] specifies spectral compatibility requirements for North America. This includes a list of legacy systems with which new technologies must be compatible, as well as methods for determining if this compatibility exists, and spectrum management deployment rules. Spectrum management (and related topics) is described in more detail in Volume 2 of this series.

A recent development in DSL technology is dynamic spectrum management (DSM). This has arisen from the recognition that the existing fixed spectrum management guidelines are predicated on DSL systems operating in worst-case conditions. However, DSL systems may well operate in more benign conditions most of the time. DSM attempts to allow DSL systems to achieve the maximum possible performance, while remaining spectrally compatible. In particular, instead of treating each DSL line in isolation, DSM looks at all of the lines in a given binder cable as a multi-user system and performs joint optimization to ensure maximum performance of all systems. Dynamic spectrum management and related issues are covered in detail in Volume 2 of this series and are also described in [Song 2002].

5.4.12.2 Deployment and Testing

As with any communications system, testing and related deployment issues are extremely important; these topics are covered in depth in Volume 2 of this series, which includes chapters on wire line channel simulation, evolution of test procedures from POTS to DSL, and loop qualification and planning for DSL deployment.

5.4.12.3 End-to-End Architectures

This chapter has largely focused on the physical layer of DSL systems, but clearly DSL is of no benefit unless it carries useful services. A large number of different types of service may be carried over DSL, including synchronous data, ATM data, and packet-based services. Figure 5.3 suggested that the physical part of the DSL largely exists on the local loop between the user and the nearest central office. However, from the service perspective, the situation is much broader. The chapters in Volume 2 of this series will address how the DSL physical layer may be used as part of end-to-end system architectures, for example, how DSL may be used to carry services such as packet-based data, and also voice services (so-called voice over DSL, or VoDSL). Volume 2 will also cover topics such as how the flexibility that exists in DSL systems may be leveraged to provide service differentiation, as well as issues in DSL operations and maintenance and security.

5.4.12.4 Ethernet in the First Mile (EFM)

Recent work in IEEE Task Force 802.3ah is aimed at the provision of "full" Ethernet services all the way to a customer premises; this concept is referred to as "Ethernet in the first mile" (EFM). Although copper-based Ethernet standards currently exist for the provision of bit rates up to 1 Gbit/s, there are restrictions on the length and quality of line that may be supported. For example, at higher speeds, Ethernet typically requires multiple pairs of wires of a certain minimum quality.

There is some interest in the use of DSL as a potentially ideal physical layer for the provision of EFM services over copper. In particular, IEEE 802.3ah has standardized VDSL as the transmission technology for use over short loops (≥ 10 Mbit/s symmetric over 750 m), and G.shdsl has been selected for transmission over longer distances (≥ 2 Mbit/s over loops of length 2.7 km); the latter is of particular interest in North America, where the average loop length tends to be longer.

Further information on this project may be found in [IEEE 802.3ah].

5.5 Representative DSL Transceivers

The previous section gave a broad overview of the different classes of DSL technology and briefly covered their salient features. This section attempts to dig more deeply into how DSL systems are actually realized, as a prelude to later chapters that describe the different technologies in more detail (both from the point of view of the fundamental underlying algorithms, as well as implementation-related issues). Design of DSL physical layers encompasses many different technologies, particularly digital signal processing (DSP), advanced coding techniques for error correction, and high-performance analog technology.

Functional block diagrams of "typical" DSL systems are introduced, and the various blocks described in later chapters are highlighted. DSL systems use one of three basic modulation techniques: PAM, QAM/CAP, or DMT. Each modulation type has its own unique characteristics; at the same time, there is some commonality in functional blocks between DSL systems, particularly in functions like error correction. Therefore, in the interests of simplicity, "representative" systems for each modulation method will be presented, and each of these systems will be deemed to be representative of all of the DSLs that use this modulation. For example, PAM is used in HDSL, HDSL2, HDSL4, and G.shdsl. This is clearly a gross simplification, because many differences exist even between systems that use the same basic modulation technique; however, it will suffice for the purposes of this chapter.[4] Later chapters will deal with each transceiver function in more detail and will also point out the differences between the multiple systems that use the same basic modulation technique.

5.5.1 Symmetric DSLs

A block diagram of a DSL transceiver to provide symmetric bit rates using PAM is shown in Figure 5.6. Although this particular block diagram is closest to a G.shdsl transceiver, the basic functions are also common to other symmetric DSLs. The major functional blocks are described in the following subsections.

5.5.1.1 Scrambler and Descrambler

The function of the scrambler is to randomize the transmit data bits that are provided to the DSL transceiver. This is necessary in order to ensure optimum performance of several transceiver components, for example, the equalizer and the echo canceller. At the receiver side, the descrambler carries out the inverse operation of the scrambler in order to regenerate the payload data.

5.5.1.2 Trellis-Coded Modulation (Encoder and Decoder)

Trellis-coded modulation is used in many systems to provide increased robustness to errors, and hence increase the achievable bit rate or reach. The basic principle of this technique

---

[4] Also, it is important to point out that the block diagrams as presented here are simply examples used for illustrative purposes; real systems may differ in some respects.

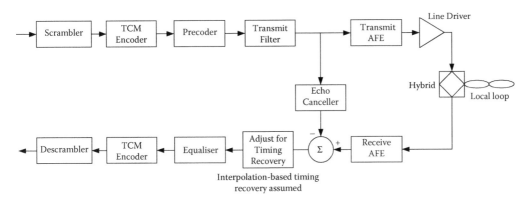

FIGURE 5.6
Block diagram of a "generic" symmetric DSL transceiver (baseband).

is to add redundancy to the transmitted data in order to enable the receiver to correct errors. The TCM decoder is typically implemented using the Viterbi algorithm, though other possibilities exist. Trellis coding is described in more detail in Chapter 8. HDSL2, HDSL4, and G.shdsl make use of TCM; HDSL does not.

5.5.1.3 *Equalizer and Precoder*

Equalization is required in DSL receivers in order to ensure reliable performance in the presence of channel distortion and interference (crosstalk, background noise, etc.). Effectively, equalization attempts to compensate for these noises and distortion; put simply, it "cancels out" signal distortion. When used with TCM, equalization is most often implemented with the aid of a precoder in the transmitter. In essence, part of the equalization function that would normally be carried out at the receiver is instead transferred to the transmitter; *i.e.*, the transmit signal is "pre-equalized" before it is sent through the channel. Obviously the transmitter needs to have some knowledge of the equalization that is required, and this information is usually transferred from receiver to transmitter as part of the transceiver startup procedure. Equalization is described in detail in Chapter 11.

5.5.1.4 *Transmit Filter*

Each DSL standard has particular limitations on the PSD of the transmitted signal, which implies that the transmit signal must somehow be "shaped" so that it conforms to this PSD specification. This shaping is typically carried out in the transmit filter (which is also commonly referred to as a "spectral shaper"). For example, for HDSL2 or HDSL4, the transmit filter needs to be able to impose quite complex PSDs on the transmit signal, with different PSDs required for the upstream and downstream directions.

5.5.1.5 *Transmit Analog Front End (AFE)*

The transmit AFE provides the interface between the DSL "datapump" (which is generally implemented using DSP technology) and the analog transmission medium. In particular, this function will encompass digital-to-analog conversion, analog filtering, and line driver functionality to ensure sufficient signal power is supplied to the line. This circuit interfaces with the hybrid circuit, which connects the four-wire circuit of the DSL transmitter and receiver to the two-wire subscriber line.

5.5.1.6   Receive AFE

The receive AFE provides similar functionality to the transmit AFE, but in the opposite direction. Indeed, both functions would normally be implemented in the same physical piece of hardware, but they are treated here as different entities for clarity. The receive AFE would normally include analog filtering to limit noise and minimize aliasing, and an analog-to-digital converter (ADC). Normally, this circuit would also include an analog automatic gain control (AGC) circuit to optimize the signal level for the ADC input.

5.5.1.7   Echo Canceller

Most of the symmetric bit rate DSLs overlap the downstream and upstream signals in frequency. Because of the imperfect nature of the hybrid interface, impedance mismatches exist, resulting in a significant echo of the transmit signal being received by the near-end receiver. This is essentially a source of interference to the received signal, albeit one for which the interference source is known. The hybrid circuit itself tries to reduce the amount of echo that is received; however, this is usually insufficient, and an adaptive echo canceller must be employed to ensure that the received echo does not dominate the noise in the received signal.

5.5.1.8   Timing Recovery

Because a DSL transmitter and receiver for a given transmission direction are located at opposite ends of a subscriber loop, there will naturally be differences in the frequencies of the symbol (and other) clocks between transmitter and receiver, which is a significant source of error if not corrected. Correction of these differences is the function of the timing recovery (or synchronization) block. In essence, this function tries to regenerate the clock that would be used in the transmitter and hence generate the data that would have been sent from the transmitter. There are many different techniques for timing recovery; some techniques attempt to vary the clock used to control the time instants at which the ADC actually samples the data, whereas others do not modify the ADC sampling clock, but instead attempt to carry out "all digital" timing recovery, *e.g.*, using signal interpolation. Techniques for synchronization are described in Chapter 12.

5.5.1.9   Additional Functions

The previous discussion has identified the key functional blocks of symmetric-rate DSL transceivers, most of which are described in more detail in subsequent chapters. However, there are many other functions that have not been covered in this section. For example, transceivers must implement functions to control the startup or initialization procedure when a connection is established between two DSL modems. Furthermore, many DSL technologies are multi-rate, which means that many parameters of the functional blocks must be properly configured (for example, different symbol clocks may be used for different bit rates, different PSDs may need to be imposed by the transmit filter, etc.). Although the various standards specify many details of functions such as these (for example, the procedures to be used for startup), the details of implementation of these functions are very much vendor-specific.

5.5.2   Passband Single-Carrier Systems

A block diagram of a typical DSL system using passband single-carrier modulation (QAM/CAP) is shown in Figure 5.7. This could be representative of a transceiver used for VDSL1 (though CAP has also been specified for use in providing HDSL service in Europe).

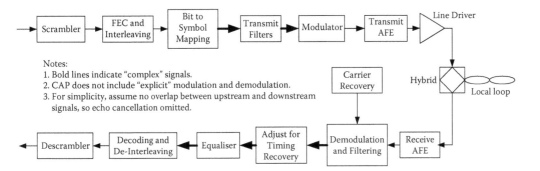

FIGURE 5.7
Block diagram of a generic passband single-carrier DSL transceiver (QAM/CAP).

Many of the blocks used in CAP/QAM systems have similar functions to those used in PAM systems (although many of the details are different), so this subsection will only focus on those functions where significant differences exist. One point to note is that in CAP/QAM systems, the datapath consists of in-phase and quadrature (real and imaginary) components, so most of the filters (equalizer, etc.) can be viewed as processing signals in the form of complex numbers.

5.5.2.1 Error Correction and Interleaving

Only TCM is used in most symmetric rate DSL systems. With other systems such as VDSL1, additional methods for mitigating errors are used. In particular, such systems make use of Reed–Solomon coding and interleaving (see Chapter 9).

5.5.2.2 Modulation and Demodulation

Unlike baseband PAM systems, CAP/QAM systems "shift" the transmit signal frequency up to some defined frequency, depending on the particular DSL service being provided. Hence, CAP/QAM systems use some form of modulation to shift the spectrum. In the case of QAM, this generally takes the form of multiplication of the baseband signal (in complex form) by in-phase and quadrature (cosine and sine) carriers. At the receiver, another multiplication by the carriers must be carried out, in order to shift the signal back down to baseband again.

In the case of CAP modulation, there is no "explicit" multiplication by carriers; instead, the modulation is "implicit" in the impulse responses of the transmit filters, thus avoiding the need for additional carrier generation. Many systems can operate in "dual-mode"; *i.e.*, they can transmit and receive both QAM and CAP signals.

5.5.2.3 Carrier Recovery

For effective demodulation and recovery of the transmitted data, it is important that the frequency and phase of the demodulator carrier in the receiver match those of the carrier used in the transmitter. Because there will always be some inherent difference between the two carriers (for the same reasons that there are differences in the symbol clock frequencies, as discussed above), it is necessary for the receiver to attempt to "recover" the characteristics of the carrier used in the transmitter; this is normally carried out by the carrier recovery block (the function of carrier recovery is similar in many ways to the function of timing recovery discussed earlier).

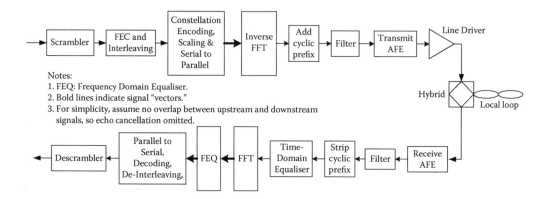

FIGURE 5.8
Block diagram of "generic" DMT-based DSL transceiver (multi-carrier).

5.5.2.4 Additional Points

Although QAM and CAP technology have been classified here as passband single-carrier systems, it should be pointed out that in some DSL systems, QAM and CAP technology may actually make use of multiple carriers. For example, VDSL1 systems may use two disjoint frequency bands for each direction of transmission, with each band utilizing a separate carrier. So, although such systems could be viewed as multi-carrier, it is more commonly the case that these systems are referred to as multi-band single carrier.

Single-carrier modulation is described in detail in Chapter 6.

5.5.3 Multi-Carrier Systems

The third principal type of modulation that is widely used in DSL systems is multi-carrier modulation, specifically, discrete multi-tone (DMT) modulation. A block diagram of a generic DMT-based transceiver is shown in Figure 5.8.

Many of the blocks have the same function as for PAM and CAP/QAM systems; however, a significant difference lies in the modulation. In simple terms, the transmitted data stream is divided among a large number of subcarriers, with the system as a whole operating at a low symbol rate.[5] In the case of DMT however, modulation and demodulation are carried out using Fourier transform, resulting in orthogonal carriers. Furthermore, DMT systems generally use carriers that are reasonably closely spaced, which implies that the magnitude response of the channel is almost "flat" across the bandwidth of each carrier; this means that equalization may be simplified. Chapter 7 describes multi-carrier modulation and DMT in detail.

5.5.4 Closing Remarks

This subsection has introduced representative block diagrams of various types of DSL systems, identifying the major functional blocks that are used, and also pointing out major similarities and differences between different classes of systems. Although the discussion has focused on the functionality, implementation is clearly an important aspect of DSL systems development as well. Chapters in Volume 2 of this series address a number of implementation aspects.

---

[5] In one sense, this is like a multi-band single-carrier system, though with many more carriers than would normally be the case.

5.6 Summary

This chapter has briefly discussed some of the motivations for the development of DSL systems, including the benefits and disadvantages of alternative local access mechanisms that may be used. It then continued to describe the principal DSL systems that are in use or under development, many of the details of which are described in later chapters. Finally, representative block diagrams of the main classes of DSL transceiver were presented, in order to highlight the most important functional blocks that form part of a typical DSL transceiver.

Later chapters will describe DSL technology in greater detail, both from the point of view of the fundamental technologies used in these systems, as well as in the specifics of particular DSL standards.

References

[ANSI T1.601 1992] ANSI T1.601. *ISDN Basic Access Interface for Use on Metallic Loops for Application on the Network Side of the NT (Layer 1 Specification)*. 1992.

[ANSI T1 1994] ANSI Committee T1. *High Bit-Rate Digital Subscriber Line (HDSL) Technical Report*. 1994.

[ANSI T1 1998] ANSI Committee T1. *Asymmetric Digital Subscriber Lines Metallic Interface*. Issue 2, T1.413-1998.

[ANSI T1 2000] ANSI Committee T1. *High Bit-Rate Digital Subscriber Line—2nd Generation*. T1.418-2000.

[ANSI T1 2001] ANSI Committee T1. *Spectrum Management for Loop Transmission Systems*. T1.417-2001.

[Ayanoglu 1998] E. Ayanoglu, N.R. Dagdeviren, G.D. Golden, and J.E. Mazo. *An Equalizer Design Technique for the PCM Modem: A New Modem for the Digital Public Switched Network*. IEEE Trans. on Communications, 46(6), pp. 763–774, June 1998.

[Boelcskei 2001] H. Boelcskei, A.J. Paulraj, K.V.S. Harhi, R.U. Nabar, and W.L. Lu. *Fixed Broadband Wireless Access: State of the Art, Challenges and future Directions*. IEEE Communications Magazine, 39(1), pp. 100–108, January 2001.

[Cioffi 1999] J.M. Cioffi, V. Oksman, J.-J. Werner, T. Pollet, P.M.P. Spruyt, J.S. Chow, and K.S. Jacobsen. *Very-High-Speed Digital Subscriber Line*. IEEE Communications Magazine, 37(4), pp. 72–79, April 1999.

[Cramer 2002] G. Cramer and G. Pesavento. *Ethernet Passive Optical Network (EPON): Building a Next-Generation Optical Access Network*. IEEE Communications Magazine, 40(2), pp. 66–73, February 2002.

[ETS 300 403] ETSI Specification ETS 300 403. *Integrated Services Digital Network*.

[Forney 1984] G.D. Forney, R.G. Gallager, G.R. Lang, F.M. Longstaff, and S.U. Qureshi. *Efficient Modulation for Band-Limited Channels*. IEEE Journal on Selected Areas in Communications, 2(5), pp. 632–647, September 1984.

[Forney 1996] G.D. Forney, L. Brown, V. Eyuboglu, and J.L. Moran. *The V.34 High-Speed Modem Standard*. IEEE Communications Magazine, 34(12), pp. 28–33, December 1996.

[Harashima 1972] H. Harashima and H. Miyakawa. *Matched Transmission Technique for Channels with Intersymbol Interference*. IEEE Trans. on Communications, 20(8), pp. 774–780, August 1972.

[IEEE Com. Mag. Aug. 1992] *IEEE Communications Magazine*, 30(8), Special Issue on ISDN, August 1992.

[IEEE Com. Mag. May 2000] *IEEE Communications Magazine*, 38(5), Special Section on "Very High-Speed Digital Subscriber Line," May 2000.

[IEEE Com. Mag. June 2001] *IEEE Communications Magazine*, 39(6), Special Section on "The Emergence of Integrated Broadband Cable Networks," June 2001.

[IEEE Com. Mag. Sept. 2001] *IEEE Communications Magazine*, 39(9), Special Section on "Broadband Wireless Access Technologies and Applications," September 2001.

[IEEE Com. Mag. Dec. 2001] *IEEE Communications Magazine*, 39(12), Special Section on "Topics in Broadband Access," December 2001.

[IEEE Com. Mag. 2002] *IEEE Communications Magazine*, 40(4), Special Section on "Topics in Broadband Access," April 2002.

[IEEE Com. Mag. March 2003] *IEEE Communications Magazine*, 41(3), Special Section on "Optical Wireless Communications," March 2003.

[IEEE Com. Mag. May 2003] *IEEE Communications Magazine*, 41(5), Special Section on "Internet Access Through the Power Line Network," May 2003.

[IEEE 802.16] IEEE 802.16 Working Group on Broadband Wireless Access Standards, http://ieee802.org/16/.

[IEEE 802.3ah] IEEE 802.3ah Ethernet in the First Mile Task Force, http://www.ieee802.org/3/efm/.

[ITU-T G.991.1 1999] ITU-T Recommendation G.992.1. *Asymmetrical Digital Subscriber Line (ADSL) Transceivers*. July 1999.

[ITU-T G.991.2 2001] ITU-T Recommendation G.991.2. *Single-Pair High-Speed Digital Subscriber Line (SHDSL) Transceivers*. February 2001.

[ITU-T G.992.2 1999] ITU-T Recommendation G.992.2. *Splitterless Asymmetrical Digital Subscriber Line (ADSL) Transceivers*. July 1999.

[ITU-T G.992.3 2002] ITU-T Recommendation G.992.3. *Asymmetrical Digital Subscriber Line (ADSL) Transceivers-2 (ADSL2)*. July 2002 (Draft).

[ITU-T G.994.1 2003] ITU-T Recommendation G.994.1. *Handshake Procedures for Digital Subscriber Line (DSL) Transceivers*. May 2003.

[ITU-T PF-R15 2003] ITU-T Temporary Document PF-R15. *G.shdsl.bis: Draft Text*. Perros-Guirec, France, August 2003.

[Song 2002] K.B. Song, S.T. Chung, G. Ginis, and J.M. Cioffi. *Dynamic Spectrum Management for Next-Generation DSL Systems*. IEEE Communications Magazine, 40(10), pp. 101–109, October 2002.

[Starr 1999] T. Starr, J.M. Cioffi, and P.J. Silverman. *Understanding Digital Subscriber Line Technology*. Upper Saddle River, NJ: Prentice-Hall, 1999.

[Starr 2003] T. Starr, M. Sorbara, J.M. Cioffi, and P.J. Silverman. *DSL Advances*. Upper Saddle River, NJ: Prentice-Hall, 2003.

[Stokesberry 1993] D. Stokesberry and S. Wakid. *ISDN in North America*. IEEE Communications Magazine, 31(5), pp. 88–94, May 1993.

[Tomlinson 1971] M. Tomlinson. *New Automatic Equaliser Employing Modulo Arithmetic*. IEE Electronics Letters, 7, pp. 138–139, March 1971.

[TS 101 135 2000] ETSI Specification TS 101 135. *Transmission and Multiplexing: High Bit-Rate Digital Subscriber Line (HDSL) Transmission Systems on Metallic Local Lines*. Version 1.5.3, September 2000.

[TS 101 388 2002] ETSI Technical Specification TS 101 388. *Transmission and Multiplexing (TM); Access Transmission Systems on Metallic Access Cables; Asymmetric Digital Subscriber Line (ADSL)*. Version 1.3.1, May 2002.

[TS 101 524 2001] ETSI Technical Specification TS 101 524. *Transmission and Multiplexing: Access Transmission System on Metallic Access Cables: Symmetrical Single Pair High Bitrate Digital Subscriber Line (SDSL)*. Version 1.1.3, November 2001.

6

Fundamentals of Single-Carrier Modulation

Vladimir Oksman

CONTENTS

6.1 Overview 144
6.2 Basics of QAM, PAM, and CAP Modulation 144
    6.2.1 QAM 144
        6.2.1.1 Transmitter 146
        6.2.1.2 Constellation Encoder (Mapper) 148
        6.2.1.3 Spectrum Shaping Filters 150
        6.2.1.4 Receiver 152
        6.2.1.5 Demodulator 153
        6.2.1.6 Equalizer 154
        6.2.1.7 Decision Circuits (Slicers) and Constellation Decoder 155
        6.2.1.8 Carrier and Timing Recovery 157
    6.2.2 PAM 157
    6.2.3 CAP 159
6.3 Main Parameters of SCM Signals 161
6.4 Performance 164
    6.4.1 Error Performance 164
    6.4.2 Transport Capability (Bit Rate) 166
6.5 Special Techniques Used in SCM Applications 167
    6.5.1 Blind Equalization 167
    6.5.2 Multi-Symbol Constellation Encoding 168
    6.5.3 Spectrum Allocation and PSD Shaping 169
    6.5.4 CAP/QAM Dual Mode Operation 170
    6.5.5 Initialization 170
    6.5.6 Burst Mode SCM Transceivers 171
    6.5.7 Multi-Band SCM Transceivers 172
        6.5.7.1 Operation of the Transceiver 172
        6.5.7.2 Frame Splitting and Recovering 173
        6.5.7.3 Transport Capability 174
        6.5.7.4 Constellation Assignment (Bit Loading) 175
6.6 Annex I: Constellation Diagrams 175
References 178

ABSTRACT This chapter reviews the fundamentals of the QAM, PAM, and CAP modulation techniques and describes the main principles of modulation, spectrum shaping, filtering, equalization, decoding, link initialization, and performance estimation of single-carrier modulation (SCM) transceivers. In addition, several special techniques used in SCM are briefly presented. Operation of multi-band SCM transceivers, developed for VDSL1, is described in detail. Examples clarify such issues as distribution of transmission capabilities, main parameters, and performance figures.

6.1   Overview

There are two main scenarios usually considered in DSL deployments. In the first scenario, a DSL system is deployed over unused copper pairs (dark copper) or replaces a digital loop carrier (DLC) system (T1 or E1). Because only DSL uses the wire, a baseband transmission can be used to utilize frequencies almost down to zero, benefiting from low attenuation of the twisted pair on low frequencies. The popular modulation type for baseband transmission is pulse amplitude modulation (PAM) used in BRI (Basic Rate ISDN), HDSL, HDSL2, and SHDSL. In the second scenario, a DSL system has to share a wire with POTS or basic rate ISDN. To ensure minimal impact on their operation, DSL systems use only frequencies above the spectrum of these baseband services. Accordingly, passband transmission is preferable for this scenario, and the most popular modulation technique is the well-known quadrature amplitude modulation (QAM), currently used in VDSL1 [ETSI TS 101 270-2] [G.993.1 2004], Etherloop [Etherloop 1998], and for subchannel modulation in ADSL [G.992.3 2002]. QAM is also used in many non-DSL applications such as voiceband modems, cable modems, and home phone line networking (for example, HPNA modems [G.989.1 2001]). Carrierless amplitude/phase modulation (CAP) is similar to QAM. It is specified as an alternative line code for HDSL [G.991.1 1998] and was used in rate-adaptive DSL (RADSL) [T1.TR.59 1999] as a pre-standard ADSL solution, and in some high-speed networking applications [Im 1995a].

The term "single-carrier modulation (SCM)" was originally introduced to denote VDSL1 systems utilizing QAM or CAP. This term is convenient to describe all DSL systems using QAM, CAP, or PAM, all of which are techniques that make use of a single carrier. Some implementations of SCM transceivers work in dual mode, allowing either QAM or CAP modulation to be used for a particular connection [ETSI TS 101 270-2].

6.2   Basics of QAM, PAM, and CAP Modulation

Modulation is the process of mapping the digits of transmitted data into signal waveforms that can be successfully propagated over a transmission medium. Usually, a group of one or more bits is mapped to a given waveform, called a symbol. Different groups of bits use different waveforms, distinguished by one or several parameters. The transmission time for each symbol is usually fixed and referred to as the symbol period, T. The symbol rate is given by $\frac{1}{T}$. It expresses the number of symbols transmitted during each second. Various modulation types described in this section mostly differ by the used signal waveforms.

6.2.1   QAM

As in many other modulation techniques, QAM uses sine waveforms of a specific frequency, called the carrier frequency f_c. The amplitude and phase of the sine wave distinguish symbols one from another, which makes QAM a two-dimensional modulation method. To increase bandwidth efficiency, the QAM waveform is constructed from two quadrature

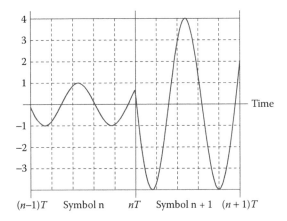

FIGURE 6.1
Example of a QAM symbol waveform.

carriers $S_I(t)$ and $S_Q(t)$ occupying the same frequency band. If the amplitudes of the in-phase carrier S_I and the quadrature carrier S_Q in the ith symbol period equal I_i and Q_i, respectively, the QAM waveform will be:

$$
\begin{aligned}
S_{QAM_i}(t) &= S_{I_i}(t) + S_{Q_i}(t) \\
&= I_i \cos(2\pi f_c t) + Q_i \sin(2\pi f_c t) \\
&= A_i \sin(2\pi f_c t + \theta_i),
\end{aligned}
\tag{6.1}
$$

with

$$
A_i = \sqrt{I_i^2 + Q_i^2}, \quad \tan(\theta_i) = \frac{I_i}{Q_i},
\tag{6.2}
$$

where A_i and θ_i are, respectively, the amplitude and the phase of the QAM sine wave in the ith symbol period. An example of a QAM symbol is presented in Figure 6.1.

Usage of sine waveforms provides the passband nature of QAM. As is shown in [Proakis 2001], the spectrum of the QAM-modulated signal is located around the carrier frequency f_c. Thus, by changing f_c, the QAM spectrum can be placed in the desirable frequency range, provided that $f_c \geq \frac{1}{2T}$. Some examples of QAM spectra are presented in Figure 6.6 and Figure 6.7.

Usage of quadrature carriers to form the QAM signal allows their separation at the receive side because the carriers are orthogonal. Two functions φ_i and φ_j are called orthogonal if they satisfy the following condition.

$$
\int_{-\infty}^{\infty} \varphi_i(t) \cdot \varphi_j(t) \cdot dt = \delta_{ij}, \quad \text{where } \delta_{ij} = \begin{cases} 1 & i = j \\ 0 & i \neq j \end{cases}.
\tag{6.3}
$$

If there is a mixture of k signals, all based on orthogonal functions $\varphi_1 \ldots \varphi_k$:

$$
S = a_1\varphi_1 + a_2\varphi_2 + \cdots + a_k\varphi_k,
\tag{6.4}
$$

any signal a_i can be filtered out by using its basis function φ_i:

$$
\int_{-\infty}^{\infty} S \cdot \varphi_i \cdot dt = \int_{-\infty}^{\infty} \left(\sum_{n=1}^{k} a_n\varphi_n \right) \cdot \varphi_i \cdot dt = a_i.
\tag{6.5}
$$

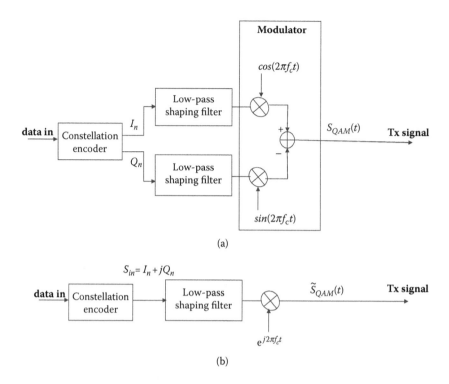

FIGURE 6.2
(a) Functional diagram of a QAM transmitter and (b) its complex-analytic representation.

Indeed, because $\sin(2\pi f_c t)$ and $\cos(2\pi f_c t)$ (the basis functions for QAM carriers) are orthogonal, the same sin and cos functions have to be used for their separation, as shown further in Equation 6.14. Similarly, the fact that sin and cos basis functions using different carrier frequencies (say, $f_{c1} \neq f_{c2}$) are orthogonal is used to separate QAM-modulated signals. In particular, it is used in multi-band SCM transmission described in this chapter and in multi-carrier transmission described in Chapter 7.

6.2.1.1 Transmitter

A functional diagram of a QAM transmitter is shown in Figure 6.2a. The input data stream is divided into groups of M consecutive bits $\{b_1, b_2, \ldots, b_M\}$, and each group is encoded into one QAM symbol. Consecutive groups (of M bits each) are encoded into consecutive QAM symbols as illustrated in Figure 6.3, and each QAM symbol is transmitted during a time of duration T. The total number of different QAM symbols required to represent all possible combinations of M bits is 2^M.

The particular QAM symbol to be transmitted during the nth symbol period is determined by the values I_n and Q_n, which are the in-phase and quadrature components, respectively. The size of the group M and the values of I_n and Q_n for the particular group of bits are assigned by the constellation encoding (mapping) procedure. The number of possible settings k_I of I_n and k_Q for Q_n should be such that $k_I \times k_Q \geq 2^M$. The full set of possible QAM symbols is usually presented in the form of a constellation diagram (signal-space diagram), where each point represents a symbol with specific values of I and Q. Figure 6.3a shows an arrangement of groups of bits and two examples of constellation diagrams for $M = 3$ (8-QAM). In this example $k_I = k_Q = 4$, and both I_n and Q_n take values $\pm 1, \pm 3$. Other examples of constellation diagrams can be found in the annex of this chapter and also in

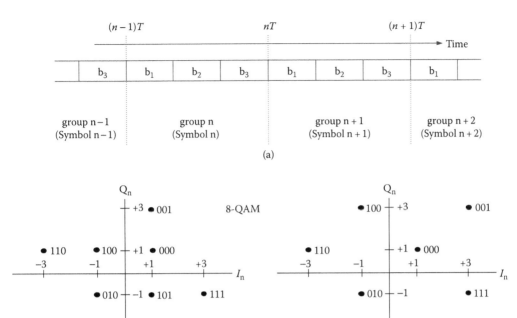

FIGURE 6.3
(a) Bit-to-symbol mapping and (b) examples of 8-QAM constellation diagrams.

[Gitlin 1995], [Ungerboeck 1987], and [Proakis 2001]. Different systems use a wide variety of different constellations, compromising between the complexity of the implementation and the efficiency of the constellation for the given transmit power. For most practical cases, the values of k_I and k_Q are even (to get a symmetric signal space) and can be calculated as: $k_I = k_Q = 2 \times \lceil (2^{M/2-1} \rceil$, where the ceiling function $\lceil \cdot \rceil$ rounds up to the nearest integer. The ceiling function is necessary for noninteger values of $M/2$. For instance, in the case of QAM-32, $M = 5$, $k_I = k_Q = 6$. The application of this equation can be easily verified using the examples of constellation diagrams in Figure 6.3 and in Figures 6.23 to 6.26 of the annex.

The output signal of a QAM modulator can be written as:

$$S_{QAM}(t) = \left[\sum_n I_n g(t - nT) \right] \cos(2\pi f_c t) - \left[\sum_n Q_n g(t - nT) \right] \sin(2\pi f_c t), \qquad (6.6)$$

where $g(t)$ is the impulse response of the two identical low-pass shaping filters.

Because QAM is a two-dimensional signal, it is sometimes convenient to describe it using a complex-analytic notation, representing the nth symbol as a complex vector $S_n = I_n + j Q_n$. The complex-analytic representation of a QAM transmitter is shown in Figure 6.2b. After S_n passes through the shaping filter and is multiplied by $e^{j2\pi f_c t}$, a complex output signal results:

$$\tilde{S}_{QAM}(t) = \sum_n (I_n + j Q_n)g(t - nT)e^{j2\pi f_c t}. \qquad (6.7)$$

The orthogonal components I and Q are, respectively, the real and imaginary parts of the complex QAM signal. It is easy to show that $Re(\tilde{S}_{QAM}(t)) = S_{QAM}(t)$. This approach is also used in Chapter 7.

6.2.1.2   Constellation Encoder (Mapper)

The first step in the modulation process is the encoding of bits into symbols. First, the encoder divides the incoming data into equal groups of bits. The number of bits in the group is determined by the value of SNR at the receiver (for example, see Table 6.6) and is usually estimated during the initialization of the link. After the number of bits in the group (bits per symbol) is determined, the encoder maps[1] each group to a symbol. In different applications, different types of mapping are used. Some examples of mappings are shown in Figure 6.3 and in Figures 6.21 to 6.26. The main requirements for the encoder are to allow a simple decoding procedure and to minimize the number of bit errors when symbols are decoded incorrectly.

Because the phase of the signal changes as it traverses the loop, the absolute phase in the received symbols is unknown. The phase change may be represented as a rotation of the constellation diagram around the origin. For example, both mappings presented in Figure 6.3 are not rotation invariant: if either of them is rotated by 90°, for instance, the decoded mapping will be different from the original one. The absolute phase reference has to be known in the receiver to decode these types of mappings.

Obtaining the absolute phase reference is sometimes complex. In applications where absolute phase reference is not available, differential encoding [Gitlin 1995] solves the problem by utilizing relative values of phase, referring to the phase of the previously transmitted symbol. Thus, there is no need for the receiver to know the absolute phase reference in the decoder. In addition, the decoder is resistant to occasional phase transients. If a transient happens during a particular symbol, only this symbol and perhaps the next one (if the relative phase of the impaired symbol cannot be recovered) will be received in error.

In differential encoding, the first two bits of the encoded group determine a shift between the quadrant used by the current symbol and the quadrant used by the previous symbol. The actual quadrant is determined by the combination of the signs of I_n and Q_n of the previously received symbol. An example of a differential encoding table is presented in Figure 6.4.

The remaining $(M - 2)$ bits of the encoded group are mapped over the constellations inside the specified quadrant. To simplify the encoding, mappings in adjacent quadrants usually have 90° rotational symmetry. Thus, the mapping in the first quadrant actually represents the entire constellation diagram, because other quadrants may be derived by rotation of the first quadrant mapping counter-clockwise by 90°, 180°, and 270°, respectively.

Gray mapping [Proakis 2001] minimizes the number of bit errors caused by a symbol error by mapping groups of bits that differ by only one bit on adjacent constellation points. Thus, the most probable symbol error (detection of the closest symbol instead of the original one sent) causes only one bit error. Unfortunately, for large constellations, Gray mapping is often not possible, especially if the constellation diagram is required to be 90° rotationally symmetric. However, inside each quadrant, Gray coding is possible and effectively used. One simple rule providing quadrant Gray mapping for square constellations 2M-QAM (M = even) is presented in Table 6.1. This type of mapping is used in SCM VDSL1 and yields a 1-bit difference between the encoded groups mapped over adjacent constellation points and a 2-bit difference between the encoded groups mapped over diagonal points for all square constellations (see examples in Figure 6.23 and Figure 6.25).

To clarify the usage of Table 6.1, consider 64-QAM ($M = 6$), and a group with the four last bits $b_3b_4b_5b_6 = 0001$. The mappings are $X_1 = X_2 = 0$, $Y_1 = 0$, $Y_2 = 1$, and $I_n = 001$, $Q_n = 011$. Using decimal values for I_n and Q_n, the constellation point with $I_n = 1$ and $Q_n = 3$ is coded 0001. This can be easily verified from Figure 6.25.

---

[1] In some references, the term "labelling" is used instead of "mapping," because the mapping procedure may be represented as labelling each constellation point by the value of the bit group this point is assigned.

$b_1 b_2$	Previous quadrant	Sign of previous symbol I_{n-1} Q_{n-1}	Current quadrant	Sign of current symbol I_n Q_n
00	1st	+ +	1st	+ +
00	2nd	− +	2nd	− +
00	3rd	− −	3rd	− −
00	4th	+ −	4th	+ −
01	1st	+ +	4th	+ −
01	2nd	− +	1st	+ +
01	3rd	− −	2nd	− +
01	4th	+ −	3rd	− −
10	1st	+ +	2nd	− +
10	2nd	− +	3rd	− −
10	3rd	− −	4th	+ −
10	4th	+ −	1st	+ +
11	1st	+ +	3rd	− −
11	2nd	− +	4th	+ −
11	3rd	− −	1st	+ +
11	4th	+ −	2nd	− +

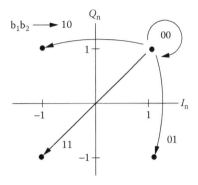

FIGURE 6.4
Differential encoding table and quadrant rotation scheme.

Another type of quadrant Gray mapping is widely used for cross-shaped constellations (odd values of M, $M > 3$). The quadrant is divided into eight sections, as shown in Figure 6.5, and each section is coded by a 3-bit section code using bits $b_3 b_4 b_5$ of the mapped group. The rest of the bits (applicable for $M > 5$ and denoted "XXX") are mapped inside each section by the rule shown in Table 6.1. Furthermore, some of the coded sections are flipped horizontally, or vertically, or both, as shown in Figure 6.5. The resulting 32-QAM and 128-QAM mappings are illustrated in Figure 6.24 and Figure 6.26, respectively.

Figure 6.4 is built so that a misdetection to any point of any adjacent quadrant results in one error in two bits used for quadrant coding. Therefore, if, instead of the point {1, 1}, the receiver will detect any of two adjacent points belonging to adjacent quadrants, only one bit error will occur. On the other hand, up to $M - 1$ errors may occur if, instead of any other constellation point along the {I,Q} axes, the closest point in the alien quadrant is detected. The likelihood of this event, however, is relatively low, because only 1 adjacent point from 4 possible belongs to the alien quadrant. In the diagram presented in Figure 6.23, among 16 points in each quadrant, there are 10 points for which misdetection to any of 4 closest

TABLE 6.1

Quadrant Gray Mapping for 2^M-QAM (square constellations)

I_n (binary) = $[X_1 X_2 \ldots X_{M/2-1} 1]$	Q_n (binary) = $[Y_1 Y_2 \ldots Y_{M/2-1} 1]$
$X_1 = b_3$	$Y_1 = b_{M/2+2}$
$X_2 = X_1 + b_4$	$Y_2 = Y_1 + b_{M/2+3}$
$X_3 = X_2 + b_5$	$Y_3 = Y_2 + b_{M/2+4}$
. . . .	. . . . .
$X_{M/2-1} = X_{M/2-2} + b_{M/2+1}$	$Y_{M/2-1} = Y_{M/2-2} + b_M$

TABLE 6.2

In-section Gray Mapping for Cross-Shaped Constellations ($M > 5$)

I_{n-sec} (binary) = $[X_1 \, X_2 \ldots X_{(M-5)/2} \, 1]$	Q_{n-sec} (binary) = $[Y_1 \, Y_2 \ldots Y_{(M-5)/2} \, 1]$
$X_1 = b_6$	$Y_1 = b_{(M-5)/2+6}$
$X_2 = X_1 + b_7$	$Y_2 = Y_1 + b_{(M-5)/2+7}$
....	.....
$X_{(M-5)/2} = X_{(M-5)/2-1} + b_{(M-5)/2+5}$	$Y_{(M-5)/2} = Y_{(M-5)/2-1} + b_M$

points will cause one bit error (9 points with $I, Q > 1$ and point $\{1, 1\}$). For the other 6 points, misdetection to any of 3 closest points will cause 1 bit error. However, for points $\{1, 3\}$, $\{1, 7\}$, $\{3, 1\}$, $\{7, 1\}$ misdetection to the fourth closest point will cause 3 bit errors, and for points $\{1, 5\}$ and $\{5, 1\}$ misdetection to the fourth closest point will cause 5 bit errors. Therefore, if all symbols are equally probable, the average probability of a 1-bit error per symbol error is $(10 \times 4 + 6 \times 3)/(16 \times 4) = 0.906$. The probabilities of a 3-bit error and a 5-bit error, respectively, are $4/64 = 0.063$, and $2/64 = 0.031$. Neither 2-bit nor 4-bit errors will occur.

In DSL, additional data encoding schemes, such as forward error-correction (FEC) coding or trellis coding, are usually used prior to the constellation encoding. Referring to the functional diagram in Figure 6.2, these types of encoding are assumed to have been applied already to the incoming data stream. These and other encoding techniques are described in Chapters 8, 9, and 10 of this book.

6.2.1.3 Spectrum Shaping Filters

Two low-pass filters are intended for limiting and shaping of the transmit signal spectrum. If the passband of the filter is much wider than $\frac{1}{T}$, the amplitude and phase of the carrier changes almost instantly at the transition from one symbol to another (Figure 6.1). The

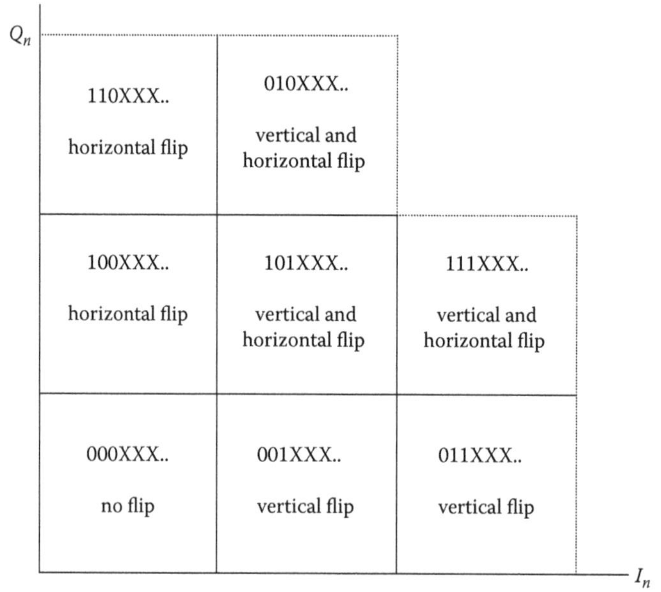

FIGURE 6.5
Mapping sections for crossed constellations.

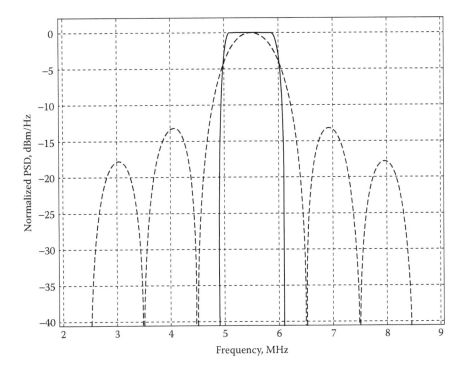

FIGURE 6.6
Spectrum of QAM signal with $1/T = 1$ Mbaud, $f_c = 5.5$ MHz (wideband shaping—dashed line, square-root raised cosine shaping with $\alpha = 0.2$—solid line).

normalized spectral magnitude function of the signal may be expressed as:

$$|S_{QAM}(f)| \approx \left| \frac{\sin(\pi T(f - f_c))}{\pi T(f - f_c)} \right|. \tag{6.8}$$

Usage of this sinc-type spectrum (see Figure 6.6) is inefficient, because its slowly decaying side-lobes cause significant crosstalk into signals occupying neighboring frequency bands, except those that use narrow frequency bands spaced by $1/T$ from the carrier frequency f_c. The latter advantage is used in multi-carrier modulation, described in Chapter 7.

Shaping improves the efficiency of the spectrum usage. A popular example is a square-root raised cosine shaping filter having a spectral magnitude function given by

$$|G(f)| = \begin{cases} 1 & , |f| \le f_1 \\ \cos\left(\frac{\pi \cdot T}{2\alpha}[|f| - f_1]\right) & , f_1 \le |f| \le f_2 \\ 0 & , \text{elsewhere} \end{cases}, \quad f_1 = \frac{1 - \alpha}{2T}, f_2 = \frac{1 + \alpha}{2T}. \tag{6.9}$$

Accordingly, the spectral magnitude function of the transmit signal is given by

$$|S_{QAM}(f)| = |G(f - f_c)|. \tag{6.10}$$

Parameter $0 \le \alpha \le 1$ in Equation 6.9 is called the excess bandwidth. The frequency boundaries of the QAM signal spectrum with excess bandwidth α are:

$$f_{min} = f_c - \frac{(1 + \alpha)}{2T}, \quad f_{max} = f_c + \frac{(1 + \alpha)}{2T}. \tag{6.11}$$

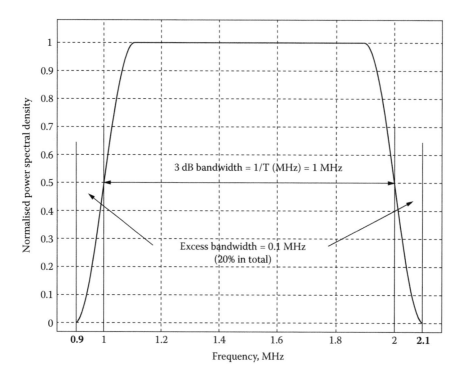

FIGURE 6.7

Spectrum of a QAM signal with square-root raised cosine shaping ($1/T = 1$ Mbaud, $f_c = 1.5$ MHz, $\alpha = 0.2$).

Respectively, the bandwidth occupied[2] by the QAM signal equals $\frac{(1+\alpha)}{T}$. In practical applications, the PSD of the signal at frequencies below f_{min} and above f_{max} drops at least 20 dB relative to its value at frequency f_c due to the square-root raised-cosine function. The 3-dB bandwidth, W_{3dB}, representing the width of the QAM spectrum, is equal to the QAM symbol rate $\frac{1}{T}$:

$$f_{max,3dB} = f_c + 1/2T,$$
$$f_{min,3dB} = f_c - 1/2T,$$
$$W_{3dB} = f_{max,3dB} - f_{min,3dB} = \frac{1}{T}.$$

(6.12)

Typically, the value of α varies from 0.1 to 0.2. Some systems targeted for transmission media with poor and barely predictable characteristics, such as home wiring, may operate with $\alpha = 1$ [G.989.1 2001]. The normalized PSD of a QAM signal using square-root raised-cosine shaping with $\alpha = 0.2$ is presented in Figure 6.7. The frequency boundaries are: $f_{min} = 1.5 - 1 \times (1 + 0.2)/2 = 0.9$ MHz, $f_{max} = 1.5 + 1 \times (1 + 0.2)/2 = 2.1$ MHz.

6.2.1.4 Receiver

The functional diagram of a QAM receiver is presented in Figure 6.8a. It includes a demodulator with low-pass shaping filters, an equalizer and decision circuits (slicers) for both in-phase and quadrature components, and a constellation decoder. The timing recovery

---

[2] As was mentioned above, the symbol rate of a QAM signal must be less than $2f_c$ to avoid modulation distortion. In the case that $1/2T < f_c \leq (1+\alpha)/2T$, the signal bandwidth is slightly less than $(1+\alpha)/T$ and equals $f_c + (1+\alpha)/2T$. The 3-dB bandwidth remains the same.

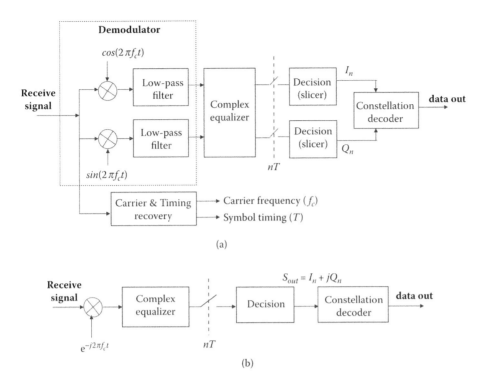

FIGURE 6.8
(a) Functional diagram of a QAM receiver and (b) its complex-analytic representation.

circuit extracts the carrier frequency for demodulation purposes and symbol timing for the rest of the processing. The fact that after demodulation the signal has a baseband format is sometimes an advantage, because it simplifies implementation of QAM transceivers operating with high carrier frequencies. A complex-analytic representation of the receiver is presented in Figure 6.8b.

6.2.1.5 Demodulator

The demodulation process can be described by the following equations for the demodulated in-phase and quadrature signal components.

$$S_{inph}(t) = S_{QAM}(t)\cos(2\pi f_c t + \theta), \quad S_{quad}(t) = S_{QAM}(t)\sin(2\pi f_c t + \theta), \quad (6.13)$$

where θ is the phase shift in the recovered carrier. Substituting Equation 6.6 and eliminating the high-frequency components, which are filtered out by the low-pass filters (see Figure 6.8), yields

$$S_{inph}(t) = \left[\sum_n I_n g(t - nT)\right]\frac{\cos\theta}{2} - \left[\sum_n Q_n g(t - nT)\right]\frac{\sin\theta}{2},$$

$$S_{quad}(t) = \left[\sum_n I_n g(t - nT)\right]\frac{\sin\theta}{2} - \left[\sum_n Q_n g(t - nT)\right]\frac{\cos\theta}{2}. \quad (6.14)$$

Equation 6.14 shows that demodulation requires a very accurate phase adjustment of the recovered carrier, and the demodulation is referred to in the industry as *coherent* demodulation. If θ is nonzero, the signal components are not orthogonal, and interference from the quadrature component affects the detection of the in-phase component, and vice versa.

The required accuracy of carrier phase recovery for QAM transmission may be derived directly from Equation 6.14. Assume that operation with the desired constellation requires a signal-to-noise ratio (SNR) of a dB. The inaccuracy of phase recovery will not cause significant SNR reduction if interference from the quadrature component is at least 12 dB below the level of noise corresponding to the required SNR.[3] Because the used constellation diagrams are symmetric ($k_I = k_Q$), the average power of in-phase and quadrature components is usually the same,[4] and the maximum allowed value of θ may be obtained from the following equation.

$$20 \log_{10} \left(\frac{\cos \theta}{\sin \theta} \right) = a + 12. \tag{6.15}$$

Consider, for instance, that one uses 256-QAM transmission with a required BER of 10^{-7} and a noise margin of 6 dB. The minimum required SNR is $a = 31.7$ dB $+ 6$ dB $= 37.7$ dB. Solving Equation 6.15 yields $\theta < 0.188°$, which is remarkably accurate. Usage of higher-order constellations requires even higher accuracy. To reach this high accuracy, special techniques of carrier frequency and phase recovery are used. Some of them are mentioned below.

6.2.1.6 *Equalizer*

The equalizer attempts to reduce intersymbol interference (ISI) in the receive signal and to maximize the SNR at the input of the decision circuit. Equalization techniques used in QAM receivers are described in Chapter 11. Typically, a fractionally spaced linear equalizer (FSLE) or decision feedback equalizer (DFE) is used. Both are usually built using finite impulse response (FIR) digital filters with adjustable complex coefficients, which allow simultaneous processing of the in-phase and quadrature components. In addition, the equalizer can automatically adapt its impulse response to the characteristics of the particular loop. The equalizer is usually adjusted to the loop during the link initialization. During normal operation, the filter coefficients are finely tuned in response to environmental changes.

The main impairments complicating DSL signal detection are ISI and noise (mostly crosstalk) accumulated in the line. ISI is caused by reflections and bandwidth limitation introduced in the loop and shaping filters. As the equalizer attempts to reduce ISI by amplifying the signal in the suppressed frequency ranges (sometimes this procedure is called "inverting the channel"), the noise tends to grow as a result of the amplification. Therefore, in the aim to increase the SNR, the equalizer attempts to find a compromise that allows low ISI without significant noise enhancement. This compromise is usually reached when the impulse response of the transmission channel is close to satisfying the Nyquist criterion, resulting in zero ISI at the instants $t = nT$. If a square-root raised-cosine shaping filter is used in the transmitter, and the rest of the transmission channel (from the input of the modulator in Figure 6.2 to the output of the equalizer) has the frequency/phase response that is the same square-root raised-cosine, then the transmission channel has the well-known

---

[3] Assume the noise due to the interference of the quadrature component is additive, uncorrelated with the other noise components, and is below the level of the power sum of all other noise components by 12 dB. Then, impact of the interference from the quadrature component can be estimated as: $10 \times \log_{10}(1 + 10^{-1.2}) = 0.26$ dB, which is usually considered insignificant.

[4] To be more precise, the mentioned average power assumes averaging over the whole set of symbols. For any particular symbol, the value of quadrature noise may be either above the averaged (if its quadrature component power is larger than the average) or below the average (if its quadrature component power is smaller than the average). The peak value of noise is above the average by $20 \times \log_{10}(Q_{max}/I_{min})$; the minimum value is below the average by the same value.

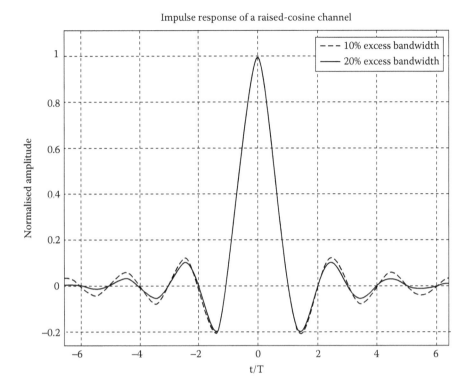

FIGURE 6.9
Impulse response of a channel with raised-cosine transfer function.

raised-cosine transfer function with the impulse response

$$g(t) = \frac{\sin\left(\frac{\pi \cdot t}{T}\right) \cdot \cos\left(\frac{\pi \alpha \cdot t}{T}\right)}{\pi \cdot \left(\frac{t}{T}\right) \cdot \left(1 - 4\left(\frac{\alpha \cdot t}{T}\right)^2\right)}. \tag{6.16}$$

The impulse response in Equation 6.16 is presented in Figure 6.9. It obviously satisfies the Nyquist criterion because $g(0) = 1$ and $g(nT) = 0$, the latter due to the sin component in the equation. The figure shows that higher values of α result in impulse responses that decay faster and have lower ISI between the decision points nT, which relaxes restrictions on jitter in symbol timing. On the other hand, higher values of α obviously reduce the efficiency of QAM transmission by introducing additional bandwidth overhead. The optimal compromise is usually found in the range $\alpha = 0.1$ to 0.2.

A superposition of traces of different waveforms appearing at the output of the equalizer is usually referred to as an eye-diagram. An example of an eye-diagram for either the in-phase or quadrature component of 4-QAM, 8-QAM, and 16-QAM is presented in Figure 6.10. The traces show some residual ISI, which is close to zero at the instants of decision (sampling points).

6.2.1.7 *Decision Circuits (Slicers) and Constellation Decoder*

Two multi-threshold slicers are used to decide on the values of I_n and Q_n being received. The decision technique is rather simple: if the received value I'_n or Q'_n at the instant of the decision differs from the possible transmit value I_n or Q_n, respectively, less than the slicing threshold does, the decision $I'_n = I_n$ or $Q'_n = Q_n$ is made. For the relative values of I_n and Q_n in the transmit signal equal to $\pm 1, \pm 3, \pm 5, \ldots, \pm(k_I - 1)$, the thresholds of the slicers are

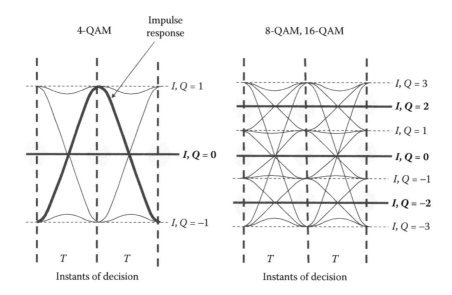

FIGURE 6.10

Illustration of eye-diagrams for in-phase and quadrature components (centers of decision regions are shaded).

set to $0, \pm2, \pm4, \pm6, \ldots, \pm(k_I - 2)$, *i.e.*, exactly in the middle between two possible values of I_n. The instant of the decision should be as close as possible to the center of the symbol period, where the eye has its widest opening (see Figure 6.10) and the highest signal-to-ISI ratio is achieved.

It may be shown ([Gitlin 1995], [Proakis 2001]) that this decision technique, usually referred to as maximum-likelihood detection, provides maximum probability of correct decisions if all transmitted symbols are equiprobable. An error (misdetection) will occur when the value of the disturbing noise at the decision instant exceeds the slicer threshold. For instance, in Figure 6.10 a noise with a relative value of 1.1 at the decision instant will cause the receive signal to exceed the threshold $I = -2$ when the original value of the transmitted signal is $I = -3$. Considering additive Gaussian noise with the variance σ^2 and the probability distribution function $P(x) = \frac{e^{-x^2/(2\sigma^2)}}{\sqrt{2\pi\sigma^2}}$, the probability of misdetection for either symbol component may be expressed using a Q-function, [Proakis 2001]:

$$P_e = Pr\left(x > \frac{d}{2}\right) = \frac{1}{\sigma\sqrt{2\pi}} \int_{d/2}^{\infty} e^{-x^2/2\sigma^2}dx = \frac{1}{\sqrt{2\pi}} \int_{d/2\sigma}^{\infty} e^{-y^2/2}dy = Q\left(\frac{d}{2\sigma}\right) \approx \frac{1}{2}e^{-d^2/8\sigma^2},$$

$$(6.17)$$

where $Pr(\cdot)$ denotes the probability of an event, and d is the distance between the two adjacent levels of either I_n or Q_n. For the presumed relative values of I_n and Q_n, $d = 2$.

The accuracy of settings for thresholds and instances of the decisions is critical: any inaccuracy is equivalent to a certain SNR reduction. The inaccuracy of settings of I_n and Q_n at the transmit side is equivalent to improper threshold settings as well. For example, one can assume 1 percent inaccuracy of I_n-threshold setting when 256-QAM is used ($k_I = 16$). For symbols with $I_n = \pm15$, this inaccuracy reduces the distance to the closest threshold by $14 \times 0.01 = 0.14$, which is equivalent to an SNR reduction of about $20 \times \log_{10}(1/(1 - 0.14)) = 1.31$ dB. This reduction is considerable, and it is even worse for higher constellations. Fortunately, the high resolution of modern signal processing devices (at least 16-bit representation) and A/D and D/A converters usually allows sufficiently accurate and stable settings of transmit signal amplitudes, thresholds, and timing instants.

Decoding of the received signal is performed directly, by using the resolved values I_n and Q_n. In the case of differential encoding, decoding begins with quadrant recognition by comparing signs of the obtained I_n and Q_n with their signs in the previous symbol (I_{n-1} and Q_{n-1}) as presented in the table in Figure 6.4. Other bits are decoded directly using the given constellation encoding rule (see Table 6.1 as an example).

6.2.1.8 Carrier and Timing Recovery

The carrier at the receive side is necessary for coherent demodulation. It could be either received from the transmitter (for example, as a pilot tone) or recovered from the received data signal. Both options are used in practical implementations, although carrier recovery is usually more attractive because it does not require any additional signals to be transmitted. However, it is also more complex.

The required high accuracy of carrier phase recovery was shown during the description of the demodulation process. Proper symbol timing is a key requirement for successful signal processing and decision making. The accuracy of symbol timing is characterized by possible fluctuations of the symbol period (jitter), usually estimated as a percentage. Obviously, these fluctuations influence equalization and reduce accuracy of the decision instants. Typically, jitter in symbol timing is maintained to be below a few percent.

Methods of carrier recovery and timing recovery, including acquisition and precise tracking of phase, are discussed in Chapter 12, and additional information can be found in [Gitlin 1995]. High precision of the carrier and symbol timing recovery is achieved by using adjusting algorithms directed by decisions on the received data. In many implementations, a joint carrier and timing recovery mechanism is used. This is especially convenient for systems like VDSL1 where the carrier frequency period and symbol period are multiples of the same timing reference; thus, the carrier recovery circuit may source the timing recovery circuit and vice versa.

6.2.2 PAM

PAM uses pulses with duration T as a signal waveform. Different PAM symbols are distinguished by their amplitudes. In particular, a symbol of 2^M-PAM can have 2^M possible values of amplitude (also called "levels") representing 2^M possible values of an M-bit group. An example of a 4-PAM (also known as 2B1Q) signal is presented in Figure 6.11. Sometimes, a passband PAM is also considered. Passband PAM refers to a modulation technique that uses a sine waveform of a particular frequency (carrier frequency) with different amplitudes. All conclusions regarding PAM technology obtained in this section hold for passband PAM.

The functional diagrams of the PAM transmitter and receiver are presented in Figure 6.12. They are very similar to those of QAM and include the same functional elements, except the multiplication by the carrier. Referring to Figure 6.2 and Figure 6.8, the PAM transceiver may be represented as a result of multiplication of the QAM in-phase component by 1 ($\cos(2\pi f_c t) = 1$ if $f_c = 0$) and the QAM quadrature component by 0 ($\sin(2\pi f_c t) = 0$ if $f_c = 0$). Thus, PAM can be directly interpreted as QAM with a zero carrier frequency. Consequently, all the main components of the PAM transceiver shown in Figure 6.12 operate in the same way as those described for QAM.

The constellation diagram of PAM is one-dimensional, as shown in Figure 6.11b. The number of possible settings of $I_n(k_I)$ equals the number of transmit signal levels. Gray mapping is typically used, so that groups of bits mapped into adjacent levels of amplitude differ by one bit only. A simple rule providing Gray mapping for 2^M-PAM is similar to the one in Table 6.1 and is presented in Table 6.3.

Differential encoding allows the PAM decoder to be invariant to crossing of the wires in the twisted pair. Similar to QAM, the first bit of the encoded group of bits (bit b_1 missing in

(a). 4-PAM (2B1Q) signal

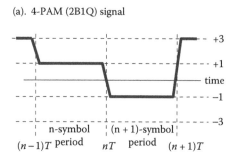

(b). PAM constellation diagrams and encoding examples

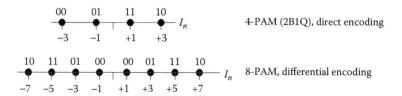

FIGURE 6.11
Examples of PAM signal and constellation diagrams.

the second column of Table 6.3) defines whether the polarity of the current symbol coincides with or differs from the polarity of the previous symbol. An example of bit mapping for 8-PAM when differential encoding is used is presented in Figure 6.11. In practical applications, however, a "trial and error" method is also used in the decoder to identify crossing of the wires instead of differential encoding.

Low-pass shaping filters, equalizers, and decision circuits of a PAM transceiver operate in the same way as in a QAM transceiver and have similar characteristics; decoding is a

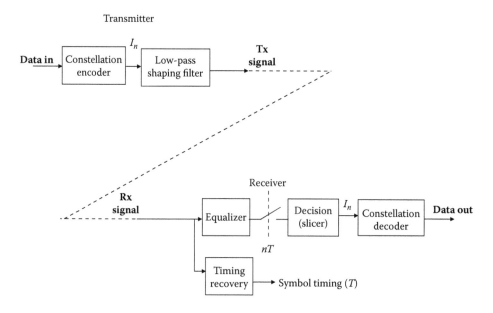

FIGURE 6.12
Functional diagram of a PAM transceiver.

TABLE 6.3

Gray Mapping for PAM

Direct encoding I_n (binary) = $[X_1 X_2 \ldots X_M\ 1] - 2^M$ (binary)	Differential encoding I_n (binary) = $[X_1 X_2 \ldots X_{M-1}\ 1]$
$X_1 = b_1$	$X_1 = b_2$
$X_2 = X_1 + b_2$	$X_2 = X_1 + b_3$
$X_3 = X_2 + b_3$	$X_3 = X_2 + b_4$
....	....
$X_M = X_{M-1} + b_M$	$X_{M-1} = X_{M-2} + b_M$

simple inversion of the mapping rule (as in Table 6.3, for instance). Figure 6.13 shows some examples of PAM signal spectra. The eye-diagrams of 4-QAM and 8/16-QAM presented in Figure 6.10 are also relevant for 2-PAM (commonly known as NRZ) and 4-PAM (2B1Q), respectively.

6.2.3 CAP

The carrierless amplitude-phase modulation [Im 1995b], [Haykin 1998] was proposed as an alternative to QAM. CAP creates a passband transmit signal with characteristics very similar to QAM but uses digital filtering instead of multiplication by the carrier frequency.

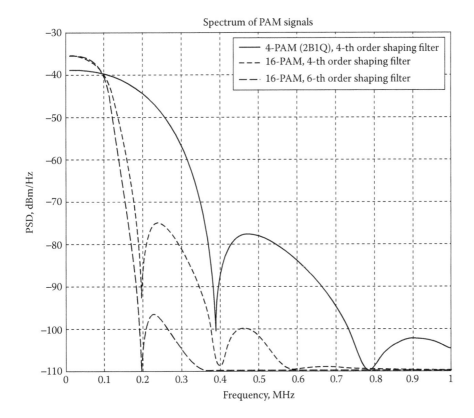

FIGURE 6.13

Spectrum of 4-PAM ($1/T = 396$ kBaud) and 16-PAM ($1/T = 198$ kBaud); ($\alpha \approx 0.8$ for the 4th order shaping filter, and $\alpha \approx 0.45$ for the 6th order shaping filter).

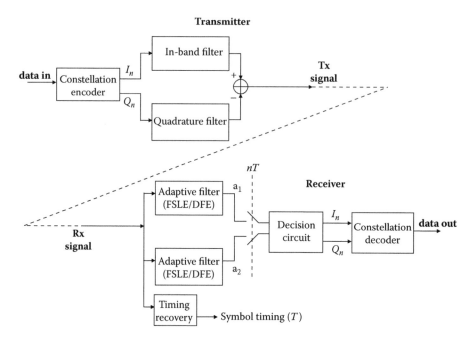

FIGURE 6.14
Functional diagram of a CAP transceiver.

This sometimes simplifies the implementation and avoids carrier recovery in the receiver. The penalty is a relatively high sampling rate, which makes CAP inconvenient for applications when the ratio of the carrier frequency to the symbol rate is small. Thus, CAP was used in pre-standard ADSL and in HDSL but is not used in VDSL1, which utilizes several rather narrow frequency bands (although some standards such as [ETSI TS 101 270-2] allow QAM/CAP dual mode operation). The rest of the properties of CAP are mostly the same as for QAM.

The functional diagrams of a CAP transmitter and receiver are presented in Figure 6.14. The transmitter is similar to the one in Figure 6.2a, and creates a passband signal by the in-phase and quadrature filters with impulse responses

$$f(t) = g(t)\cos(2\pi f_c t)q(t) = g(t)\sin(2\pi f_c t), \tag{6.18}$$

where f_c is the "virtual carrier," which is simply the center frequency of the transmit signal spectrum.[5] The CAP transmit signal, accordingly, equals

$$
\begin{aligned}
S_{CAP}(t) &= \sum_n [I_n p(t - nT) - Q_n q(t - nT)] \\
&= \sum_n [I_n p(t - nT) - Q_n \tilde{p}(t - nT)] \\
&= \sum_n [I_n g(t - nT)\cos(2\pi f_c(t - nT))] - \sum_n [Q_n g(t - nT)\sin(2\pi f_c(t - nT))], \tag{6.19}
\end{aligned}
$$

---

[5] In practical implementations, the actual time span where the channel impulse responses accurately follow Equation 6.18 should be at least $8T$ ($\pm 4T$ from the center lobe of the impulse response).

where I_n and Q_n are outputs from the constellation encoder, and $\tilde{p}$ denotes a Hilbert transformation of p.

Equation 6.19 shows that the CAP signal is actually very similar to a QAM signal. It occupies the same spectrum (determined by the shaping component $g(t)$ of the in-phase and quadrature filters (see Figure 6.7)), may use the same constellation diagrams, but experiences additional rotation by a fixed phase increment of $2\pi f_c T$ in each symbol period. This rotation is the same in magnitude but opposite in sign to the rotation of the QAM carrier during the symbol period. Therefore, a CAP signal looks like a QAM signal in which the phase of the carrier is set to zero at the beginning of each symbol period. In the case when $f_c \times T$ is equal to any integer, CAP and QAM signals are exactly the same, and thus the same receiver (either QAM or CAP) works for both signals. This feature is used in some combined CAP/QAM SCM transceivers to simplify the startup process.

The receiver (see Figure 6.14) contains two adaptive filters, a decision circuit, and a constellation decoder. Adaptive filters are intended to combat the ISI and to provide a channel response that satisfies the Nyquist criterion. They are built as an inverted Hilbert pair with impulse responses $g_{a_1}(t) = -\tilde{g}_{a_2}(t)$, respectively. As the paths between the transmitter and the adaptive filters are linear, the signals a_1, a_2 on the outputs of the adaptive filters are:

$$a_1 = \sum_n [I_n x(t - nT) - Q_n \tilde{x}(t - nT)], \tag{6.20}$$

$$a_2 = \sum_n [Q_n x(t - nT) + I_n \tilde{x}(t - nT)], \tag{6.21}$$

where $x(t) = g(t) \star h(t) \star g_a(t)$ is the impulse response of the signal path (where $\star$ denotes the convolution operator), including the shaping filters $g(t)$, the loop $h(t)$, and the adaptive filters $g_a(t)$. Because the channel response $x(t)$ is equalized to satisfy the Nyquist criterion, i.e., $x(kT) = 1$ and $\tilde{x}(kT) = 0$, the decision circuit recovers the values of I_n and Q_n using output signals of the adaptive filters at the decision instants.

6.3   Main Parameters of SCM Signals

The main parameters of modulated SCM signals are the average power per symbol, peak-to-average ratio (PAR), and the minimum Euclidean distance. All these parameters can be found analyzing the signal constellation diagram. A 2^M-QAM (equivalently, a 2^M-CAP) signal is initially considered.

The relative average power per symbol is

$$P = \frac{1}{2^M} \sum_{k=1}^{2^M} (I_k^2 + Q_k^2). \tag{6.22}$$

The PAR, in dB, is calculated as a ratio between the peak and the average power:

$$PAR = 10 \log_{10} \left[\frac{\max_k (P_k)}{P} \right] + 3 = 10 \log_{10} \left[\frac{\max_k (I_k^2 + Q_k^2)}{P} \right] + 3 \quad [\text{dB}], \tag{6.23}$$

TABLE 6.4

Main Parameters of QAM (square and cross-shaped constellations)

	2^M-QAM									
Parameter	$M = 1$	$M = 2$	$M = 3$	$M = 4$	$M = 5$	$M = 6$	$M = 7$	$M = 8$	$M = 9$	$M = 10$
P	2	2	10	10	20	42	82	170	330	682
PAR [dB]	3	3	5.5	5.5	5.3	6.68	6.16	7.23	6.59	7.49
d_N	2	1.41	0.894	0.632	0.447	0.309	0.221	0.153	0.110	0.0766
η [dB]	0	−3.0	−7.0	−10.0	−13.0	−16.2	−19.1	−22.3	−25.2	−28.3

where 3 dB is due to the PAR of a sine wave.[6] In both equations, k is the number of the given constellation point, $k = 1, 2, \ldots, 2^M$.

The Euclidian distance $d(k, i)$ is the geometric distance between points k and i of the constellation diagram. In accordance with the decision technique used, the value $0.5 \times d(k, i)$ determines how much the received vector $\{I_k, Q_k\}$ of symbol k can deviate from its original point before it will be misdetected as symbol i and vice versa. According to Equation 6.17, the probability that a symbol is detected in error decays exponentially with d; thus the minimum Euclidian distance actually determines the symbol error probability. Consequently, the minimum Euclidian distance normalized to the averaged symbol power P indicates the relative noise immunity of the particular constellation: $d_N = \min_{k,i}[d(k, i)]/\sqrt{P}$. That value, in turn, expresses the difference in SNR necessary to ensure the same average probability of symbol error when different constellations are used. Taking 2-QAM as a convenient reference, the noise immunity of other constellations can be estimated, in dB, as:

$$\eta = 20 \log_{10} \frac{d_N[2^M - QAM]}{d_N[2 - QAM]} \text{ [dB]}. \tag{6.24}$$

The following example illustrates calculation of the parameters for the square-shaped constellation of 16-QAM presented in Figure 6.23. Substituting the values of I and Q from the first quadrant of Figure 6.23 into Equations 6.22 to 6.24, and noting that all the quadrants are symmetric, the following results are obtained.

$$P = \frac{1}{16} \times 4 \times [(1^2 + 1^2) + (1^2 + 3^2) + (3^2 + 1^2) + (3^2 + 3^2)] = 10$$

$$PAR = 10 \log_{10} \frac{\max_k(P_k)}{P} + 3 = 10 \log_{10} \frac{(3^2 + 3^2)}{10} + 3 = 5.5 \text{ dB}$$

$$d_N = \frac{\min_{k,i}[d(k, i)]}{\sqrt{P}} = \frac{2}{\sqrt{P}} = 0.632,$$

$$\eta = 20 \log_{10} \left(\frac{d_N}{d_N(M = 1)} \right) = -10 \text{ dB}.$$

The values of the parameters for some common 2^M-QAM constellations are presented in Table 6.4. It is assumed that all constellations, except 8-QAM, are square-shaped for even values of M and cross-shaped for odd values of M. The constellation for 8-QAM is shaped as shown in Figure 6.22.

For PAM, the same parameters can be easily derived from Equations 6.22 to 6.24, assuming $Q_n = 0$ and dropping the sine wave PAR of 3 dB. Considering the constellation diagram for

---

[6] Equation 6.23 doesn't account for the impact of shaping filters; see the explanation under Table 6.5.

2^M-PAM similar to those presented in Figure 6.11, the results are:

$$P = \frac{1}{2^M} \cdot \sum_k I_k^2 = \frac{d^2}{12}(2^{2M} - 1), \tag{6.25}$$

$$PAR = 10 \log \frac{P_{max}}{P} + 3 = 10 \log \frac{\max_k \left(I_k^2 \right)}{P} \text{ [dB]}, \tag{6.26}$$

$$\eta = 20 \log \frac{d_N[2^M - PAM]}{d_N[2 - PAM]} \text{ [dB]}, \tag{6.27}$$

where, as previously, k is the number of the given pulse level (constellation point in one-dimensional interpretation), $k = 1, 2, \ldots, 2^M$, and d is the distance between two adjacent constellation points. For 8-PAM ($M = 3$), having a constellation diagram presented in Figure 6.11, the results are:

$$P = \frac{1}{8} \times 2 \times (1^2 + 3^2 + 5^2 + 7^2) = 21,$$

$$PAR = 10 \log \frac{\max_k \left(I_k^2 \right)}{P} = 10 \log \frac{7^2}{21} = 3.7 \text{ dB},$$

$$d_N = \frac{\min_{k,i} [d(k,i)]}{\sqrt{P}} = \frac{2}{\sqrt{21}} = 0.4364,$$

$$\eta = 20 \log_{10} \left(\frac{d_N}{d_N(M = 1)} \right) = -10 \text{ dB}.$$

The values of parameters for 2^M-PAM are presented in Table 6.5.

It should be noted that the values of the PAR presented in Table 6.4 and Table 6.5 do not account for the impact of the shaping filters. Shaping filters increase the peak power due to ISI between adjacent symbols. They can increase the PAR by about 3 dB when the symbol waveform reaches its maximum value both at the end of one symbol interval and at the beginning of the next one.

The last line in Table 6.4 clearly demonstrates the "3 dB rule" widely used in QAM and CAP engineering practice: every additional bit per symbol requires about 3 dB of additional SNR. The similar "6 dB rule" well known for PAM could be concluded from the last line of Table 6.5. Both rules work more accurately for higher constellations, starting from 16-QAM (4-PAM).

TABLE 6.5

Main Parameters of PAM

Parameter	2-PAM	4-PAM	8-PAM	16-PAM	32-PAM
P	1	5	21	85	341
PAR [dB]	0	2.6	3.7	4.23	4.50
d_N	2	0.894	0.436	0.217	0.108
η [dB]	0	−7	−13.2	−19.3	−25.3

6.4 Performance

The performance of an SCM transceiver is characterized by the average probability of symbol error and the bit rate it is capable of transmitting (transport capability).

6.4.1 Error Performance

The error performance is usually estimated under the assumption that the noise has a Gaussian distribution. Consider first 2^M-PAM, which has $k_I = 2^M$ possible values of I_k: $\pm 1, \pm 3, \ldots, \pm(k_I - 1)$. The probability of a symbol error due to a misdetection between two adjacent constellation points is expressed by Equation 6.17. Because all symbols, except those two with maximum absolute values of I_k, are surrounded by two possible values for which they could be mistaken, and those with maximum absolute value of I_k have only one such value, the total number of possible cases of misdetection equals $2 \times (k_I - 1)$. Thus, there are, on average, $2 \times (k_I - 1)/k_I$ possible cases of misdetection, and the average error probability per symbol equals

$$P_e = \frac{2\,(k_I - 1)}{k_I} \cdot Q\left(\frac{d}{2\sigma}\right), \tag{6.28}$$

where σ^2 is the variance of the noise, and d is the Euclidean distance between two adjacent PAM levels (which equals 2 in this case). Substituting the value of SNR $= P/\sigma^2$, and the average symbol power P from Equation 6.25 yields

$$P_e = \frac{2\,(k_I - 1)}{k_I} \cdot Q\left(\frac{d}{2\sqrt{P/SNR}}\right) = 2\left(1 - \frac{1}{2^M}\right) \cdot Q\left(\sqrt{\frac{3 \cdot SNR}{(2^{2M} - 1)}}\right). \tag{6.29}$$

In the case of QAM (or CAP), the probability of a symbol error is a superposition of incorrect decisions for either I or Q. Assuming that the probability of an incorrect decision is the same for I and Q (due to symmetry of the constellations) and that the probabilities of incorrect decisions for I and Q are independent, the total averaged error probability P_e equals

$$P_e = P_e(I) + P_e(Q) = 2P_e(I). \tag{6.30}$$

For 2^M-QAM, I_k and Q_k may have k_I and k_Q possible values: $\pm 1, \pm 3, \ldots, \pm(k_I - 1)$, where k_I and k_Q equal $2^{M/2}$ rounded to the closest even integer (see the constellation diagrams presented in Figures 6.23 to 6.26). Using this approach, the probability of an erroneous decision for I_k and Q_k can be calculated using the same equation as for PAM, with appropriate modification of the value k_I. Each of the QAM components carries half of the total symbol power, so the SNR for each component equals

$$SNR = \frac{P}{\sigma^2} \quad \text{and} \quad \sigma = \sqrt{\frac{P}{2 \cdot SNR}}. \tag{6.31}$$

Substituting this into Equation 6.28 results in

$$P_e = 2P_e(I) = \frac{4(k_I - 1)}{k_I} \cdot Q\left(\frac{d}{2\sqrt{P/(2 \cdot SNR)}}\right)$$

$$= 4\left(1 - \frac{1}{k_I}\right) \cdot Q\left(\frac{d}{2} \times \sqrt{\frac{2 \cdot SNR}{P}}\right), \tag{6.32}$$

where $k_I = 2 \times \lceil(2^{M/2-1})\rceil$.

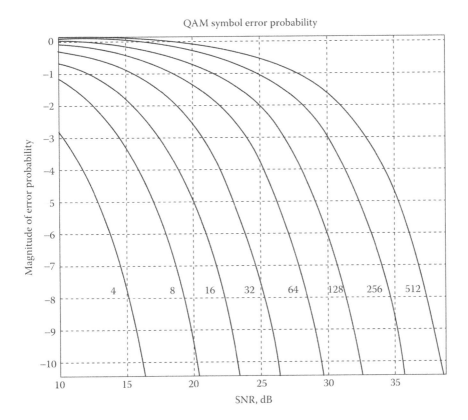

FIGURE 6.15
Symbol error probability of QAM with different constellation sizes.

Equation 6.33 may be used for both square and nonsquare constellations. For all square and cross-shaped constellations, $d = 2$; for the 8-QAM constellation presented in Figure 6.22, $d = 2\sqrt{2}$. The value of P is obtained from Equation 6.22, and for some constellations can be found in Table 6.4.

The results of the computation of QAM symbol error probability versus SNR, as given by Equation 6.33, are presented in Figure 6.15. 2^M-PAM symbol error probability curves are very close to those of 2^{2M}-QAM. For instance, the 16-QAM curve in Figure 6.15 can be also used for 4-PAM (2B1Q), and 256-QAM can be used for 16-PAM.

The SNR values required to keep the average symbol error probability equal to 10^{-7} are presented in Table 6.6.

In practice, engineers often keep in mind the 14.5 dB reference (4-QAM) or 21.5 dB reference (16-QAM) and the "3 dB rule" to estimate SNR requirements for other constellations.

TABLE 6.6

SNR Requirements for $P_e = 10^{-7}$ in the Presence of Gaussian Noise

	2^M-QAM								
Parameter	$M = 2$	$M = 3$	$M = 4$	$M = 5$	$M = 6$	$M = 7$	$M = 8$	$M = 9$	$M = 10$
	2-PAM		4-PAM		8-PAM		16-PAM		32-PAM
SNR [dB]	14.5	18.6	21.5	24.5	27.7	30.8	33.8	36.9	39.9

For instance, using 4-QAM as a reference, 2-QAM requires $SNR \approx 14.5 - 3 = 11.5$ dB and 256-QAM requires $SNR \approx 14.5 + [\log_2(256) - \log_2(4)] \times 3 = 14.5 + 18 = 32.5$ dB. Similarly, the "6 dB rule" is usually used with reference to 4-PAM (2B1Q). Both the "3 dB" and "6 dB" rules are more accurate for larger constellations and are quite accurate for constellations larger than 16-QAM and 4-PAM.

6.4.2   Transport Capability (Bit Rate)

The transmit bit rate R provided by a 2^M-PAM or a 2^M-QAM transceiver operating at a symbol rate $1/T$ is

$$R = M \cdot \frac{1}{T} \quad \text{[bit/s]}. \tag{6.33}$$

Because the symbol rate $1/T$ determines the bandwidth occupied by the signal, the same bit rate in a wider band can be obtained using a smaller constellation. Use of a wider bandwidth with a smaller constellation may be beneficial for error performance if the noise immunity gain due to reduction of the constellation size (see Table 6.4 and Table 6.5) balances the loss of SNR due to the increase in the bandwidth. This issue matters particularly when high-order constellations are used. For instance, expanding the bandwidth by a factor of two and switching from 256-QAM to 16-QAM will lead to extra noise immunity of $\Delta\eta = \eta\,[16 - QAM] - \eta\,[256 - QAM] = 22.3 - 10 = 12.3$ dB.

In DSL applications, however, signal propagation loss increases exponentially as frequency increases. Accordingly, for a loop of a particular length, all signal components transmitted above a certain frequency will be received at levels below the level of the noise, thus limiting the useful bandwidth expansion. Furthermore, the useful bandwidth decreases as the loop length increases, so systems operating on long loops need to confine their transmissions to the lower frequencies. Additionally, some DSLs are strictly limited to specific frequency bands because of spectrum compatibility rules. With a limited bandwidth, a higher bit rate can be reached only by using higher-order constellations. This, however, requires higher SNR, which may be not available in longer loops. For QAM, support of the next higher value of M requires about 3 dB of excess SNR. One way to utilize SNR margin less than 3 dB is to use the multi-symbol (multi-dimensional) coding briefly described below. Another possibility is to use more powerful encoding at upper layers, partially utilizing the additional bit rate gained due to the use of a higher-order constellation.

As shown in Section 4.6, the bit rate may also be calculated using the famous Shannon formula for channel capacity (see Equation 4.58):

$$C = \omega \cdot \log_2\left(1 + \frac{SNR}{\Gamma(P_e)}\right) \quad \text{[bit/s]}, \tag{6.34}$$

where ω is the bandwidth of the channel and $\Gamma(P_e)$ is the Shannon gap (see Chapter 4). Because the Shannon capacity C obtained using a PAM or a QAM signal with the bandwidth of ω shall be the same as R in Equation 6.33, the value of the gap may be estimated as

$$\Gamma(P_e)\text{[dB]} = 10 \cdot \log_{10}\left(\frac{SNR}{2^M - 1}\right) = SNR\text{[dB]} - 10 \cdot \log_{10}(2^M - 1). \tag{6.35}$$

Calculation of $\Gamma(P_e)$ for the larger constellation sizes and their corresponding SNR values in Table 6.6 for a symbol error rate of $P_e = 10^{-7}$ yields $\Gamma(P_e) = 9.75$ dB for PAM and QAM with square constellations and $\Gamma(P_e) = 9.57$ dB for QAM with cross-shaped constellations. A gap value of 9.8 dB is commonly used in the industry in calculations for QAM or PAM. For channels with frequency-dependent SNR, Equation 6.34 may be used for a steplike approximation of the SNR figure. This leads to the well-known equation for the channel capacity (and the bit rate) achievable at the given error probability (10^{-7}):

$$C = R = \int_{f_{min}}^{f_{max}} \log_2\left(1 + \frac{SNR(f)}{\Gamma(P_e)}\right) df, \qquad (6.36)$$

where f_{max} and f_{min} are the maximum and the minimum frequency bounds, respectively, of the used spectrum. Equation 6.36 is widely used in engineering practice.

6.5 Special Techniques Used in SCM Applications

6.5.1 Blind Equalization

Blind equalization usually refers to a startup procedure in which no specific or known a priori signal is required to train the equalizer of the receiver. In particular, the equalizer may be trained as it receives "live" data. Blind equalization usually takes more time than training with a specific signal, but it sometimes simplifies implementation of the transmitter because no training signal is required. Another feature of blind equalization is that there is no need to stop the service if the link fails for a relatively short time, because the equalizer can recover using the incoming data as a training signal. This is a key requirement for point-to-multi-point connections, in particular for video broadcasting used in pre-standard CAP VDSL1 systems [Harman 1995].

As the link is turned on, the equalizer is adjusted to adapt to the loop. The stochastic gradient algorithm, usually used for the adjustment, updates the transfer function (tap coefficients) of the equalizer to minimize the cost function CF selected for the particular adaptation procedure. The equalizer is declared to be converged when the cost function is minimized; *i.e.*, $\nabla(CF) = 0$.

With a pre-defined training sequence, the cost function usually reflects the difference between the received symbol values and the known transmit values. This decision-directed adaptation (see Chapter 11) cannot be used for blind equalization, because the transmitted values are unknown. Instead, algorithms of blind equalization minimize the cost function using the known statistics of the transmit symbol values, mostly based on the assumption that in an equalized channel, all the constellation points are located at some distance around the origin. The two simple and well-known algorithms are the reduced constellation algorithm (RCA) and the constant modulus algorithm (CMA), which are shown in Figure 6.16.

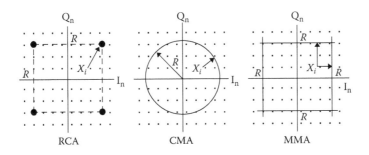

FIGURE 6.16
Principles of blind equalization.

With the reduced constellation algorithm (RCA), the cost function is defined as an averaged distance between the received value X_i and the reference corner point inside the same quadrant:

$$CF_{RCA} = E\left[|X_i - r \cdot (sgn(X_i) + j \cdot sgn(\tilde{X}_i))|^2\right], \tag{6.37}$$

where $E(\cdot)$ denotes expectation, and r is the radius of a circle within which the received signal values lie. The reference points reflect the expected constellation points of the equal-power 4-QAM. With the constant modulus algorithm (CMA), similarly, the cost function reflects the dispersion of the received signal values around the circle of radius r:

$$CF_{CMA} = E\left[\left(|X_i|^2 - r^2\right)^2\right]. \tag{6.38}$$

RCA is the simplest for implementation but does not provide reliable convergence, even for square constellations, above 16-QAM. CMA is much more reliable, but it is still not very effective for non-square and dense constellations (and it becomes almost impractical for constellations larger than 64-QAM).

The multi-modulus algorithm (MMA) uses a more comprehensive cost function

$$CF_{MMA} = E\left[\left(X_i^2 - r^2\right)^2 + \left(\tilde{X}_i^2 - r^2\right)\right], \tag{6.39}$$

which allows reliable convergence with constellations up to 128-QAM. For more dense constellations, a generalized MMA (GMMA) algorithm was recently proposed. The idea behind GMMA is to divide the constellation diagram into smaller regions using a multi-dimensional cost function combining the MMA cost functions built for different regions. A detailed description of GMMA and other blind equalization algorithms can be found in [Werner 1999].

The reader should be aware that almost all blind equalization methods have a probability of false convergence, especially when high-density constellations are used [Werner 1999]. This problem can be overcome by restarting the receiver. Also, all the described blind equalization algorithms are applicable only for implementations using linear equalization (FSLE), but not DFE. Due to the feedback in DFE, converging under blind equalization becomes unstable, even for simple constellations. Typically, the equalizer is blindly trained with its feedback disabled, and then finely adjusted with the feedback.

In point-to-point configurations, a "semi-blind" equalization procedure can be used to improve convergence, which is very efficient in operation with nonsquare and dense constellations. In semi-blind mode (also called "two-step mode"), the receiver is first trained with a 4-QAM pseudo-random signal and then switched to higher constellations providing the required data bit rate. This mode is often used in SCM VDSL1 implementations.

6.5.2 Multi-Symbol Constellation Encoding

Multi-symbol encoding can increase spectral efficiency. Considering QAM, it was shown that increasing the number of bits per symbol by 1 bit while still maintaining the same error probability and noise margin requires at least 3 dB of additional SNR. Multi-symbol coding can utilize smaller quantities of excess SNR in order to increase the bit rate, thus improving the system performance. In particular, usage of N-symbol coding allows a reduction of the required excess SNR by approximately $1/N$. Consider a simple case of $N = 2$. Assume that instead of encoding 4-bit groups of data into 16-QAM symbols, one encodes a 9-bit group into two sequential 24-QAM symbols. The total number of symbols needed to transmit a 9-bit group is $2^9 = 512$. Two 24-QAM symbols provide $24^2 = 576$ possible combinations. As a result, two 24-QAM symbols can be used to transmit a 9-bit group, providing a transport capability of 4.5 bit/symbol.

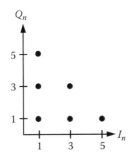

FIGURE 6.17
First quadrant of a QAM-24 symbol.

The relative noise immunity η of 24-QAM can be calculated using Equations 6.22 and 6.24. Considering the 24-QAM constellation presented in Figure 6.17, the values of P and d_N can be calculated, and then averaged over two 24-QAM symbols under an assumption that none of the possible symbol combinations when both symbols have their maximum power is used. This assumption is based on the fact that the 24-QAM constellation includes 8 points (two in each quadrant) with maximum relative power $P = 26$ ($|I| = 1, |Q| = 5$ and $|I| = 5, |Q| = 1$). There are, accordingly, 64 possible combinations when both symbols use one of those constellation points. However, use of those combinations is unnecessary, because there are enough other symbol combinations to encode a 9-bit group ($576 - 64 = 512$). As a result, $P = 12.4$, $d_N = 2/\sqrt{14.4} = 0.568$, and $\eta = -10.95$ dB. Thus, it takes only 0.95 dB of additional SNR to change from 16-QAM to 24-QAM. The next step, to 32-QAM, requires about 2 dB.

Multi-symbol coding requires additional synchronization to recognize relevant pairs of symbols.

6.5.3 Spectrum Allocation and PSD Shaping

The center frequency f_c and the symbol rate $1/T$ define the allocation of the QAM (CAP) signal spectrum and the occupied bandwidth, as shown in Figure 6.7. By varying these two parameters, the spectrum can be moved into any part of the available frequency range. Besides shifting of the spectrum and changing of the bandwidth, in-band shaping of the transmit spectrum is sometimes required in order to prevent radio-frequency (RF) egress or to reduce crosstalk in the frequency region occupied by the QAM signal.

Spectral shaping is performed at the transmitter by using finite impulse response (FIR) or infinite impulse response (IIR) filters. The FIR filters are preferable for gradual smooth PSD shaping; IIR are preferable for notching out the transmit spectrum at frequencies used by amateur radio services, which is a requirement for DSL systems. A frequency response of a typical fourth-order IIR notching filter used to reduce egress into an amateur radio band of 1.8 to 2 MHz is presented in Figure 6.18.

All types of spectral shaping cause additional ISI, which is compensated by the equalizer. In some cases, excessively complex shaping may cause performance reductions and even error propagation in the DFE. In the case where error propagation is a concern, Tomlinson–Harashima pre-coding, described in Chapter 11, is usually the best way to resolve the problem.

The reduction in performance due to notching does not follow exactly the channel capacity calculation presented in Equation 6.36. A good engineering rule for deep notches, following the investigation done in [Salz 1973], says that the average SNR reduction when the notch occupies X percent of the band may be calculated as $X \times SNR[\text{dB}]/100$. For example, if a 256-QAM signal transported with 6 dB noise margin (SNR = 33.8 dB + 6 dB =

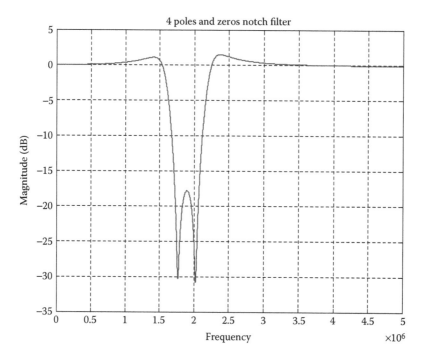

FIGURE 6.18
IIR notching filter of fourth order.

39.8 dB) has 5 percent of the band notched, the SNR (and the noise margin, respectively) will drop by $0.05 \times 39.8 = 1.96$ dB, *i.e.*, from 6 dB to 4.04 dB. If a minimum 6 dB margin has to be maintained, the signal constellation should be dropped to 128-QAM, which results in losing channel capacity of 1 bit/symbol/Hz.

6.5.4 CAP/QAM Dual Mode Operation

Dual mode operation can be used to allow CAP and QAM modems to interoperate. When dual mode operation is supported, the transmitter may be either CAP or QAM, and the receiver must be capable of receiving both signals. Because CAP and QAM signals differ only by rotation, it is sometimes practical to implement both options in the receiver, recognizing whether CAP or QAM is being transmitted by analyzing the received constellation diagram, which will be rotating if the receiver does not match the transmitted signal. The details of this method are explained in [Oksman 2000].

In other applications, the link is started with a specifically selected symbol rate and center frequency, so that the same receiver can work for both CAP and QAM as described above. After startup, the transmitter informs the receiver whether CAP or QAM will be further transmitted, and the receiver reconfigures itself for either CAP or QAM operation.

6.5.5 Initialization

Initialization is a process to establish an SCM link. Generally, transceivers at the two sides of the loop cannot communicate until they both use the same set of transmission parameters. Those are at least the center frequencies, symbol rates, and constellation sizes. Because initially transceivers may have different setups, a special handshake procedure is first activated over the loop to deliver the desired transmission parameters from the master transceiver, located at the service originating side (usually CO or ONU), to the slave transceiver at the customer premises. Furthermore, the transceivers on both sides of the loop are trained, and the SCM link is established after echo-cancellers (if used) and equalizers at both sides converge, and data frames are synchronized. If transmission parameters have to be modified for some reason (to increase the noise margin, for instance, or in order to get a higher bit rate), a new set of transmission parameters is downloaded to the slave transceiver via a specially arranged management channel. After the new set of parameters is available, the transceivers on both sides of the line are trained with the new parameters, and the link is established again.

The handshake procedure may be initialized from either side of the loop, and uses a dedicated (default) set of transmission parameters, which is the same for all modems of the particular type. The settings for default parameters are usually picked to provide low attenuation of the handshake signal, but sufficiently wideband to cope with bridged taps, RF ingress, and other loop irregularities. Handshake protocols are designed to be extremely robust and usually include multiple retransmissions to improve reliability. After handshake, depending on the implementation, transceivers may use a special training sequence based on the relevant transmission parameters, or may be trained blindly or semi-blindly.

In some implementations, line probing is used to discover characteristics of the loop prior to link activation. Line probing is usually based on measurements of the attenuation of handshake signals in the upstream and downstream directions, although other signals with special settings of transmission parameters can be used additionally or instead.

6.5.6 Burst Mode SCM Transceivers

Burst mode operation was originally defined to support packet transport (IP and Ethernet). Perhaps the best-known system of this kind is Etherloop [Etherloop 1998], which uses QAM. In burst mode, the modem transmits in one direction at a time, only when a packet is sent (not continuously as a standard DSL does). This mode, also known as "ping-pong," has considerable advantages in the implementation complexity of the transceiver, because signals are not transmitted and received simultaneously. Consequently, all duplexing facilities such as the hybrid, echo canceller, and band-separating filters are unnecessary. The dynamic range of the receive signal decreases because no transmit echo is involved, which simplifies the A/D and D/A converters. Burst systems also benefit from the idle time, which is usually significant in packet-based systems (50 to 80 percent and more). Silence during the idle periods reduces power consumption and decreases the crosstalk in multi-pair cables. The drawbacks are potentially reduced efficiency due to the idle time, and nonstationary characteristics of the generated crosstalk. In burst systems using high frequencies, NEXT generated by the packet transmitted into one pair can completely suppress the received packet in another pair. This effect (also called "NEXT collision") is similar to collision in Ethernet LAN and requires re-transmission of the packet. Burst systems have not been standardized internationally, although they are successfully deployed in some regions, especially over very long loops where duplexing requires very high implementation accuracy.

Typically, a modem accumulates one or several data packets to be transmitted and encapsulates them into a frame (usually several thousand bytes long). The frame is prepended by an 8- to 16-byte preamble with pre-defined contents. This short preamble is still sufficient for the QAM receiver with a small constellation (typically, 16-QAM or smaller) to recover the timing and update the equalizer. During transmission of the frame, the other side is silent, accumulating data packets for the response. As the frame is received, the response is sent back. The communication protocol usually includes a "token back" frame to respond if there is no actual data to send.

By selection of a proper carrier frequency and symbol rate, the burst mode system is capable to allocate its spectrum and adjust the transmit power so that for the given loop

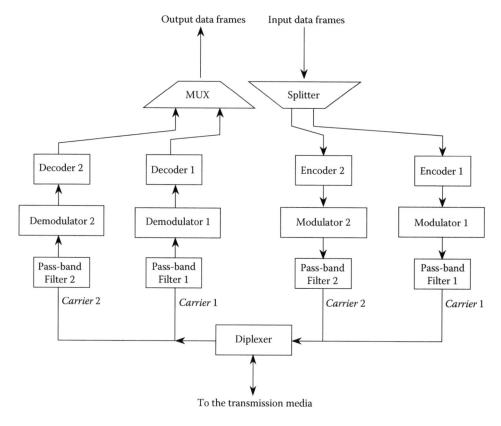

FIGURE 6.19
Functional diagram of an SCM VDSL transceiver using two carriers.

characteristics, the crosstalk to and from other DSL operating in the same binder can be minimized [Telcordia 1999].

6.5.7 Multi-Band SCM Transceivers

High-speed DSL instantiations, in particular VDSL, often use frequency division duplexing (FDD). With FDD, signals propagating in the upstream and downstream directions utilize distinct frequency bands, and NEXT between systems deployed in the same cable is avoided. Furthermore, a multi-band structure improves performance when loops of different lengths operate in the same cable, and is especially effective for deployment of high-speed symmetric services (with the same upstream and downstream bit rates). Currently a nonoverlapping band allocation has been adopted for VDSL systems.

SCM transceivers operating over multiple frequency bands combine several QAM (or CAP) transmitters and receivers, separated by bandpass filters, each intended for a particular band. A standard SCM VDSL1 transceiver, for instance, uses four frequency bands: two in the transmit direction and two in the receive direction. Accordingly, it includes two transmitters and two receivers, as shown in Figure 6.19.

6.5.7.1 Operation of the Transceiver

The four-band transceiver shown in Figure 6.19 operates in the following way. In the transmit direction, the input frame of data is split into two bit streams. Each stream is encoded, modulated, and sent onto the transmission line via the bandpass filter. The signal transmitted in a particular band is usually called a carrier. For the transceiver presented in Figure 6.19, either one or both carriers[7] (depending on the required data bit rate) can be transmitted in the upstream or downstream direction. At the receiver site, signals of both carriers are demodulated, decoded, and multiplexed into the output data frame.

The bandpass filters restrict the out-of-band power of the transmit signal in order to prevent crosstalk between the opposite transmission directions, and may also provide in-band spectral shaping, if necessary. The diplexer reduces the transmit signal echo in the receiver.

During operation, the transmitters and receivers of all carriers at both sides of the link are fully synchronized, independently of the transported data. The transceiver at the network node (CO or ONU) operates as a master: its transmitters are both synchronized to the master clock, which is usually derived from the local DSLAM. At the customer side of the loop, transmitters use the clock recovered from the received signal of either carrier (loop timing). A local clock source is often used as a backup if the clock recovery is lost, or when the transceiver at the customer site activates the link.

In VDSL1, the center frequencies and symbol rates of the transceivers are scalable, and both are multiples of the basic symbol rate (*BSR*):

$$\frac{1}{T} = s \cdot BSR \quad \text{and} \quad f_c = k \cdot BSR, \tag{6.40}$$

where s and k are integers. This simplifies clock recovery and allows the same phase-locked loop (PLL) to serve both carriers, although each receiver can be equipped to use its own PLL.

6.5.7.2 Frame Splitting and Recovering

The frame splitter disassembles the input data frame and composes a data frame for each of the transmitters involved (see Figure 6.19). This frame (called also the PMD-frame) helps to accommodate the difference in propagation delays between the carriers when transmitted over the line. Operation of a two-way frame splitter used in standard SCM VDSL1 is described in Figure 6.20. A similar splitting process can also be used if more than two carriers are involved.

The SCM VDSL1 input frame contains a 2-octet frame alignment header and a 403-octet payload. A PMD-frame has the same format. The splitting cycle in Figure 6.20 starts at the beginning of input frame #1. First, frame alignment octets (syncwords) are mapped into the PMD-frames of both carriers. Then the first N_1 data octets of the input frame #1 are mapped into carrier-1 and the N_2 following data octets are mapped into carrier-2. Repetition of this process forms the payload of the PMD-frame. In total, the splitting cycle involves $(N_1 + N_2)$ input frames, from which N_1 frames are mapped into carrier-1, and N_2 frames are mapped into carrier-2. Inverted syncwords inside the splitting cycle assist fast delineation of the PMD frame in the receiver.[8]

The two PMD-frames transmitted by two carriers have a data rate ratio of N_1/N_2. The particular values of N_1 and N_2 depend on the transmit data bit rates R_1, R_2 provided by each of the carriers, and may be calculated as $N_1 = R_1/GCD$ and $N_2 = R_2/GCD$, where GCD is the greater common divisor of the bit rates R_1 and R_2. The described splitting procedure works for any combination of integers N_1, N_2, however, some combinations may result in extremely long splitting cycles, which complicate frame delineation in the receiver. Those

---

[7] Both the frame splitter and the multiplexer are bypassed if only one of the carriers is used.
[8] The data transported over the link is usually randomized prior to transmission to reduce undesired impact of its internal correlation couplings on the PMD-frame delineation in the receiver.

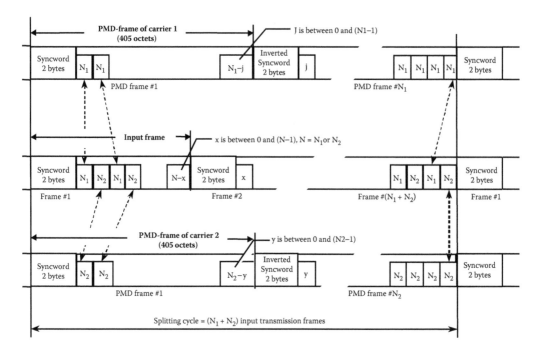

FIGURE 6.20
Composing PMD-frames from the input frame (SCM VDSL1, two carriers).

combinations are usually replaced with more convenient ones, despite a slight impact on the transmission bit rate granularity.

The multiplexer (MUX in Figure 6.19) combines the PMD-frames of the received carriers back into the original data frame format. Prior to multiplexing, PMD-frames of all carriers are aligned in time. The delay difference between the carriers is accommodated by appropriate buffering in the receiver. The absolute value of the delay difference can be estimated as

$$\Delta t = |m_1 \cdot T_1 - m_2 \cdot T_2| + \tau, \tag{6.41}$$

where τ is the difference in the propagation time over the line and analog front end, such as twisted pair line, POTS/ISDN splitters, etc., usually less than 5 μs; m_1, m_2 is the number of taps in the shaping and bandpass filters of each carrier, respectively; and T_1, T_2 are the symbol periods of carrier-1 and carrier-2, respectively. A typical value of Δt in standard SCM VDSL1 applications does not exceed 20–30 μs.

6.5.7.3 Transport Capability

The transport capability of a multi-band transceiver is the sum of transport capabilities over all utilized bands. In particular, the transmit bit rate of a two-band transceiver with symbol rates $1/T_1$, $1/T_2$ using 2^{M_1}-QAM, 2^{M_2}-QAM is:

$$R = M_1 \cdot \frac{1}{T_1} + M_2 \cdot \frac{1}{T_2}. \tag{6.42}$$

The minimum available bit rate (when 2-QAM is used for both carriers) equals $(\frac{1}{T_1} + \frac{1}{T_2})$, and the granularity of bit rate adjustment is either $R \cdot \min(M_1, M_2)$ (by symbol rate adjustment) or $\min(\frac{1}{T_1}, \frac{1}{T_2})$ (by constellation adjustment). Both types of adjustment allow to fit the required bit rate, unless its value is not too high for the particular loop and noise environment.

6.5.7.4 Constellation Assignment (Bit Loading)

When more than one carrier is used for transmission, there is always a question of how to distribute the data to be transmitted among the carriers. In other words, what should be the symbol rate and the constellation for each of the transmitters to maximize the throughput over the particular link? Currently there is no standard solution for this problem, although some rather simple proprietary algorithms are successfully used.

Generally, the optimization algorithm tries first to select the appropriate symbol rate. After this is complete, the best constellation assignment is one that provides the same value of noise margin for each receiver. One of the possibilities is similar to the bit loading technique described in Chapter 7: the constellation size is assigned in accordance with the SNR measured in the particular band.

The optimization algorithm usually requires one or two iterations during the link initialization. First a training signal with low constellation and maximum available symbol rate is applied to collect the SNR measurements in each band, and then the appropriate symbol rate and constellation size are selected. A semi-blind equalization described above facilitates this algorithm. Further, after optimized values are available, either semi-blind or blind equalization may be used to start the link.

For most loops (straight loops or loops with bridged taps, and crosstalk noise environment), selection of the symbol rate and constellation assignment is straightforward: it is usually beneficial to use the maximum available symbol rate, as was explained earlier, and set the constellation size appropriate for the SNR measured in the particular band.[9] In some special cases, however, selection of the symbol rate can be challenging and requires a detailed analysis of loop characteristics. Further, adjustments in both symbol rate and constellation size may be necessary. Those usually require several iterations during the link initialization and may not be possible when blind equalization is a requirement.

6.6 Annex I: Constellation Diagrams

This annex includes several square-shaped and cross-shaped constellation diagrams and mappings currently used in SCM and DMT VDSL1. Square-shaped and cross-shaped constellations are simple for implementation because the slicing levels are the same for both in-phase and quadrature components. Special constellation shapes, such as 8-QAM, for instance, are more efficient and provide performance advantages.

All mappings for SCM VDSL1, except for 2-QAM and 8-QAM, are built using the rules explained by Table 6.1 and Table 6.2 for the first quadrant; mappings in the second, third, and fourth quadrants are equal to the mapping in the first quadrant rotated counter-clockwise by 90, 180, and 270 degrees, respectively. (See Figures 6.21 to 6.26.) At the time of writing, most SCM transceivers use constellations that do not exceed 1024 points (10 bits per symbol). In some implementations (generally those targeting very short reach or utilizing low-frequency bands), constellations as large as 4096 (12 bits per symbol) may be useful. However, most SCM modems do not implement constellations this large because the very high required SNR to support 12 bits per symbol is not realistic in most of the access network; generally only short loops benefit. The latter follows from loop capacity analysis of DSL frequency allocations [Ungerboeck 2000].

---

[9] For straight loops the constellation size can also be easily predicted. Therefore, many SCM systems use predefined fixed transmission profiles.

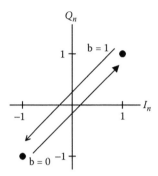

FIGURE 6.21
2-point constellation with differential bit encoding.

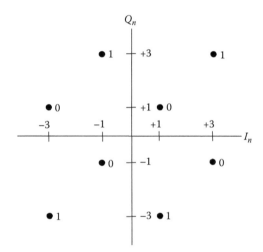

FIGURE 6.22
8-point constellation and bit mapping.

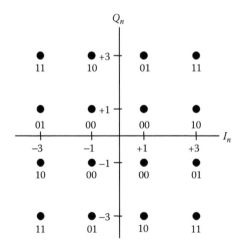

FIGURE 6.23
16-point constellation and bit mapping (first quadrant).

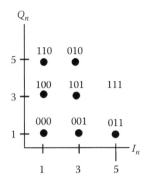

FIGURE 6.24
32-point constellation and bit mapping (first quadrant).

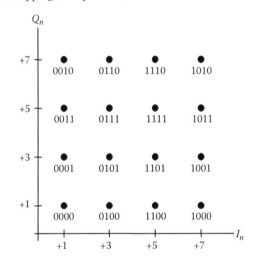

FIGURE 6.25
First quadrant of 64-point constellation and bit mapping.

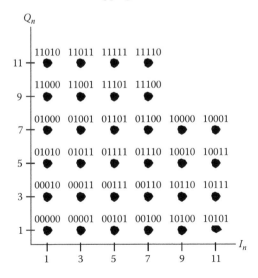

FIGURE 6.26
First quadrant of 128-point constellation and bit mapping.

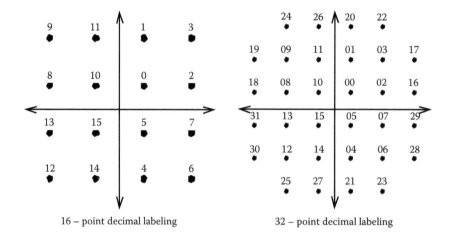

16 – point decimal labeling   32 – point decimal labeling

FIGURE 6.27
Examples of decimal labelling used in standard DMT constellations.

In DMT transceivers, due to the very narrow subchannel bandwidths, constellations up to 32,768 points (15 bits per symbol) are used in ADSL and VDSL, and even larger constellations are being considered. Similar square-shaped and cross-shaped constellations are mostly used. However, mappings for DMT are not rotation-invariant due to use of pilot tones (see Chapter 12). For convenience, decimal format is used for labels [G.992.3 2002]. Some examples of constellation diagrams used in DMT are shown in Figure 6.27.

References

[Etherloop 1998] Elastic Networks. *Etherloop*. Technology White Paper. January 1998.

[ETSI TS 101 270-2] ETSI Technical Specification 101 270-2, V 1.2.1. *Very-high-speed digital subscriber line (VDSL). Part 2: Transceiver specification.*

[Gitlin 1995] R.D. Gitlin, J.F. Hayes, and S.B. Weinstein. *Data communications principles.* Plenum Press, New York, 1992.

[G.989.1 2001] ITU-T Recommendation G.989.1 (2001). *Phoneline Networking Transceivers—Foundation.* 2001.

[G.991.1 1998] ITU-T Recommendation G.991.1 (1998). *High-speed digital subscriber line (HDSL) transceivers.* 1998.

[G.992.3 2002] ITU-T Recommendation G.992.3 (2002). *Asymmetric digital subscriber line (ADSL) transceivers.* 2002.

[G.993.1 2004] ITU-T Recommendation G.993.1 (2004). *Very high speed digital subscriber line.*

[Harman 1995] D.D. Harman, G. Huang, G.H. Im, M.H. Nguyen, J.J. Werner, and M.K. Wong. *Local distribution for interactive multimedia TV.* IEEE Multimedia Magazine, pp. 14–23, Fall 1995.

[Haykin 1998] S.S. Haykin. *Communications systems.* 4th edition. John Wiley & Sons, New York, 1998.

[Im 1995a] G.H. Im, D.D. Harman, G. Huang, A.V. Mandzik, M.H. Nguyen, and J.J. Werner, *51.84 Mb/s 16-CAP ATM LAN standard,.* IEEE J. Select. Areas Commun., Vol.13, No.4, pp. 620–632, May 1995.

[Im 1995b] G.H. Im and J.J. Werner. *Bandwidth-efficient digital transmission over unshielded twisted pair wiring.* IEEE JSAC, Vol. 13, No. 9, Dec. 1995, pp. 1643–55.

[Oksman 2000] V. Oksman and J.J. Werner. *Single-carrier modulation technology for very-high-speed subscriber line.* IEEE Communications Magazine, May 2000, pp. 82–89.

[Proakis 2001] J.G. Proakis. *Digital Communications.* Fourth edition. McGraw Hill Series, New York, 2001.

[Salz 1973] J. Salz. *Optimum mean-square decision feedback equalization.* The BSTJ, Vol. 52, No. 8, October 1973.

[Telcordia 1999] Telcordia Technologies. *Spectral compatibility and EtherLoop's Spectrum Manager.* Contribution T1E1.4/1999-191, Arlington, VA, April 1999.

[T1.TR.59 1999] ATIS Committee T1. *Single-Carrier Rate Adaptive Digital Subscriber Line (RADSL).* ATIS Committee T1 technical report T1.TR.59-1999.

[T1.TRQ-12 2004] ATIS Committee T1 technical requirements document T1.TRQ-12-2004. *Interface between Networks and Customer Installation; Very-high-speed digital subscriber lines (VDSL) metallic interface (QAM-based).*

[Ungerboeck 1987] G. Ungerboeck. *Trellis-coded modulation with redundant signal sets. Part I: Introduction; Part II: State of the art.* IEEE Commun. Mag., Vol. 25, No. 2, pp. 5–21, Feb. 1987.

[Ungerboeck 2000] G. Ungerboeck. *Spectral efficiency advantage of SCM over MCM.* Contribution ITU-T SG15/Q4 contribution FI-083, Fiji Islands, January 2000.

[Werner 1999] J.J. Werner, J. Yang, D.D. Harman, and G.A. Dumont. *Blind equalization for broadband access.* IEEE Communications Magazine, Vol. 37, No. 4, April 1999.

7

Fundamentals of Multi-Carrier Modulation

Krista S. Jacobsen

CONTENTS

7.1 Introduction 181
7.2 Basics of Multi-Carrier Modulation 182
7.3 Discrete Multi-Tone (DMT) 186
    7.3.1 Digital Duplexing 193
        7.3.1.1 Cyclic Suffix 194
        7.3.1.2 Timing Advance 196
        7.3.1.3 Cyclic Extension 196
    7.3.2 Peak-to-Average Ratio 197
        7.3.2.1 PAR Reduction Techniques 198
7.4 Initialization 200
    7.4.1 Activation 201
    7.4.2 Channel Discovery 201
    7.4.3 Transceiver Training 201
    7.4.4 Channel Analysis 202
        7.4.4.1 Channel Identification 202
        7.4.4.2 Noise Identification 203
        7.4.4.3 Bit Allocation 204
        7.4.4.4 Bit Allocation to Maximize the Bit Rate
                at a Target Noise Margin 204
        7.4.4.5 Bit Allocation to Maximize the Noise Margin
                at a Target Bit Rate 206
    7.4.5 Parameter Exchange 206
7.5 Steady-State Adaptation 206
    7.5.1 Bit Swapping 207
    7.5.2 FEQ Adaptation 208
7.6 Summary 208
7.7 Acknowledgments 209
References 209

ABSTRACT This chapter discusses the basics of multi-carrier modulation. The emphasis is on discrete multi-tone (DMT) modulation, which is the form of multi-carrier modulation that has been standardized for ADSL and VDSL.

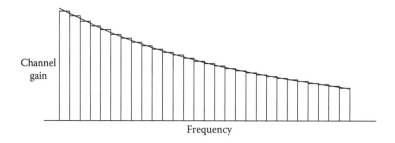

FIGURE 7.1
Partitioning of a channel bandwidth into subchannels.

7.1   Introduction

As described in Chapter 6, single-carrier systems support a constant spectral efficiency, in terms of bits per second per Hertz, within the bandwidth in which they operate. Neglecting excess bandwidth, a system that supports four bits per symbol, for instance, supports four bits per second per Hertz. The spectral efficiency of a single-carrier system does not follow the SNR variations over the channel bandwidth. Instead, the receiver aggregates signal power and noise over the entire bandwidth, and the overall SNR, which depends on the chosen demodulation and equalization methods, determines how many bits per symbol the channel can accommodate at the desired symbol error probability.

Multi-carrier systems take a different approach to the distribution of information bits in the available bandwidth. Unlike single-carrier systems, they do not typically transmit a constant number of bits across the bandwidth. Instead, multi-carrier systems partition the available bandwidth into a set of subchannels, and each subchannel supports a number of bits proportional to its SNR and the desired symbol error probability. Figure 7.1 illustrates the concept of partitioning a channel's bandwidth into subchannels, and Figure 7.2 shows how this partitioning results in a distribution of bits that varies with frequency based on the channel SNR. The "subsymbols" corresponding to the subchannels are aggregated during each symbol period to compose a multi-carrier symbol. In total, each multi-carrier symbol may support hundreds or even thousands of bits, but at a low symbol rate. Therefore, a multi-carrier system can be considered to be equivalent to many single-carrier systems operating in parallel, each at its particular carrier frequency (the subcarrier) and at a low symbol rate. (See Chapter 6 for details on single-carrier modulated systems.)

A key benefit of partitioning the channel into a set of parallel and independent subchannels is that transmission in noisy or highly attenuated regions of the channel can be avoided. Figure 7.3 illustrates the avoidance of low-SNR regions on a channel that can support transmission below 300 kHz and above 550 kHz, but not between 300 and 550 kHz. The subchannels in the low-SNR region of the bandwidth are simply turned off, whereas those in the regions with higher SNR support data transmission.

7.2   Basics of Multi-Carrier Modulation

This section presents a high-level overview of multi-carrier modulation. Readers interested in a more detailed discussion are referred to [Cioffi EE379c notes], which provides a thorough, mathematically rigorous treatment of multi-carrier modulation.

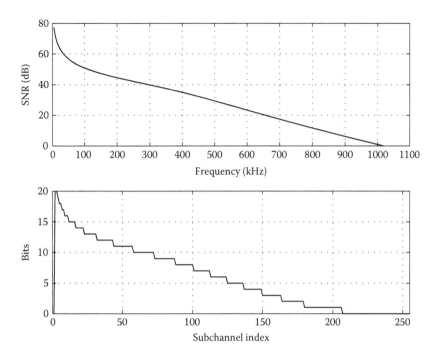

FIGURE 7.2
Multi-carrier systems support a variable number of bits across the frequency band.

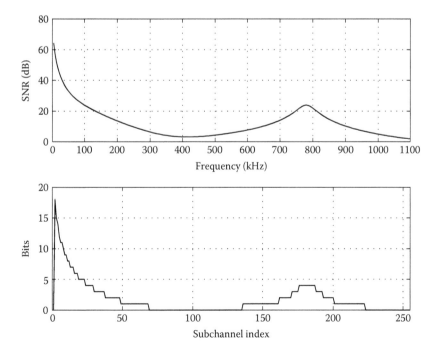

FIGURE 7.3
Multi-carrier systems avoid transmitting in low-SNR regions of the spectrum. (Calculation of bits per subchannel assumes 6 dB of noise margin and 5 dB of coding gain.)

A multi-carrier modulator partitions a channel into a set of $\overline{N}+1$ parallel subchannels[1] that are, ideally, independent. The subchannels are generated by a single transmitter and demodulated by a single receiver. Each subchannel has an associated subcarrier, and thus each subchannel can be considered an independent single-carrier channel that supports a number of bits per symbol, b_k, appropriate for its SNR and the desired symbol error probability, P_e. In standardized DSL systems that use multi-carrier modulation, the subchannel bandwidths are equal. If the subchannels are considered independent single-carrier modulated systems, having equal bandwidths means the subchannels operate at the same symbol rate.[2] In total, the channel supports a number of bits per symbol of

$$B = \sum_{k=1}^{\overline{N}-1} b_k,$$

assuming neither the subchannel at zero nor the one at the Nyquist frequency is used for transmission, which is the case in DSL. Denoting the symbol period as T and the symbol rate as $1/T$, the bit rate supported by the system is $R = B/T$.

To simplify equalization, multi-carrier modulation systems generally attempt to partition the channel bandwidth into a large enough number of subchannels that the channel frequency response is effectively flat across the bandwidth of each subchannel. Under such a condition, the Nyquist criterion is satisfied and intersymbol interference (ISI) from one symbol to the next is eliminated. The receiver can then demodulate the signal in a symbol-by-symbol manner, without complex equalization.

In practice, of course, the frequency response across the finite bandwidth of each subchannel will never be completely flat, and some amount of ISI will be unavoidable. Generally, the ISI is proportional to the subchannel bandwidth. Although multi-carrier systems can apply the same types of channel equalization techniques that are used in single-carrier modulation, Section 7.3 describes how the multi-carrier systems used in DSL eliminate (or at least dramatically reduce) ISI.

Furthermore, the partitioning of the channel into subchannels probably will not result in perfectly independent subchannels. Time/frequency duality dictates that partitioning into subchannels with "brick wall" characteristics in the frequency domain would require infinitely long symbols (and infinite processing delay) in the time domain. Therefore, practical partitioning methods strive to construct a set of subchannels that are mostly independent.

The partitioning of the channel into subchannels can be accomplished using any set of orthonormal basis functions. Mathematically, a set of functions $\{\varphi_i(t)\}$ represents an orthonormal basis if

$$\int_{-\infty}^{\infty} \varphi_m(t)\varphi_n(t)dt = \delta_{mn}, \text{ where } \delta_{mn} = \begin{cases} 1 & m = n \\ 0 & m \neq n \end{cases}. \tag{7.1}$$

---

[1] In a baseband multi-carrier system, one of the subchannels is located at zero, one is located at the Nyquist frequency, and the rest are passband subchannels. If the multi-carrier signal is a discrete-time signal (prior to digital-to-analog conversion and application to the channel), the spectrum of the signal is periodic, so the Nyquist-frequency subchannel also "appears" about zero. Therefore, the "DC" (zero frequency) and Nyquist-frequency subchannels can be considered to compose a single subchannel centered about zero frequency. For this reason, and because neither is used to support data in DSL systems, sometimes the DC and Nyquist subchannels are taken together as a single subchannel, and the total number of subchannels is given as $\overline{N}$ rather than $\overline{N}+1$. In the development of the explanation of the multi-carrier modulator and demodulator, it is assumed there are $\overline{N}+1$ subchannels. In the subsequent discussions of discrete multi-tone modulation, the DC and Nyquist subchannels are taken as one, and the total number of subchannels is assumed to be $\overline{N}$.

[2] If the subchannel bandwidths differ, the concept of "a symbol" no longer makes sense, because different subchannels will correspond to different symbol periods. In this chapter, it is assumed that all subchannels have equal bandwidths, which is the case in all standardized DSL systems that use multi-carrier modulation.

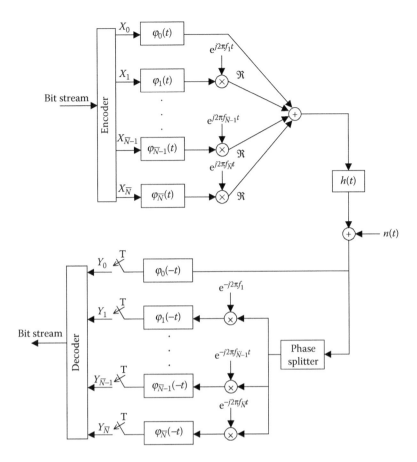

FIGURE 7.4
Block diagram of generic multi-carrier modulator and demodulator.

Figure 7.4 illustrates a generic multi-carrier modulator and demodulator. In the transmitter, the bit stream is converted from serial to parallel format, such that the input to the modulator can be considered a vector denoted

$$\mathbf{X} = \begin{bmatrix} X_{\overline{N}} \\ X_{\overline{N}-1} \\ \vdots \\ X_0 \end{bmatrix}, \tag{7.2}$$

where $\overline{N}+1$ is the number of subchannels into which the channel bandwidth is partitioned. The subsymbols X_0 and $X_{\overline{N}}$ are one-dimensional and real, and correspond, respectively, to the zero and Nyquist frequency subcarriers. The subsymbols X_1 through $X_{\overline{N}-1}$ are two-dimensional and correspond to the $\overline{N} - 1$ passband subchannels.

Each subsymbol X_k is modulated by a basis function $\varphi_k(t)$. For X_0 and $X_{\overline{N}}$, the modulation is one-dimensional, and the modulators can be considered to be pulse amplitude modulated (PAM). The rest of the subsymbols are quadrature amplitude modulated (QAM). The subcarrier frequencies for subchannels 1 through $\overline{N} - 1$ are $f_k = k \cdot \frac{f_s}{N}$, where f_s is the sampling rate of the system. It should be noted that the expression for f_k assumes a baseband (or baseband-equivalent) system.

In one possible multi-carrier receiver structure, a phase splitter is applied to decouple the quadrature components of the QAM subchannels, and then each subchannel is processed as a baseband signal using a matched filter and sampler combination. A maximum-likelihood detector, which is part of the decoder block in Figure 7.4, then operates independently on each subchannel to reconstruct the transmitted subsymbols.

If the subchannel bandwidths are infinitely narrow, which means the symbol period is infinite, then the output of the channel following the sampling process can be written in terms of Fourier transforms as

$$Y_k = X_k \cdot H_k + N_k, \tag{7.3}$$

where $H_k \triangleq H(k/T) = H(f)$ for $|f - n/T| < 1/2T$, $H(f)$ is the Fourier transform of the channel impulse response $h(t)$, and N_k is the value of the noise power spectral density (PSD) at the subcarrier frequency f_k. Equation 7.3 indicates that each subsymbol is scaled by the pulse response gain of its associated subchannel. Because each H_k is a complex scalar, the receiver can estimate the transmitted subsymbol X_k by executing a single complex multiply per subchannel.

The objective in multi-carrier modulation is to choose a set of basis functions that remain orthogonal following transmission through a channel with additive Gaussian noise, so that complicated equalization at the receiver can be avoided. The eigenfunctions of the channel,[3] which are independent "modes" that allow transmission without interference from other "modes" (see [Cioffi EE379c notes]), form a basis that has this property; however, using the channel eigenfunctions as the basis functions is computationally prohibitive for practical systems and could require infinite delay. Moreover, the eigenfunctions of different channels differ, so using the eigenfunctions as the basis would significantly complicate the design of a DSL modem intended to work on any telephone line. In effect, the most basic elements of the transceiver would have to be reconfigured for every line on which the modem operated. Therefore, suboptimal partitioning methods are used in practical multi-carrier implementations. It can be shown that system performance using these suboptimal methods approaches optimal performance when the number of subchannels into which the bandwidth is partitioned approaches infinity.

Several partitioning methods have been proposed for multi-carrier transmission. This chapter focuses on the technique that has been standardized for DSL: discrete multi-tone.

7.3 Discrete Multi-Tone (DMT)

Although the partitioning of a channel bandwidth into subchannels can be achieved using any orthonormal basis, one particular basis has found favor for DSL applications. That basis is the Fourier transform, and particularly the inverse discrete Fourier transform (IDFT). The use of the IDFT and discrete multi-tone modulation for DSL was originally described in [Bingham 1990] and [Cioffi 1991]. One key advantage of the IDFT is the availability of efficient computational methods, such as the fast Fourier transform (FFT). The FFT of an N-point sequence can be computed in $N \log_2(N)$ operations rather than the N^2 operations that would be required for most matrix multiplications, thus greatly reducing computational complexity. An additional advantage of using the IDFT for multi-carrier modulation is that the basis functions are fixed and independent of the channel.

---

[3] An eigenfunction of the channel is any function $\varphi_n(t)$ that, when convolved with the matched-filtered channel over a specific interval, reproduces itself scaled by a constant (the eigenvalue).

The discrete Fourier transform (DFT) of an N-dimensional sequence $\mathbf{x}$, where

$$\mathbf{x} \triangleq \begin{bmatrix} x_{N-1} \\ x_{N-2} \\ \vdots \\ x_0 \end{bmatrix}, \tag{7.4}$$

is given by

$$\mathbf{X} \triangleq \begin{bmatrix} X_{N-1} \\ X_{N-2} \\ \vdots \\ X_0 \end{bmatrix}, \tag{7.5}$$

where

$$X_k = \frac{1}{\sqrt{N}} \sum_{n=0}^{N-1} x_n \cdot e^{-j(2\pi/N)kn} \ \forall \, k \in [0, N-1]. \tag{7.6}$$

The IDFT of $\mathbf{X}$ is then given by

$$x_n = \frac{1}{\sqrt{N}} \sum_{k=0}^{N-1} X_k \cdot e^{j(2\pi/N)kn} \ \forall \, n \in [0, N-1]. \tag{7.7}$$

The DFT can be written in matrix form as $\mathbf{X} = Q\mathbf{x}$, where the matrix Q is defined as

$$Q = \frac{1}{\sqrt{N}} \begin{bmatrix} e^{-j(2\pi/N)(N-1)(N-1)} & e^{-j(2\pi/N)(N-2)(N-1)} & \cdots & e^{-j(2\pi/N)2(N-1)} & e^{-j(2\pi/N)(N-1)} & 1 \\ e^{-j(2\pi/N)(N-1)(N-2)} & e^{-j(2\pi/N)(N-2)(N-2)} & \cdots & e^{-j(2\pi/N)2(N-2)} & e^{-j(2\pi/N)(N-2)} & 1 \\ \vdots & \vdots & \ddots & \vdots & \vdots & \vdots \\ e^{-j(2\pi/N)(N-1)2} & e^{-j(2\pi/N)(N-2)2} & \cdots & e^{-j(2\pi/N)4} & e^{-j(2\pi/N)2} & 1 \\ e^{-j(2\pi/N)(N-1)} & e^{-j(2\pi/N)(N-2)} & \cdots & e^{-j(2\pi/N)2} & e^{-j(2\pi/N)} & 1 \\ 1 & 1 & \cdots & 1 & 1 & 1 \end{bmatrix}. \tag{7.8}$$

Defining the matrix Q^* as

$$Q^* = \frac{1}{\sqrt{N}} \begin{bmatrix} e^{j(2\pi/N)(N-1)(N-1)} & e^{j(2\pi/N)(N-2)(N-1)} & \cdots & e^{j(2\pi/N)2(N-1)} & e^{j(2\pi/N)(N-1)} & 1 \\ e^{j(2\pi/N)(N-1)(N-2)} & e^{j(2\pi/N)(N-2)(N-2)} & \cdots & e^{j(2\pi/N)2(N-2)} & e^{j(2\pi/N)(N-2)} & 1 \\ \vdots & \vdots & \ddots & \vdots & \vdots & \vdots \\ e^{j(2\pi/N)(N-1)2} & e^{j(2\pi/N)(N-2)2} & \cdots & e^{j(2\pi/N)4} & e^{j(2\pi/N)2} & 1 \\ e^{j(2\pi/N)(N-1)} & e^{j(2\pi/N)(N-2)} & \cdots & e^{j(2\pi/N)2} & e^{j(2\pi/N)} & 1 \\ 1 & 1 & \cdots & 1 & 1 & 1 \end{bmatrix}, \tag{7.9}$$

the IDFT can be written as $\mathbf{x} = Q^*\mathbf{X}$.

Note that each entry of Q and Q^* can be written in the form $\cos(x) + j\sin(x)$, which means the DFT and IDFT are sums of sinusoids.

The length-N (where $N = 2\overline{N}$) sequence $\mathbf{X}$ has the property of Hermitian symmetry if it can be written in terms of a set of entries X_k, $k \in [1, \overline{N}]$, such that

$$X_{k,Herm} = \begin{cases} Re\{X_{\overline{N}}\} & k = 0 \\ X_k & k = 1, \dots, \overline{N} - 1 \\ Im\{X_{\overline{N}}\} & k = \overline{N} \\ X_{N-k}^* & k = \overline{N} + 1, \dots, N - 1 \end{cases}, \tag{7.10}$$

where * denotes complex conjugation. Similarly, given an initial complex sequence $\mathbf{Z}$, a Hermitian symmetric sequence $\mathbf{Z_h}$ is easily constructed using Equation 8.10.

In the DMT transmitter, when the input sequence $\mathbf{X}$ is Hermitian symmetric, the output following application of the IDFT will be real. In practice, this property means an N-point (where $N = 2\overline{N}$) complex-to-real IDFT can be used at the DMT transmitter to generate the $\overline{N}$ subchannels so the output of the transform is real and can be applied directly to the channel after digital-to-analog conversion.

Although the objective in multi-carrier transmission is to partition the channel bandwidth into independent subchannels, in a practical system with reasonable processing delay, the subchannels will overlap. In DMT, the subchannels overlap in such a way that they remain orthogonal at the subcarrier frequencies.

Each symbol the DMT transmitter applies to the channel, where each symbol is the result of an IDFT operation, can be considered to be windowed in the time domain by a rectangular pulse, which is caused by the finite duration of each symbol. Denoting the subcarrier spacing as Δf, the nth DMT symbol is the sum of components that can be written in the time domain in the form of

$$x_{n,k}(t) = \left[X_k \cdot e^{j2\pi \Delta f k t} + X_k^* \cdot e^{-j2\pi \Delta f k t} \right] \cdot w(t), \tag{7.11}$$

where $x_{n,k}(t)$ represents the components of the nth symbol due to the kth subchannel. The rectangular window is defined as

$$w(t) = \begin{cases} 1 & t \in \left(0, \dfrac{1}{\Delta f}\right) \\ 0 & t \notin \left(0, \dfrac{1}{\Delta f}\right) \end{cases}. \tag{7.12}$$

The Fourier transform of $w(t)$ is

$$W(f) = \mathrm{sinc}\left(\frac{f}{\Delta f}\right), \tag{7.13}$$

which is a sinc function with its peak at 0 Hz and zeros at multiples of Δf, as shown in Figure 7.5.

Because multiplication in time corresponds to convolution in frequency, and because $e^{j2\pi \Delta f k t}$ can be written as $\cos(2\pi \Delta f k t) + j \sin(2\pi \Delta f k t)$, the Fourier transform of $x_{n,k}(t)$ is the convolution of signals of the form $X_k \cdot \delta(f - \Delta f)$ and $W(f)$. (If Hermitian symmetry is imposed, the sin component of $e^{j2\pi \Delta f k t}$ is eliminated.) This convolution simply corresponds to copies of $W(f)$ that are centered at multiples of Δf and scaled by the X_k corresponding to the subchannels. In ADSL and DMT-based VDSL1, the copies of $W(f)$ appear at multiples of 4.3125 kHz. (See [G.992.1], [G.992.3], [G.992.5], and [T1.424].)

Note that for any selected copy of $W(f)$, its value at any other integer multiple of Δf is zero due to the properties of the sinc function. Therefore, at any subcarrier frequency, the value of the aggregate signal—the sum of all the sinc functions corresponding to the subchannels—is due only to the signal on that subchannel. Thus, the signal at any multiple of Δf is independent of all other subchannels.

The DMT transmitter can be viewed in a less mathematical manner as follows. The bit stream to be transmitted can be considered to be a frequency-domain variable. The number of bits each subchannel can support has been determined by some bit distribution algorithm (see Subsection 7.4.4.3), so the total number of bits per symbol is known. The subchannels at zero and the Nyquist frequency are not used in practice, so each of the other $\overline{N} - 1$ subchannels supports an independent QAM constellation that represents b_k bits. During each symbol period, a sequence of B bits is assigned to the subchannels. The assigned bits

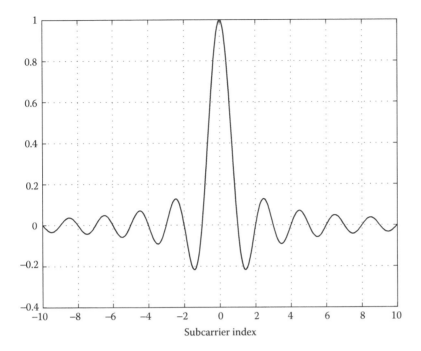

FIGURE 7.5
Result of rectangular windowing of symbols is sinc functions at the subcarrier frequencies. Zero crossings correspond to indices of other subcarriers.

are mapped, using the appropriate constellation diagram, to constellation points, which then become the X_k input to an IDFT. The result of the IDFT yields a composite signal whose frequency spectrum is a sum of sinc functions that are centered at integer multiples of the subchannel spacing (the subcarrier frequencies), with zeros at all other subcarrier frequencies. Figure 7.6 shows a few of the component sinc functions that are then scaled by the X_k and added to form the composite signal. In the time-domain, the block of samples representing B bits is converted from digital to analog format and applied to the channel. Figure 7.7 illustrates the DMT transmitter.

Ideally, $\overline{N}$ is sufficiently large that the subchannels are memoryless; that is, ISI is negligible. In practice, however, the frequency response across any selected subchannel is not likely to be perfectly flat, so ISI is not completely eliminated by the partitioning process. Instead, if the noise on each subchannel is assumed to be white and Gaussian, intersymbol and inter-subchannel interference caused by the channel's impulse response length being greater than one can be eliminated by use of a *cyclic prefix*. The cyclic prefix is a copy of the last v time-domain samples (that is, the samples after the IDFT has been applied) of each DMT symbol, as shown in Figure 7.8. As its name implies, the transmitter prepends the cyclic prefix to each symbol. If the channel impulse response length is no more than $v + 1$ samples in duration (at the selected sampling rate), then ISI caused by each symbol is confined to the cyclic prefix of the following symbol. Consequently, by discarding the cyclic prefix samples in the receiver prior to demodulation using the DFT, ISI can be eliminated completely.

Following the addition of the cyclic prefix, each symbol contains $N + v$ samples. Clearly, the cyclic prefix carries redundant information and is transmission overhead. From an efficiency standpoint, therefore, it is desirable to minimize its duration. However, the duration of the cyclic prefix required to eliminate ISI is a function of the channel impulse response duration, which is a function not only of the loop length, but also of whether the loop is of

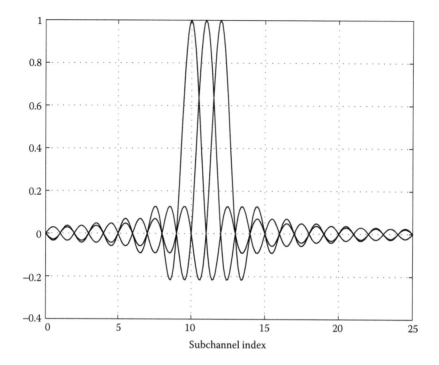

FIGURE 7.6
Individual sinc functions that are then multiplied by the subchannel gains and summed to form the composite signal.

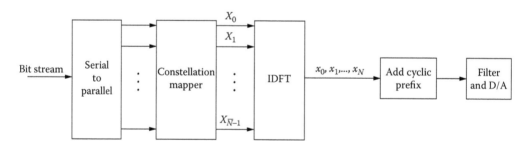

FIGURE 7.7
DMT transmitter block diagram.

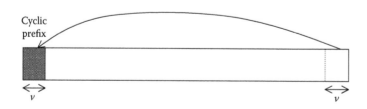

FIGURE 7.8
Illustration of cyclic prefix.

a configuration that causes signal reflections, such as one with bridged taps. The required length cannot be known precisely a priori. One could choose to design an efficient system by choosing a "safe" cyclic prefix length and then partitioning the channel into an enormous number of subchannels by using a very large IDFT. However, the system complexity, in terms of memory requirements, would increase due to the need for very large (I)DFTs at the transmitter and receiver. Furthermore, the latency of the system would increase, because the subchannels would be narrow, and the symbol rate would be very low. (However, each symbol would carry a lot of bits!)

In real-world DMT systems, the cyclic prefix length and (I)DFT size are selected so the system efficiency is over 90 percent. (See [G.992.1], [G.992.3], [G.992.5], and [T1.424], for example.) To compensate for residual ISI that occurs if the channel impulse response duration exceeds the chosen cyclic prefix duration, a time-domain equalizer (TEQ) may be used in the receiver. The TEQ is designed to shorten the channel impulse response length to within the cyclic prefix duration. Chapter 11 describes the TEQ in detail.

It is worth noting that the presence of the cyclic prefix results in the symbol rate of a DMT system not being the same as the subcarrier spacing. Given a critically sampled system with sampling rate f_s equal to twice the Nyquist frequency, the subcarrier spacing is calculated as $\Delta f = \frac{f_s}{N}$. In contrast, the rate at which data-carrying symbols are transmitted is $\frac{1}{T} = \frac{f_s}{N+\nu}$, which excludes the "excess time" required to transmit the cyclic prefix. Thus, $\Delta f > \frac{1}{T}$.

In DMT VDSL1 [T1.424], a valid combination of system parameters is a sampling rate of 35.328 MHz with a DFT size of 8192. Consequently, the subcarrier spacing is $\Delta f = 4.3125$ kHz. The cyclic extension[4] corresponding to this combination of sampling rate and DFT size is 640 samples in duration. Therefore, the symbol rate is $\frac{1}{T} = \frac{35,328}{8192+640} = 4$ kHz. The reader can easily verify that the other valid combinations of the VDSL1 sampling frequency, DFT size, and cyclic extension duration also yield 4.3125 kHz subcarrier spacing and a 4 kHz symbol rate [T1.424].

In ADSL [G.992.1] [T1.413] and ADSL2 [G.992.3], the value of N in the downstream direction is 512, the cyclic prefix duration is 32 samples, and the sampling rate is 2208 kHz. As in VDSL1, the subcarrier spacing is 4.3125 kHz. However, the calculation of the data symbol rate is not as straightforward as in VDSL. To aid ADSL receiver synchronization, a special symbol known as the synchronization symbol is transmitted after every 68 data symbols. Because this symbol does not transport any of the data stream, it is additional overhead, and its presence must be taken into account when the ADSL data symbol rate is computed. Each ADSL symbol, including the synchronization symbol, is 544 samples long. In calculating the ADSL data symbol rate, the impact of the synchronization symbol can be considered to be distributed among the 68 data-carrying symbols it follows. As a result, the synchronization symbol adds $544/68 = 8$ samples of additional overhead to each data symbol. Therefore, the data symbol rate is calculated as $\frac{1}{T} = \frac{2208}{512+32+8} = 4$ kHz, which is the same as in VDSL1. ADSL2plus [G.992.5] maintains the same subcarrier spacing and data symbol rate as in ADSL (and ADSL2) but defines 512 subchannels in the downstream direction.

That the subcarrier spacing and symbol rate are the same in ADSL(2)(plus) and VDSL is not a coincidence. The 4.3125 kHz subcarrier spacing and 4 kHz data symbol rate were chosen in VDSL1 to facilitate the VDSL1 modem entering a mode in which it can establish a connection with an ADSL modem. VDSL2 is expected to exploit this feature and enable the convergence of ADSL and VDSL hardware.

In both ADSL and VDSL, the cyclic prefix (and, in the case of VDSL, the cyclic suffix; see Subsection 7.3.1) results in an overhead penalty of approximately 8 percent. Often, the excess time associated with the cyclic prefix in a DMT system is said to be analogous to

---

[4] The cyclic extension, which is composed of a cyclic prefix and a cyclic suffix, is unique to DMT VDSL and is described later in the chapter.

the excess bandwidth of a single-carrier system, as both represent a fundamental system overhead.

One might wonder why the last v data samples are used as the "pad" between DMT symbols. After all, the (presumed corrupted) samples corresponding to the cyclic prefix are discarded in the receiver prior to demodulation, so zeros, or even random samples, could be used to eliminate ISI. In fact, the cyclic prefix serves another purpose. It is well known that with continuous-time signals, convolution in the time domain corresponds to multiplication in the frequency domain. Thus, the output of a channel can be determined either by performing a convolution of the time-domain input signal and the channel impulse response, $h(t)$, or by multiplying the Fourier transform of the signal by the channel frequency response and computing the inverse Fourier transform of the result.

In discrete time, the product of the DFTs of two finite-length sequences corresponds to the circular convolution of the two sequences. (Recall that in circular convolution, one of the sequences is time reversed and circularly shifted with respect to the other sequence.) Therefore, to take advantage of the relation that multiplication in frequency corresponds to convolution in time, it is necessary in the DMT system to force periodicity on one of the time-domain sequences. Clearly, because a modem has no control over the channel impulse response, the only practical option is to force the input sequence to appear periodic over the length of the convolution.

To illustrate, consider a discrete-time channel with four samples in its impulse response: $\mathbf{h} = [h_0\ h_1\ h_2\ h_3]$. Assume an 8-point complex-to-real IDFT is used in the DMT transmitter to partition the channel bandwidth into four subchannels.[5] Because the channel impulse response constraint length (that is, the impulse response length in samples minus one sample) is 3 samples, the required duration of the cyclic prefix is also 3. For ease of explanation, it is assumed the channel is noiseless.[6] Under these assumptions, the output of the channel is the linear, discrete-time convolution of the channel impulse response and the cyclically prefixed input sequence, which can be written as:

$$\mathbf{y} = \mathbf{x} \star \mathbf{h} = [\cdots\ \tilde{x}_5\ \tilde{x}_6\ \tilde{x}_7\ x_0\ x_1\ x_2\ x_3\ x_4\ x_5\ x_6\ x_7\ \cdots] \star [h_0\ h_1\ h_2\ h_3],$$

where $\tilde{x}_i$ denotes the ith sample of the previous symbol and $\star$ represents convolution. Denoting the received samples corresponding to the cyclic prefix samples as y_{-3}, y_{-2}, and y_{-1}, the convolution of the channel and the input sequence can be written using matrices and vectors as

$$\begin{bmatrix} y_{-3} \\ y_{-2} \\ y_{-1} \\ y_0 \\ y_1 \\ y_2 \\ y_3 \\ y_4 \\ y_5 \\ y_6 \\ y_7 \end{bmatrix} = \begin{bmatrix} h_3 & h_2 & h_1 & 0 & 0 & 0 & 0 & 0 & h_0 & 0 & 0 \\ 0 & h_3 & h_2 & 0 & 0 & 0 & 0 & 0 & h_1 & h_0 & 0 \\ 0 & 0 & h_3 & 0 & 0 & 0 & 0 & 0 & h_2 & h_1 & h_0 \\ 0 & 0 & 0 & h_0 & 0 & 0 & 0 & 0 & h_3 & h_2 & h_1 \\ 0 & 0 & 0 & h_1 & h_0 & 0 & 0 & 0 & 0 & h_3 & h_2 \\ 0 & 0 & 0 & h_2 & h_1 & h_0 & 0 & 0 & 0 & 0 & h_3 \\ 0 & 0 & 0 & h_3 & h_2 & h_1 & h_0 & 0 & 0 & 0 & 0 \\ 0 & 0 & 0 & 0 & h_3 & h_2 & h_1 & h_0 & 0 & 0 & 0 \\ 0 & 0 & 0 & 0 & 0 & h_3 & h_2 & h_1 & h_0 & 0 & 0 \\ 0 & 0 & 0 & 0 & 0 & 0 & h_3 & h_2 & h_1 & h_0 & 0 \\ 0 & 0 & 0 & 0 & 0 & 0 & 0 & h_3 & h_2 & h_1 & h_0 \end{bmatrix} \cdot \begin{bmatrix} \tilde{x}_5 \\ \tilde{x}_6 \\ \tilde{x}_7 \\ x_0 \\ x_1 \\ x_2 \\ x_3 \\ x_4 \\ x_5 \\ x_6 \\ x_7 \end{bmatrix}. \qquad (7.14)$$

---

[5] Such a system probably would not be designed in reality because it would be too inefficient. However, for the purpose of showing the effect of the cyclic prefix, this system is adequate.
[6] If noise is not neglected, and that noise is Gaussian, the analysis and conclusions do not change, but the notation becomes more cumbersome.

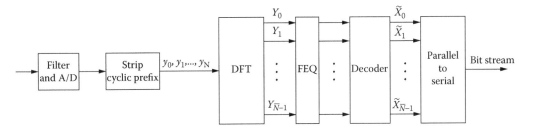

FIGURE 7.9
DMT receiver block diagram.

One can verify by inspection of Equation 7.14 that although the samples y_{-3}, y_{-2}, and y_{-1} have components due to samples from the previous symbol, the samples $[y_0, y_1, y_2, \ldots, y_7]$ are dependent only on the input samples of the current symbol $[x_0, x_1, x_2, \ldots, x_7]$. Thus, the last N samples of the signal emerging from the channel are exactly what they would have been if the input signal had been truly periodic. Therefore, use of the cyclic prefix ensures that what is actually a linear convolution appears to be circular in the period of interest to the receiver.

Because the subchannels are independent and are transmitted at different subcarrier frequencies, recovery of the input subsymbols is straightforward. Figure 7.9 shows a block diagram of the DMT receiver. Following analog-to-digital conversion, the cyclic prefix, which is corrupted by ISI from the previous symbol, is discarded. The receiver then takes an N-point real-to-complex DFT of the received sampled symbol, which converts the signal back to the frequency domain. If the channel is assumed to be noiseless, denoting the samples after the DFT as $Y_k = X_k \cdot H_k$, the input points of X_k can be recovered exactly simply by dividing each of the Y_k by the corresponding H_k. If the channel is not noiseless, then the points recovered through division by H_k do not correspond precisely to constellation points. Instead, they are noisy points that lie within the constellation diagram.

In an implementation, typically division is not convenient for signal processing hardware. Consequently, Y_k is scaled and rotated to remove the effect of H_k; this operation can be achieved using a single-tap filter. The set of taps, one per subchannel, is known as the frequency-domain equalizer (FEQ). Application of the FEQ requires one complex multiply per subchannel per DMT symbol, which is a negligible implementation complexity relative to the multi-tap equalization that might be required if ISI were not confined to the cyclic prefix.

When the channel is not noiseless, in the absence of coding, the noisy received points are recovered using a simple decision process (symbol-by-symbol detection), as described in Chapter 6. If trellis encoding is used, then a trellis decoder is needed in the receiver to reconstruct the transmitted bit stream, as discussed in Chapter 8.

7.3.1 Digital Duplexing

One of the benefits of DMT is the flexibility that results from partitioning the channel into a large number of subchannels. If a modem can generate and receive a full set of sub-channels (*i.e.*, spanning the entire available bandwidth) in both the upstream and downstream directions, then by allocating disjoint sets of subchannels to the downstream and upstream directions, a wide variety of frequency plans (*i.e.*, allocations of frequency bands to the downstream and upstream directions) can be accommodated by a single modem pair. For example, different regional frequency plans can be supported by a single modem. Furthermore, the complexity of DSL modems can be reduced if band separation filters are not required to mitigate interference (typically echo) from transmitted signals to received signals.

However, simply allocating disjoint sets of subchannels to the two directions does not eliminate the need for filters. Unsynchronized transmissions on adjacent and nearby subchannels have sidelobes, which are caused by the rectangular windowing (see Figure 7.5), that can interfere with received subchannels. If adjacent subchannels support transmission in opposite directions, it is necessary to synchronize the symbol periods of the transmitted and received signals so the zeros of the sinc functions of the transmitted signal are coincident, in frequency, with the zeros of the sinc functions of the received signal. Therefore, some clever construction is necessary to fully eliminate the need for band separation filters.

Several techniques to synchronize the transmitted and received symbols are described in [Mestdagh 2000] and [Sjoberg 1999]. Instantiations of DMT that use these enhancements are sometimes called *digitally duplexed* DMT or *Zipper*.[7] This section overviews the method; interested readers are referred to [Mestdagh 2000] and [Sjoberg 1999] for additional details on Zipper.

If the two DMT modems on either end of a line somehow synchronize the transmitted and received signals such that their symbol boundaries are aligned, then the received subchannels will be unaffected by the transmitted subchannels. In other words, in the frequency domain, the zeros of the sinc functions of the transmitted signal will all lie at all other subchannels' center frequencies at the sampling instants, so there is no interference with the received subchannels. Under this condition, bidirectional transmission is possible without the use of filters to separate transmit and receive bands. Furthermore, when groups of adjacent subchannels are allocated to the downstream and upstream directions, the need for large guard bands between bands used in opposite transmission directions is eliminated.

The purpose of the *cyclic suffix*, which is used in VDSL, is to allow the transmitted and received symbol boundaries to be aligned at both ends of the line. A *timing advance* can be applied to reduce the required length of the cyclic suffix and thus minimize the additional overhead required to support digital duplexing. The following subsections explain these concepts.

7.3.1.1 Cyclic Suffix

The cyclic suffix is the same idea as the cyclic prefix. However, as its name implies, the cyclic suffix is a copy of some number of samples from the beginning of the symbol, and it is appended to the end of the symbol, as illustrated in Figure 7.10. Initially, let the length of the cyclic suffix be denoted as 2Δ, where Δ is the phase (propagation) delay of the channel. To enable digital duplexing, the sum of the cyclic prefix and cyclic suffix[8] durations must be large enough that the transmitted and received symbol boundaries can be aligned at both ends of the line [Mestdagh 2000]. It is assumed here that the cyclic prefix is not overdimensioned and is thus entirely corrupted by ISI. Therefore, the symbols are properly aligned if there are, at the same time, sets of N samples in both the transmitted and received data streams that do not include any part of the cyclic prefix.

---

[7] The technique was nicknamed "Zipper" because, ideally, it is possible to allocate alternating subchannels to the two transmission directions. In such a configuration, if the transmitted power spectra are the same in both directions, and the noise spectra are the same in both directions, approximately symmetrical transmission is supported on every line, regardless of length. This desirable behavior is generally not achievable through any other means without causing spectral compatibility problems, unless time-division duplexing is used.

[8] The cyclic suffix of symbol i is immediately followed by the cyclic prefix of symbol $i + 1$. Clearly the cyclic prefix and cyclic suffix of a particular symbol are not adjacent. Nevertheless, combination of the cyclic prefix and cyclic suffix is often referred to as the *cyclic extension*.

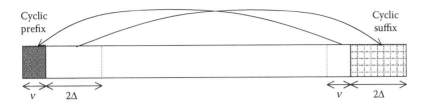

FIGURE 7.10
Illustration of cyclic suffix.

Alignment at one end of the line is easily achieved simply by observing the symbol boundaries of received symbols, and then transmitting symbols such that their boundaries are aligned with the received symbol boundaries. However, achieving alignment at one end of the line generally results in misalignment at the other. Figure 7.11 illustrates the simple case of a system with a DFT size of 12 (*i.e.*, 6 subchannels) and a channel with phase delay Δ of 2 samples and a constraint length v of 3 samples. Including the cyclic prefix, the length of each transmitted symbol is 15 samples. (With 20 percent overhead due to the cyclic prefix alone, this system is not very efficient, but it is only for illustration purposes.) Note that to achieve alignment of the transmitted and received symbol boundaries on the left-hand side of the figure, the transmitter on the right-hand side has to advance transmission of its symbols by 2 samples. This action results in significant misalignment of the transmitted and received symbol boundaries on the right-hand side, because the valid transmit symbol overlaps the cyclic prefix of the received symbol. In addition, when more than one symbol period is considered, it is clear that the valid received symbol will overlap the cyclic prefix of the subsequent transmit symbol.

Now assume a cyclic suffix of length 4 samples (corresponding to twice the phase delay of the channel) has been appended to all symbols so each transmitted symbol is 19 samples in duration. Figure 7.12 illustrates the result of using the cyclic suffix. With the cyclic suffix, when the transmitted and received symbol boundaries are perfectly aligned on the left-hand side, there are 12 consecutive samples (from 5 through 16) in both the transmitted and received symbols on the right-hand side that are not corrupted by ISI. Note that the transmitted symbol corresponding to samples 5 through 16 is a shifted version of the original symbol. However, because a shift in time corresponds to a rotation in frequency, the receiver on the left-hand side will simply use FEQ taps that are correspondingly rotated. The rotation of the FEQ taps happens naturally when the modems are initialized, because the cyclic suffix is always appended. The receiver cannot and need not distinguish between the shifted symbol and the original symbol.

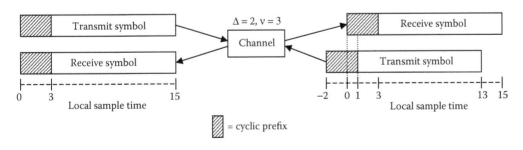

FIGURE 7.11
Without a cyclic suffix, alignment of transmitted and received symbols at one end of the line results in misalignment at the other end of the line.

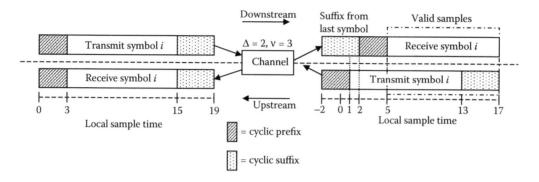

FIGURE 7.12
Addition of the cyclic suffix allows alignment of transmitted and received symbol boundaries at both ends of the line.

As the example illustrates, use of the cyclic suffix allows the transmitted and received symbol boundaries to be aligned, thereby eliminating interference to the received signal caused by the transmitted signal [Mestdagh 2000]. The penalty for eliminating the interference is the addition of the cyclic suffix, which, as with the cyclic prefix, results in overhead. One method to minimize the required length of the cyclic suffix is the timing advance.

7.3.1.2   *Timing Advance*

Based on the example in the previous section, dimensioning the cyclic suffix so that its duration is twice the phase delay of the channel facilitates achieving the desired condition of time-synchronized valid transmit and receive symbols at both ends of the line. However, as with the cyclic prefix, the cyclic suffix is redundant information and thus results in a bit rate penalty. (In reality, the penalty is not as severe as the example would suggest.) Therefore, it is desirable to minimize the length of the cyclic suffix.

Referring again to Figure 7.12, it is clear that with a cyclic suffix of duration twice the channel phase delay, there are actually several sets of valid transmit and receive symbols at the modem on the left-hand side. (Recall that, unlike the cyclic prefix, the cyclic suffix is not corrupted by ISI.) Any contiguous set of 12 samples starting from sample 3, 4, 5, 6, 7, or 8 constitutes a valid symbol in both the transmit and receive directions on the left-hand side of the line. To minimize overhead, it is desirable to reduce the number of sets of valid symbols at both ends of the line to exactly one.

The *timing advance* provides a way to achieve this objective [Sjoberg 1999]. Rather than precisely synchronizing the boundaries of the transmitted and received symbols on the left-hand side after appending the cyclic suffix, the transmitted symbol is advanced by a number of samples equal to the phase delay of the channel, Δ. The cyclic suffix length can then be halved, as Figure 7.13 illustrates. When the transmitted symbol on the left-hand side is advanced by 2 samples, a single set of valid transmit and receive samples, from 3 through 14 inclusive, results on the left-hand side. On the right-hand side, because the cyclic suffix length has been halved, the valid samples are now also 3 through 14. Through use of the cyclic suffix and timing advance, the desired condition of aligned transmit and receive symbol boundaries is achieved with minimum additional overhead, which yields an efficient system that does not require filters to separate the transmit and receive bands.

7.3.1.3   *Cyclic Extension*

When discussing digital duplexing, often the cyclic prefix and cyclic suffix samples together are referred to as the *cyclic extension*. The length of the cyclic extension is often denoted as N_{CE}. In some systems, such as VDSL, the cyclic extension is over-dimensioned so that

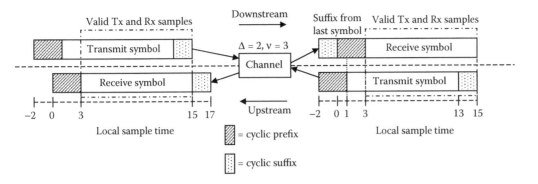

FIGURE 7.13
Use of timing advance allows the required length of the cyclic suffix to be halved.

its duration is longer than the sum of the channel constraint length and the propagation delay. Some of the cyclic extension samples are then windowed to reduce the subchannel sidelobes, which decreases the susceptibility of the system to radio-frequency interference (RFI). Chapter 13 discusses in detail the use of windowing to improve the rejection of RFI in DMT systems.

7.3.2 Peak-to-Average Ratio

A DMT system can be considered to be an aggregation of narrow-band QAM systems. One issue that must be considered in modem design is how the summation of a large number of QAM signals influences the power requirements. If the data stream is such that each subcarrier transmits a subsymbol farthest from the origin of the constellation diagram, then the aggregate transmit power will be at its peak. On average, of course, the power requirements will be much lower than the peak value, but the peak value must be taken into account when modems are designed.

The peak-to-average ratio (PAR) of a signal is defined as $\frac{V_{peak}}{V_{RMS}}$, where V_{peak} is the peak voltage of the signal, and V_{RMS} is the root-mean-square (RMS) voltage. Typically, the PAR is expressed as a power ratio in dB as $20 \log_{10}(\frac{V_{peak}}{V_{RMS}})$. The PAR of a sinusoid is easily derived as $\sqrt{2}$, or 3 dB.

The PAR of a signal is important because it has an impact on the required precision of various components in a modem, including the digital-to-analog and analog-to-digital converters and the signal processing circuitry. Signals with higher PAR require components with higher precision, and they also require analog components, such as amplifiers, filters, and transformers, to remain linear over a wide range of input power levels. Furthermore, a system with a high PAR requires that the line driver be designed to accommodate the peak power. All these requirements can increase component cost and power consumption.

When the amplitude of a signal exceeds the maximum signal level the implementation has been designed to accommodate, the signal is "clipped" by the digital-to-analog converter. Clipping is a nonlinear distortion that causes intermodulation distortion, which in turn causes the bit error rate of the (presumed) linear receiver to increase. Although a system could avoid clipping altogether by using a larger quantization step size in the digital-to-analog and analog-to-digital converters, the quantization process would then lead to significant quantization noise. In addition, most of the bits in the converters would often be unused, because the signal amplitude is more likely to be near its mean value than at its peak value most of the time. Thus, finding a reasonable trade-off between clipping and quantization noise is fundamental in the design of DMT modems.

The signal applied to the channel by a DMT transmitter can be considered the sum of up to $\overline{N}$ sinusoids. From a PAR perspective, the best case is when each subchannel supports only two bits, and the constellation supported by each subcarrier is quadrature phase-shift keyed (QPSK). In this case, the PAR is always the same, regardless of which constellation point is transmitted. If $\overline{N}$ such sinusoids are added together, the PAR is $\sqrt{2\overline{N}}$.

The PAR of a subchannel that supports a constellation larger than 2 bits is greater than $\sqrt{2}$. For large constellations, if all points are assumed to be equally likely to be transmitted, the reader can verify that the value of the PAR of a single QAM signal approaches $\sqrt{3}$. (Alternatively, the reader can consult Table 6.4 of Chapter 6.) An estimate of the PAR of a DMT signal composed of many subchannels, each of which supports a large constellation, is $\sqrt{6\overline{N}}$ [Bingham 2000].

In ADSL, $\overline{N} = 256$ in the downstream direction, which corresponds to a PAR of nearly 32 dB. Accommodating such a high PAR in a practical system would require very expensive (possibly even nonexistent) components, or it would significantly increase the noise caused by the quantization process. Therefore, to reduce the required precision of modem components while maintaining a reasonable quantization noise, typically systems are designed to accommodate a PAR that is significantly lower than 32 dB. ADSL standards require systems to accommodate a PAR of 16 dB [G.992.1]. Clipping may occur whenever the PAR of the actual signal exceeds 16 dB.

The probability of a clip can be expressed as [Bingham 2000]

$$\text{Pr\{clip\}} = \sqrt{\frac{2}{\pi}} \int_y^\infty e^{-x^2/2} dx, \tag{7.15}$$

where y is the PAR, as a linear ratio (*i.e.*, not in dB), that the system can accommodate without clipping. The probability that a symbol contains at least one clip is [Bingham 2000]

$$\text{Pr\{clipped symbol\}} = 1 - (1 - \text{Pr\{clip\}})^N. \tag{7.16}$$

For an ADSL system that accommodates a PAR of 16 dB, $\text{Pr\{clip\}} = 2 \times 10^{-10}$, and, in the downstream direction, $\text{Pr\{clipped symbol\}} = 10^{-7}$, which is the same as the target symbol error probability.

Clearly, reducing the PAR of the transmitted signal is desirable to improve system performance and reduce the probability that a clip occurs and causes errors. Consequently, several methods to reduce the PAR of a signal have been developed.

7.3.2.1 PAR Reduction Techniques

Many PAR reduction methods have been developed over the years. Interested readers are referred to [Tellado 2000] for a detailed discussion of a number of methods. In this chapter, the primary focus is on some of the newer PAR reduction techniques that have been developed with DSL as the intended application. All the methods considered incur some sort of penalty in exchange for reducing clipping. Some result in an increased probability of bit errors, others reduce the bit rate that can be supported, and still others increase the computational complexity of the transmitter or receiver.

It is important to note that application of these methods in a practical system generally does not entirely eliminate the probability of a clip. Instead, these methods work to lower the clipping probability.

7.3.2.1.1 PAR Reduction Methods that Cause Distortion

Distortion is a consequence of clipping. Several PAR reduction methods have been developed to mitigate the impact of this distortion, although none completely eliminates the resultant increase in the bit error rate. With these methods, the output of the IDFT is typically monitored to identify samples that will clip.

In clip windowing (see [Pauli 1997], [van Nee 1998], and [Pauli 1998]), a window is applied to the clipped time domain signal to smooth discontinuities that would otherwise cause significant out-of-band energy.

In clip shaping (see [Chow 1997a] and [Chow 1997b]), the energy of samples that will cause a clip is redistributed to samples around the clip sample, thus changing the spectrum of the clipping noise. Energy is added to or subtracted from neighboring samples so the new signal does not clip. For example, a filter can be applied to the sample that will clip and the samples preceding and following that sample. The receiver then applies an inverse filter to remove the clip shaping.

7.3.2.1.2 *Distortionless PAR Reduction Methods*

If the level of a signal can be reduced before digital-to-analog conversion, then clipping, and the resultant signal distortion and bit error rate increase, can be avoided altogether. Several methods of "distortionless" PAR reduction have been developed. In [Tellado 2000], these methods are classified in three groups: coding methods, discrete parameter optimization, and continuous parameter optimization. The latter two groups include the newest methods of PAR reduction.

Coding Methods — Given that the DMT signal is a sum of sinusoids, manipulation of the phases of the subchannels would seem to be a promising strategy to reduce the PAR of the overall signal. Indeed, coding methods strive to modify the phases of the subchannels such that the PAR of the resulting time-domain signal is low. However, optimizing the phases such that the PAR is minimized is a nonlinear problem, and there is no global optimum. Furthermore, the computational complexity required for these methods is high, because a large set of symbols with low PAR must be found to maximize the bit rate. For additional details on coding methods, the reader is referred to [Tellado 2000].

Discrete Parameter Optimization — The second class of distortionless PAR reduction methods is discrete parameter optimization, in which a reversible transformation is applied to either the frequency-domain vector **X** or the time-domain sequence **x**. The transformation, which may be applied only to some subchannels, is chosen such that it reduces the PAR of the resulting signal relative to what it would have been otherwise. Methods in this class of PAR reduction algorithms exploit the fact that clipping is a rare occurrence, so the likelihood that the transformed symbol and the original symbol both clip is low. The receiver can accurately decode the signal with information about what sort of modification was applied by the transmitter.

The simplest transformation method is dynamic clip scaling [Bingham 1996], in which the transformation is simply a scaling of the symbol. When the peak sample in a symbol is detected to exceed a threshold, the entire symbol is scaled by an amount that reduces the peak sample to below the threshold. The amount by which the symbol is scaled is then transmitted to the receiver, typically using a reserved subchannel, so demodulation can be performed correctly. Because a subchannel is reserved to transmit the scaling, dynamic clip scaling results in a small loss in the bit rate. In addition, the SNR on each subchannel is reduced because the transmitted energy of the entire symbol is decreased to avoid a clip. However, if the noise margin on each subchannel is higher than the amount by which the symbol is scaled, the scaling just results in a constant (but temporary) reduction in the noise margin; it does not result in any bit errors.

More sophisticated methods employ phase shifts of the subsymbols, including pseudo-random phase terms [Mestdagh 1996], [Muller 1997a], [Muller 1997b]. Additional methods that reduce computational complexity by exploiting the mathematical properties of the DFT are described in [Muller 1997c], [Verbin 1997], [Muller 1997a], [Muller 1997b], and [Zekri 1999].

A newer method of discrete parameter optimization is Tellado's tone injection method, which reduces the PAR without a degradation to the bit rate [Tellado 2000]. This method expands the size of the constellation and then maps each point from the original constellation to several equivalent points in the expanded constellation. The bit error probability does not increase if the minimum distance between duplicate points is at least as large as in the original constellation. Because several points in the expanded constellation represent the same point from the original constellation, the point with the appropriate frequency and phase can be selected by the modulator to reduce the PAR of the signal. The penalty of the tone injection method is the increased transmitter complexity required to choose which of the equivalent points in the expanded constellation results in the greatest PAR reduction. For a more detailed description of the tone injection method, readers are referred to [Tellado 2000].

Continuous Parameter Optimization — In this class are the PAR reduction methods that compute the transmitted symbol as a function of the original symbol and a set of design parameters that are optimized [Tellado 2000]. Generally, these methods define a subset of the available subchannels that are used to reduce or, ideally, to cancel peaks that occur due to data on the remaining subchannels. Subchannels in a single DMT symbol can be grouped into two sets, $\{D\}$ and $\{P\}$, where each subchannel is a member of either the set of data subchannels $\{D\}$ or the set of peak-reducing subchannels $\{P\}$. A peak-reducing pulse that will be transmitted on the subchannels in $\{P\}$ is then computed. The objective is to define a pulse such that the IDFT of the composite signal $D + P$ has a lower PAR than it would have had otherwise.

During each symbol, bits are allocated to the subchannels in set $\{D\}$. If the resulting symbol is expected to result in clipping, a pulse is constructed by finding amplitudes and phases on the subchannels in $\{P\}$ that reduce or cancel the peaks in the symbol composed of the $\{D\}$ subchannels. The two signals are then added together in the time domain, and the signal is applied to the channel.

Because continuous parameter optimization methods require some subchannels in each symbol to be allocated for peak reduction, they may reduce the data-carrying capability of the channel. Sometimes only the subchannels near the band edge will be allocated to the set $\{P\}$, because these subchannels typically suffer from high attenuation, particularly on long lines, and would not be able to support data. Utilizing unused subchannels allows PAR reduction without any loss in data rate; however, the amount by which the PAR can be reduced using only these (often poor-quality) subchannels is limited.

Two examples of continuous parameter optimization methods are the Gatherer/Polley method [Gatherer 1997] and Tellado's tone reservation method. Using Tellado's method, the PAR of the transmitted signal can be decreased by as much as 10 dB. Readers are referred to [Tellado 2000] for additional details.

7.4 Initialization

Thus far, this chapter has described how DMT transceivers modulate and demodulate a bit stream. Before steady-state transmission can occur, the modems on either side of a line must execute a coordinated initialization procedure. An important initial step is to determine symbol boundaries in the received signal. The modems must also work together to determine the number of bits that can be allocated to the downstream and upstream subchannels. Finally, they must exchange information necessary to enable accurate modulation and demodulation after a connection has been established.

Initialization of a pair of DMT modems can be described as five phases:

1. Activation,
2. Channel discovery,
3. Transceiver training,
4. Channel analysis, and
5. Parameter exchange.

The following sections describe these phases in some detail. The focus is on the processes that need to be completed by a pair of generic DMT modems to establish a connection. A discussion of the detailed protocols used in standardized DMT-based DSL modems is outside the scope of this chapter. The reader is referred to [G.992.1], [T1.413], [ETSI TS 101 388], [G.992.3], [G.992.5], and [G.994.1] to learn about protocols in standardized DMT-based systems and additional initialization procedures that are not covered here.

7.4.1 Activation

Activation of DSL transceivers is accomplished using a handshake procedure. During this process, basic capabilities are exchanged so specific initialization procedures can follow. Activation takes place before synchronization has been established and well before the modems have learned anything about the line on which they are operating.

A detailed explanation of handshaking is beyond the scope of this chapter; interested readers are referred to [G.994.1]. In the following subsections, it is assumed the two DMT modems on either side of the line have completed the handshake process and need to perform DMT-specific initialization procedures.

7.4.2 Channel Discovery

After two modems have completed the activation procedure, they enter the channel discovery phase of initialization. During this phase, they may perform coarse timing recovery to establish symbol boundaries, and they may apply transmitter power cutback. The receiver may also identify a subchannel suitable as a pilot tone (see Chapter 12).

Transmitter power cutback is required to ensure the receiver of the modem on the other end of the line is not saturated on short loops. In addition, the modem may need to implement power back-off in order to prevent high-level far-end crosstalk from appearing on other loops. The power is adjusted using an estimate of the loop attenuation, which is computed based on the power of the received signal. If the line is long, the received power will be low, and the transmitter may not need to apply any power cutback.

7.4.3 Transceiver Training

During the transceiver training phase, the automatic gain control (AGC) settings at both receivers are adjusted. The receiver time-domain equalizers and, if present (in ADSL, for instance), echo cancellers are also trained.

The receiver AGC must be set to the appropriate value to provide the best SNR at the receiver without saturating. The AGC may be initialized to its lowest setting at the start of the initialization process and then be adjusted based on the received power so that the received signal power is some number of dB below the maximum dynamic range of the analog-to-digital converter.

The receiver time-domain equalizers, which are used to shorten the effective length of the channel impulse response, must also be trained. Methods to train the TEQ are discussed in Chapter 11.

Some modem instantiations use echo cancellers, either to overlap the downstream and upstream spectra or, in nonoverlapped ADSL, to reduce the required width of the transition region between the downstream and upstream channels. (In VDSL, which uses digital duplexing, the use of a cyclic suffix results in orthogonal transmit and receive signals, so in theory use of an echo canceller is not necessary.) If an echo canceller is present, it must be trained during the initialization process. Training is simplest when the modem on the far end is not transmitting, which allows the modem on the near end to estimate the echo due to transmitted signals. Thereafter, echo due to the transmitted signal can be subtracted from the received signal.

7.4.4 Channel Analysis

Typically, the management system controlling the modem at the central office side of the line provides information that dictates what the downstream and upstream bit rates will be. The desired performance might be a specific data rate combination, a set of acceptable data rate combinations, or simply the best rates that can be accommodated while maintaining a required noise margin.

During the channel analysis phase, the subchannel gain values H_k and the noise variance on each subchannel are estimated so the subchannel SNRs can be computed. Based on the desired configuration dictated by the management system, the modems then determine how many bits will be supported by each subchannel.

7.4.4.1 Channel Identification

In order to determine the values of the FEQ taps and calculate the subchannel SNRs that are needed to compute the appropriate allocation of bits to the subchannels, the modems must perform channel identification. A signal known to both the transmitter and receiver is transmitted repeatedly over the channel, and the receiver simply averages the samples of the subsequent received symbols to estimate the channel gain. The use of an averaging process assumes the noise is additive and zero-mean. Denoting by M the number of DMT symbol periods during which channel estimation is performed, the receiver averages M received time-domain symbols,[9] computes the DFT of the averaged signal, and divides the result by the DFT of the known transmitted signal.

The signal input to the channel can be periodic or cyclically prefixed. For ease of explanation, here it is assumed the input signal is truly periodic. The period of the signal must be greater than or equal to the constraint length v of the unknown channel impulse response. $M + 1$ symbols of the training signal are applied to the channel. The first received symbol "clears" the channel and is discarded by the receiver. The last M symbols are then averaged:

$$\bar{\mathbf{y}} = \frac{1}{M} \sum_{m=1}^{M} \mathbf{y_m},\tag{7.17}$$

where $\bar{\mathbf{y}}$ and $\mathbf{y}$ are vectors containing samples of the time-domain symbols, and m is the symbol index. Denoting the DFT of $\mathbf{y}$ as $\mathbf{Y}$ and indexing the subchannels by k, the estimate

---

[9] In reality, it is the samples of the collection of received symbols that are averaged. To avoid cumbersome (but more precise) verbiage, the symbols are said to be averaged.

of the channel frequency response can be written as

$$\hat{H}_k = \sum_{m=1}^{M} \frac{Y_{m,k}}{X_{m,k}}, \tag{7.18}$$

where $X_{m,k}$ is the constellation point transmitted on subchannel k during training symbol period m.

As one would expect, under the assumption that the noise is zero-mean, the accuracy of the channel estimate improves as M increases.

7.4.4.2 Noise Identification

Noise identification is similar to channel identification. Following completion of the estimate of the channel gain, the noise variance can be estimated from the sequence of symbols obtained by subtracting the product of the known transmitted signal and the estimate of the channel gain from the received signal. Denoting the residual (error) signal on the kth subchannel following subtraction of the product of the channel and input signal as E_k, the objective is to estimate the variance of several instances of

$$E_{m,k} = Y_{m,k} - X_{m,k}\hat{H}_k, \tag{7.19}$$

where m is the symbol index. The variance of the noise on the kth subchannel is estimated as

$$\hat{\sigma}_k^2 = \frac{1}{M} \sum_{m=1}^{M} |E_{m,k}|^2. \tag{7.20}$$

Assuming the noise is stationary, the mean of Equation 7.20 is the variance of the zero-mean estimate. The variance of the estimate is

$$var\{\hat{\sigma}_k^2\} = E\{\hat{\sigma}_k^4\} - E\{\sigma_k^2\}^2, \tag{7.21}$$

where σ_k^2 is the true variance of the noise on the kth subchannel, and $E\{\cdot\}$ denotes the expected value. Proceeding,

$$E\{\hat{\sigma}_k^4\} = E\left\{\left[\frac{1}{M} \cdot \sum_{m=1}^{M} |E_{m,k}|^2\right]^2\right\}, \tag{7.22}$$

which simplifies to

$$E\{\hat{\sigma}_k^4\} = \frac{1}{M}\left[E\left\{|E_k|^4\right\} + (M-1)E\left\{|E_k|^2\right\}^2\right]. \tag{7.23}$$

Assuming the noise on each subchannel is Gaussian, $E\{|E_k|^4\} = 3\sigma_k^4$, and the variance of the estimate is

$$var\{\hat{\sigma}_k^2\} = \frac{1}{M}\left[3\sigma_k^4 + (M-1)\sigma_k^4\right] - \sigma_k^4, \tag{7.24}$$

which simplifies to

$$var(\hat{\sigma}_k^2) = \frac{2}{M}\sigma_k^4. \tag{7.25}$$

Thus, the excess noise in the estimate is $\sqrt{\frac{2}{M}}\sigma_k^2$.

To ensure the excess noise is no more than 0.1 dB at three standard deviations from the mean, M must be at least 269. To have no more than 0.1 dB excess noise at 4σ, the number of symbols that must be averaged in the noise estimate is 478.

7.4.4.3 *Bit Allocation*

A key step in the initialization process of multi-carrier modems is the determination of the so-called *bit distribution*, that is, how many bits should be allocated to each subchannel. An entire book could be written to describe the various bit allocation algorithms that have been invented over the years. (See [Cioffi EE379c notes], [Bingham 2000], and [Starr 2002] for examples.) Here, the focus is on bit allocation within the constraints imposed on DSL modems.

Given a desired symbol error probability P_e on each subchannel, an expression for the number of bits that can be supported by the kth subchannel is

$$b_k = \log_2 \left(\frac{\text{SNR}_k \cdot \gamma_c}{\Gamma(P_e) \cdot \gamma_m} + 1 \right), \tag{7.26}$$

where SNR_k is the signal-to-noise ratio of the kth subchannel, γ_c is the coding gain due to the use of forward error correcting or other codes (see Chapters 8, 9, and 10), γ_m is the noise margin, and $\Gamma(P_e)$ is the Shannon gap at the desired symbol error probability. The Shannon gap is defined by $\Gamma(P_e) = \frac{d_{\min}^2}{12\sigma^2}$, where $d_{\min}$ is the minimum distance between points in a QAM constellation at the receiver, and σ^2 is the received noise variance per dimension. The gap is a constant for a given symbol error probability and constellation type. (See Chapter 4.) For QAM with $P_e = 10^{-7}$, the Shannon gap is 9.75 dB.

One observes from Equation 7.26 that for a given P_e, the maximum number of bits a subchannel can support is dependent on its SNR, coding gain, and noise margin. The (uncoded) SNR of the kth subchannel can be written as

$$\text{SNR}_k = \frac{\varepsilon_k |H_k|^2}{2\sigma_k^2}, \tag{7.27}$$

where ε_k is the average input signal energy per two-dimensional QAM subsymbol, $|H_k|^2$ is the gain of the kth subchannel, and σ_k^2 is the received noise variance per dimension on the kth subchannel.

The variables over which a modem designer has control are ε_k, γ_c, and γ_m. The coding gain is determined by the code selection and framing (see Chapters 8, 9, and 10), which leaves only the input energy per subchannel and the noise margin available for adjustment by the bit allocation algorithm.

There are two basic types of bit allocation algorithms: those that maximize the data rate R while maintaining a particular noise margin γ_m, and those that maximize the noise margin while supporting a particular bit rate.

In DSL, both the transmitted PSD and total power are constrained by standards. For DSL systems based on DMT, a PSD template and PSD mask are usually specified. The mask specifies the upper limit on the transmit spectrum, and the template specifies the nominal (average) PSD. In ADSL, the integral of the template over the maximum bandwidth corresponds to the maximum allowed total power. This restriction greatly simplifies bit allocation. In VDSL1, the integral of the maximum allowed PSD over the bandwidth exceeds the total power, which means bit allocation algorithms allowing more latitude in parameters can be pursued. Here, the focus is on the simplest bit allocation algorithms that are appropriate when a PSD constraint corresponding to the total allowed transmit power is imposed, such as those that might be used in ADSL. For information on other bit allocation approaches, the reader is referred to [Starr 1999] and [Cioffi EE379c notes].

7.4.4.4 *Bit Allocation to Maximize the Bit Rate at a Target Noise Margin*

In cases in which the bit rate is to be maximized, a target noise margin is specified. In DSL, a noise margin target of 6 dB is typically required. Equation 7.26 can then be applied directly to determine the values of b_k. However, the b_k calculated in this manner will generally be noninteger. Decoder implementations are greatly simplified when integer-sized constellations are used, and therefore it is desirable to round the b_k to the nearest integers. However, to maintain a constant error probability on each subchannel (so that no subchannel dominates the system's error performance), the ε_k must then be adjusted. The ε_k are scaled by the gain factors

$$g_k = \frac{2^{[b_k]} - 1}{2^{b_k} - 1},\qquad(7.28)$$

where $[b_k]$ is the rounded value of b_k. Note that for $b_k > 2.5$,

$$g_k \approx 2^{[b_k] - b_k}.\qquad(7.29)$$

Using Equation 7.28, one can compute that the maximum value of g_k is 3.8 dB, which is the gain adjustment necessary to maintain the same error rate when a b_k value of 0.5 is rounded to 1. For $b_k > 2.5$, the swing in g_k is approximately ±1.5 dB. As a result, if a constant PSD is assumed, the energy distribution at the transmitter often has a sawtooth shape with up to 7.6 dB variation in amplitude. Figure 7.14 illustrates the sawtooth nature of g_k for a simple channel with monotonically decreasing gain and a constant transmitted energy across the bandwidth. The upper plot shows the rounded bit distribution, and the lower plot shows the necessary values of the gains (in dB) to support the rounded numbers of bits. This sawtooth characteristic is one reason why both a PSD mask and a PSD template are specified for DMT systems. The template provides the values of ε_k, and the mask provides flexibility for the transmitted PSD following application of the g_k. In ADSL, the PSD mask lies 3.5 dB above the template in the passband [G.992.1]. Of this 3.5 dB, 2.5 dB is assumed to be to accommodate application of the g_k. (A g_k value of 2.5 dB corresponds to rounding from 0.64 bits to 1 bit, so the mask does not allow rounding of $b_k < 0.64$ to 1 bit.) Referring

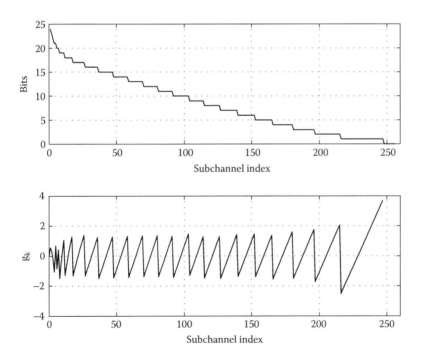

FIGURE 7.14
Rounding of bits on subchannels requires use of nonunity gain values.

again to Figure 7.14, it is clear that the gains required on the highest subchannels in this example would not be feasible in a real ADSL system.

Note that the total power constraint in a DSL system must always be obeyed, so in cases where the PSD template corresponds to the maximum allowed total power, a system would not be able to round up the bits on every subchannel.

7.4.4.5   Bit Allocation to Maximize the Noise Margin at a Target Bit Rate

Given that

$$B = \sum_{k=1}^{\overline{N}-1} b_k,$$

and

$$b_k = \log_2 \left(\frac{\text{SNR}_k \cdot \gamma_c}{\Gamma(P_e) \cdot \gamma_m} + 1 \right),$$

one might consider solving for γ_m and then computing the b_k to find a bit distribution that maximizes the noise margin while supporting a fixed bit rate. However, the rounding necessary to enable decoding simplicity is a nonlinear operation. Therefore, an iterative algorithm is necessary to maximize the noise margin at a target bit rate. With the values of γ_c and SNR_k known, the following steps may be used to maximize the noise margin γ_m while supporting a target bit rate of B_T [Bingham 2000].

1. Denote the minimum acceptable noise margin, for example, 6 dB, as $\gamma_{m,i}$. Using the method used for bit allocation to maximize the bit rate for a desired noise margin, calculate the achievable bit rate B_i using $\gamma_{m,i}$. Denote the number of subchannels allocated bits as $Ncar_i$.

2. If $B_i < B_T$, the desired bit rate is not sustainable at even the minimum acceptable noise margin. In this case, the bit allocation procedure should be aborted so a lower bit rate can be negotiated with the management system.

3. If $B_i > B_T$, calculate a new noise margin using $\gamma_{m,i+1} = \gamma_{m,i} \cdot 2^{(B_i - B_T)/Ncar_i}$.

4. Iterate through steps 1 through 3 until $B_i = B_T$.

It is possible that B_i never equals B_T, in which case the designer of the bit allocation algorithm must incorporate a metric to terminate execution of the iterations.

7.4.5   Parameter Exchange

Upon completion of the bit allocation algorithm, the receivers at both ends of the line have determined the number of bits each subchannel should support. In the final phase of initialization, this information and other calculated parameters, such as framing and coding parameters, must be exchanged so the modems can transition to steady-state operation.

The standards define various signals and protocols to enable parameter exchange. Interested readers are referred to [G.992.1], [G.992.2], [G.992.3], [G.992.5], and [T1.424] for additional information.

7.5   Steady-State Adaptation

DSL is potentially an "always on" service, and certainly most connections are likely to persist for long periods of time. Therefore, it is necessary to provide mechanisms that allow modems to adapt to changing channel and noise conditions during a connection.

During steady-state operation, the channel characteristics may change. In general, how-ever, it is the noise PSD that changes, possibly because of other DSL systems in the binder turning on or off, or perhaps because of the appearance of or a change in radio-frequency ingress. To maintain stability of the system, DMT modems must adapt to these changing conditions.

Several components of the DMT modem can be adapted to accommodate changes in the channel or noise. Among these components are the bit allocation and the FEQ settings.

7.5.1 Bit Swapping

The bit allocation established during initialization provides the preliminary distribution of bits on the downstream and upstream subchannels. During a connection, the receiver at each end of the line monitors its error performance. For example, a receiver might track the mean-squared error (MSE) between the received (noisy) subsymbols and the decoded subsymbols. When the receiver detects that the SNR on a specific subchannel has degraded beyond a certain threshold, it attempts to find another subchannel that can accommodate one or more additional bits. Using a control channel, the local receiver then sends instructions to change the bit allocation to the far-end transmitter, where the change is executed at a time known to the receiver.[10]

When the number of bits on a subchannel is changed, the power on that subchannel may also need to be modified to maintain the same noise margin and error performance. For large constellations, support of an additional bit requires an additional 3 dB of power. (To add a bit to a small constellation, the incremental power required is larger. For example, approximately 4.8 dB of additional power is required to accommodate an extra bit on a subchannel that was previously supporting only one bit. See Chapter 6 for more details.) Likewise, if a bit is removed from a subchannel that was supporting a large constellation, the power can be reduced by 3 dB without changing the noise margin on that subchannel. Therefore, bit swapping must also include a mechanism to change the power allocated to the subchannels.

As an example, assume subchannel i, which supports a 7-bit constellation, begins to degrade. The local receiver determines that the noise margin is not the desired 6 dB, but instead only 4 dB. Fortunately, the receiver identifies another subchannel, j, that currently supports only 5 bits and has a noise margin of 7 dB. Through an overhead channel, the local receiver then instructs the far-end transmitter to

1. Move one bit from subchannel i to subchannel j,
2. Reduce the power on subchannel i by 1 dB, and
3. Increase the power on subchannel j by 2 dB.

Assuming the PSD mask allows the power on subchannel j to be increased sufficiently, after the bit swap the bit rate of the system is preserved, and the subchannels both have 6 dB noise margins.

To facilitate coordination of changes in the bit allocations, the modems on both sides of a line can operate with synchronized counters. In ADSL, for example, time is tracked using superframe counters, where a superframe is a consecutive set of 69 DMT symbols [G.992.1]. When the modems transition from initialization to steady-state operation, they set their counters to zero. Times at which bit swaps are executed are then given in terms of the superframe counter values.

---

[10] Obviously, the "receiver" does not send anything to the far end without help from the local transmitter.

The bit swap protocol used in both ADSL and VDSL1 modems has three steps [G.992.1] [T1.424]. The first step is the transmission of the bit swap request from Modem A, whose receiver would like to implement a bit swap. The request is transmitted on an overhead channel. The second is the transmission, also on the overhead channel, of an acknowledgment from Modem B, whose transmitter will have to modify its bit allocation to support the request. Finally, Modems A and B implement the swap synchronously when their counters reach the designated value.

In ADSL as specified by ITU-T Recommendation G.992.1 [G.992.1], a bit swap request is composed of 4 two-byte fields. The first byte of each pair designates the index of the subchannel involved in the bit swap, and the second byte indicates what action should be taken. The action could be to add a bit to or remove a bit from the subchannel, or it could be to increase or decrease the power on that subchannel by 1, 2, or 3 dB. One can see that a typical bit swap request message might consist of a modification in bits plus a change in power to one subchannel, followed by a modification in bits plus a change in power to another subchannel.

Upon receiving the bit swap request, the transmitter in Modem B determines the superframe counter value at which the change will be implemented. The transmitter then sends an acknowledgment message to the Modem A receiver. The message echoes the request and appends the superframe counter value at which the change will be executed. When the superframe counter reaches the value contained in the acknowledgment message, the transmitter in Modem B and the receiver in Modem A execute the swap, and the Modem B transmitter makes the changes to the subchannel powers requested by the receiver.

The second-generation ADSL specification, ADSL2 (which is defined in ITU-T Recommendation G.992.3 [G.992.3]), allows additional flexibility in bit swapping, including a mechanism to change the bit rate on the fly, *i.e.*, either to increase or decrease the bit rate in one or both transmission directions without dropping a connection to re-initialize. DMT-based VDSL1 defines an additional mode of bit swapping called express swapping, details of which can be found in [Starr 2002].

7.5.2   FEQ Adaptation

The FEQ is used by the receiver to compensate for the scaling and rotation of subchannel constellation points due to the subchannel gains. During steady-state operation, the FEQ taps must be updated to compensate for small timing errors and to allow proper detection following bit swaps.

The mechanism used to update the FEQ is similar to the mechanism used by the receiver to determine that a bit swap is required, namely, tracking the error between the points in the constellation diagram and the decoded points. An error signal can be defined for each subchannel as the difference between the input and output of the decision device. Based on this error, the FEQ taps can be updated using standard adaptation algorithms, such as least mean-squares (LMS).

7.6   Summary

This chapter explained the fundamentals of multi-carrier modulation. The chapter began with an overview of multi-carrier modulation and a discussion of the concept of channel partitioning. Discrete multi-tone modulation, the instantiation of multi-carrier modulation that has been standardized for ADSL and VDSL, was then introduced. Two of the mechanisms used in VDSL—cyclic suffix and timing advance—were presented. It was shown that the addition of a cyclic suffix allows the transmitted and received symbol boundaries to be aligned at both ends of the line, which eliminates interference in the received signal from subchannels allocated for transmission in the opposite direction. Under this condition, the use of filters to separate the downstream and upstream bands is unnecessary. Timing advance was then shown to reduce the required length of the cyclic suffix by half.

The issues of the peak-to-average power ratio and clipping were addressed. Methods to reduce the PAR, including those that allow distortion and those that eliminate it, were then briefly overviewed.

The chapter continued with an overview of the steps required to initialize a pair of DMT modems. The overview included an explanation of how the subchannel gains and noise variances are estimated, from which the subchannel signal-to-noise ratios are computed. A detailed discussion of processes that can be used to allocate bits to the subchannels followed. Two strategies were considered: allocation of bits when a target noise margin is provided and the bit rate is to be maximized, and allocation of bits when the target rate is provided and the noise margin is to be maximized.

Finally, the need to adapt various components of the modems to maintain reliable and robust performance during steady-state transmission was discussed. Bit swapping, the process of moving bits from a degraded subchannel to a subchannel with excess margin, was presented. The need to adapt the frequency-domain equalizer to compensate for timing inaccuracies and bit swaps was also described.

7.7 Acknowledgments

I am blessed to find myself consistently in the company of exceptional individuals and talented colleagues, and I would be remiss not to thank them for their direct and indirect contributions to this chapter. Without John Cioffi's mentorship and influence, I wouldn't know much about DMT, and this chapter surely would not exist. I am also grateful to several people for providing thoughtful and helpful feedback on the chapter, including Brian Wiese and George Ginis, my standards friend and foe Vladimir Oksman, and the "official" reviewers of the chapter (some known, some unknown). The rigorous review of the chapter definitely improved its quality. I would also like to thank Jose Tellado for sending me his dissertation, from which I was able to learn more than I ever wanted to know about PAR reduction! Finally, to Jim, for putting up with the countless weekends I spend working on chapters for books, thank you for your patience and support.

References

[Baum 1996] R.W. Baum, R.F.H. Fischer, and J.B. Huber. *Reducing the Peak-to-Average Ratio of Multicarrier Modulation by Selected Mapping.* Electronics Letters, October 1996, pp. 2056–2057.

[Bingham 1990] J.A.C. Bingham. *Multicarrier Modulation for Data Transmission: An Idea Whose Time Has Come.* IEEE Commun. Mag., May 1990.

[Bingham 1996] J.A.C. Bingham and J.M. Cioffi. *Dynamic Scaling for Clip Mitigation in the ADSL Standard Issue 2.* T1E1.4 contribution number 96-019, Los Angeles, CA, January 1996.

[Bingham 2000] J.A.C. Bingham. *ADSL, VDSL and Multicarrier Modulation.* Wiley-Interscience, New York, 2000.

[Chow 1997a] J.S. Chow, J.A.C. Bingham, and M.S. Flowers. *Mitigating Clipping Noise in Multicarrier Systems.* In Proceedings of IEEE Int. Conf. Commun., pp. 715–719, Montreal, Canada, 1997.

[Chow 1997b] J.S. Chow, J.A.C. Bingham, M.S. Flowers, and J.M. Cioffi. *Mitigating clipping and quantization effects in digital transmission systems.* U.S. Patent No. 5,623,513, April 22, 1997.

[Cioffi EE379c notes] J.M. Cioffi. *EE379C Course notes.* Stanford University, Stanford, CA. Available at http://www.stanford.edu/class/ee379c/.

[Cioffi 1991] J.M. Cioffi. *A Multicarrier Primer.* T1E1.4 contribution number 91-157, 1991.

[ETSI TS 101 388] *Asymmetric Digital Subscriber Line (ADSL)—European specific requirements [ITU-T G.992.1 modified].* ETSI TS 101 388 (2002).

[Gatherer 1997] A. Gatherer and M. Polley. *Controlling Clipping Probability in DMT Transmission.* In Proceedings of the Asilomar Conference, November 1997.

[G.992.1] *Asymmetric digital subscriber line (ADSL) transceivers.* ITU-T Recommendation G.992.1 (1999).

[G.992.2] *Splitterless asymmetric digital subscriber line (ADSL) transceivers.* ITU-T Recommendation G.992.2 (1999).

[G.992.3] *Asymmetric digital subscriber line (ADSL) transceivers—2 (ADSL2).* ITU-T Recommendation G.992.3 (2002).

[G.992.5] *Asymmetric digital subscriber line (ADSL) transceivers—extended bandwidth ADSL2 (ADSL2plus).* ITU-T Recommendation G.992.5 (2003).

[G.994.1] *Handshake procedures for Digital Subscriber Line (DSL) transceivers.* ITU-T Recommendation G.994.1 (2003).

[Mestdagh 1996] D.J.G. Mestdagh and P. Spruyt. *A Method to Reduce the Probability of Clipping in DMT-Based Transceivers.* IEEE Trans. Commun., COM-44(10):1234–1238, 1996.

[Mestdagh 2000] D.J. Mestdagh, M.R. Isaksson, and P. Odling. *Zipper VDSL: A Solution for Robust Duplex Communication over Telephone Lines.* IEEE Commun. Mag., No. 5: 90–96, May 2000.

[Muller 1997a] S.H. Muller and J.B. Huber. *A Comparison of Peak Power Reduction Schemes for OFDM.* In Proceedings IEEE Globecom, Vol. 1, pp. 1–5, Phoenix, AZ, 1997.

[Muller 1997b] S.H. Muller and J.B. Huber. *A Novel Peak Power Reduction Scheme for OFDM.* In Proceedings IEEE PIMRC, pp. 1090–1094, Helsinki, Finland, 1997.

[Muller 1997c] S.H. Muller and J.B. Huber. *OFDM with Reduced Peak-to-Average Power Ratio by Optimum Combination of Partial Transmit Sequences.* Electronics Letters, 33(5):368–369, Feb. 1997.

[Pauli 1997] M. Pauli and H.P. Kuchenbecker. *Minimization of the Intermodulation Distortion of a Nonlinearly Amplified OFDM Signal.* Wireless Personal Commun., 4(1):93–101, 1997.

[Pauli 1998] M. Pauli and H.P. Kuchenbecker. *On the Reduction of the Out-of-Band Radiation of OFDM Signals.* In Proceedings IEEE Int. Conf. Commun. pp. 1304–1308, Atlanta, Georgia, 1998.

[Sjoberg 1999] F. Sjoberg, M. Isaksson, R. Nilsson, P. Odling, S.K. Wilson, and P.O. Borjesson. *Zipper: A Duplex Method for VDSL Based on DMT.* IEEE Trans. Comm., No. 8:1245–1253, August 1999.

[Starr 1999] T. Starr, J.M. Cioffi, and P.J. Silverman. *Understanding Digital Subscriber Line Technology.* Prentice-Hall, Upper Saddle River, NJ, 1999.

[Starr 2002] T. Starr, M. Sorbara, J.M. Cioffi, and P.J. Silverman. *DSL Advances.* Prentice-Hall, Upper Saddle River, NJ, 2002.

[Tellado 1997] J. Tellado and J.M Cioffi. *PAR Reduction in Multicarrier Transmission Systems.* T1E1.4 contribution number 1997-367, Sacramento, CA, December 1997.

[Tellado 2000] J. Tellado. *Multicarrier Modulation with Low Peak-to-Average Power: Applications to xDSL and Broadband Wireless.* Kluwer Academic, Boston, 2000.

[T1.413] *Network and Customer Installation Interfaces – Asymmetric Digital Subscriber Line (ADSL) Metallic Interface.* ANSI Standard T1.413-1998.

[T1.424] *Very-high-bit-rate Digital Subscriber Lines (VDSL) Metallic Interface (DMT based).* ANSI Standard T1.424 (2003).

[van Nee 1998] R.D.J. van Nee and A. de Wild. *Reducing the Peak-to-Average Power Ratio of OFDM.* In Proc. IEEE Vehicular Tech. Conf., Vol. 3, pp. 2072–2076, Ottawa, Canada, 1998.

[Verbin 1997] R. Verbin. *Efficient Algorithm for Clip Probability Reduction.* T1E1.4 contribution 1997-323, Sep 1997.

[Zekri 1999] M. Zekri and L. Van Biesen. *Super Algorithm for Clip Probability Reduction of DMT Signals.* In Proceedings of the 13th International Conference on Information Networking (ICOIN-13), Vol. 2, pp. 8B-3.1–6, Cheju Island, Korea, 1999.

8

Trellis-Coded Modulation in DSL Systems

Gottfried Ungerboeck

CONTENTS

8.1 Introduction 211
8.2 Principles of Trellis-Coded Modulation 212
    8.2.1 Uncoded Modulation and Coded Modulation 212
    8.2.2 Trellis-Based Coded Modulation and Trellis-Coded
           Modulation (TCM) 213
    8.2.3 Set Partitioning 214
    8.2.4 General Structure of TCM Encoders 217
    8.2.5 Free Euclidean Distance 218
    8.2.6 Code Search and Optimum Codes 220
    8.2.7 Performance of TCM Schemes with Different Symbol Dimensions . . . 222
8.3 Trellis Coding in SHDSL 223
    8.3.1 TCM Encoder 223
    8.3.2 16-PAM Symbol Mapping 224
    8.3.3 Convolutional Encoder 224
8.4 Trellis Coding in ADSL 225
    8.4.1 Bit Allocations and Tone Ordering 225
    8.4.2 Convolutional Encoding and Bit Conversion 226
    8.4.3 2-D Symbol Mapping 226
           8.4.3.1 Even Values of b 226
           8.4.3.2 Odd Values of b 227
    8.4.4 Discussion 227
8.5 Conclusions 230
References 231

ABSTRACT This chapter reviews the principles of trellis-coded modulation and discusses the trellis-coding schemes used in SHDSL and ADSL transceivers.

8.1 Introduction

Trellis-coded modulation (TCM) evolved in the early 1980s as a combined coding and modulation technique for band-limited channels. Prior to TCM, it was widely believed that coding gains could only be achieved at the expense of compromising bandwidth efficiency by sending redundant code symbols along with information-bearing symbols. With TCM, this barrier was overcome by the introduction of redundancy for coding in the form of expanded constellations of modulation symbols and optimizing coding and modulation jointly to achieve good distance properties of code sequences in Euclidean signal space. Thus, significant gains in robustness against noise are achieved, compared to uncoded modulation, for given values of signal power and spectral efficiency. TCM can be viewed as an extension of convolutional coding with binary modulation to convolutional encoding with nonbinary modulations such as 8-PSK or higher-order QAM.

The first TCM schemes were proposed in 1976 [Ungerboeck 1976]. Following a more detailed publication [Ungerboeck 1982] in 1982, an explosion of research and actual implementations of TCM took place. Today, TCM is employed in most digital transmission systems whenever higher-order modulations are used for the sake of bandwidth efficiency. In many cases, TCM serves as an inner channel coding scheme in combination with interleaving and outer algebraic error-control coding with Reed–Solomon codes. In DSL systems, TCM employed in SHDSL [G.991.2] and ADSL [G.992.1] transceivers. TCM has also been added to VDSL in the VDSL standards.

8.2   Principles of Trellis-Coded Modulation

Assume an ideal discrete-time channel with additive white Gaussian noise (AWGN) and zero intersymbol-interference (ISI). The sequence of received noisy signals is $\{r_n\} = \{a_n + w_n\}$, where a_n represents the transmitted modulation symbol, and w_n denotes additive noise at symbol time n. In baseband transmission systems, the signals are naturally one-dimensional (real-valued). In carrier modulated systems employing phase-shift keying (PSK) or quadrature amplitude modulation (QAM), the signals are usually two-dimensional (complex-valued). To obtain higher-dimensional signals, lower-dimensional signals may be grouped together into N-dimensional (N-D) signal blocks.

Generally, the modulation symbols a_n are selected from an N-D symbol constellation $\mathcal{A}$ with average energy σ_a^2 per dimension. The noise samples w_n consist of independent Gaussian noise components with zero mean and variance σ_w^2 per dimension. The signal-to-noise ratio is

$$SNR = \sigma_a^2/\sigma_w^2.$$

8.2.1   Uncoded Modulation and Coded Modulation

Let $M = |\mathcal{A}| = 2^m$ be the number of N-D symbols in $\mathcal{A}$. In uncoded modulation systems, m information bits are mapped into a modulation symbol a_n independently for each time n. Thus, information is represented with a spectral efficiency of $\eta = m/N$ bits per dimension. An optimum symbol-by-symbol detector determines independently for each time n, the most likely transmitted symbol $\hat{a}_n = \arg \min_{\alpha_n \in \mathcal{A}} |r_n - \alpha_n|^2$. Let $d_{\min}$ denote the smallest Euclidean distance between the symbols of $\mathcal{A}$, and $K_{\min}$ be the average number of nearest neighbor symbols at distance $d_{\min}$ from each symbol in $\mathcal{A}$. For higher SNR, symbol-error probability is well approximated by

$$P_{se} \cong K_{\min} Q(d_{\min}/2\sigma_w),$$

where $Q(.)$ is the Gaussian error integral.

In coded modulation systems, the succession of modulation symbols is constrained by coding rules. With block coding, the code constraints apply to symbols within finite windows. With sequence coding, such as nonterminated convolutional coding, the code constraints act in a sliding-window fashion. Taking the viewpoint of infinite sequences, the modulation code C is the set of permitted symbol sequences $\{a_n\} \in C \subset \cdots A \times A \times A \cdots$. An optimum *soft* sequence decoder operates on the unquantized sequence $\{r_n\}$ and determines the most likely transmitted sequence

$$\{\hat{a}_n\} = \arg \min_{\{\alpha_n\} \in C} \sum_n |r_n - \alpha_n|^2. \tag{8.1}$$

Let $\{\alpha_n\}$ and $\{\alpha'_n\}$ be two sequences of modulation symbols. The squared Euclidean distance (SED) between these sequences is given by $d^2(\{\alpha_n\}, \{\alpha'_n\}) = \sum_n |\alpha_n - \alpha'_n|^2$. The smallest SED between sequences $\{\alpha_n\} \neq \{\alpha'_n\}$, both in C, is referred to as squared *free* Euclidean distance d_{free}^2.

The average number of code sequences departing at time n from a given code sequence and remerging later with this sequence with accumulated distance d_{free}^2 is denoted by K_{free}. At high SNR, the *error-event* probability per symbol time that the optimum sequence decoder decides in favor of an erroneous sequence $\{\hat{a}_n\} \neq \{a_n\}$ is well approximated by

$$P_{ee} \cong K_{free} Q(d_{free}/2\sigma_w).$$

8.2.2 Trellis-Based Coded Modulation and Trellis-Coded Modulation (TCM)

A wide class of coded modulation schemes can be appropriately described in terms of finite-state machines. A state transition diagram depicted in a temporally rolled-out fashion is called a *trellis* diagram. Transitions in the trellis diagram are associated with a current state s_n, an input symbol x_n, an output symbol a_n, and a next state s_{n+1}. The modulation code C is the set of output symbols associated with connected transitions, or *paths*, through the trellis diagram. Unless the number of states is excessively large, sequence decoding can practically be performed by the Viterbi algorithm.

The most well-known trellis-based coded modulation scheme is convolutional coding with binary modulation. Convolutional codes have time-invariant trellis diagrams. However, the class of trellis-based coded modulation schemes also includes block codes with binary or nonbinary modulation and dense lattice constellations, which have time-varying trellis descriptions.

The narrower form of trellis-coded modulation (see [Ungerboeck 1982], [Wei 1984], [Ungerboeck 1987], [Wei 1987], [Forney 1988a], [Forney 1988b], [Wei 1989], and [Forney 1989]) discussed in this chapter extends the concept of convolutional coding with binary modulation to convolutional coding with higher-order modulation. A characteristic feature of TCM is the use of symbol set redundancy for coding. Among various forms of coded modulation, TCM has in practice been most successful.

For transmission of m bits per N-D modulation symbol, TCM schemes use an expanded symbol constellation A of $M = 2^{m+1}$ symbols. The redundancy available for coding diminishes with an increasing number of dimensions. The sufficiency of using twice as many constellation symbols as needed for uncoded modulation has been shown in [Ungerboeck 1982] for 1-D and 2-D symbol constellations. Doubling the size of 4-D constellations provides enough redundancy for achieving good coding gains at moderate code complexities. However, the reduced redundancy per dimension becomes a limiting factor for obtaining significant further gain improvements with codes of higher complexity.

Example 8.1: Four-State Coded 8-PSK

Figure 8.1 illustrates four-state coded 8-PSK, which was the first useful TCM scheme discovered in 1975. Transitions in the trellis diagram are associated with 8-PSK symbol labels. Because four transitions stem from each state, two information bits can be encoded per transmitted 8-PSK symbol. The spectral efficiency is the same as in uncoded 4-PSK modulation.

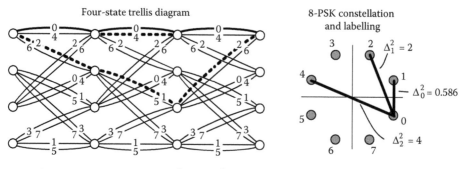

SED between symbol 0 and 4 : $\Delta_2^2 = 4 = d_{free}^2$

SED between paths with symbols 0 0 0 and 212 : $\Delta_1^2 + \Delta_0^2 + \Delta_1^2 = 4.586$

FIGURE 8.1
A simple example of TCM: four-state coded 8-PSK modulation.

By not showing in Figure 8.1 input symbols associated with transitions, the realization of an encoder is left open. The trellis diagram exhibits *parallel transitions* between current states and next states. The 8-PSK sequences can differ in one symbol along parallel transitions or otherwise must differ in at least three symbols. In this example, the squared free distance $d_{free}^2 = 4$ is found between symbols assigned to parallel transitions, for example, the symbols with labels 0 and 4. Sequences differing in three or more symbols must at least have SED 4.586. Compared to uncoded 4-PSK with $d_{min}^2 = 2$, the four-state coded 8-PSK scheme achieves a coding gain of $\log_{10}[d_{free}^2/d_{min}^2] = 3$ dB asymptotically at high SNR.

8.2.3   Set Partitioning

An important element in the construction of TCM schemes is the concept of *set partitioning* [Ungerboeck 1982], [Ungerboeck 1987]. A symbol constellation $\mathcal{A}$ of size 2^{m+1} is partitioned into 2^{k+1} equal-sized subsets, where $1 \leq k \leq m$. Hence, the constellation is partitioned at least into 4 subsets. The partitioning is accomplished in $k + 1$ successive two-way partitioning steps. The objective is to increase uniformly and as much as possible in every partitioning step the minimum squared subset distance (MSSD) Δ_j^2, $1 \leq j \leq k + 1$, among the symbols of the jth level subsets. Each two-way selection is identified by a label bit $y^j \in \{0, 1\}$. The jth level subsets are labelled with binary $j + 1$ tuples $[y^j y^{j-1} \cdots y^0]$, or equivalently integers $i = 2^j y^j + \cdots 2y^1 + y^0$. This indexing of subsets is referred to as set partitioning (SP) labelling.

$$\mathcal{A} \xrightarrow{y^0} \mathcal{B}_{y^0} \xrightarrow{y^1} \mathcal{C}_{[y^1,y^0]} \xrightarrow{y^2} \cdots \xrightarrow{y^k} \mathcal{S}_{[y^k \cdots y^1, y^0]},$$

$$\text{MSSD} \quad \Delta_0^2 \leq \Delta_1^2 \leq \Delta_2^2 \leq \cdots \leq \Delta_{k+1}^2,$$

$$\text{Size} \quad 2^{m+1} \quad 2^m \quad 2^{m-1} \quad\quad 2^{m-k}.$$

For $k < m$, the final subsets $\mathcal{S}_i$, $0 \leq i \leq 2^{m-k} - 1$, contain 2^{m-k} symbols with MSSD Δ_{k+1}^2. The symbols of these subsets become associated with parallel transitions in a trellis diagram. A decoder decides among the symbols of each final subset in the same way as symbol decisions are made for uncoded modulation. Free Euclidean distance is limited either by the free distance between paths involving nonparallel transitions or the MSSD of the final subsets; *i.e.*, Δ_{k+1}^2 upperbounds d_{free}^2. The set partitioning depth is chosen such that no other partitioning depth can give a larger value of d_{free}^2. If $k = m$, the final subsets contain only one symbol and by convention $\Delta_{k+1}^2 = \infty$.

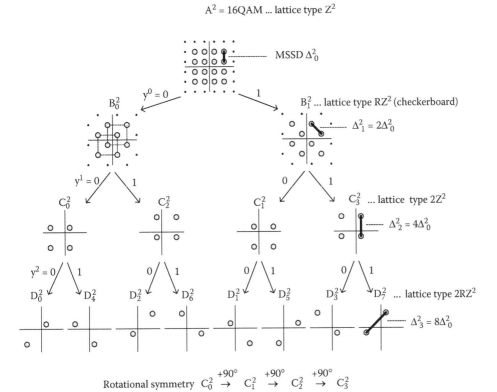

FIGURE 8.2
Set partitioning of 16-QAM constellation.

Set partitioning of 1-D and 2-D constellations is straightforward. Figure 8.2 illustrates set partitioning of a 16-QAM constellation ($N = 2$). Henceforth, the symbol dimensions of symbol sets are indicated by a superscript, when this contributes to clarity.

The 16-QAM constellation is congruent to a compact subset of the 2-D integer lattice Z^2, which is the infinite set of points (a, b) with integer coordinates. The class of constellations based on Z^2 includes square constellations (4-QAM = 4-PSK = QPSK, 16-QAM, 64-QAM, ...), cross constellations (32-QAM-CR, 128-QAM-CR, ...), and checkerboard or double-square constellations (8-QAM-DS, 32-QAM-DS, 128-QAM-DS, ...). The latter constellations are based on the rotated lattice RZ^2. Every two-way partitioning of these constellations increases the MSSD by a factor of two; *i.e.*, $\Delta_j^2 = 2\Delta_{j-1}^2$.

Higher-dimensional constellations can be viewed as the Cartesian set product of lower-dimensional constituent constellations. Figure 8.3 depicts the partitioning of a 4-D constellation $\mathcal{A}^4 = \mathcal{A}^2 \times \mathcal{A}^2$ based on Z^4.

The partitioning of the constituent 2–D constellation $\mathcal{A}^2$ is known from Figure 8.2. One can write $\mathcal{A}^4 = \mathcal{A}^2 \times \mathcal{A}^2 = (\mathcal{B}_0^2 \cup \mathcal{B}_1^2) \times (\mathcal{B}_0^2 \cup \mathcal{B}_1^2)$. The first two-way partitioning of $\mathcal{A}^4$ leads to the first-level subsets $\mathcal{B}_0^4 = (\mathcal{B}_0^2 \times \mathcal{B}_0^2) \cup (\mathcal{B}_1^2 \times \mathcal{B}_1^2)$ and $\mathcal{B}_1^4 = (\mathcal{B}_0^2 \times \mathcal{B}_1^2) \cup (\mathcal{B}_1^2 \times \mathcal{B}_0^2)$ with MSSD $\Delta_1^2 = 2\Delta_0^2$. In $\mathcal{B}_0^4$, the distance Δ_1^2 occurs between 2-D symbol components in subsets $\mathcal{B}_0^2$ or $\mathcal{B}_1^2$ and between 4-D symbols, one in $\mathcal{B}_0^2 \times \mathcal{B}_0^2$ and the other in $\mathcal{B}_1^2 \times \mathcal{B}_1^2$. The first-level subsets are then partitioning into second-level subsets $\mathcal{B}_0'^4 = \mathcal{B}_0^2 \times \mathcal{B}_0^2$, $\mathcal{B}_2'^4 = \mathcal{B}_1^2 \times \mathcal{B}_1^2$ and $\mathcal{B}_1'^4 = \mathcal{B}_0^2 \times \mathcal{B}_1^2$, $\mathcal{B}_3'^4 = \mathcal{B}_1^2 \times \mathcal{B}_0^2$. This leaves the MSSD at $\Delta_2^2 = 2\Delta_0^2$. The third- and fourth-level partitioning steps are conceptually similar to the first- and second-level partitioning steps, respectively. The third-level subsets $\mathcal{C}_i^4$, $0 \le i \le 7$, and the fourth-level subsets $\mathcal{C}_i'^4$, $0 \le i \le 15$, have the same MSSD $\Delta_3^2 = \Delta_4^2 = 4\Delta_0^2$.

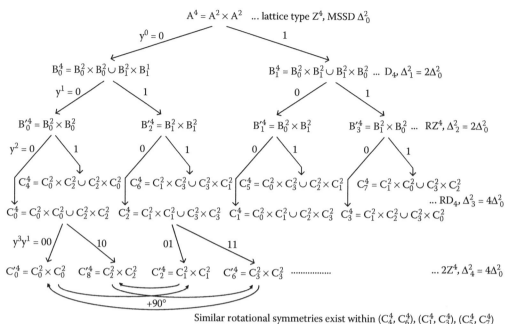

FIGURE 8.3
Set partitioning of 4-D constellation of lattice type $\mathcal{Z}^4$. [Ungerboeck 1987]

Inspection of Figure 8.3 reveals the following relation between the 4-D labels $i = [y^3 y^2 y^1 y^0]$ and the 2-D labels p and q in the 4-D subsets $C_i'^4 = C_p^2 \times C_q^2$:

$$p = (2y^3 + y^1), \quad q = p + (2y^2 + y^0) \mod 4. \tag{8.2}$$

Cyclic incrementation of $2y^3 + y^1$ by 1 modulo 4, *i.e.*, $y^3 y^1 = 00, 01, 10, 11, 00, \ldots$, causes the 2-D symbols in C_p^2 and C_q^2 to go in lockstep through successive $+90°$ rotations. The 4-D label bits $y^3 y^1$ are therefore called the *rotation-sensitive* bits. The modulo 4 operation in Equation 8.2 makes the conversion from 4-D labels to 2-D labels nonlinear in terms of modulo-2 arithmetic. Rotational symmetries of this kind have been the basis for designing rotationally invariant 4-D trellis codes [Wei 1987], [Ungerboeck 1987] with differential encoding of the rotation-sensitive bits. The Wei code has been employed in voiceband modems since V.34 [V.34].

A lattice Λ is an infinite set of discrete N-D points, which form an algebraic group under vector addition. The origin serves as the zero element of the group. Let $\Lambda' \subset \Lambda$ be a sublattice of Λ. The notation Λ/Λ' denotes the set of subsets of Λ (= quotient group), in which Λ' is the zero element containing the origin, and the other subsets are translated versions of Λ' (= cosets). The union of Λ' and its cosets is the lattice Λ.

For TCM schemes with lattice-type symbol constellations, the task of set partitioning can be separated from the determination of the actual size and shape of finite symbol constellations [Forney 1988a], [Forney 1988b]. The employed lattice Λ is almost always the integer lattice $\mathcal{Z}^N$. For $N = 1, 2$, and 4, $\mathcal{Z}^N$ is partitioned as follows.

$$\mathcal{Z}(1)/2\mathcal{Z}(4)/4\mathcal{Z}(16)/8\mathcal{Z}(64) \cdots$$
$$\mathcal{Z}^2(1)/R\mathcal{Z}^2(2)/2\mathcal{Z}^2(4)/2R\mathcal{Z}^2(8)/4\mathcal{Z}^2(16) \cdots$$
$$\mathcal{Z}^4(1)/\mathcal{D}_4(2)/R\mathcal{Z}^4(2)/R\mathcal{D}_4(4)/2\mathcal{Z}^4(4)/2\mathcal{D}_4(8)/2R\mathcal{Z}^4(8) \cdots.$$

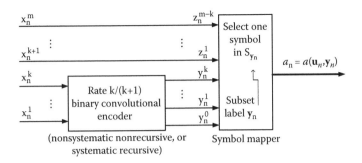

FIGURE 8.4
General structure of TCM encoder.

The values in parentheses are the lattice MSSDs. R denotes a rotation-and-expansion operator. $\mathcal{D}_4$ is the Schaefli lattice, the densest lattice in four dimensions.

From Λ, a finite symbol constellation is obtained by the operation $\mathcal{A} = \alpha(* \cap \mathcal{R})$, where $\mathcal{R}$ is a compact region of desired shape containing the desired number of lattice points, and $\alpha(.)$ denotes an appropriate scaling, translation, or rotation operation. The subsets of $\mathcal{A}$ are obtained by replacing in $\alpha(\Lambda \cap \mathcal{R})$ the lattice Λ by the corresponding sublattices.

8.2.4 General Structure of TCM Encoders

The general structure of a TCM encoder is depicted in Figure 8.4. At time n, m information bits $\mathbf{x}_n = [x_n^m, x_n^{m-1}, \ldots, x_n^1]$ are encoded. Of these, $k \leq m$ bits $[x_n^k, x_n^{k-1}, \ldots, x_n^1]$ enter a rate $k/(k+1)$ binary convolutional encoder. The encoder produces $k+1$ coded bits $\mathbf{y}_n = [y_n^k, \ldots, y_n^1, y_n^0]$. For $k < m$, the remaining $m - k$ uncoded bits are denoted by $\mathbf{u}_n = [x_n^m, \ldots, x_n^{k+1}]$. The symbol mapper converts $\mathbf{u}_n$ and $\mathbf{y}_n$ into a modulation symbol $a_n = a(\mathbf{u}_n, \mathbf{y}_n) \in \mathcal{A}$. The $(k+1)$-tuple $\mathbf{y}_n$ determines a subset $S_{\mathbf{y}_n}$, and the $(m-k)$-tuple $\mathbf{u}_n$ selects the symbol a_n among the 2^{m-k} symbols of $S_{\mathbf{y}_n}$. Thus the coded bits play the role of subset labels, and the existence of uncoded bits gives rise to parallel transitions in the trellis diagram. The labelling of symbols in the subsets $S_{\mathbf{y}_n}$ has no effect on the distance properties of the code. However, it should exhibit desirable properties of regularity and symmetry, and facilitate simple implementations. If $k = m$, then there are no uncoded bits and hence no parallel transitions.

A convolutional code is most compactly defined in terms of its parity-check equation(s). Specifying a convolutional code in terms of generator polynomials or rational generator functions generally requires more parameters because a generator description not only defines a code but also the input/output relation of a specific encoder.

Let $\mathcal{C}_\mathbf{y}$ denote the set of binary code sequences, or label sequences, produced by the convolutional encoder. Expressed in polynomial sequence notation $\xi(D) = \Sigma_n \xi_n D^n$, the label sequences $\mathbf{y}(D) \in \mathcal{C}_\mathbf{y}$ satisfy the parity-check equation of a rate $k/(k+1)$ convolutional code:

$$\mathbf{y}(D)\mathbf{H}(D)^T = [y^k(D), \ldots, y^1(D), y^0(D)] . [H^k(D), \ldots, H^1(D), H^0(D)]^T = 0(D). \quad (8.3)$$

The maximum degree v of the parity-check polynomials $H^j(D), 0 \leq j \leq k$, is the *constraint length* of the code. Minimal encoders can be realized with v binary storage elements, and hence the trellis diagram has 2^v states. Good TCM codes have at least 4 states ($v \geq 2$) and parity-check polynomials of the form

$$H^j(D) = 0 + h_{v-1}^j D^{v-1} + \cdots h_1^j D + 0, \quad 1 \leq j \leq k,$$
$$H^0(D) = D^v + h_{v-1}^0 D^{v-1} + \cdots h_1^0 D + 1. \quad (8.4)$$

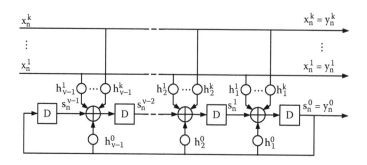

FIGURE 8.5
Systematic recursive rate-$k/(k+1)$ convolutional encoder with restricted parity-check coefficients, shown in observer canonical form [Kailath 1980].

The restriction to $h_\nu^0 = h_0^0 = 1$ and $h_\nu^j = h_0^j = 0, 1 \le j \le k$, will be explained in the next subsection.

Minimal encoders can be realized in different ways [Forney 1970], [Johannesson 1999]. The preferred encoders are either *nonsystematic nonrecursive* or *systematic recursive*. Figure 8.5 depicts a systematic recursive encoder for restricted parity-check polynomials as in Equation 8.4. The encoder can be understood as the circuit that produces from unconstrained binary sequences $y^k(D), \ldots, y^1(D)$ the sequence $y^0(D)$ required to satisfy the parity-check equation. Convolutional encoders are essentially linear filters employing modulo-2 arithmetic instead of ordinary number arithmetic. The encoder of Figure 8.5 generates the sequence of coded bits $y^0(D)$ by a recursive "filter" in observer canonical form [Kailath 1980] with k inputs and one output.

8.2.5   Free Euclidean Distance

It will now be shown that free Euclidean distance of a TCM code can be determined in a manner similar to finding free Hamming distance in a binary convolutional code. The following definitions and lemmas are needed to prove Theorem 8.1 below.

DEFINITION 8.1
Let $v^2(\mathbf{y}, \mathbf{y}') = \min d^2(a(\mathbf{u}, \mathbf{y}), a(\mathbf{u}', \mathbf{y}'))$ be the SED between one symbol in subset $S_\mathbf{y}$ and another symbol in subset $S_{\mathbf{y}'}$ minimized over all $\mathbf{u}, \mathbf{u}' \in \{0, 1\}^{m-k}$.

DEFINITION 8.2
Let $w^2(\mathbf{e}) = \min v^2(\mathbf{y}, \mathbf{y} \oplus \mathbf{e})$ be the Euclidean weight of the label error $\mathbf{e}$ minimized over all $\mathbf{y} = [y^k, \ldots, y^1, y^0] \in \{0, 1\}^{k+1}$.

LEMMA 8.1 **(Set Partitioning)**
Let $q(\mathbf{e})$ be the number of trailing zeros in $\mathbf{e} = [e^k, \ldots, e^1, e^0]$. Then

$$w^2(\mathbf{e}) = \min_{[y^k, \ldots, y^1] \in \{0,1\}^k} v^2(\mathbf{y}, \mathbf{y} \oplus \mathbf{e}) \ge \Delta_{q(\mathbf{e})}^2, \qquad (8.5)$$

where $\oplus$ denotes bit-wise modulo 2 addition. Note that $w^2(\mathbf{0}) = \Delta_{q(\mathbf{0})}^2 = 0$.

The lemma consists of two parts. The first part states that instead of minimizing $v^2(\mathbf{y}, \mathbf{y} \oplus \mathbf{e})$ over all $\mathbf{y} = [y^k, \ldots, y^1, y^0] \in {0, 1}^{k+1}$ to obtain $w^2(\mathbf{e})$ as in Definition 2, it suffices to minimize $v^2(\mathbf{y}, \mathbf{y} \oplus \mathbf{e})$ over all $[y^k, \ldots, y^1] \in \{0, 1\}^k$. The value of y^0 does not matter because of the symmetry of the first-level subsets $\mathcal{B}_0$ and $\mathcal{B}_1$ in all practical cases.

The second part provides a lower bound on $w^2(\mathbf{e})$. The bound follows immediately from inspection of Figures 8.2 and 8.3. For example, if $\mathbf{e} = [\dots 1,\ 0]$, then $q(\mathbf{e}) = 1$. Two symbols in subsets with labels differing by $q(\mathbf{e}) = 1$ must be in the same first-level subset B_0 or B_1 because of $e^0 = 0$, but in different second-level subsets because of $e^1 = 1$. Therefore, the two symbols must have at least SED Δ_1^2. The lower bound in Equation 8.5 is always achieved for larger lattice-type constellations, such that $w^2(\mathbf{e}) = \Delta_{q(\mathbf{e})}^2$. Only for small constellations such as 8-PSK and 16-QAM, the lower bound is not achieved for some values of $\mathbf{e}$ [Ungerboeck 1982].

LEMMA 8.2 **(Rate $k/(k+1)$ Code)**
For every unconstrained sequence $y^k(D), \dots, y^1(D)$ there exists a sequence $y^0(D)$ such that the parity-check equation of the rate $k/(k+1)$ code is satisfied. The lemma is trivially true in light of Figure 8.5.

THEOREM 8.1 **(Free Euclidean Distance** [Ungerboeck 1982]**)**
The squared free Euclidean distance of a TCM code is given by

$$d_{free}^2 = \min\left(\Delta_{k+1}^2, d_{free}^2(k)\right), \tag{8.6}$$

where

$$d_{free}^2(k) = \min_{\substack{\mathbf{e}(D) \in \mathcal{C_y} \\ \neq 0}} \sum_{i=n}^{n+L} w^2(\mathbf{e}_i) \;\geq\; \min_{\substack{\mathbf{e}(D) \in \mathcal{C_y} \\ \neq 0}} \sum_{i=n}^{n+L} \Delta_{q(\mathbf{e}_i)}^2. \tag{8.7}$$

PROOF Because Δ_{k+1}^2 is the minimum SED between symbols assigned to parallel transitions, Δ_{k+1}^2 is an obvious upper bound to d_{free}^2. The quantity $d_{free}^2(k)$ is the squared free SED between symbol sequences with diverging nonparallel paths through the trellis diagram. These paths remerge after more than one transition. Let $\mathbf{y}(D)$ and $\mathbf{y}'(D) = \mathbf{y}(D) \oplus \mathbf{e}(D)$ be the label sequences of such paths. The label error sequence

$$\mathbf{e}(D) = \mathbf{e}_n D^n + \mathbf{e}_{n+1} D^{n+1} + \cdots \mathbf{e}_{n+L} D^{n+L}, \quad \mathbf{e}_n, \mathbf{e}_{n+L} \neq \mathbf{0}, \quad L > 0 \tag{8.8}$$

is a code sequence because of code linearity. One can now express $d_{free}^2(k)$ as

$$
\begin{aligned}
d_{free}^2(k) &= \min_{\substack{\mathbf{e}(D) \in \mathcal{C_y} \\ \neq 0}} \min_{\mathbf{y}(D) \in \mathcal{C_y}} \sum_{i=n}^{n+L} \min_{\mathbf{u}_i, \mathbf{u}'_i} d^2(a(\mathbf{u}_i, \mathbf{y}_i), a(\mathbf{u}'_i, \mathbf{y}_i \oplus \mathbf{e}_i)) \\
&= \min_{\substack{\mathbf{e}(D) \in \mathcal{C_y} \\ \neq 0}} \min_{\mathbf{y}(D) \in \mathcal{C_y}} \sum_{i=n}^{n+L} v^2(\mathbf{y}_i, \mathbf{y}_i \oplus \mathbf{e}_i).
\end{aligned}
\tag{8.9}
$$

From Lemma 8.2, the minimization in Equation 8.9 over all $\mathbf{y}(D) \in \mathcal{C_y}$ involves a minimization over unconstrained sequences $[y^k(D), \dots, y^1(D)]$ with $y^0(D)$ chosen to satisfy the parity-check equation. Lemma 8.1 states that minimizations of $v^2(\mathbf{y}_i, \mathbf{y}_i \oplus \mathbf{e}_i)$ over $[y_i^k, \dots, y_i^1] \in \{0, 1\}^k$ do not depend on y_i^0. Therefore, the minimization in Equation 8.9 over all $\mathbf{y}(D) \in \mathcal{C_y}$ can be replaced by individual minimizations over unconstrained k-tuples $[y_i^k, \dots, y_i^1]$ for each i for arbitrary y_i^0. Thus, $d_{free}^2(k)$ is obtained exactly as a sum of Euclidean weights $w^2(\mathbf{e}_i)$ minimized over nonzero sequences $e(D) \in \mathcal{C_y}$. The lower bound in Equation 8.7 follows trivially from the second part of Lemma 8.1.

For large lattice-type constellations, the lower bound in Equation 8.7 is always achieved because $w^2(\mathbf{e}) = \Delta^2_{q(\mathbf{e})}$ for all $\mathbf{e}$. The lower bound is usually also attained in TCM codes with small symbol constellations, even if $w^2(\mathbf{e}) > \Delta^2_{q(\mathbf{e})}$ for some $\mathbf{e}$.

The property of the parity-check polynomials of good TCM codes with coefficients $h^0_0 = h^0_\nu = 1$ and $h^j_\nu = h^j_0 = 0, 1 \le j \le k$, for $\nu \ge 2$, can now be explained. It can be verified from Figure 8.5 that a label error sequence $\mathbf{e}(D)$ can only depart from the all-zero sequence with a first label error of $\mathbf{e}_n = [e^k_n, \ldots, e^1_n, 0]$ and remerge with the all-zero sequence only with a last label error of $\mathbf{e}_{n+L} = [e^k_{n+L}, \ldots, e^1_{n+L}, 0]$ for some $L > 0$. Lemma 8.1 states that Euclidean weights associated with the first and last label errors are at least Δ^2_1, and hence $d^2_{free}(k) \ge 2\Delta^2_1$. In other words, transitions originating from or joining into one state can only be associated with symbols either from the first-level subset $\mathcal{B}_0$ or from $\mathcal{B}_1$. In most cases, the value of Δ^2_1 in a TCM code is equal to the value d^2_{min} of uncoded modulation at the same spectral efficiency. Therefore, good TCM codes achieve a coding gain of at least 3 dB with at least 4 states over uncoded modulation, provided the first-level set partitioning is characterized by $\Delta^2_1 \ge 2\Delta^2_0$. This is always the case, except for some poor choices of higher-dimensional symbol constellations. ∎

THEOREM 8.2 **(Equal Free Euclidean Distance [Ungerboeck 1982])**
Two TCM codes with parity-check polynomials $\mathbf{H}(D)$ and $\mathbf{H}'(D)$ have the same value of $d^2_{free}(k)$, if $w^2(\mathbf{e}) = \Delta^2_{q(\mathbf{e})}$ for all $\mathbf{e}$, and the parity-check polynomials are related by

$$\mathbf{H}'(D) = [H^k(D), \ldots, H^t(D), \ldots, H^t(D) \oplus H^s(D), \ldots, H^0(D)], 0 \le s < t \le k. \quad (8.10)$$

PROOF Let $\mathbf{H}(D)$ and $\mathbf{H}'(D)$ define the codes of label sequences $C_\mathbf{y}$ and $C'_\mathbf{y}$, respectively. If a label error sequence $\mathbf{e}(D)$ satisfies $\mathbf{e}(D)\mathbf{H}^T(D) = 0(D)$, then

$$\mathbf{e}'(D) = [e^k(D), \ldots, e^t(D) \oplus e^s(D), \ldots, e^s(D), \ldots, e^0(D)] \quad (8.11)$$

satisfies $\mathbf{e}'(D)\mathbf{H}'^T(D) = 0(D)$. The $(k+1)$-tuples $\mathbf{e}'_i$ and $\mathbf{e}_i$ have the same number of training zeros, and hence $\Delta^2_{q(\mathbf{e}'_i)} = \Delta^2_{q(\mathbf{e}_i)}$. If symbol sequences in the first code defined by $\mathbf{H}(D)$ differ by a label error sequence $\mathbf{e}(D)$, then there exist symbol sequences in the second code defined by $\mathbf{H}'(D)$ differing by a label error sequence $\mathbf{e}'(D)$. The smallest Euclidean distances between such symbol sequences in the first code and in the second code are the same. ∎

8.2.6 Code Search and Optimum Codes

Let a symbol constellation $\mathcal{A}$ and constraint length ν be given. Finding a TCM code with largest free Euclidean distance $d^2_{free} = \min(\Delta^2_{k+1}, d^2_{free}(k))$ involves making a tradeoff between Δ^2_{k+1} and $d^2_{free}(k)$. With increasing set-partitioning depth, Δ^2_{k+1} increases or remains unchanged for certain partitioning steps in the case of higher-dimensional constellations. On the other hand, for given ν, the largest achievable value of $d^2_{free}(k)$ generally decreases with increasing k. If for a given value of k a code with $d^2_{free}(k) > \Delta^2_{k+1}$ is found, then the code search should be extended to the next larger value of k.

Theorem 8.1 provides the basis for determining the value of $d^2_{free}(k)$ for given $\mathbf{H}(D)$. Finding the free Euclidean distance in a TCM code is similar to finding free Hamming distance in binary convolutional codes. One only needs to replace Hamming weights of error $(k+1)$-tuples $\mathbf{e}$ by the Euclidean weights $w^2(\mathbf{e})$ defined in Equation 8.5. The search order for parity-check polynomials and the employed rejection rules are described in [Ungerboeck 1982]. Theorem 8.2 can be used as a powerful additional rejection rule to avoid checking the distance of codes, when a distance-equivalent code has already been tested earlier.

The optimum TCM codes for symbol constellations based on the lattices $\mathcal{Z}, \mathcal{Z}^2$, and $\mathcal{Z}^4$ are listed in Tables 8.1, 8.3, and 8.4, respectively. Table 8.2 gives the optimum codes for 8-PSK.

TABLE 8.1

Optimum TCM Codes for 1-D Modulation Based on Lattice $\mathcal{Z}$ [Ungerboeck 1982], [Ungerboeck 1987]. MSSDs: Δ_0^2, $4\Delta_0^2$, $16\Delta_0^2$, ...

ν	k	h^1	h^0	d^2_{free}/Δ_0^2	$\gamma_{4AM/2AM}$ $m=1$	$\gamma_{8AM/4AM}$ $m=2$	γ_{clu} $m\to\infty$	K_{free} $m\to\infty$
2	1	2	5	9.0	2.55	3.31	3.52	4
3	1	04	13	10.0	3.01	3.77	3.97	4
4	1	04	23	11.0	3.42	4.18	4.39	8
5	1	10	45	13.0	4.15	4.91	5.11	12
6	1	024	103	14.0	4.47	5.23	5.44	36
7	1	126	235	16.0	5.05	5.81	6.02	66
8	1	362	515	16.0*	—	5.81	6.02	2
9	1	0342	1017	16.0*	—	5.81	6.02	2

TABLE 8.2

Optimum TCM Codes for 8-PSK Modulation [Ungerboeck 1987]. MSSDs: $\Delta_0^2 = 1$, $\Delta_1^2 = 0.586$, $\Delta_2^2 = 4$

ν	k	h^2	h^1	h^0	$d^2_{free}\Delta_0^2$	$\gamma_{8-PSK/4-PSK}$ $m=2$	K_{free}
2	1	—	2	5	4.000*	3.01	1
3	2	04	02	11	4.586	3.60	2
4	2	16	04	23	5.172	4.13	≈2.3
5	2	34	16	45	5.758	4.59	4
6	2	066	030	103	6.343	5.01	≈5.3
7	2	122	054	277	6.586	5.17	≈0.5
8	2	130	072	435	7.515	5.75	≈1.5

TABLE 8.3

Optimum TCM Codes for 2-D Modulation Based on Lattice $\mathcal{Z}^2$ [Ungerboeck 1987]. MSSDs: Δ_0^2, $2\Delta_0^2$, $4\Delta_0^2$, ...

ν	k	h^2	h^1	h^0	$\dfrac{d^2_{free}}{\Delta_0^2}$	$\gamma_{\frac{16-QAM}{8-PSK}}$ $m=3$	$\gamma_{\frac{32-CR}{16-QAM}}$ $m=4$	$\gamma_{\frac{64-QAM}{32-CR}}$ $m=5$	γ_{clu} $m\to\infty$	K_{free} $m\to\infty$
2	1	—	2	5	4.0*	4.36	3.01	2.80	3.01	4
3	2	04	02	11	5.0	5.33	3.98	3.77	3.98	16
4	2	16	04	23	6.0	6.12	4.77	4.56	4.77	56
5	2	10	06	41	6.0	6.12	4.77	4.56	4.77	16
6	2	064	016	101	7.0	6.79	5.44	5.23	5.44	56
7	2	042	014	203	8.0	7.37	6.02	5.81	6.02	344
8	2	304	056	401	8.0	7.37	6.02	5.81	6.02	44
9	2	0510	0346	1001	8.0*	7.37	6.02	5.81	6.02	4

The parity-check polynomials are specified in octal notation; that is, $\mathbf{h}^j = 103 \triangleq H^j(D) = D^6 + D + 1$. An asterisk (*) indicates cases where $d^2_{free} = \Delta^2_{k+1} < d^2_{free}(k)$. In these codes, free Euclidean distance occurs only between symbols assigned to parallel transitions. Code performance is given in terms of asymptotic coding gains at high SNR over uncoded modulation,

$$\gamma_{coded/uncoded} = 10 \log_{10}\left(d^2_{free}/d^2_{min}\right), \tag{8.12}$$

where d^2_{min} is the minimum squared Euclidean in a symbol constellation used for uncoded modulation with equal signal power and spectral efficiency. The code search has been

TABLE 8.4

Optimum TCM Codes for 4-D Modulation Based on Lattice $\mathcal{Z}^4$ [Ungerboeck 1987]. MSSDs: $\Delta_0^2, 2\Delta_0^2, 2\Delta_0^2, 4\Delta_0^2, 4\Delta_0^2, \ldots$

ν	k	h^4	h^3	h^2	h^1	h^0	$\dfrac{d^2_{free}}{\Delta_0^2}$	$\gamma_{clu} m \to \infty$	$K_{free}\, m \to \infty$
3	2	—	—	04	02	11	4.0	4.52	88
4	2			14	02	21	4.0*	4.52	24
5	3		30	14	02	41	4.0*	4.52	8
6	4	050	030	014	002	101	5.0	5.48	144
7	4	120	050	022	006	203	6.0	6.28	

performed in lexicographical order of the parity-check polynomials. The parity-check coefficients of the first code encountered with the indicated free Euclidean distance is listed.

8.2.7 Performance of TCM Schemes with Different Symbol Dimensions

Figure 8.6 shows the effective coding gains of TCM codes for large lattice-type constellations of different dimensions [Forney 1998]. The effective coding gains have been computed from the distance spectrum of these codes and are plotted versus decoding complexity measured by a detailed operation count. The following codes are considered.

1. 1-D trellis codes "Ung 1D" [Ungerboeck 1982], [Ungerboeck 1987] with large $\mathcal{Z}$-type constellations and convolutional codes for $k = 1$ and $2 \le \nu \le 9$.

2. 2-D trellis codes "Ung 2D" [Ungerboeck 1982], [Ungerboeck 1987] with large $\mathcal{Z}^2$-type constellations and convolutional codes for $k = 1$, $\nu = 2$ and $k = 2$, $3 \le \nu \le 9$.

3. 4-D-trellis codes "Wei 4D" [Wei 1987] with large $\mathcal{Z}^4$-type constellations and convolutional codes for $k = 2$, $\nu = 3, 4$ and $k = 3$, $\nu = 5$, and $k = 4$, $\nu = 6$.

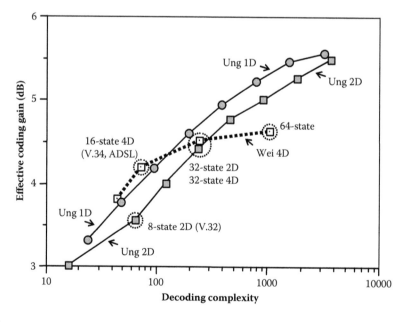

FIGURE 8.6

Effective coding gain versus decoding complexity for 1-D, 2-D, and 4-D TCM codes. (Adapted from [Forney 1998].)

From Figure 8.6 one can see that 4-D trellis codes outperform the 2-D codes at lower decoding complexities. The 4-D 16-state code of Wei has first been adopted in V.34 modems [V.34]. A variation of this code is also used in ADSL transceivers [G.992.1]. The V.34 modem specification includes also 4-D codes with 32 and 64 states, but most manufacturers have only implemented the 16-state code. The incremental coding gains of the additional codes in V.34 have not justified their significantly higher decoding complexities. At higher decoding complexities, the 1-D and 2-D codes outperform the 4-D codes. The lower redundancy per dimension of 4-D constellations limits the achievement of higher coding gains.

It is noteworthy that the performance/complexity tradeoffs of the original 1-D and 2-D codes of Ungerboeck [Ungerboeck 1982] and the 4-D codes of Wei [Wei 1987] and Ungerboeck [Ungerboeck 1987] have not been improved since the publication of these codes.

8.3 Trellis Coding in SHDSL

SHDSL transceivers are defined in ITU-T Recommendation G.991.2 [G.991.2]. SHDSL transceivers are designed primarily for duplex operation over two-wire twisted-pair subscriber lines. For this mode of operation, selected symmetric user data rates in the range of 192 kbit/s to 2312 kbit/s are supported. One-dimensional TCM with 16-PAM modulation is employed and referred to as 16-level trellis-coded PAM (16-TCPAM). G.991.2 also specifies operating modes for asymmetric data rates and for four-wire operation at higher data rates. A recent update to G.991.2 added 32-TCPAM to SHDSL.

Figure 8.7 illustrates the SHDSL transmitter operations in data mode. The TCM encoder generates from scrambled data bits trellis-coded 16-PAM symbols. The time indices m and n represent bit time and symbol time, respectively. The PAM symbols enter a Tomlinson–Harashima precoder [Forney 1998]. The precoded signal is then conditioned by a spectral shaping filter such that the power spectral density (PSD) of the transmitted signal meets the requirements of a given PSD mask. During link initialization, the receivers determine precoder coefficients and transfer them to the corresponding transmitters.

8.3.1 TCM Encoder

The TCM encoder is shown in Figure 8.8. The serially received bits s_m from the scrambler are converted to a 3-bit parallel word $[x_n^3 \, x_n^2 \, x_n^1]$ at the nth symbol time. The bit x_n^1 enters the rate-1/2 convolutional encoder. The coded output bits $y_n^1 y_n^0$ together with the uncoded bits $x_n^3 = y_n^3$ and $x_n^2 = y_n^2$ are used in the 16-PAM symbol mapper to select an output symbol a_n.

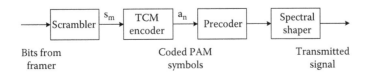

FIGURE 8.7
Transmitter operations in data mode.

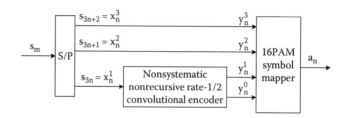

FIGURE 8.8
TCM (16-TCPAM) encoder.

8.3.2   16-PAM Symbol Mapping

The set partitioning of the 16-PAM constellation and the symbol mapping are illustrated in Figure 8.9. The 16-PAM constellation is partitioned into four subsets C_i, $0 \leq i \leq 3$, where $i = 2y^1 + y^0$. The bits $y^1 y^0$ represent subset labels of SP type. The four symbols within each of these subsets are Gray-labeled with the bits $y^3 y^2$.

8.3.3   Convolutional Encoder

The nonsystematic nonrecursive rate-1/2 convolutional encoder is as shown in Figure 8.10. The generator polynomials

$$g^1(D) = b_0 + b_1 D + \cdots b_{20} D^{20}, \quad g^0(D) = a_0 + a_1 D + \cdots a_{20} D^{20} \tag{8.13}$$

are programmable. Codes with constraint length up to $\nu = 20$ can be specified. Vendor-specific generator coefficients are sent during link initialization from the receivers to the transmitters.

For rate-1/2 convolutional codes, the generator polynomials are related to the parity-check polynomials by $g^1(D) = H^0(D)$ and $g^0(D) = H^1(D)$. The SP-type labelling of the subsets of the 16-PAM constellation permits using the parity-check polynomials as specified in Table 8.1 to obtain optimum one-dimensional TCM encoding for $2 \leq \nu \leq 9$. With

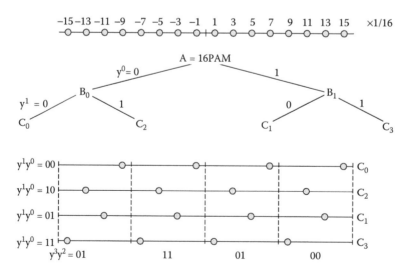

FIGURE 8.9
Set partitioning of 16-PAM constellation and symbol mapping.

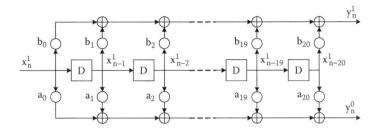

FIGURE 8.10
Nonsystematic nonrecursive rate-1/2 convolutional encoder.

generator polynomials $g'^1(D) = H^1(D) \oplus H^0(D)$ and $g^0(D) = H^1(D)$ distance equivalent codes according to Theorem 8.2 are obtained.

8.4 Trellis Coding in ADSL

ADSL transceivers are defined in ITU-T Recommendations G.992.1 [G.992.1], G.992.2 [G.992.2], G.992.3 [G.992.3], G.992.4 [G.992.4], and G.992.5 [G.992.5]. The transceivers employ discrete multi-tone (DMT) modulation with 256 subcarriers for ADSL [G.992.1], [G.992.2] and ADSL2 [G.992.3], [G.992.4], and 512 subcarriers for ADSL2plus [G.992.5].

G.992.1 specifies a 4-D 16-state trellis code as an optional coding method to obtain an asymptotic coding gain of 4.5 dB over uncoded modulation. Terminated trellis coding is used to generate a block of coded 2-D modulation symbols for modulating the subcarriers during one DMT frame. Encoding starts in the zero state and is terminated in state zero. The same code is also employed for ADSL2 and ADSL2plus. The ADSL code is similar to Wei's 4-D 16-state TCM code [Wei 1987] first introduced in V.34 modems [V.34], but uses a different symbol mapping. The free Euclidean distance of the code is $d_{free}^2 = 4\Delta_0^2$ and occurs between 4-D symbols assigned to parallel transitions. The number of nearest neighbors is $K_{free} = 24$ for large constellations.

8.4.1 Bit Allocations and Tone Ordering

During link initialization, each receiving entity determines a list of bit and gain allocations (b_i, g_i), $i = 1, 2, \ldots, \overline{N} - 1$, where $\overline{N} - 1$ is the number of employed subcarriers. The value of b_i indicates the number of bits to be mapped into 2-D symbols for the ith subcarrier, and g_i is a gain factor to be applied to these symbols. For ADSL, the number of bits per 2-D symbol is either $b_i = 0$ or $2 \leq b_i \leq 15$. For ADSL2 and ADSL2plus, $b_i = 1$ is also permitted. According to G.992.1, the b_i values are sorted in ascending order to obtain a tone ordering list $(t[k])$, $k = 1, 2, \ldots, N_{sc} - 1$, such that $b_{t[1]} \leq b_{t[2]} \leq b_{t[3]} \cdots$. The original list (b_i, g_i) and the tone ordering list $(t[k])$ are sent to the transmitting entity. In this section, trellis encoding is described only for the case of all $b_i \geq 2$.

In the transmitting entity, after scrambling, Reed–Solomon encoding and interleaving, the bits to be transmitted during one DMT frame are held in a data frame buffer. Bits are extracted from the data frame buffer according to the tone ordering list; *i.e.*, bit retrieval and encoding into 2-D modulation symbols begins for carriers with the lowest numbers of bits. The 4-D nature of the code requires reading two entries $b_{t[k]} = x$ and $b_{t[k+1]} = y$ from the ordered list of bit allocations to generate two 2-D modulation symbols. Because

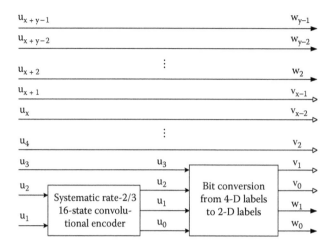

FIGURE 8.11
Convolutional encoding and bit conversion.

4-D trellis coding adds one bit per two 2-D symbols, $x + y - 1$ bits are extracted from the data frame buffer. The bits are assembled, least significant bit first, in a binary word $\mathbf{u} = [u_{x+y-1}, u_{x+y-2}, \ldots, u_2, u_1]$.

8.4.2   Convolutional Encoding and Bit Conversion

Figure 8.11 illustrates the encoding and bit conversion operations leading from $\mathbf{u}$ to symbol labels $\mathbf{v} = [v_{x-1}, \ldots, v_1, v_0]$ and $\mathbf{w} = [w_{y-1}, \ldots, w_1, w_0]$. The x-bit label $\mathbf{v}$ is mapped into a 2-D symbol for the $t[k]$th subcarrier, and the y-bit label $\mathbf{w}$ is mapped into a 2-D symbol for the $t[k+1]$th subcarrier.

The systematic rate-2/3 16-state convolutional encoder is depicted in Figure 8.12(a), as given in [Wei 1987] and [G.992.1]. Figure 8.12(b) shows the equivalent encoder in observer canonical form. At the beginning of a DMT frame, the convolutional encoder is initialized to the all-zero state $s_3' = s_2' = s_1' = s_0' = 0$. At the end of a DMT frame, the encoder is returned to the all-zero state by entering into the convolutional encoder for the last two 4-D symbol periods instead of data bits u_2 and u_1 constrained bits $u_2 = s_2'$ and $u_1 = s_1' \oplus s_3'$.

The bit conversion from 4-D labels to 2-D labels is accomplished by

$$v_1 = u_1 \oplus u_3 \qquad , \quad v_0 = u_3,$$
$$w_1 = u_0 \oplus u_1 \oplus u_2 \oplus u_3, \quad w_0 = u_2 \oplus u_3. \tag{8.14}$$

8.4.3   2-D Symbol Mapping

Because symbol constellations can be very large, an algorithmic symbol mapper is employed to map b bits $v = [v_{b-1}, \ldots v_1, v_0]$ into 2-D symbols $a = (a_R, a_I)$ located on the 2-D grid of odd integer values. In lattice notation, $a \in 2Z^2 + 1$.

8.4.3.1   Even Values of b

For even values of b, a_R and a_I are the odd integers with two's complement representations $(v_{b-1}, v_{b-3}, \ldots, v_1, 1)$ and $(v_{b-2}, v_{b-4}, \ldots, v_0, 1)$, respectively. The bits v_{b-1} and v_{b-2} are the sign bits. Figure 8.13 shows the symbol constellations for $b = 2$ and $b = 4$.

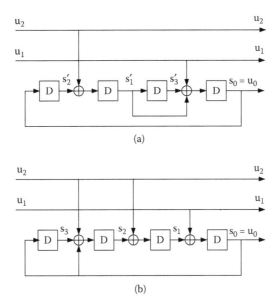

FIGURE 8.12
Convolutional encoder for 4-D trellis-coded modulation in ADSL: (a) 16-state systematic convolutional encoder as in G.992.1 [Wei 1987]; (b) equivalent encoder in observer canonical form, as in V.34 [V.34].

8.4.3.2 Odd Values of b

Figure 8.14 depicts the symbol constellations for $b = 1$, $b = 3$, and $b = 5$. For $b > 3$, a_R and a_I are the odd integers with two's complement representations (X_c, X_{c-1}, v_{b-4}, v_{b-6}, ..., v_3, v_1, 1) and (Y_c, Y_{c-1}, v_{b-5}, v_{b-7}, ..., v_2, v_0, 1), respectively. The bits X_c and Y_c are the sign bits. The relationship between X_c, X_{c-1}, Y_c, Y_{c-1} and v_{b-1}, v_{b-2}, ..., v_{b-5} is given in Table 8.5.

8.4.4 Discussion

The ADSL trellis code has been concisely defined in the preceding subsections by describing the encoding operations. Convolutional encoding and conversion from 4-D labels to 2-D labels are accomplished by linear operations in terms of modulo-2 arithmetic. The question arises how the ADSL code relates to other known 4-D 16-state trellis codes. For this, the labelling of 4-D and 2-D symbol subsets requires further examination.

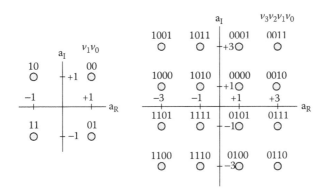

FIGURE 8.13
Two-dimensional symbol constellations for $b = 2$ and 4.

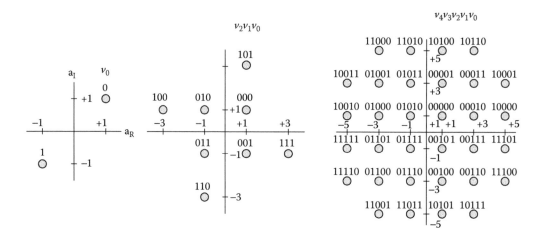

FIGURE 8.14

Two-dimensional symbol constellations for $b = 1, 3,$ and 5 ($b = 1$ not allowed for trellis coding in G.992.1).

TABLE 8.5

Determining $X_c, X_{c-1}, Y_c, Y_{c-1}$

$\nu_{b-1}\nu_{b-2}\nu_{b-3}\nu_{b-4}\nu_{b-5}$	$X_c X_{c-1}$	$Y_c Y_{c-1}$
0 0 0 0 0	0 0	0 0
0 0 0 0 1	0 0	0 0
0 0 0 1 0	0 0	0 0
0 0 0 1 1	0 0	0 0
0 0 1 0 0	0 0	1 1
0 0 1 0 1	0 0	1 1
0 0 1 1 0	0 0	1 1
0 0 1 1 1	0 0	1 1
0 1 0 0 0	1 1	0 0
0 1 0 0 1	1 1	0 0
0 1 0 1 0	1 1	0 0
0 1 0 1 1	1 1	0 0
0 1 1 0 0	1 1	1 1
0 1 1 0 1	1 1	1 1
0 1 1 1 0	1 1	1 1
0 1 1 1 1	1 1	1 1
1 0 0 0 0	0 1	0 0
1 0 0 0 1	0 1	0 0
1 0 0 1 0	1 0	0 0
1 0 0 1 1	1 0	0 0
1 0 1 0 0	0 0	0 1
1 0 1 0 1	0 0	1 0
1 0 1 1 0	0 0	0 1
1 0 1 1 1	0 0	1 0
1 1 0 0 0	1 1	0 1
1 1 0 0 1	1 1	1 0
1 1 0 1 0	1 1	0 1
1 1 0 1 1	1 1	1 0
1 1 1 0 0	0 1	1 1
1 1 1 0 1	0 1	1 1
1 1 1 1 0	1 0	1 1
1 1 1 1 1	1 0	1 1

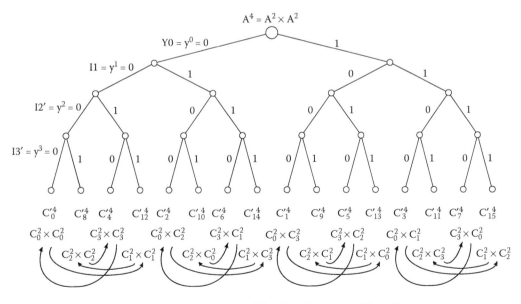

$\begin{matrix} c & b \\ (01) & (10) \end{matrix}$	$\begin{matrix} c & b \\ (01) & (10) \end{matrix}$

$Z = (Z1, Z0)$

Wei labeling
(V.34)(Wei 1987)

$\mu = (\mu^1\mu^0)$

SP labeling of Figure 8.2
(Ungerboek 1982),
(Ungerboek 1987)

$v = (v_1v_0)$

Gray labeling
ADSL (G992.1)

$$\mu^1 = Z1 \oplus Z0 = v_0, \mu^0 = Z0 = v_1 \oplus v_0$$

FIGURE 8.15
Wei labelling, SP labelling of Figure 8.2, and Gray labelling of the four second-level subsets of $\mathcal{Z}^2$-type constellations.

The constellations of constituent 2-D symbols need to be partitioned into four subsets, denoted in Figure 8.2 by C_i^2, $0 \le i \le 3$. First, one should notice that in the ADSL code and the code of Wei [Wei 1987], [V.34] the 2-D subsets are labelled differently. The differences are illustrated in Figure 8.15. The Wei labelling is also of SP-type, but differs from the SP labelling in Figure 8.2 in the sign of 90° rotations. For ADSL, Gray labelling is employed. The three labelling schemes are related by simple linear transformations.

Differences in the labelling of 4-D subsets can be exposed more clearly by replacing in the Wei and ADSL schemes the 2-D subset labels uniformly with corresponding SP labels as in Figure 8.2. The set partitioning of 4-D constellations for the Wei code and the ADSL code is shown in Figure 8.16 and Figure 8.17, respectively. The 4-D labels $i = [y^3 y^2 y^1 y^0]$,

+90° rotations: I3'I2' = 00 → 11 → 10 → 01 → 00

FIGURE 8.16
Set partitioning of 4-D constellations as in Wei's code [Wei 1987] [V.34] with equivalent SP labelling of 2-D subsets as in Figure 8.2.

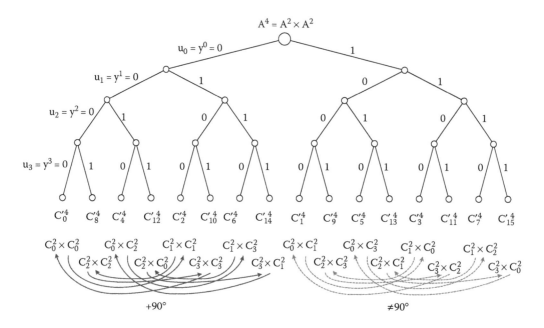

FIGURE 8.17
[G.992.1] Set partitioning of 4-D constellations as in ADSL code [G992.1] with equivalent SP labelling of 2-D subsets as in Figure 8.2.

$0 \leq i \leq 15$, and the 2-D labels p and q, $0 \leq p, q \leq 3$ of the 4-D subsets $C_i'^4 = C_p^2 \times C_q^2$ are related as follows.

$$\text{Figure 8.3: } p = (2y^3 + y^1), \qquad q = p + (2y^2 + y^0) \bmod 4, \qquad (8.15)$$

$$\text{Wei, V.34: } p = -(2y^3 + y^2) \bmod 4 \qquad q = p - (2y^1 + y^0) \bmod 4, \qquad (8.16)$$

$$\text{ADSL: } p = (2y^3 + y^1) \qquad q = 2(y^3 \oplus y^2 \oplus y^1 \oplus y^0) + (y^1 \oplus y^0). \qquad (8.17)$$

Compared to the set partitioning of Figure 8.3, in Wei's set partitioning the bits $y^2(= u_2)$ and $y^1(= u_1)$ are swapped and $90°$ rotations of the 2-D subsets occur with a different sign. The difference may be reduced to the sign difference of the $90°$ rotations by swapping bits y^2 and y^1 in Wei's 16-state convolutional encoder. Consider now Wei's convolutional encoder in observer canonical form as in Figure 8.12(b), with the two bits swapped. In the notation of the code tables of Section 8.2, the Wei's encoder generates a convolutional code with parity check polynomials $\mathbf{h}^2 = 14$, $\mathbf{h}^1 = 02$, $\mathbf{h}^0 = 31$. Table 8.4 specifies for $\nu = 4$ a code with the same Euclidean distance properties as the Wei code, but with parity check polynomials $\mathbf{h}^2 = 14$, $\mathbf{h}^1 = 02$, $\mathbf{h}^0 = 21$. The two codes differ only in the values of $\mathbf{h}^0$. The code in Table 8.4 precedes the Wei code in lexicographical order among distance-equivalent codes, and has therefore been listed.

The 4-D set partitioning for the ADSL code does not possess the same symmetries under $90°$ rotations as the partitioning of Wei and Figure 8.3. The ADSL partitioning is therefore not suited for the construction of a rotationally invariant 4-D trellis code.

8.5 Conclusions

The principles of TCM have been described in Section 8.2 with the intent to provide a concise summary of the theoretical underpinnings of TCM. These theoretical foundations and lists of optimal TCM codes have been available for quite some time. Not surprisingly, however, the use of different notation by various authors may have prevented a widespread common understanding of TCM. There may have also been other reasons. As a result, compared to the application of Reed–Solomon codes for forward error correction, the use of TCM in standardized telecommunication equipment is less homogeneous. Many variations have been created and successfully promoted. In the area of DSL, the developers of standards have not agreed on a single TCM code or a family of such codes for SHDSL and ADSL.

For SHDSL, one-dimensional TCM with a programmable convolutional encoder in non-systematic nonrecursive form has been chosen. SHDSL allows for codes with up to 2^{20} (!) states. Although codes with that high number of states may never practically be employed, permitting codes with a high number of states fits well with the fact that one-dimensional TCM codes offer better performance than higher-dimensional TCM codes, when high decoding complexities are allowed.

For ADSL, a single 4-D 16-state TCM code with a convolutional encoder in systematic recursive form has been specified (although not in canonical form). At the moderate decoding complexity of this code, the ADSL code offers a higher coding gain than 1-D or 2-D TCM schemes at comparable decoding complexity. The choice of the ADSL code may have been motivated by coding expertise gained from voiceband modems. The 4-D 16-state Wei code specified for V.34 and V.90/V.92 voiceband modems is invariant under $90°$ carrier phase rotations. This code attribute has been considered a necessary feature for voiceband modems. The ADSL code is also based on Wei's 16-state convolutional encoder, but uses a different symbol mapping. The property of rotational invariance is not needed in ADSL, because receivers recover absolute carrier phase from pilot carriers and known synch symbols.

There will probably be no significant new developments of TCM codes as defined in the narrower sense of this chapter. During the last decade, the focus of coding research has shifted to capacity-approaching concatenated coding schemes with interleaving and iterative decoding. Achieving diversity gains in addition to good performance in the presence of stationary noise has become a major focus for wireless communication. The newer coding concepts may find applications also in DSL systems.

References

[Forney 1970] G.D. Forney, Jr., *Convolutional codes I: Algebraic structure*, IEEE Trans. Inform. Theory, vol. IT-16, pp. 720–738, Nov. 1970.

[Forney 1988a] G.D. Forney, Jr., *Coset codes — Part I: Introduction and geometrical classification*, IEEE Trans. Inform. Theory, vol. 34, pp. 1123–1151, Sept. 1988.

[Forney 1988b] G.D. Forney, Jr., *Coset codes — Part II: Binary lattices and related codes*, IEEE Trans. Inform. Theory, vol. 34, pp. 1152–1187, Sept. 1988.

[Forney 1989] G.D. Forney, Jr. and L.-F. Wei, *Multidimensional constellations Part I: Introduction, figures of merit, and generalized cross constellations*, IEEE J. Select. Areas Commun., vol. 7, pp. 877–892, Aug. 1989.

[Forney 1998] G.D. Forney, Jr. and G. Ungerboeck, *Modulation and coding for linear Gaussian channels*, IEEE Trans. Inform. Theory, vol. 44, pp. 2384–2415, Oct. 1998.

[G.991.2] ITU-T Recommendation G.991.2, *Single-pair high-speed digital subscriber line (SHDSL) transceivers*, Feb. 2001.

[G.992.1] ITU-T Recommendation G.992.1, *Asymmetric digital subscriber line (ADSL) transceivers*, June 1999.

[G.992.2] ITU-T Recommendation G.992.2, *Splitterless asymmetric digital subscriber line (ADSL) transceivers*, June 1999.

[G.992.3] ITU-T Recommendation G.992.3, *Asymmetric digital subscriber line (ADSL) transceivers — 2 (ADSL2)*, July 2002 (pre-published).

[G.992.4] ITU-T Recommendation G.992.4, *Splitterless asymmetric digital subscriber line transceivers 2 (splitterless ADSL2)*, July 2002 (pre-published).

[G.992.5] ITU-T Recommendation G.992.5, *Asymmetric digital subscriber line (ADSL) transceivers — Extended bandwidth ADSL2 (ADSL2plus)*, May 2003 (pre-published).

[Johannesson 1999] R. Johannesson and K.S. Zigangirow, *Fundamentals of Convolutional Coding*, IEEE Press, Los Alamitos, CA, 1999.

[Kailath 1980] T. Kailath, *Linear Systems*, Prentice Hall, Upper Saddle River, NJ, 1980.

[Ungerboeck 1976] G. Ungerboeck and I. Csajka, *On improving data-link performance by increasing the channel alphabet and introducing sequence coding*, Proc. 1976 IEEE Int. Symposium on Information Theory, p. 53, June 1976.

[Ungerboeck 1982] G. Ungerboeck, *Channel coding with multilevel/phase signals*, IEEE Trans. Inform. Theory, vol. 28, pp. 55–67, Jan. 1982.

[Ungerboeck 1987] G. Ungerboeck, *Trellis-coded modulation with redundant signal sets — Part I: Introduction; — Part II: State of the art*, IEEE Commun. Mag., vol. 25, no. 2, pp. 5–21, Feb. 1987.

[V.34] ITU-T Recommendation V.34, *A modem operating at data signaling rates of up to 33 600 bit/s for use on the general switched telephone network and on leased point-to-point 2-wire telephone-type circuits*, Feb. 1998, replacing first version of 1994.

[Wei 1984] L.-F. Wei, *Rotationally invariant convolutional channel encoding with expanded signal space, Part II: Nonlinear codes*, IEEE J. Select. Areas Commun., vol. 2, pp. 672–686, Sept. 1984.

[Wei 1987] L.-F. Wei, *Trellis-coded modulation using multidimensional constellations*, IEEE Trans. Inform. Theory, vol. 33, pp. 483–501, July 1987.

[Wei 1989] L.-F. Wei, *Rotationally invariant trellis-coded modulations with multidimensional M-PSK*, IEEE J. Sel. Areas Commun., vol. 7, pp. 1281–1295, Dec. 1989.

9

Error Control Coding in DSL Systems

Cory S. Modlin

CONTENTS

9.1   Introduction 234
9.2   Background on Error Control Codes 235
    9.2.1   Galois Fields 235
    9.2.2   Cyclic Redundancy Check (CRC) 238
    9.2.3   An Upper Bound on the Probability of an Undetected Error 241
    9.2.4   Exact Calculation of the Probability of Undetected Error 244
    9.2.5   Reed–Solomon Codes 246
    9.2.6   Decoding Reed–Solomon Codes 248
    9.2.7   Consequences of Uncorrectable Errors 249
        9.2.7.1   Application of Reed–Solomon Codes 253
9.3   Forward Error Correction Coding Gain 253
9.4   Interleaving 258
    9.4.1   Optimum Memory Implementation Using Tong's Method 261
    9.4.2   Forney's Triangular Interleaver 262
    9.4.3   Error Correction Comparison of DMT
        and Single-Carrier Modulation 265
    9.4.4   Erasures 265
9.5   Concatenated Coding 266
9.6   Summary 269
9.7   Acknowledgments 269
References 269

ABSTRACT   Error control codes are designed to detect or correct errors caused by noise sources that are usually nonstationary. The often aging wires that carry DSL signals are subject to noise conditions that change sometimes on the order of milliseconds and other times over days. Cyclic redundancy check (CRC) codes are used to detect errors without correcting them. Error detection is crucial for the management of DSL systems. This chapter analyzes the ability of CRC codes to detect errors. Reed–Solomon forward error correction (FEC), often in combination with interleaving, can detect and correct errors. This chapter looks at the error rate and burst error correction capability with a realistic implementation of a Reed–Solomon decoder and examines two classes of memory-optimal interleavers. It also discusses how error control codes concatenated with trellis codes behave somewhat differently from when there is no trellis code.

9.1 Introduction

With DSL systems running over existing telephone lines, environmental conditions are not easily controlled. Changing weather, varying loop quality and age, inconsistent indoor wiring, radio interference, and DSL services running on adjacent wire pairs, to name a few, can all cause errors with different characteristics. Some errors are the result of persistent thermal noise. Noise of this sort is typically modelled as Gaussian and causes short, infrequent errors. Other noise events, for example, caused by the switching on and off of electrical appliances or by quickly varying radio frequency interference, can cause relatively longer bursts of errors.

Depending on the application running over the DSL service, errors can cause varying degrees of disruption. For data (for example, Internet) applications, information is typically delivered in packets. If there are errors anywhere in a particular packet, the entire packet is discarded and a new packet is requested. In this case, the correction of error bursts at the DSL layer may not be required or may be limited to correcting short error bursts. However, for video applications where there typically is not time or memory to retransmit packets, errors can show up as annoying picture distortion. In this case, it is desirable to correct as many error bursts as possible.

Regardless of the application running over DSL, users demand the data they receive be reliable and be delivered at high speed. They also demand that the DSL transceivers be inexpensive. Given the environmental, cost, and high-speed constraints, it is unrealistic to assume that DSL will be as consistent as fiber or magnetic storage. However, all DSL systems employ error control coding to limit the impact of errors by detecting and sometimes correcting them.

In DSL systems, there are usually a number of different codes whose primary function is to detect and possibly correct errors with different characteristics. Unlike trellis codes, whose primary function is to increase the overall data rate in the presence of Gaussian-like noise, error control codes do not necessarily provide for higher data rates and are designed to combat error bursts of varying lengths.

In DSL transceivers, error detection is done using cyclic redundancy check (CRC) codes. Essentially, the CRC is extra bits appended to a block of data that are calculated based on the content of the data. If there are errors anywhere in the block of data (including in the CRC itself), the CRC calculated at the receiver will not match the received CRC, and an error is detected.

Information about detected errors is used for a couple of different purposes. Persistent errors in transmitted data can indicate a problem with the transmission that might require the transceivers to re-train or adjust somehow. A CRC can also be used in the modem overhead channel. In ADSL2 [G.992.3 1993], for example, a high-level data link control (HDLC)-based [ISO/IEC 1993] modem overhead channel is provided to exchange physical layer status information between transceivers. If errors are detected, the message will be re-transmitted. This type of error control coding is referred to as automatic re-transmission request (ARQ).

Finally, some DSL transceivers include forward error correction (FEC) based on Reed–Solomon codes. These codes, frequently used together with interleaving, can detect and then correct a certain number of errors within a block of data. This is the primary tool used to combat long bursts of errors.

This chapter presents a brief introduction to linear, cyclic codes of the type used in DSL. Forward error correction in particular offers an advantage by correcting errors; this coding gain for single-carrier and multi-carrier DSL is investigated. The chapter then discusses interleaving, often used in conjunction with forward error correction, to correct bursts of errors. Requirements for interleaving, which are different for single- and multi-carrier systems, are then presented. Finally, concatenated trellis and forward error correction codes, as they are used in ADSL, are examined.

9.2   Background on Error Control Codes

Error control coding is applied to binary data just outside of the physical layer encoding. Error control codes used in DSL are systematic block codes, meaning that data is partitioned into blocks of k symbols and r redundancy symbols are appended to the block without altering the original k symbols before the entire codeword of $k + r$ symbols is transmitted over a channel. The channel in this case is the combination of the physical layer DSL transceiver and the line. The redundancy of the error control code allows the receiver to detect errors and possibly correct errors. This is illustrated in Figure 9.1.

When the modulo-2 sum of any two codewords forms another valid codeword, the code is a linear code. When the cyclic rotation of any codeword forms another valid codeword, the code is a cyclic code. Linear cyclic codes describe the subclass of codes used in DSL systems for error control.

9.2.1   Galois Fields

Galois fields, named for the French mathematician who first identified them [Toti Rigatelli 1996], form the foundation of error control codes. Loosely defined, a field is a set of elements over which addition, subtraction, multiplication, and division are possible without leaving the set. There must exist two field elements, 0 and 1, that satisfy $a + 0 = a$ and $a \cdot 1 = a$, and the commutative, associative, and distributive rules of algebra must apply. A Galois field is a finite field, meaning that the set has a finite number of elements. All finite fields have p^m elements, where p is a prime number. For practical communication applications with binary data inputs, fields with $p = 2$ are of interest. Galois fields that are based on $p = 2$ are denoted by GF(2^m). For a more complete discussion on Galois fields, the reader is referred to [Lin 1983], [Blahut 1983], [Berlekamp 1984], and [Blahut 1990].

To complete the description of the GF(2^m) Galois field, addition and multiplication must be defined. The inverse operations of subtraction and division are implicitly defined given the definition of a Galois field. Because a GF(2^m) Galois field has 2^m elements, one way to

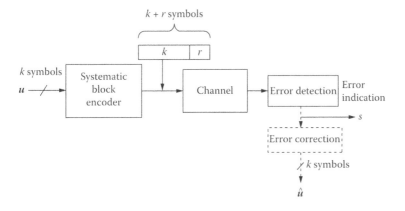

FIGURE 9.1

Illustration of block coding. k information symbols are encoded with r redundant symbols and transmitted over a channel. At the receiver, errors can be detected or possibly corrected.

represent the elements of the field is as an m-bit number where each bit is an element of GF(2), either 0 or 1. Addition is then defined as the bit-by-bit, modulo-2 addition of the elements. For example, in GF(2^8), it is possible to add (10110101)+(01100110) = (11010011).

Multiplication is more complicated. It might be tempting to simply multiply the values. For example, one could say that (01100110) · (00000100) = (0110011000)(102 · 4 = 408). But then the result is no longer in GF(2^8) because the value is greater than 255. To get around this problem, one could try taking the result modulo-2^8 by subtracting 256. In this case, (01100110) · (00000100) = (0110011000) − (100000000) = (10011000). However, this method of multiplication also violates the properties of a field, because, for example, there is no value when multiplied by (01100000) that equals (00000001).

Multiplication that does satisfy the properties of a Galois field can be defined using polynomial multiplication over GF(2), modulo a pre-defined, irreducible, fixed polynomial, $p(x)$. A polynomial is said to be over GF(2) if all of the coefficients are in GF(2) (0 or 1). The degree of a polynomial, $p(x)$, is the largest power of x with a nonzero coefficient. The polynomial, $p(x)$, is irreducible if it cannot be factored by any other polynomial of degree greater than zero.

To multiply (01100110) · (00000100) in GF(2^8), first each octet would be represented as a polynomial. Polynomial representation is another way to represent elements in a Galois field. The multiplication becomes

$$c(x) = (x^6 + x^5 + x^2 + x)(x^2) = x^8 + x^7 + x^4 + x^3.$$

One irreducible polynomial in GF(2^8) is $p(x) = x^8 + x^4 + x^3 + x^2 + 1$. Using this polynomial, $c(x)$ modulo $p(x)$, which is the remainder of $c(x)/p(x)$, is $x^7 + x^2 + 1$ or (10000101).

In the last example, the polynomial $p(x) = x^8 + x^4 + x^3 + x^2 + 1$ is of a special class of irreducible polynomials called a primitive polynomial. An irreducible polynomial of degree m is primitive if the smallest positive integer, n, for which $p(x)$ divides $x^n + 1$ is $n = 2^m - 1$. In general, there is more than one primitive polynomial for any degree m. A transmitter and a receiver generating a code using Galois field arithmetic must be using the same polynomial. When using a primitive polynomial $p(x)$ to construct a Galois field, the element x is a primitive element; every nonzero element of the Galois field can be represented as a power of x. In general, x is not the only primitive element in a Galois field and so the variable α is often used to denote a primitive element. When $\alpha = x$, it is possible to construct a table of the powers of α, as shown in Table 9.1.

Having defined addition and multiplication, subtraction and division are straightforward. Because any element added to itself is zero, every element is its own additive inverse. Therefore, addition and subtraction amount to the same operation. The multiplicative inverse can be found by noting that $\alpha^{q-1} = 1$ in the GF(q) field constructed by a primitive polynomial. Therefore, $\alpha^n · \alpha^{q-n-1} = 1$ and α^n and α^{q-n-1} form an inverse pair. In other words, division by α^n is equivalent to multiplication by α^{q-n-1}.

Example 9.1   (Data Scrambler Using Primitive Polynomials)

Nearly all digital communications systems scramble transmitted data before it is line coded and transmitted over the channel. At the receiver, a descrambler recovers the transmitted data if there are no errors in between. Data scramblers are used to prevent patterned transmit sequences, such as all ones, all zeros, or some short periodic pattern, such as asynchronous transfer mode (ATM) idle cells, from creating problems with the steady-state operation of the modem. Problems could include a signal with a very high peak-to-average ratio that would cause transmit and receive clipping or could also involve the destruction of adaptive algorithms that depend on random input data. Scramblers can also generate a pseudo-random binary sequence (PRBS) that can be used during modem training.

TABLE 9.1

Elements of GF(256) Generated by
$p(x) = x^8 + x^4 + x^3 + x^2 + 1$

Power Representation	Polynomial	Octet
0	0	0000 0000
1	1	0000 0001
α	x	0000 0010
α^2	x^2	0000 0100
α^3	x^3	0000 1000
α^4	x^4	0001 0000
α^5	x^5	0010 0000
α^6	x^6	0100 0000
α^7	x^7	1000 0000
α^8	$x^4 + x^3 + x^2 + 1$	0001 1101
α^9	$x^5 + x^4 + x^3 + x$	0011 1010
$\vdots$		
α^{254}	$x^7 + x^3 + x^2 + x$	1000 1110
α^{255}	1	0000 0001

A simple and widely used way to create a training sequence is by using a primitive polynomial to generate a sequence. For example, a training sequence can be generated using a primitive polynomial $p(x)$ of degree m, using a linear feedback shift register (LFSR) as shown in Figure 9.2. To generate a training signal, there is no input stream. In Figure 9.2, p_n is the coefficient of x^n in $p(x)$. Using an LFSR, the m bits in the shift register are initialized

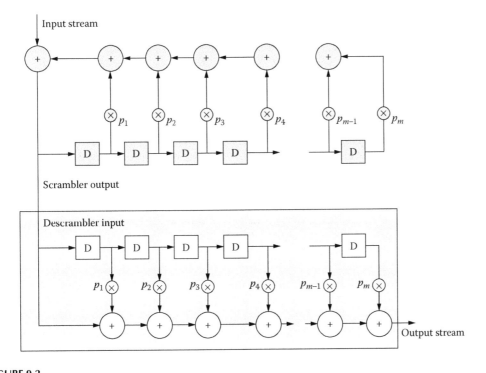

FIGURE 9.2
Self-synchronized scrambler and descrambler. The scrambler also serves as a PRBS generator when there is no input sequence and the delay elements are initialized to any nonzero state. For binary sequences over GF(2), all operations are modulo-2.

to some nonzero value. If the receiver and transmitter both know what the initial value is, the receiver can create the same training sequence. Then, the LFSR effectively multiplies the value in the shift register by two (left shift) and gives the remainder of the division by the $p(x)$. Because $p(x)$ is a primitive polynomial, the pattern will not repeat until after $2^m - 1$ bits.

Used as a scrambler, the LFSR (as shown in Figure 9.2) takes a binary stream as input. For example, if the scrambler is based on the primitive polynomial

$$p(x) = x^{23} + x^5 + 1,$$

as it is in HDSL downstream and ISDN downstream, the output of the scrambler could be written

$$a(x) = m(x) + a(x - 5) + a(x - 23), \tag{9.1}$$

where $m(x)$ is the input message. The descrambler can be written as

$$\hat{m}(x) = y(x) + y(x - 5) + y(x - 23), \tag{9.2}$$

where $y(x)$ is the received sequence, possibly with errors.

This scrambler is called a self-synchronized scrambler because no information about the state of the transmitter is required at the receiver. Using a self-synchronized scrambler simplifies the transitions from initialization to data exchange. Also, if data is ever lost by the receiver, the descrambler will re-acquire synchronization on its own. However, as evident in Equation 9.2, a single error in the received sequence will cause three errors in the received message for a polynomial with three nonzero taps. This multiplication of errors is called error propagation.

Especially in block processing systems like DMT, where loss of synchronization between the transmitter and receiver is very unlikely, this problem of error propagation can be avoided by using a frame-synchronized scrambler [Starr 1999]. As shown in Figure 9.3, a frame-synchronized scrambler generates a PRBS and then adds this, modulo-2, to the transmit sequence at the transmitter and to the received sequence at the receiver. The state of the LFSR must remain synchronized between the transmitter and receiver throughout data transmission, but error propagation is avoided.

9.2.2   Cyclic Redundancy Check (CRC)

Nearly all DSL systems and nearly all transport protocols running over DSL such as TCP/IP and ATM use a cyclic redundancy check to detect errors. A CRC is an example of a linear, cyclic, block code where r symbols of redundancy are appended to a block of k symbols to form a codeword of $n = k + r$ symbols. A code with a codeword size $n = k + r$ symbols is often referred to as an (n, k) code. A code with k information symbols is linear if and only if the sum of any two codewords is also a codeword. A cyclic code has the additional property that the cyclic shift of any codeword is itself a codeword. Normally, a CRC is used in systematic form, meaning that r redundant symbols are appended to k unaltered data symbols.

A bit-level CRC in $GF(2^r)$ is encoded as the remainder of a message polynomial, $d(x)$, multiplied by x^r and divided by the polynomial used to construct the Galois field, $p(x)$, of degree r. This can be implemented as a linear feedback shift register as in Figure 9.4. The complete codeword, $d(x) \cdot x^r + r(x)$, is evenly divisible by $p(x)$. Therefore at the receiver, if there are no errors, the received codeword can be divided by $p(x)$ using the same procedure and the remainder will be zero.

For example, a Galois field can be constructed using the primitive polynomial $p(x) = x^3 + x^2 + 1$, which can also be written in binary form as (1101). For this code, $r = 3$ because it is a third-order polynomial. Assume a length $k = 2$ message is encoded, $d(x) = x$, or (10) in binary form, using this code. Using polynomial division, $d(x) \cdot x^3 = x^4$ is divided

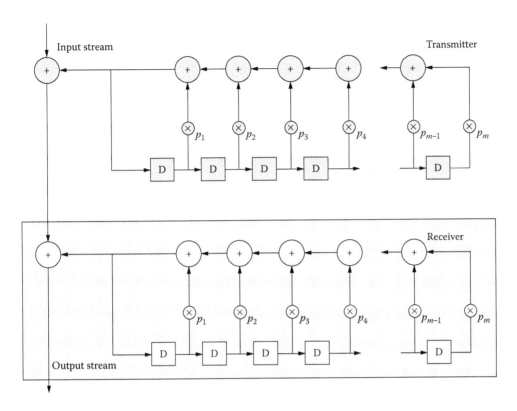

FIGURE 9.3
Frame-synchronized scrambler and descrambler. The initial states of the scrambler and descrambler must be identical, and the states must remain synchronized throughout the connection. For binary sequences over GF(2), all operations are modulo-2.

by $p(x)$, which can be done using long division as shown in Figure 9.5. The remainder is $x^2 + x + 1$ (111) and the complete message is then (10111) of length $n = 5$.

Because this is a linear code, it can also be represented as a matrix multiplication. The message, $d(x)$, can be multiplied by a generator matrix, G, to compute the message. The generator matrix has dimensions $k \times n$. Given the generator polynomial $p(x)$, the generator

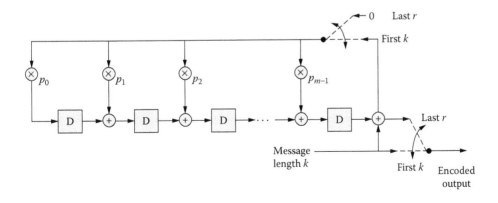

FIGURE 9.4
Cyclic encoder for a systematic code with k information symbols and r symbols of redundancy. Addition and multiplication are over the Galois field.

```
                    Quotient
        111
   1101 | 10000
          1101
          ----
          1010
          1101
          ----
           111    Remainder
```

FIGURE 9.5
Polynomial division.

matrix G can be constructed as [Pless 1998]

$$G = \begin{bmatrix} p(x) & 0 & 0 & \cdots & 0 \\ 0 & p(x) & 0 & \cdots & 0 \\ 0 & 0 & p(x) & \cdots & \vdots \\ \vdots & & & \ddots & 0 \\ 0 & 0 & \cdots & 0 & p(x) \end{bmatrix}_{k \times n}. \tag{9.3}$$

To create a systematic code using the generator matrix, G must be put in systematic form so that

$$G_{sys} = [I_{k \times k} | P_{r \times k}]. \tag{9.4}$$

The matrix G can be converted to systematic form using row operations. For example, using the code from the previous example with $p(x) = x^3 + x^2 + 1$, the generator matrix is equal to

$$G = \begin{bmatrix} 1 & 1 & 0 & 1 & 0 \\ 0 & 1 & 1 & 0 & 1 \end{bmatrix},$$

which can be put in systematic form by replacing the first row with the sum of the first and second rows:

$$G_{sys} = \begin{bmatrix} 1 & 0 & 1 & 1 & 1 \\ 0 & 1 & 1 & 0 & 1 \end{bmatrix}.$$

It is clear that $[10] \cdot G_{sys} = (10111)$ as it is using polynomial division.

If there are errors during transmission, the remainder calculated at the receiver (equal to the received message divided by the generator polynomial $p(x)$) will not, in general, be zero. However, if errors occur in such a way as to create another valid codeword, the remainder of the division at the receiver will be zero and the errors will not be detected. Considering the received codeword (with possible errors) as $w(x) = u(x) + e(x)$ where $u(x)$ is the noiseless transmitted codeword, it is clear that $u(x) \bmod p(x)$ is zero because $u(x)$ is a valid codeword. Therefore,

$$w(x) \bmod p(x) = (u(x) + e(x)) \bmod p(x)$$
$$= u(x) \bmod p(x) + e(x) \bmod p(x)$$
$$= e(x) \bmod p(x).$$

When selecting a CRC, the probability of having undetected errors is of interest. The types of error patterns in $e(x)$ that cannot be detected depend on the polynomials used and the message length, k. To calculate the probability of an undetected error, let A_i be the number of codewords with i ones in it or with Hamming weight i. The set of A_i is known as a weight distribution of the code [Peterson 1972]. If, for example, there exists a codeword

TABLE 9.2

CRC Generator Polynomials Used in DSL and Related Standards

Standard	Number of Check Bits m	CRC Polynomial	Comment
ADSL/ADSL2/ADSL2+/ DMT VDSL1[1]	8	$x^8 + x^4 + x^3 + x^2 + 1$	Primitive
CRC-6: SHDSL/ HDSL/ HDSL2	6	$x^6 + x + 1$	Primitive
QAM VDSL1	4	$x^4 + x + 1$	Primitive
CRC-16: HDSL2/ HDLC, IP	16	$x^{16} + x^{12} + x^5 + 1 =$ $(x^{15} + x^{14} + x^{13} + x^{12} + x^4$ $+ x^3 + x^2 + x + 1) \cdot (x + 1)$	Primitive times $(x + 1)$

[1] DMT VDSL has been selected as the American national standard and as the 802.3ah standard. In the ITU, the first generation of VDSL, called VDSL1, is DMT-based with QAM defined in an annex. Moving forward, VDSL2 in the ITU will be DMT-based only.

with i ones in it, this also means that there exists an error pattern with i ones in it that cannot be detected. Therefore, the probability of an undetected error is

$$P_{ue} = \sum_{i=1}^{n} A_i p^i (1 - p)^{n-i}, \tag{9.5}$$

where p is the probability of a single bit error. This assumes that all errors are independent and equally likely.

One way to calculate A_i is to form all 2^k codewords and to keep track of the weight of each codeword. For small values of k, this is possible on a computer. However, most CRCs are used to detect errors in many hundreds or thousands of message bits, for which forming all possible codewords is not practical. In this case, two strategies can be used to calculate the probability of undetected errors. The first is to place an upper bound on this probability using some characteristics of the generator polynomial. The second is to use the dual code to calculate this probability exactly.

9.2.3 An Upper Bound on the Probability of an Undetected Error

To place an upper bound on the probability of an undetected error, one can draw general conclusions about the polynomial used to generate the CRC code. A partial list of CRC polynomials used in DSL and related protocols is given in Table 9.2. In all of these cases, the CRC polynomial is either a primitive polynomial, $p_1(x)$, or a primitive polynomial multiplied by $(x + 1)$. The order of the primitive polynomial $p_1(x)$ is defined as e and the order of the entire CRC polynomial, $p(x)$, as r. If $p(x)$ is primitive, then $e = r$.

Looking at these polynomials, it is clear that [Peterson 1972], [Starr 1999], [Stallings 1997], [Peterson 1961]:

- All single bit errors will be detected, because $e(x) = x^a$ is never divisible by $p(x)$ for any a.
- If e is the degree of the primitive polynomial, $p_1(x)$, only double errors spaced a multiple of $2^e - 1$ bits apart will not be detected. By definition of a primitive polynomial, the smallest number, a, for which $x^a + 1$ is a multiple of $p_1(x)$ is $a = 2^e - 1$. This means that if $x^{2^e} - 1$ is divided by $p_1(x)$, the remainder will be 1. Because $x^a + 1$ is not a multiple of $p_1(x)$ for any value of a less than $2^e - 1$, the next smallest power

of x that will have a remainder of one when divided by $p_1(x)$ must be $x^{2(2^e-1)}$. This demonstrates that double errors must be a multiple of 2^e-1 samples apart to avoid detection. Any double error also divides the factor $(x+1)$, so this factor provides no additional help. In many applications, the length of the block of data will be longer than 2^e-1 bits and so it is possible, although rare, to have undetectable double bit errors. However, when the block is equal to or shorter than 2^e-1 symbols, all double errors can be detected and, frequently, even longer errors can also always be detected. This is the basis for Hocquenghem, Bose, and Chaudhuri (known as "BCH") and Reed–Solomon codes, which are discussed later.

- One can easily verify that any polynomial multiplied by $(x+1)$ has an even number of terms. Therefore, if $(x+1)$ is a factor in $p(x)$, no errors with an odd number of bits will evenly divide $p(x)$ and therefore all errors with an odd number of bits will be detected.

- Any burst of length r or less will be detected. A burst of length r is a sequence $e(x)$ of r bits for which the first and last bits in the sequence are in error as well as any combination of errors in between. A burst of length r starting at bit j can be written as $e(x) = x^j \cdot e_0(x)$, where $e_0(x)$ is of order $r-1$. Clearly, $e_0(x)$ will not divide $p(x)$, of order r, because $e_0(x)$ is a lower-order polynomial than $p(x)$. And because x^j will not divide $p(x)$, the burst will always be detected.

- Most bursts that are longer than r can also be detected. The probability of an undetected burst is 2^{-r+1} for a burst of length $r+1$ and 2^{-r} for any length burst greater than $r+1$.

PROOF [Peterson 1961] A burst of length b starting at bit j can be written as $x^j f(x)$, where $f(x)$ is of order $b-1$ and has a constant term equal to 1. Because $p(x)$ is never a multiple of x^j, $p(x)$ must be a multiple of $f(x)$ in order for the burst to be mistaken for a valid codeword. Among the 2^{b-2} random polynomials of order $b-1$ with a constant term equal to 1, the ones that are multiples of $f(x)$ can be written as $p(x) \cdot q(x)$, where $q(x)$ is of order $b-r-1$ and has a constant term equal to 1. For $b = r+1$, the only choice for $q(x)$ is $q(x) = 1$ and so only 1 of the $2^{b-2} = 2^{r-1}$ sequences will be valid codewords. For $b > r+1$, there are 2^{b-r-2} possible values of $q(x)$ out of the total of 2^{b-2} random polynomials, making the overall probability of error 1 in $2^{b-2-(b-r-2)} = 2^r$. ∎

- If $(x+1)$ is a factor in $p(x)$ and $p_1(x)$ is a primitive polynomial factor of degree e, any two bursts, each of length two or less, will be detected as long as the distance between the start of the two bursts is less than or equal to 2^e-1.

- Of the 2^n possible sequences, only 2^k are valid codewords. Therefore, given a completely random sequence, only one in 2^r are valid, because $r = n-k$, and the probability of an undetected burst is 2^{-r} exactly as it is for a burst of length $r+1$ or greater.

Example 9.2 (ADSL)

Given these observations, it is possible to place bounds on the probability of an undetected error. In ADSL, for example, an eighth-order primitive polynomial is used. Each CRC is associated with a superframe of data which is 69 DMT frames and spans 17 ms. At an average data rate of 1.5 Mbps, for example, a superframe will carry about $n = 25{,}500$ bits, which is much greater than $2^8 - 1 = 255$, meaning that some double errors will go undetected. The number of ways a double error will not be detected is the number of ways two bits can be a multiple of 255 bits apart in n bits, which is

$$n_{und\_2bit\_errs} = \sum_{k=1}^{\lfloor n/(2^8-1)\rfloor} n - k(2^8 - 1).$$

The primitive polynomial used in ADSL can detect all single errors and all but $n_{und\_2bit\_errs}$ double errors, and it will fail to detect, at most, $1/2^r$ of the errors length three or greater. Therefore, an upper bound on the probability of undetected errors is given by

$$P(\text{undetected error}) \leq n_{und-2bit-errs}\, p^2(1 - p)^{n-2} + \frac{1}{2^r}P(3 \text{ or more errors})$$

$$= n_{und-2bit-errs}\, p^2(1 - p)^{n-2} + \frac{1}{2^r}\left[\sum_{k=3}^{n} \binom{n}{k} p^k(1 - p)^{n-k}\right],$$

(9.6)

where $\binom{n}{k} = \frac{n!}{k!(n-k)!}$ means the number of combinations of k errors in n bits, P indicates the probability of an event, and p is the probability of a bit error. For ADSL with a bit error rate of $p = 10^{-7}$ the probability of an undetected error in a superframe is less than or equal to 1.5×10^{-8}.

Because the superframe is 17 ms in duration, the mean time between failures is

$$17 \text{ ms} \cdot \frac{1}{1.5 \times 10^{-8}} \approx 13 \text{ days,}$$

which means that, on average, there will be a superframe with undetected errors once every 13 days. Of course, ADSL operates with a noise margin, so that the bit error rate is typically much lower than 10^{-7} and undetected errors will essentially never happen.

It is also interesting to know the probability of an undetected error given that there are errors. By Bayes rule,

$$P(\text{undetected error} \mid \text{there are errors}) = \frac{P(\text{undetected error and there are errors})}{P(\text{there are errors})}.$$

Undetected errors always coincide with having errors, and therefore P(undetected error and there are errors) is just P(undetected error). The probability that there are errors is given by the expression

$$P(\text{there are errors}) = 1 - P(\text{no errors})$$

$$= 1 - \binom{n}{k} p^0(1 - p)^n.$$

(9.7)

For the ADSL example, the probability that there are errors in a block of 25,500 bits is 2.5×10^{-3}. The probability of having undetected errors in a block of n bits, given that there are errors, is about $1.5 \times 10^{-8} \div 2.5 \times 10^{-3} = 6 \times 10^{-6}$.

Finally, if there are so many errors that the entire string of bits is basically random, the probability of an undetected error is $\frac{1}{2^r} = \frac{1}{256} = 3.9 \times 10^{-3}$, regardless of the length of the block.

Detected errors are reported to the management entity to indicate a possible maintenance problem. Errors detected at the customer premises equipment (CPE) are sent to the central office (CO) equipment via indicator bits in the ADSL frame. Errors detected at the CO are communicated directly to the management entity. Generally, DSL transceivers do not use CRC errors to request retransmission of data.

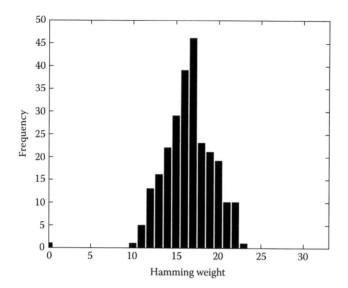

FIGURE 9.6
Weight distribution, B_i, of the dual of the eighth-order CRC used in ADSL with $k = 25$.

9.2.4 Exact Calculation of the Probability of Undetected Error

The exact probability of undetected errors can be calculated for most practical CRC generator polynomials.[1] As mentioned previously, calculating the weight distribution, A_i, by computing all of the 2^k codewords in an (n, k) code is usually impractical. However, it is possible to calculate A_i by using the dual code, which has only r information bits. For most practical CRCs, r is on the order of 8 or 16, and it is rarely more than 32.

The dual code is a code generated by the matrix, H, such that $G \cdot H^T = \mathbf{0}_{k \times r}$, where $\mathbf{0}$ is an all-zeros matrix. The dual code has a blocksize n but only r information bits. If G has dimensions $k \times n$, H has dimensions $r \times n$. One can find the matrix, H, given the matrix, G, by noting that if $G_{sys} = [I_k \ P_{k \times r}]$ then $H_{sys} = [P_{k \times r}^T \ I_r]$, where I is the square identity matrix [Peterson 1972]. It is easy to verify that $G_{sys} \cdot H_{sys}^T = \mathbf{0}_{k \times r}$, because $G_{sys} \cdot H_{sys}^T = I_k \cdot P_{k \times r} + P_{k \times r} \cdot I_k^T = P_{k \times r} + P_{k \times r} = \mathbf{0}_{k \times r}$ using the rules of binary addition. Finding the weight distribution of the dual code, B_i, is practical on a computer for all CRCs used in DSL systems by creating all 2^r codewords and keeping track of the weights.

A_i can be calculated from the weight distribution of the dual code, B_i, using the MacWilliams identity [MacWilliams 1977]. For a binary code, the MacWilliams identity can be written as [Peterson 1972]

$$A_i = \frac{1}{2^r} \cdot \sum_{j=0}^{n} B_j \cdot \sum_{l=0}^{\min(i,j)} (-1)^l \cdot \binom{j}{l} \cdot \binom{n-j}{i-l}. \tag{9.8}$$

As a first example, the eighth-order CRC polynomial used in ADSL from Table 9.2 is considered. For this example, it is assumed $k = 25$ bits. The weight distribution of the dual code B_i is shown in Figure 9.6, and the corresponding weights of the code as derived from Equation 9.8 are shown in Figure 9.7.

---

[1] The author is very grateful to Idan Alrod, currently a Ph.D. student at Tel Aviv University under Professor Simon Litsyn, for his assistance in writing this section on using the MacWilliams identity to calculate the exact probability of error. The formula in Equation 9.8 and the presentation of weight distributions as in Figures 9.6 and 9.7 were the results of his work.

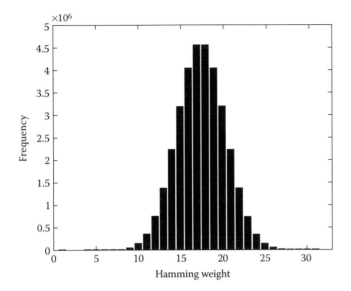

FIGURE 9.7
Weight distribution, A_i, of the eighth-order CRC used in ADSL with $k = 25$ calculated using the MacWilliams identity.

Table 9.3 lists the first few weight values. The table shows, as expected, that there are no codewords with weight one or two ones because $k < 2^8 - 1$ and the generator polynomial is primitive. There is always a codeword of weight zero because, with a linear code, an all-zeros message results in all-zeros redundancy.

TABLE 9.3

First Eight Weight Values for
Eighth-Order ADSL CRC
Polynomial with $k = 25$ bits

A_0	1
A_1	0
A_2	0
A_3	20
A_4	148
A_5	942
A_6	4347
A_7	16640

Example 9.3   (HDLC)

The HDLC protocol documented in ISO 3309 [ISO/IEC 1993] is a common way to encapsulate packets of data. An HDLC frame is shown in Figure 9.8.

Opening flag 8 bits $0 \times 7e$	Address 0, 8, or 16 bits	Control field 8 or 16 bits	Information $8 \times P$ bits	CRC 16 bits $x^{16} + x^{12} + x^5 + 1$	Closing flag 8 bits $0 \times 7e$

FIGURE 9.8
HDLC encapsulation. Shaded area is encoded by CRC.

The CRC in HDLC encompasses the address, control, and information fields. As shown in Figure 9.8, the CRC block is $P+(1, 2, 3,$ or 4) bytes. In ADSL2 [G.992.3 1993], for example, physical layer management "showtime" configuration commands are encapsulated in HDLC frames. Here, P is specified to be less than 1024.

Because the HDLC CRC shown in Table 9.2 contains the factor $(1+x)$, only even numbers of errors could go undetected. The block of data, a maximum of $(P + 4) \cdot 8 = 8224$ bits is shorter than $2^e - 1 = 2^{15} - 1 = 32767$, where e is the order of the primitive polynomial factor which is 15 for HDLC. Therefore, all double errors are also detected.

An upper bound on the probability of an undetected error is then

$$P(\text{undetected error}) \leq \frac{1}{2^r} P(4 \text{ or more errors})$$

$$= \frac{1}{2^r} \left[\sum_{k=4}^{n} \binom{n}{k} p^k (1 - p)^{n-k} \right].$$

For a block of 4000 bits and a probability of bit error of 10^{-7}, the probability of an undetected error is upper bounded by 1.0×10^{-15}. Using the dual code and MacWilliams identity to calculate the exact weight distribution of this code using $k = 4000$, the weight distribution is given in Table 9.4 and the actual probability of an undetected error is 3.3×10^{-20} (from Equation 9.5).

In ADSL2, the message overhead rate is between 4 kbps and 64 kbps. At 32 kbps, for example, there would be 0.125 HDLC blocks per second if all of them were 4000 bits long. The mean time between failures would then be $0.125/3.3 \times 10^{-20} = 3.8 \times 10^{18}$ seconds $=$ over 100 billion years where a failure is an HDLC frame with an undetected error. Clearly, this aspect of the HDLC protocol is sufficiently robust.

TABLE 9.4

Weight Distribution of the Code for Sixteenth-Order HDLC CRC Polynomial with $k = 4000$

A_0	1
A_1	0
A_2	0
A_3	0
A_4	3.3×10^8
A_5	0
A_6	1.8×10^{14}
A_7	0

9.2.5 Reed–Solomon Codes

It is desirable to construct a code such that in a codeword block of length n, it is possible to correct up to t errors anywhere within that block with as little redundancy as possible. This implies that even if there are t errors, the resulting block of n is still closer to the actually transmitted codeword than to any other codewords because if it were closer to another codeword, the decoder would select another codeword instead. Closeness refers to the number of positions the received block differs from the actually transmitted codeword.

The number of places in which one codeword differs from another codeword is referred to as the Hamming distance. The Hamming distance between any two codewords must be at least $2t + 1$ in order to be able to correct up to t errors. If the distance were only $2t$, for example, a codeword with t errors would be a distance t from two different codewords, and the receiver would not be able to decide which was correct.

Reed–Solomon codes provide powerful error correction capability for relatively little overhead. Introduced in 1960 by Reed and Solomon [McEliece 1997], Reed–Solomon codes are used widely in DSL for error correction. Reed–Solomon codes are a subclass of codes developed by Hocquenghem, Bose, and Chaudhuri, now referred to as BCH codes, for which the generator polynomial has non-binary coefficients. As with binary BCH codes, the generator polynomial of a Reed–Solomon code is

$$g(X) = LCM(m_0(X), \ m_1(X), \ m_2(X), \ m_3(X), \ldots, m_{2t-1}(X)),$$

where $m_i(x)$ is a polynomial of minimum degree with α^i as its root and LCM means "least common multiple." Here α is a primitive element in GF(q), where q typically equals 2^m. Over GF(q), it is possible to define $m_i(X) = (X + \alpha^i)$ with degree 1. Therefore,

$$g(X) = (X + \alpha^0) \cdot (X + \alpha^1) \cdot (X + \alpha^2). \ldots .(X + \alpha^{r-1}),$$

where the coefficients of $g(X)$ are in GF(q).

For a Reed–Solomon code:

- The minimum distance, $d_{\min} = r + 1$
- The number of redundant q-ary symbols, $n - k = r$
- The block length, $n = q - 1$
- Error correction capability $t = \lfloor r/2 \rfloor$

Therefore, this code can correct t, q-ary errors. The most common application of a Reed–Solomon code is $q = 2^8$ meaning that the code can correct up to t octets in a block length of 255. Reed–Solomon codes are a type of forward error correction, meaning they can correct errors without needing to re-transmit the data.

For example, a single-error-correcting Reed–Solomon code is given by

$$\begin{aligned} g(X) &= (X + \alpha^0)(X + \alpha^1) \\ &= X^2 + (\alpha^1 + 1) \cdot X + \alpha^1 \\ &= X^2 + (x + 1) \cdot X + x, \end{aligned}$$

where $\alpha = x$ and X, in this case, is over GF(q). Multiplication by the coefficient $(\alpha^1 + 1)$, for example, is defined by the polynomial used to construct the GF(q) field. In many DSL systems that use Reed–Solomon codes, $p(x) = x^8 + x^4 + x^3 + x^2 + 1$ is used to construct the GF(256) field[2] and $\alpha = 00000010_2 = 2_{16}$. The cyclic encoder used to encode a CRC is also used to encode a Reed–Solomon code as shown in Figure 9.4. In this case, however, the multipliers and adders are over GF(q) instead of GF(2) as with a binary encoder.

The block size n of a Reed–Solomon code can be shortened from $q - 1$ while maintaining the same error correcting capabilities. Consider all of the codewords for which the l high-order symbols are zero. This set of codewords forms a linear subcode of the original code. By restricting the valid set to only these codewords and by deleting the l high-order zeros, a shortened code with a codeword size of $n - l$ symbols and $k - l$ information symbols is formed. Any cyclic code can be shortened and maintain its error correction capability. However, the new code is not cyclic.

---

[2] One of the reasons this polynomial is also used by the CRC is to simplify hardware so that only one Galois field multiplier is necessary. In more recent recommendations, such as G.992.5, for example, multiple polynomials are defined in recognition that Galois field multiplication is often done in software.

9.2.6 Decoding Reed–Solomon Codes

Many books and articles have been written on decoding Reed–Solomon codes, including [Lin 1983], [Blahut 1983], [Berlekamp 1984], [Peterson 1972], [Forney 1965], and [Chien 1964]. A Reed–Solomon code can correct up to $t = \lfloor r/2 \rfloor$ errors anywhere in the block of n symbols with r redundant symbols. If the locations of the errors are known, the errors are referred to as erasures. A Reed–Solomon code can correct up to r erasures. If there are s erasures and t errors, a Reed–Solomon code can correct the errors as long as

$$2t + s \leq r.$$

The error and erasure correcting capability are important factors when the coding gain and burst error correction capability of the Reed–Solomon code are considered.

The steps involved in a practical BCH or Reed–Solomon decoder are described in many textbooks and Internet sites. An excellent overview is given in [Morelos-Zaragoza 2002]. This chapter briefly describes the decoder process to provide a background for analyzing the performance or coding gain of Reed–Solomon codes.

The basic idea behind decoding a Reed–Solomon code (or any BCH code) is to find an error sequence with the smallest number of terms that, when added to the received sequence, makes a valid codeword. As described earlier with CRC codes, the received codeword $w(x)$ can be considered as a sum of the transmitted codeword $u(x)$ plus any errors $e(x)$; $w(x) = u(x) + e(x)$. Furthermore, if at the receiver the received codeword, $w(x)$, is multiplied by the dual code generator matrix, H, the result will be zero if there are no errors. In the case of nonbinary Reed–Solomon codes, all multiplication is done using a Galois field multiplier in GF(q).

The results of the received codeword, w, multiplied by the dual code, $s = H \cdot w^T$, are called the syndromes. The vector s has r elements, normally referred to as S_1 through S_r.

The strategy behind decoding BCH codes (including Reed–Solomon codes) is to form an *error locator polynomial* defined as

$$\sigma(x) = \prod_{i=1}^{v} (1 + \alpha^{j_i} x) = 1 + \sigma_1 x + \sigma_2 x^2 + \cdots + \sigma_v x^v, \tag{9.9}$$

where $\alpha \in GF(q)$ and the set of α^{j_i} are known as error positions with the position of the error given by j_i. The value of v is the number of errors that are being corrected, normally up to t errors if erasure decoding is not used. Therefore, the roots of the error locator polynomial determine the positions of the errors.

The coefficients of the error locator polynomial can be derived from the syndromes through the relationship (for example, see [Lin 1983]):

$$\begin{pmatrix} S_{v+1} \\ S_{v+2} \\ \vdots \\ S_{2v} \end{pmatrix} = \begin{pmatrix} S_1 & S_2 & \cdots & S_v \\ S_2 & S_3 & \cdots & S_{v+1} \\ \vdots & \vdots & \ddots & \vdots \\ S_v & S_{v+1} & \cdots & S_{2v-1} \end{pmatrix} \cdot \begin{pmatrix} \sigma_v \\ \sigma_{v-1} \\ \vdots \\ \sigma_1 \end{pmatrix}. \tag{9.10}$$

This is often referred to as the key equation. Solving this equation for σ_i is typically the most computationally intensive piece of a BCH/Reed–Solomon decoder. The two most common algorithms for solving this equation are known as the Berlekamp–Massey algorithm and the Euclidean algorithm. Both algorithms find an error locator polynomial of the smallest degree. In general, if the capability of the code is exceeded (*i.e.*, there are more than t errors), the error locator polynomial will be corrupted, and it is usually not possible to detect error positions at this stage in the decoder process.

Once the error locator polynomial is derived, the roots must be determined, because the roots indicate the positions of the errors. This is commonly done using the Chien search procedure. The Chien search simply cycles through all 2^q possible roots, α^i, and determines if $\sigma(\alpha^i) = 0$.

For a binary BCH code, only the error position is needed because it is then obvious that error correction involves flipping the bit. In nonbinary Reed–Solomon codes, it is necessary also to know the error value, e_{j_i}. The error value is the corrected value at location j_i. The location is determined by the roots of the error locator polynomial. The error values are typically computed using a method developed by Forney [Forney 1965] with the error values defined as

$$e_{j_i} = \frac{(\alpha_{j_i})^2 \Lambda(\alpha^{-j_i})}{\sigma'(\alpha^{-j_i})},\qquad(9.11)$$

where the Reed–Solomon code has $(\alpha^0 \cdots \alpha^{r-1})$ as its zeros, $\sigma'(x)$ is the derivative of $\sigma(x)$, and the polynomial $\Lambda(x)$ is known as an error evaluator polynomial and is defined as

$$\Lambda(x) = \sigma(x) \cdot s(x) \bmod x^{r+1},\qquad(9.12)$$

where $s(x)$ is the syndrome polynomial. Finding the error evaluator polynomial is an intrinsic part of the Euclidean algorithm; alternatively, it can be calculated after the Berlekamp–Massey algorithm.

A typical implementation of a Reed–Solomon decoder is shown in Figure 9.9. In this implementation, received codewords are corrected on-the-fly as the roots of the error locator polynomial are being calculated using the Chien search.

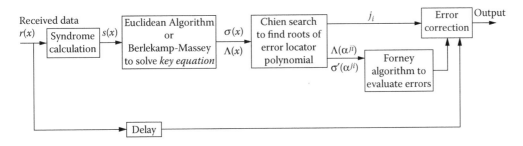

FIGURE 9.9
Typical implementation of a Reed–Solomon decoder.

9.2.7 Consequences of Uncorrectable Errors

If the number of errors exceeds the capability of the code, the errors cannot be corrected. For example, in a t error correcting code, if there are more than t errors, the code cannot correct them. Uncorrectable errors can result in either a decoder error where the errors happen to be within t symbols of another valid codeword or a decoder failure where the received codeword is not within t symbols of another valid codeword. It is impossible to detect a decoder error. A decoder failure, however, will be detected when the number of unique roots of the error locator polynomial is less than its degree, v.

It is not until after all of the roots of the error locator polynomial have been calculated that it is possible to know whether the capability of the code has been exceeded. In a typical implementation, error corrections are already completed before it has been determined whether the number of roots of the error locator polynomial is equal to the degree. This avoids the extra storage complexity and delay that would be required to put off correction until after the Chien search is completed. But it does mean that errors can be inserted by the decoder where there were no errors in the received message.

To get a feel for the number of additional errors that will be inserted, assume that the error locator polynomial is corrupt and random when the number of errors exceeds the capability of the code.[3] In this event, in the worst case, every first-order factor (also known as a root or linear factor) of the corrupt error locator polynomial will indicate a unique location that is not currently in error. Assume that the error value is also random and that therefore, on average, half of the bits inserted will be errors. Therefore, instead of only $v+1$ errors before decoding, for a code that can correct up to v errors, there can be $v+1+a$ errors after decoding and the a additional errors will have a $1/2$ probability of bit errors.

Under the assumption that the error locator polynomial is random, one simple approach is to calculate the probability of a decoder error, that is, the probability that the decoder finds a valid codeword other than the transmitted codeword. In this case, the error locator polynomial will have exactly v unique linear factors that point to locations within the codeword size, n.[4] For a code in GF(q), the number of ways to have exactly v linear factors is $\binom{n}{v}$, whereas the total number of possible polynomials of degree v is $(q-1)^v$. For $n = q-1$, it is easy to show that

$$\text{probability of decoder error} \approx \frac{\binom{q-1}{v}}{(q-1)^v}$$

$$< \frac{\left[\frac{(q-1)^v}{v!}\right]}{(q-1)^v}$$

$$< \frac{1}{v!}.$$

This result agrees very closely with one derived in [McEliece 1986]. Although this does not prove that the error locator polynomial is random when the error pattern is random (in [McEliece 1986], the authors assumed that the error pattern was random), it does at least give some confidence that the assumption is valid.

The average number of additional q-ary errors inserted by the Reed–Solomon decoder when the capability of the code is exceeded can be calculated assuming:

- The error locator polynomial is completely random when there are too many errors.

- The locations of the roots of the error locator polynomial do not coincide with locations that are already in error; that is, these are additional errors.

- There is no indication that there might be too many errors until after all correction is complete.

- The codeword size, n, is equal to $q-1$. This assumption is implicit in the following derivation, because it is assumed that all linear factors lead to errors. But if $n < q-1$, then any linear factors pointing to a location greater than n would not lead to an error.

This calculation includes both the decoder error and decoder failure conditions, because all random polynomials are considered.

---

[3] The author is not aware of any proof that the error locator polynomial is in any sense random when the capability of the code is exceeded. But the following derivation offers some understanding of how many additional errors to expect, even if it is not strictly exact.
[4] If the error location is greater than n, there is surely a decoder failure because the location would fall outside the sphere of the codeword.

The number of additional errors that can be inserted, a, cannot exceed v, the largest size of the error locator polynomial. But, on average, a will typically be less than v. To see this, consider an error locator polynomial of order v written as (Equation 9.9 repeated):

$$\sigma(x) = \prod_{i=1}^{v}(1 + \alpha^{j_i}x) = 1 + \sigma_1 x + \sigma_2 x^2 + \cdots + \sigma_v x^v.$$

Under the assumptions listed above, the average number of additional q-ary errors inserted can be calculated by finding the average number of unique linear factors of the error locator polynomial. If any linear factors are repeated, they will not cause more errors. The average number of additional errors, $\bar{a}$, is given by the expression

$$\bar{a} = \sum_{u=1}^{v} u \cdot P\left(u \text{ unique linear factors}\right)$$

$$= \sum_{u=1}^{v} u \cdot \sum_{l=u}^{v} P(u \text{ unique linear factors}|l \text{ linear factors}) \cdot P\left(l \text{ linear factors}\right) \tag{9.13}$$

The objective is to find expressions for the probabilities in Equation 9.13. Starting with the first probability on the second line, given that there are l linear factors, the total number of ways to have exactly u unique linear factors can be written

$$\text{Ways to select } u \text{ unique linear factors} \atop \text{from } l \text{ total linear factors} = \binom{n}{u} \cdot \binom{l-1}{l-u}, \tag{9.14}$$

which is the number of ways to select u unique linear factors, multiplied by the number of ways to distribute the u unique factors among the $l - u$ remaining linear factors. In this expression, n is the codeword size assumed equal to $q - 1$.

Then, the total number of ways the error locator polynomial can be factored into v linear factors is the total number of polynomials of order l with exactly l linear factors, multiplied by the total number of polynomials of order $(v - l)$ with no linear factors. The total number of polynomials of order l with l linear factors is similar to the problem of drawing v balls from an urn with n balls in it with replacement and without concern about the order, which is

$$\text{Total number of ways to have } v \text{ linear factors} = \binom{n+l-1}{l}. \tag{9.15}$$

If the number of ways a polynomial of order k will have no linear factors is denoted as $\#nlf(k)$, the total number of polynomials of order v with exactly l linear factors is equal to:

$$\binom{n+l-1}{l} \cdot \#nlf(v-l). \tag{9.16}$$

The total number of polynomials of order v is n^v. So the average number of additional errors, from Equation 9.13, can be written, using Equations 9.14, 9.15, and 9.16, as

$$\bar{a} = \sum_{u=1}^{v} u \cdot \sum_{l=u}^{v} \frac{(\# \text{ polys w/}u \text{ unique roots, } l \text{ roots total})}{(\# \text{ polys w/}l \text{ roots})} \cdot \frac{(\# \text{ polys w/}l \text{ roots})}{(\text{total } \# \text{ of polys of order } v)}$$

$$= \sum_{u=1}^{v} u \cdot \sum_{l=u}^{v} \frac{\binom{n}{u} \cdot \binom{l-1}{l-u}}{\binom{n+l-1}{l}} \cdot \frac{\binom{n+l-1}{l} \cdot \#nlf(v-l)}{n^v}$$

$$= \sum_{u=1}^{v} u \cdot \sum_{l=u}^{v} \binom{n}{u} \cdot \binom{l-1}{l-u} \cdot \frac{\#nlf(v-l)}{n^v}. \tag{9.17}$$

The order of the summation in Equation 9.17 can be reversed and the equation re-written as

$$\bar{a} = \sum_{l=1}^{v} \sum_{u=1}^{l} u \cdot \binom{n}{u} \cdot \binom{l-1}{l-u} \cdot \frac{\#nlf\,(v-l)}{n^v} \tag{9.18a}$$

$$= \frac{1}{n^v} \sum_{l=1}^{v} \#nlf\,(v-l) \sum_{u=1}^{l} u \cdot \binom{n}{u} \cdot \binom{l-1}{l-u} \tag{9.18b}$$

$$= \frac{1}{n^v} \sum_{l=1}^{v} \#nlf\,(v-l) \left[\sum_{u=1}^{l} \binom{n}{u} \cdot \binom{l-1}{l-u} + \sum_{u=2}^{l} (u-1) \cdot \binom{n}{u} \cdot \binom{l-1}{l-u} \right] \tag{9.18c}$$

$$= \frac{1}{n^v} \left[\sum_{l=1}^{v} \#nlf\,(v-l) \binom{n+l-1}{l} + \sum_{l=1}^{v} \#nlf\,(v-l) \sum_{u=2}^{l} (u-1) \cdot \binom{n}{u} \cdot \binom{l-1}{l-u} \right] \tag{9.18d}$$

$$= \frac{1}{n^v} \left[\sum_{l=1}^{v} \#nlf\,(v-l) \cdot \binom{n+l-1}{l} + \#nlf\,(v) \right] \tag{9.18e}$$

$$= \frac{1}{n^v} n^v \tag{9.18f}$$

$$= 1. \tag{9.18g}$$

Line (d) follows from line (c) of Equation 9.18 because

$$\sum_{u=1}^{l} \binom{n}{u} \binom{l-1}{l-u} = \binom{n+l-1}{l},$$

which is the total number of ways a polynomial of degree l can be factored into linear factors.

Line (e) of Equation 9.18 results from the fact that

$$\sum_{l=1}^{v} \#nlf\,(v-l) \sum_{u=2}^{l} (u-1) \binom{n}{u} \binom{l-1}{l-u} = \#nlf\,(v),$$

which is the total number of ways a polynomial of degree v has no linear factors.

The expression in the square brackets on line (e) of Equation 9.18 is the total number of polynomials with between 1 and v linear factors plus the total number of polynomials with 0 linear factors.

The average number of errors added by the Reed–Solomon decoder, under the assumptions listed earlier, is always one. The total number of polynomials with between one and v linear factors plus the total number of polynomials with 0 linear factors includes all possible polynomials. There are n^v total polynomials and n^v/n^v is 1.

This result shows that on average, under the assumptions listed, there will be one additional q-ary error created when the capability of the code is exceeded. In practice, the actual number of additional errors will be fewer because some error conditions are detected when solving the key equation and because not all added error locations avoid existing error locations.

Also, with a shortened codeword size, $n < q - 1$, linear factors with locations greater than $n - 1$ will not cause additional errors. It is assumed that the locations indicated by the error locator polynomial roots are evenly distributed. Therefore, the average number of additional errors is expected to be

$$\bar{a} = \frac{n}{q-1}. \tag{9.19}$$

The result does show that when using a suboptimal decoder implementation, where errors are corrected before it is known if there are more than v errors, the error rate is only marginally higher.

9.2.7.1 Application of Reed–Solomon Codes

Typically, Reed–Solomon codes are paired with an interleaver to correct long bursts of errors. An interleaver spreads the data out or shuffles the data after it is encoded by the Reed–Solomon code. This way, if there is a long burst of errors, the errors will be evenly distributed over many codewords. If the data is sufficiently shuffled so that each codeword has only a small number of errors, the code will correct them. Interleaving is discussed in more detail in Section 9.4.

Among current DSL systems, Reed–Solomon coding is only used in ADSL and VDSL. There are a number of reasons for this. First, ADSL and VDSL are both intended to be able to carry video and voice in addition to Internet data traffic. In video and voice applications, there is not time to re-transmit data if errors are detected. Errors in video and voice traffic are very annoying. With video applications in particular, the round-trip delay is usually not important assuming video conferencing is not considered. When the delay can be very long, it is possible to use very large interleavers that allow the correction of long error bursts.

Reed–Solomon coding is used in nearly all DMT systems for additional reasons. As discussed in Chapter 7, a DMT signal is very nearly Gaussian and has a high peak-to-average ratio (PAR). Regardless of how the analog-to-digital and digital-to-analog converters are designed, inevitably there will be clipping. Clipping in DMT can affect an entire DMT symbol and cause errors throughout the frame. Therefore, Reed–Solomon coding with interleaving is used to combat these usually rare bursts of errors.

Another reason that Reed–Solomon coding is often used in DMT systems is that it is straightforward to vary the data rate by small units by adding bits to individual sub-channels. Therefore, it is relatively easy to add the r redundant symbols regardless of the codeword size (n) or the redundancy (r). In QAM-based systems, as discussed in Chapter 6, the granularity in selecting data rates is usually higher because every QAM symbol carries the same number of bits, and this number is typically much smaller than that carried by a DMT frame. However, it is possible to accommodate the overhead by increasing the transmission bandwidth. It is also possible, with multi-dimensional constellations [Wei 1987] [Forney 1989], to transmit fractions of a bit per symbol, which also addresses the problem of granularity. So this is not a major obstacle.

9.3 Forward Error Correction Coding Gain

Forward error correction using Reed–Solomon coding is included in DSL for two primary reasons. The first, and usually the primary reason, is to work together with an interleaver to correct bursts of errors or impulse noise. Interleaving is addressed in detail in Section 9.4. The second reason for forward error correction is to allow higher data rates with a coding gain.

As discussed in Section 9.2.5, the Reed–Solomon forward error correction code adds r redundant symbols to a block of n symbols and can correct up to $t = \lfloor r/2 \rfloor$ random errors. Because the code will correct a certain number of errors, the error rate after the code will be lower than the error rate before the code. However, because r symbols of redundancy had to be added, the lower error rate does not come for free. The coding gain, γ_c, is the net or overall advantage when considering both the lower error rate and the addition of the redundancy. Without considering the penalty for adding the r redundant symbols, the gross coding gain, γ_g, is defined.

It is possible that a Reed–Solomon code will incur a net loss or a negative coding gain. For example, a designer may choose to use a Reed–Solomon code that corrects long bursts of errors even though the ability to correct random errors is reduced. Regardless of whether the Reed–Solomon code provides a positive or a negative coding gain, it is important to know what the coding gain is to determine the data rate that a DSL transceiver pair can support.

It does not make sense to include erasures in a discussion of coding gain. Erasures are only potentially useful when the errors are not completely random. When there is a pattern to the errors, as there is when there is a burst of errors, erasures can help to correct a long burst because it is possible to predict where there are errors. However, when coding gain is considered, it is assumed that errors occur at random and, therefore, erasures can provide no additional gain.

To begin a discussion on coding gain, it is necessary to ask with respect to what is the coding gain measured. In nearly all communications systems, including DMT-based and QAM/PAM-based DSL, data is transmitted in multi-dimensional symbols. In DMT, for example, each subchannel carries a point in a two-dimensional, QAM constellation. The amount of data, *i.e.*, the number of bits, that can be carried in a symbol at a given average error rate depends on the signal-to-noise ratio (SNR). It is often assumed that the noise is Gaussian and independent from one symbol to the next. In a single-carrier system, the SNR is measured over the entire bandwidth. In DMT, the SNR is measured for each individual subchannel. A more detailed discussion on modulation techniques is given in Chapters 6 and 7.

For example, a two-bit, two-dimensional QAM constellation is illustrated in Figure 9.10. The points of the constellation are $(1, 1)$, $(1, -1)$, $(-1, 1)$, and $(-1, -1)$. Figure 9.10 shows what 1000 received points could look like if the SNR were 14.5 dB. For this constellation, the signal energy, ε, equal to the average power of the constellation points, is $\varepsilon = 1^2 + 1^2 = 2$; the energy per dimension, $\bar{\varepsilon} = \varepsilon/2 = 1$. The total noise power per dimension in this case is then $\sigma^2 = 1/10^{(14.5/10)} = 0.0355$.

Assuming Gaussian noise, the error rate is dominated by the probability that the noise pushes the received level closer to an adjacent point than to the one that was transmitted. For this example, adjacent points are separated by the distance, $d_{min} = 2$, and the probability that the noise in one dimension exceeds half the minimum distance, or 1, is $Q(2/(2\sigma))$ where $Q(a) = p(y > a)$ for y, a zero-mean, normalized Gaussian random variable. For this example, each point has two nearest neighbors, so the overall probability of a constellation error is $2Q(1/\sigma)$ which is approximately equal to 1.1×10^{-7} for $SNR = 14.5$ dB.

In general, for any constellation with energy, ε, per dimension and noise

$$\bar{\sigma} = \frac{\sum \sigma^2}{N} = \sigma^2,$$

the SNR can be defined as [Starr 1999]

$$SNR = \frac{\varepsilon}{\sigma^2}.$$

The energy $\bar{\varepsilon}$ is directly proportional to the d_{min}^2. The probability of a constellation error can be approximated as [Starr 1999]

$$p_{symbol} = N_e Q \left[\frac{d_{min}}{2\sigma} \right], \tag{9.20}$$

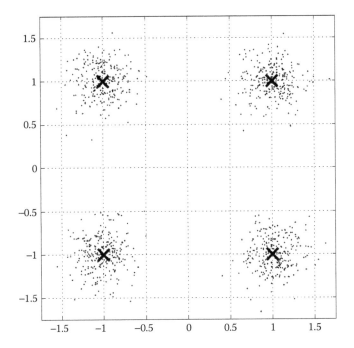

FIGURE 9.10
Illustration of a 4-QAM constellation with noise. The figure shows 1000 received points at an SNR of 14.5 dB. The points that were transmitted were $(+1, +1)$, $(+1, -1)$, $(-1, +1)$, and $(-1, -1)$ shown with an $\times$.

where N_e is the number of nearest neighbors or the average number of constellation points at a distance $d_{\min}$ from a given point.

The coding gain of a Reed–Solomon code can be quantified in terms of how much lower the SNR can be while maintaining a certain error rate. For a Reed–Solomon code defined over $GF(2^8)$, for example, the coding gain is the degree to which the SNR can be lowered while maintaining the same octet error rate. In other words, for $GF(q)$, q-ary error rates are generally of interest. However, Equation 9.20 is a constellation error rate, not a q-ary error rate. To further complicate matters, most DSL recommendations specify a target bit error rate that is neither a q-ary error rate nor a constellation error rate.

It is possible to develop an expression for the bit error rate given the constellation error rate by finding the average number of bit errors per constellation error. Let p_{ij} be the probability of mistaking constellation point j for point i. Let Δb_{ij} be the number of bit errors when point j is mistaken for point i. The average number of bit errors for every constellation error is then

$$\Delta b_{avg} = \frac{\sum_{i \in \text{ all points}} \left(\sum_{j \neq i} p_{ij} . \Delta b_{ij} \right)}{\sum_{i \in \text{ all points}} \left(\sum_{j \neq i} p_{ij} \right)}. \tag{9.21}$$

The probability of a bit error is then approximated by combining Equations 9.20 and 9.21 as

$$p_{bit} = \Delta b_{avg} \cdot N_e \cdot Q\left[\frac{d_{\min}}{2\sigma} \right]. \tag{9.22}$$

The expression in Equation 9.21 can be approximated by only considering points separated by the distance $d_{\min}$ as

$$\Delta b_{avg} \approx mean(\Delta b_{ij}) \text{ over } i, j \text{ such that } d_{ij} = d_{\min}, \tag{9.23}$$

TABLE 9.5

Average Number of Bit Errors, Using
Equation 9.23, for the 2-Dimensional QAM

Bits	Average Number of Bit Errors in a Nearest Neighbor Error
1	1.00
2	1.00
3	1.00
4	1.33
5	1.77
6	1.57
7	1.79
8	1.73
9	1.85
10	1.84
11	1.90
12	1.90
13	1.94
14	1.94
15	1.96

where d_{ij} is the distance from point i to point j. Because adjacent points are all separated by the same distance, the probability of mistaking them is always the same. The average number of bit errors in a constellation error, using Equation 9.23, for the constellations used in ADSL is shown in Table 9.5. Note that if the Reed–Solomon code is concatenated with a trellis code, discussed in Chapter 8, these equations no longer hold because nearest neighbor errors are usually not allowed with a trellis code. In order to calculate the coding gain, it is necessary also to know the q-ary (in the case of GF(q)) or octet (in the case of GF(2^8)) error rate in terms of the constellation symbol error rate. Assume, for example, that all constellations carry one bit, and let $q = 2^8$. In this case, each symbol error is $1/8$ of an octet error and, therefore, the octet error rate is eight times the symbol error rate. Similarly, with 2, 4, and 8 bits per constellation, the octet error rate is four, two, and one times the symbol error rate. On average, the q-ary error rate in terms of the symbol error rate is

$$p_{q\text{-ary}} = \frac{\log_2(q)}{b} p_{symbol},$$

$$p_{octet} = \frac{8}{b} p_{symbol} \text{ for } GF(2^8),$$

(9.24)

where b is the number of bits in the constellation. For the case of DMT where the number of bits per DMT subchannel varies from subchannel to subchannel, Equation 9.24 can be used as an approximation with b representing the average number of bits per subchannel.

The error rate after the Reed–Solomon code involves only those error events with more than t errors. As shown in Section 9.2.7, for a typical implementation when there are more than t errors, the decoder will make $n/(q-1)$ additional errors on average (Equation 9.19). Assuming that q-ary errors occur at random, the q-ary error rate can be written as

$$p_{q\text{-ary,coded}} \cong \sum_{i=t+1}^{n} \binom{n}{i} p_{q\text{-ary}}^i \cdot (1 - p_{q\text{-ary}})^{n-i} \cdot \frac{(i + n/(q-1))}{n},$$

(9.25)

where n is the codeword size and $p_{q\text{-ary}}$ comes from Equation 9.24.

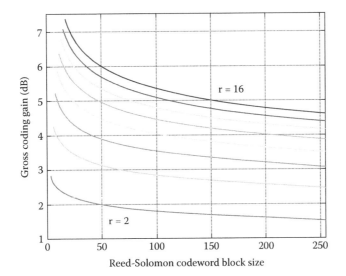

FIGURE 9.11
Gross coding gain for a Reed–Solomon code over GF(2^8) for $r = 2, 4, 6, \ldots, 16$. A four-point constellation was used with a target bit error rate of 1×10^{-7} after the scrambler.

To find the gross coding gain, γ_g, it is necessary to find the difference between the SNR required to achieve the target bit error rate both before and after the Reed–Solomon code is applied. Equation 9.22 gives the relationship between the bit error rate and the SNR because $d_{min}/(2\sigma)$ is directly related to the SNR. Then, Equation 9.24 together with Equation 9.20 yields the q-ary error rate before the code. Finally, Equation 9.25 is the q-ary error rate after the code.

To find the bit error rate from the coded q-ary error rate, it is necessary to estimate the number of bit errors in each q-ary error symbol. Denote the average number of bit errors in a q-ary symbol error caused by channel errors as $\bar{b}_{q\text{-}ary}$. Typically, $\bar{b}_{q\text{-}ary}$ can be estimated based on the characteristics of the specific system. As mentioned in Section 9.2.7, errors introduced by the decoder will have generally more bit errors: on average, $\log_2(q)/2$ bit errors. Generally, because the number of additional errors is small, this will be a second-order effect. But for small values of r, it will make a difference in the calculated coding gain. The bit error after the Reed–Solomon decoder can be approximated as

$$p_{\text{bit, coded}} \cong \sum_{i=t+1}^{n} \binom{n}{i} p_{q\text{-}ary}^{i} \cdot (1 - p_{q\text{-}ary})^{n-i} \left[\frac{i}{n} (\bar{b}_{q\text{-}ary}) + \frac{(n/(q-1))}{2n} \log_2(q) \right]. \quad (9.26)$$

The gross coding gain for a family of Reed–Solomon codeword sizes and redundancies is shown in Figure 9.11 for $r = 2, 4, 6, \ldots, 16$ and n from $r + 1$ to 255. The maximum redundancy shown is $r = 16$ because this is the maximum allowed in current DSL systems. These gross coding gains are shown for 4 bits per constellation over GF(2^8) assuming a target bit error rate of 3.3×10^{-8}. This bit error rate is selected so that the bit error rate after a three-tap, self-synchronized scrambler, such as the one used in nearly all DSL systems including ADSL, SHDSL, and HDSL, will be 10^{-7}. The number of bit errors in each octet in this case, $\bar{b}_{q\text{-}ary}$, is assumed to be 1.5.

Determining the overall net coding gain, γ_c, means taking into account the penalty for the r redundant symbols. As mentioned earlier, this depends on the modulation and system constraints. The r redundant symbols can be added by adding bits to constellations or by increasing the bandwidth, or by doing both. This is true for single- and multi-carrier systems.

9.4 Interleaving

As mentioned in the abstract of Chapter 1, one of the key reasons for the wide deployment of DSL is that it runs over existing telephone lines. Not having been designed for high-speed digital communications, these existing lines are prone to interference from external sources. This interference causes what is normally called impulse noise in DSL recommendations. Impulse noise can be caused by telephone ringing (see Section 1.6), picking up the telephone (see Section 1.7), dial pulse (see Section 1.9) [Brown 1999], or any number of other external factors such as weather, intermittent radio frequency interference (see Section 3.2), or even large appliances.

Clearly it is difficult to characterize the exact characteristics, frequency, or duration of impulse noise. As a guide, however, two "characteristic" noise impulses are captured in the ADSL standard, T1.413 Issue 2 [T1.413 1998]. These impulses last for about 22 µs. However, impulse noise can typically be 10–20 times longer or more. Interleaving is used with Reed–Solomon coding to correct bursts of errors caused by noise of this sort.

An interleaver is a device that accepts codewords from a finite alphabet and returns the identical codewords but in a different order. Combined with Reed–Solomon coding, an interleaver can spread long strings of errors over several codewords. The spreading is sufficient to correct a burst of errors if the burst causes no more than t errors in any one codeword in a t-error correcting Reed–Solomon code (or s erasures in a s-erasure correcting Reed–Solomon code if erasures are used). Long bursts of errors can result from impulse noise. But errors can also result from the concatenated coding where the first stage of decoding generates bursts of errors as in trellis decoding using a Viterbi detector.

The simplest interleaver to consider is a block interleaver. With a block interleaver, codewords are written into a rectangular array in columns and are read out in rows. There are I rows and d columns where I is the codeword size (or a divisor of the codeword size) and d is the interleaver depth. A block interleaver with $I = 7$ and $d = 4$ is shown in Figure 9.12. An example of eight consecutive errors is also shown in the figure. In this case, if this code could correct up to $t = 2$ errors and the codeword size, n is equal to I, the burst of eight errors would be corrected.

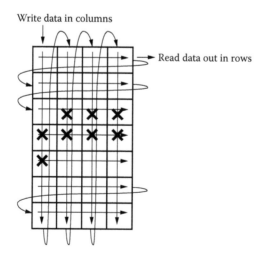

FIGURE 9.12

Block interleaver with seven rows and four columns. Codewords are written in columns and then read out in rows. As an example, each of eight consecutive errors is marked by an X. Each codeword has only two errors.

Interleaver depth
interleave read/deinterleave write

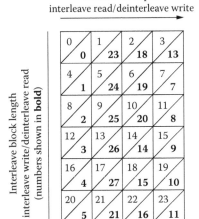

FIGURE 9.13
Convolutional interleaver with read and write addresses shown.

Because block interleavers do not lend themselves to low-memory, efficient implementations, many DSL systems use convolutional interleaving introduced originally by Ramsey [Ramsey 1970] and Forney [Forney 1971]. In general, a convolutional interleaver imposes a different delay on each input symbol. If i denotes the symbol index within a group of I symbols so that $i = 0, 1, \ldots, I - 1$, symbol i experiences a delay of $i \cdot (d - 1)$ with d the interleaver depth. The deinterleaver performs the inverse operation delaying symbol i by $(I - i - 1)(d - 1)$. The overall delay of the interleaver/deinterleaver pair is $(I - 1)(d - 1)$.

A convolutional interleaver is shown in Figure 9.13 for $I = 7$ and $d = 4$. As with the block interleaver, symbols are written into the interleaver in columns and read out in rows. However, unlike the block interleaver, it is not necessary to wait to fill the entire block before reading. As shown in Figure 9.13, the first symbol, symbol 0, is written then read immediately with no delay. Symbol 1 is written, then read at time $k = 4$ delayed by 3 samples. Symbol 2 is written then read out at time 8 delayed by 6 samples. For this example, the ordering of the input symbols in the output sequence is

$$0, 4, 8, 12, 16, 20, 24, 7, 11, 15, 19, 23, 27, 31, 14, 18, \ldots.$$

In general, for a convolutional interleaver, the ordering of the input symbols in the output sequence will be

$$0, d, 2d, \ldots, (I - 1) \cdot d, I, I + d, I + 2d, \ldots.$$

In order to ensure that an input is never repeated in the output, as required in a valid interleaver, the block length I and the depth d must be co-prime, meaning that they share no common factors aside from 1.

When I and d are not co-prime but $I + 1$ and d are co-prime, it is possible to add a dummy row to the interleaver to give it $(I + 1)$ rows instead of I rows. In this case, no input data is written into or read from the $I + 1$st row of the interleaver. The dummy row is added only to create a valid interleaver. In the ADSL and ADSL2 recommendations, the interleaver depth is restricted to be a power of 2 from 2^0 to 2^6. In this case, only odd values are co-prime with d. In ADSL and ADSL2, I is set to the Reed–Solomon codeword size, n, and a dummy row is always added whenever n is even.

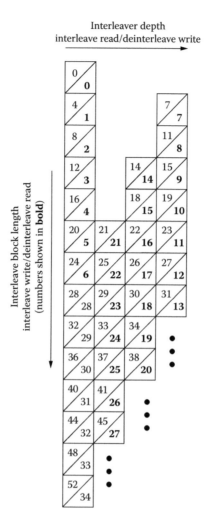

FIGURE 9.14

Cascading column representation of a convolutional interleaver with $d = 4$ and $I = 7$.

As shown in Figure 9.13, a convolutional interleaver is very similar to a block interleaver except that the blocks of I symbols are staggered. Another way to see the convolutional interleaver is as in Figure 9.14. In this view of a convolutional interleaver, blocks of I symbols are added in a constantly descending pattern; as if playing a game of Tetris with the symbol blocks coming from the bottom. In this analogy, the new block is always added to the column farthest from the bottom or to the column farthest to the left if columns share the same height. This means that rows in the interleaver do not appear in the same position in adjacent symbol blocks. However, the error-spreading properties of the block interleaver are preserved in a convolutional interleaver.

Interleavers are characterized by their delay, memory requirements, and spreading [Heegard 1999]. The delay of the interleaver is the total delay between the interleaver input and the deinterleaver output caused by the interleaver/deinterleaver. With a block interleaver, a straightforward implementation will wait for the block of $I \times d$ to fill before starting to read out the codewords. Therefore, the interleaver and de-interleaver delay combined is $2 \times I \times d$. The desire to correct long error bursts is often at odds with the requirement for small delay in DSL systems. In applications such as video conferencing, voice, and gaming, small delay is critical. Even in general Internet use, small delays yield higher throughput. In general, it is desirable to minimize the delay for a given immunity to impulse noise.

Interleaving is typically one of the largest consumers of memory on a DSL transceiver chip. Therefore, it is critical to use the smallest possible amount of interleaver memory. The smallest amount of memory required to build an interleaver/deinterleaver pair is equal to the total delay of the interleaver/deinterleaver [Heegard 1999]. Typically, for memory-optimized interleavers, the interleaver and deinterleaver memory size is nearly the same. As mentioned, the delay of a convolutional interleaver is $(I-1)(d-1)$. As will be discussed in more detail later, it is possible to design a convolutional interleaver with very nearly the theoretical minimum amount of memory equal to $(I-1)(d-1)$ symbols or $(I-1)(d-1)/2$ symbols for either the interleaver or deinterleaver alone.

Regarding spreading, the output of an (I, d) convolutional interleaver can be described as re-ordering the input sequence so that no contiguous sequence of n_2 symbols in the re-ordered sequence contains any symbols that were separated by fewer than d symbols in the original ordering [Ramsey 1970]. The value of n_2 depends on the interleaver parameters and is

$$n_2 = I - \left\lceil \frac{I}{d} \right\rceil, \tag{9.27}$$

where $\lceil . \rceil$ means "greatest integer greater than."

PROOF As mentioned earlier, the ordering of the input symbols in the output sequence is

$$0, \ d, \ 2d, \ldots, (I-1) \cdot d, \ I, \ I+d, \ I+2d, \ldots.$$

In the first I symbols and all groups of I symbols thereafter, the distance between adjacent symbols is always d. The distance between symbol I and symbol $0 \le k < I$ is $|I - k \cdot d|$. The objective is to find the smallest k such that $|I - i \cdot d| \ge d$ for $i = k \ldots I - 1$. In order for $|I - i \cdot d| \ge d$ for $i = k \ldots I - 1$, I must either be always greater than $i \cdot d$ or always less than $i \cdot d$ because if it were not, there would have to exist some value of i for which $|I - i \cdot d| < d$. If I is always greater than $i \cdot d$, this would mean that $I > (I - 1) \cdot d$, which can only happen if $I = 1$ or $d = 1$. Neither of these cases is of interest, and therefore I must be less than $i \cdot d$ for $i = k \ldots I - 1$ and must therefore be less than $k \cdot d$. The smallest k such that $I < k \cdot d$ can be found as $k = \lceil \frac{I}{d} \rceil$. Symbol k is separated from symbol I by $I - k$ symbols and $I - \lceil \frac{I}{d} \rceil$ is therefore the maximum value of n_2. ∎

Effectively, for DSL implementations, interleaver spreading amounts to error correction capabilities. For all of the interleaving methods considered here, a t-error (s-erasure) correcting Reed–Solomon code plus interleaver with depth d can correct a burst of up to $t \cdot d \cdot I/n$ or $(s \cdot d \cdot I/n)$ symbols. Bursts must be separated sufficiently so that two or more bursts do not corrupt the same codeword. As long as n/I is an integer, if bursts are periodic, the period must not exceed $n \cdot d$ symbols.

9.4.1 Optimum Memory Implementation Using Tong's Method

As mentioned earlier, a convolutional interleaver/deinterleaver pair can be implemented with $(I-1)(d-1)$ memory symbols. In [Ramsey 1970] a method for doing this is outlined using shift registers. Because interleaver memory represents such a substantial cost in DSL systems and because shift registers are very inefficient, this method is not suitable for implementation. Random access memory (RAM) is used instead. It is possible to implement a convolutional interleaver using a single RAM using the addressing as shown in Figure 9.13. However, this scheme uses more than twice the minimum amount of memory requiring

TABLE 9.6

Example of Tong's Addressing
Method with $I = 4$, $d = 3$

Symbol	0	1	2	3
Delay	1	3	5	7
Address	0	0	1	2
	0	0	3	1
	0	0	2	3
	0	0	1	2
	⋮	⋮	⋮	⋮
Period	1	1	3	3

$2 \cdot I \cdot d$ memory symbols for the complete interleaver/deinterleaver pair. A near-optimal implementation of a convolutional interleaver is described by Tong [Tong 1998]. In Tong's method, the same memory address is used for both reading and writing. For each address, data is read and then written. With this method, Tong recognizes that if a symbol, i, is delayed by $(i\%I) \cdot (d - 1)$ symbols, where $i\%I$ means "the integer remainder of i/I," this must mean that the memory location in which symbol i was written will appear again $(i\%I) \cdot (d - 1)$ symbols later. By reading and then writing the same memory location each cycle, this method introduces a fixed delay of 1 additional symbol in the interleaver and 1 additional symbol in the deinterelaver so that the total delay is actually $(I - 1)(d - 1) + 2$.

As an example, the memory addresses are shown in Table 9.6 for the case $I = 4$ and $d = 3$. In this case, $i\%I$ is 0, 1, 2, and 3 and continues to repeat as index i increments. The delay of symbol i is shown as $i \cdot (d - 1) + 1$ for $i = 0 \dots 3$. The process begins with address location 0 and then repeats address location 0, 1 symbol later because the delay is 1, then 3 symbols later because the delay is 3, and so on. Then address 1 occurs in symbol 2. Because this symbol has a delay of 5, five spaces are counted before address 1 appears again. The process of moving across the rows of the table and filling in new addresses if necessary continues until the entire sequence eventually repeats. In this case, the sequence repeats after 12 symbols. Along the first and second columns, the address sequence is 0, 0, 0, ... and in the third and fourth columns the sequence is 3, 2, 1, 3, 2, 1, ... but offset from each other.

In general, there will be I different address sequences, one for each delay value, which can always be written as decrementing, contiguous counters. Each counter can have a different start, maximum, and minimum value. In this example, the start, maximum, and minimum addresses for the first two delay values (1 and 3) are all zero. The start addresses for the last two delay values are 1 and 2, respectively, with the maximum and minimum addresses for both of these last two delays being 3 and 1, respectively. In Tong's method, three arrays of length I are stored, A, L, and U, with the current address, the lower limit on the address, and the upper limit on the address, respectively. The algorithm continues to cycle through each of the I columns, and at column i, the address used to read and write is $A(i - 1) - 1$ if $A(i) > L(i)$ or $U(i)$ otherwise.

The deinterleaver can work exactly the same way. With the deinterleaver, the delays are $(I - 1)(d - 1) + 1$, $(I - 2)(d - 1) + 1, \dots, 1$. The memory required for either the interleaver or deinterleaver is $(I - 1)(d - 1)/2 + 1$ plus the memory required to store A, L, and U which is typically much smaller than the interleaver or deinterleaver memory itself.

9.4.2 Forney's Triangular Interleaver

For convolutional interleavers, the only restriction on I and d is that they be co-prime. And even if they are not co-prime, it is usually true that $I + 1$ and d are co-prime and the interleaver can be implemented with a dummy row.

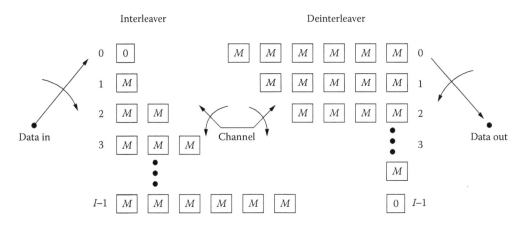

FIGURE 9.15
Forney's triangular interleaver implemented as a series of shift registers.

However, Forney [Forney 1972] and Berlekamp and Tong [Berlekamp 1985] recognized that the implementation of a memory-optimized interleaver can be simplified by constraining the relationship between I and d. In [Berlekamp 1985], Berlekamp and Tong demonstrate a helical interleaver where the interleaver depth is either $I - 1$ or $I + 1$. In [Forney 1972], Forney presents what is often referred to as a triangular interleaver named for its shape when implemented as a series of shift registers as shown in Figure 9.15. In the triangular interleaver, $d = M \cdot I + 1$ where M is any positive integer. A triangular interleaver is used in both DMT and QAM VDSL1. For DSL applications at least, the triangular interleaver is usually considered flexible enough to meet the performance requirements.

As shown in Figure 9.15, the triangular interleaver can be seen as a series of shift registers. Each box labelled with "M" represents M storage locations. The box labelled "0" means no storage; data written into a 0 storage block comes out immediately. Data is written into the interleaver on the left one row at a time in a cyclic pattern. As data is written, delayed data is shifted out the right so for every input symbol there is a delayed output symbol.

The example in Figure 9.14 showed that the columns of I symbols are shifted downward. In Figure 9.14, the fourth column is shifted down one row, then the third and second columns are shifted an additional two rows each. The triangular interleaver imposes a regular pattern on the downward shifts as shown in Figure 9.16. The interleaver in Figure 9.16 has $I = 4$, $d = 9$, and $M = 2$. A similar type of regular pattern is formed with the helical interleaver.

Because of the regular pattern, it is not necessary (but still possible) to use Tong's method to implement these interleavers using a single memory. As an example, an interleaver with $n = 4$ and $I = 5$ is shown in Table 9.7. Every "orbit" or column in the table has the same period. This will always be true with the helical or triangular interleaver. Furthermore, the address sequence can always be derived with the following algorithm assuming that y is the interleaver address and the total memory size is $\frac{(I-1)(d-1)}{2}+1$:

$$
\begin{aligned}
&y = 0 \quad // \quad starting\ address \\
&while\ (there\ is\ more\ data\ to\ process) \\
&\qquad for\ i = 0\ to\ I - 1 \\
&\qquad\qquad y = (y + i \cdot M)\ mod\ (\ total\ memory\ size) \\
&\qquad end \\
&end
\end{aligned}
$$

For the deinterleaver, i, the algorithm counts down from $I - 1$ to 0 instead of up from 0 to $I - 1$.

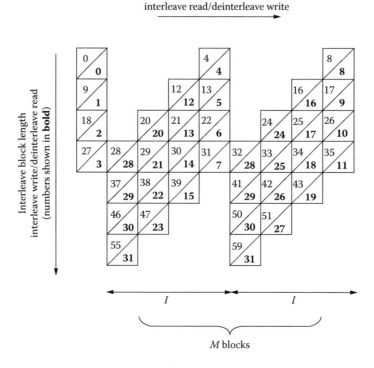

FIGURE 9.16
$M = 2$, $I = 4$, $d = 9$ triangular interleaver represented as a convolutional interleaver. Downward shifts of columns are in a regular pattern in blocks of I columns.

The advantage of the triangular or helical interleaver is the relatively simple implementation, with respect to Tong's method, of a memory optimal interleaver. However, this comes at the price of reduced flexibility. When the size of interleaver memory is not a primary concern—either because the interleaver memory is off-chip or the total amount of interleaver memory does not significantly influence the overall complexity—then it is simple and sufficient to implement a convolutional interleaver using twice the optimal amount of memory. As DSL speeds increase, the interleaver memory size requirement increases and

TABLE 9.7

Addressing for Forney's Triangular Interleaver
Using Tong's Method. $I = 4$, $d = 5$

Symbol	0	1	2	3
Delay	1	5	9	13
Address	0	0	1	3
	6	6	0	2
	5	5	6	1
	4	4	5	0
	3	3	4	6
	2	2	3	5
	1	1	2	4
	0	0	1	3
	6	6	0	2
	⋮	⋮	⋮	⋮
period	7	7	7	7

it becomes more difficult to implement the interleaver with off-chip memory. Therefore, in VDSL, having a memory optimal interleaver implementation is generally required.

9.4.3 Error Correction Comparison of DMT and Single-Carrier Modulation

As line code battles continue, debates rage on the relative merits of DMT versus single-carrier in combating impulse noise. It is fair to say that there is a time/frequency duality between the two line codes. Because DMT contains many narrow frequency subchannels, a narrowband noise impulse will corrupt only a few DMT subchannels, whereas the same impulse could corrupt all single-carrier symbols for its duration. However, a very short, wideband impulse will corrupt only a few single-carrier symbols, whereas, in the worst case, if it lies on the border between two DMT frames, it could corrupt two entire DMT frames, which is about 500 μs worth of data in all current DMT standards.

Interestingly, both the QAM and DMT VDSL1 specifications mandate 500 μs of impulse noise protection using erasures. Correcting 500 μs of data in either line code requires the same amount of memory at any given data rate. Therefore, those who drafted the competing line code proposals saw no particular advantage in terms of burst error correction for either line code.

9.4.4 Erasures

Recall from Section 9.2.5 that a Reed–Solomon code with r redundant symbols can be used to correct up to $2t + s \leq r$ symbols where s is the number of errors in known locations (erasures) and t is the number of errors in unknown locations. Because of the substantial memory requirements of interleavers and the desire to minimize the delay, it is highly desirable to take advantage of the factor of two improvement in error correction offered by using erasures. However, determining the location of errors is not trivial.

One way to try to determine the location of errors is to look for spikes in the error between the constellation points and the received, equalized input signal. Typically, these errors are computed for other purposes such as equalizer updates and trellis decoding. However, the difficulty with this method is that if erasures are flagged too frequently, the random error correcting capability of the code can be compromised. If flagged too infrequently, errors will be missed. A method that takes advantage of the inner trellis code in ADSL and ADSL2 to avoid some of the problems with too many or too few erasures is described in [Toumpakaris 2003a].

However, the VDSL1 recommendations do not use inner trellis codes, and ADSL connections are not required to use an inner trellis code.[5] Therefore, an alternate method is desired. A relatively simple method that takes advantage of the structure of convolutional interleaving is described in [Berlekamp 1985]. A very similar method is also described in [Toumpakaris 2003b]. In these methods, erasures are predicted based on detected errors in previous codewords.

Because of the cascading nature of the columns in the convolutional interleaver as shown, for example, in Figures 9.14 and 9.16, a burst of errors will always begin at the bottom of one of the columns and, therefore, at the end of a codeword. Therefore, even for a very long burst, there will be only be a few errors in the first few corrupted codewords and they can be detected and corrected without erasures. However, if more than one error is detected toward the end of a codeword, or if errors are detected in similar positions across more than one codeword, it can be assumed an error burst has started. The error positions can continue to be tracked to determine whether the burst continues.

---

[5] ADSL2 requires support for trellis codes, but the receiver can decide to disable it for any reason.

This method is simpler in the case of the triangular or helical interleavers because of the regular structure. However, it is still possible even with general convolutional interleaving.

Although this method is designed for impulse noise, it is also possible that the technique will be beneficial when considering the concatenated codes that are described in Section 9.5. With concatenated codes, errors that slip through the inner code often occur in bursts. It is possible that erasures can be used to maintain high coding gains using smaller interleaving depths than would be required without erasures. The author is not aware of any work that has been done on this topic to date.

9.5   Concatenated Coding

Concatenated coding was introduced by Forney in his doctoral thesis in 1965; the results were reprinted in [Forney 1966]. Forney presents serial concatenated codes of the form shown in Figure 9.17. A serial concatenated coding system uses two different codes, an outer code and an inner code. The two codes are encoded and decoded independently. Forney showed that their use can lead to an exponential decrease in error rate at the cost of more complexity and redundancy.

In current ADSL and ADSL2 standards, the inner code is a 16-state, Wei trellis code and the outer code is a Reed–Solomon code with interleaving. There is more discussion on trellis codes in Chapter 8 and on concatenated codes used for turbo coding in Chapter 10. Forney's work focused on inner and outer block codes. Michelson and Levesque [Michelson 1985] credit Odenwalder [Odenwalder 1972] with suggesting an inner convolutional code of the type used in ADSL.

ADSL uses concatenated codes for a number of reasons. Concatenated codes can offer higher coding gains than the trellis code or the Reed–Solomon code alone. Although it would be possible to achieve the higher level of coding gain with trellis coding alone, Reed–Solomon coding is also used to combat impulse noise, as discussed in Section 9.4. When impulse noise protection is not needed or low system delay is an important consideration, the trellis code can still be used and will achieve good performance on its own.

Evaluating the performance of concatenated codes is more challenging than evaluating the performance of any single code alone. Recall how the coding gain for the Reed–Solomon code was evaluated in Section 9.3. The development started with an expression for the error rate of a q-ary (octet in the case of GF(256)) symbol (Equation 9.24) and applied the binomial theorem (Equations 9.25 and 9.26). It was assumed that q-ary errors were independent. With concatenated codes, the q-ary symbol error rate after the inner (trellis) code depends on the performance of this inner code. Furthermore, errors following the inner code often occur in bursts. Making matters still more complicated, the length of the bursts that follow the inner

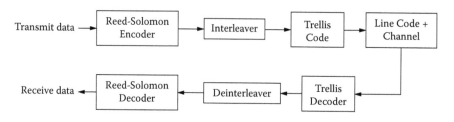

FIGURE 9.17
Serial concatenated coding used in ADSL and ADSL2. The outer code is a Reed–Solomon code, and the inner code is a 16-state trellis code.

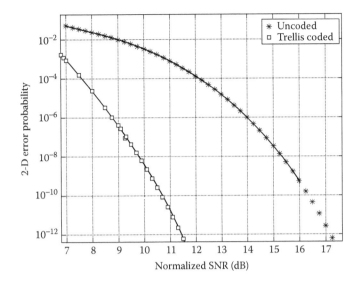

FIGURE 9.18
Two dimensional symbol error rate versus SNR for the 16-state Wei trellis code used in ADSL and ADSL2 from
[Zogakis 1994]. The SNR for an uncoded QAM constellation is also shown. These plots apply, in general, for any
large constellation.

code is usually longer when the error rate is higher. And if the outer code has a higher coding
gain, the error rate following the inner code will be higher than if the outer code is weaker.

It is possible to start by assuming that there is a "sufficiently large" interleaver between the
inner and outer codes to spread the errors over multiple codewords and to make the errors
appear independent. What is meant by "sufficiently large" depends on the inner and outer
codes being used. In this case, one can continue to use Equations 9.24 and 9.26 to determine
the coding gain of the outer, Reed–Solomon code. For the trellis code used in ADSL and
ADSL2, the expression for p_{symbol} in Equation 9.24 has no closed form expression. The error
rate versus the SNR for a trellis code can be evaluated using trellis search techniques or can
be evaluated by using computer simulation. A plot of the error rate of a two-dimensional
symbol versus the SNR is shown in Figure 9.18 [Zogakis 1994] for the 16-state Wei code
used in ADSL and ADSL2.

For the 16-state code used in both ADSL and ADSL2, the additional gross coding gain
that can be achieved by adding an outer Reed–Solomon code with sufficient interleaving
has been evaluated. This is shown in Figure 9.19 for $r = 2, 4, 6, \ldots, 16$ and n from $r + 1$ to
255. These gross coding gains are shown for 4 bits per constellation over $GF(2^8)$ assuming
a target bit error rate of 10^{-7} after the scrambler. It is interesting to compare this figure with
Figure 9.11, the gross coding gain of the Reed–Solomon code without the inner trellis code.
With the inner trellis code, the gross coding gains are considerably smaller. Intuitively, it
makes sense that simply concatenating codes does not achieve the full coding gain of the
two codes independently added. Otherwise, it would be possible to continue to achieve
higher net coding gains by continuing to concatenate more and more codes. Viewed another
way, the curve of the probability of symbol error with the trellis code is steeper than the
uncoded curve as shown in Figure 9.18. As the error rate curve gets steeper, the advantage,
in terms of SNR, between the coded and uncoded outer code is decreased. If another third
code were added, the added coding gain would be still smaller, so small that typically the
loss from the redundancy would be higher than the added gain.

Although the issue of interleaving was brushed aside earlier, it is interesting to see how
changing the interleaver depth influences the coding gain. Computer simulations using

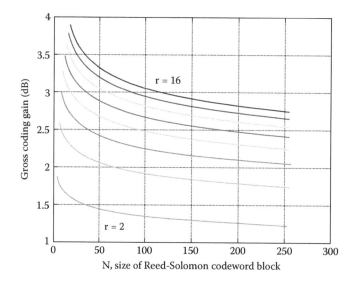

FIGURE 9.19

Additional gross coding gain provided by the Reed–Solomon code when concatenated with the trellis code used in ADSL and ADSL2. The gross coding gain is shown assuming all subcarriers use a 4-point constellation. The target bit error rate is 10^{-7} after the scrambler.

the trellis code in ADSL with the Reed–Solomon code were run to illustrate the impact of the interleaver depth. Reducing the interleaver depth does not eliminate the added coding gain, but it does reduce it. An example for the case of $n = 220$, 7 bits per subchannel, 24 tones in each DMT frame, and $r = 16$ is shown in Figure 9.20 over an interleaver depth of 2, 4, 8, 16, and 32. In this case, an interleaver depth of 16 is sufficient to achieve nearly the full coding gain of about 2.8 dB over the gross coding gain of the trellis code alone. This data point with sufficient interleaving can also be seen in Figure 9.19 at $n = 220$ and $r = 16$.

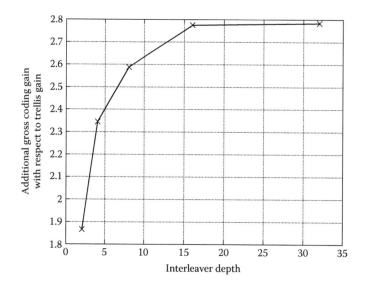

FIGURE 9.20

Gross coding gain versus the interleaver depth for concatenated trellis and Reed–Solomon codes. This example uses $n = 220$, $r = 16$, 7 bits per DMT subchannel, 24 tones per DMT frame, and an interleaver depth of 2, 4, 8, 16, and 32. Nearly all of the coding gain is recovered at a depth of 16.

Although Figure 9.19 used 4-bit constellations and Figure 9.20 used 7-bit constellations, the concatenated coding gain is not very sensitive to the size of the constellation.

9.6 Summary

The particular style and design of error control coding selected for a given application depends on the deployment environment, application, and line code. Because DSL is deployed over an aging outdoor plant, a wide variety of noise and loop conditions must be expected. Depending on whether the application is sensitive to delay or to error bursts, block coding with interleaving may or may not be appropriate. In HDSL and SHDSL where the application is primarily for low-delay data, there is no Reed–Solomon code or interleaving defined. In SHDSL, there is instead trellis coding which yields a high coding gain. ADSL is geared for residential services, which include data, voice, and also video-on-demand. In this case, both a trellis code and Reed–Solomon code plus interleaver are defined. It is left to the operators to decide whether to disable these features.

In all cases, though, DSL systems operate at high margins to account for the challenging deployment environment. In this environment, error control coding plays an essential role in creating stable connections at high speeds with error conditions appropriate to the application.

9.7 Acknowledgments

I am grateful to Professor Simon Litsyn for his generosity in reading this chapter in draft form and sharing his own ideas and research with me. The chapter is stronger for his contributions. I also wish to thank my colleague Po Tong for helping me to understand the principles and history of interleaving and Reed–Solomon coding. His suggestions and support while I was preparing this chapter were invaluable.

References

[Aslanis 1992] Aslanis, J.T., Tong, P., and Zogakis, T.N., "*An ADSL Proposal for Selectable Forward Error Correction with Convolutional Interleaving,*" T1E1.4/92-180, August 20, 1992.

[Berlekamp 1984] Berlekamp, E.R., "*Algebraic Coding Theory,*" Aegean Park Press, Laguna Hills, CA, 1984.

[Berlekamp 1985] Berlekamp, E.R. and Tong, P., "*Interleavers for Digital Communications,*" United States Patent, 4,559,625, Dec 17, 1985.

[Blahut 1983] Blahut, R.E., "*Theory and Practice of Error Control Codes,*" Addison-Wesley, Reading, MA, 1983.

[Blahut 1990] Blahut, R.E., "*Digital Transmission of Information,*" Addison-Wesley, Reading, MA, 1990.

[Brown 1999] Brown, R., "*Non-Continuous Events in the Telephone Outside Plant,*" contribution to T1E1.4, T1E1.4/99-183, April 20, 1999.

[Chien 1964] Chien, R.T., "*Cyclic Decoding Procedure for the Bose-Chaudhuri-Hocquenghem Codes,*" IEEE *Trans. on Info. Theory*, October 1964, IT-10, pp. 357–363.

[Forney 1965] Forney, G.D., "*On Decoding BCH Codes,*" IEEE *Trans. on Info. Theory*, October 1965, IT-11, pp. 549–557.

[Forney 1966] Forney, G.D., "*Concatenated Codes,*" MIT Press, Cambridge, MA, 1966.

[Forney 1971] Forney, G.D., *"Burst-Correcting Codes for the Classic Bursty Channel,"* IEEE Trans Communications Technology, Vol COM-19, October 1971, pp. 772–781.

[Forney 1972] Forney, G.D., *"Interleavers,"* United States Patent, 3,652,998, March 28, 1972.

[Forney 1989] Forney, G.D. Jr., *"Multi-dimensional constellations — Part II: Voronoi Constellations,"* IEEE J. Selected Areas Comm., August 1989.

[G.992.3 1993] ITU-T Recommendation G.992.3 (2002), *"Asymmetric Digital Subscriber Line (ADSL) Transceivers — 2 (ADSL2)."*

[Heegard 1999] Heegard, C. and Wicker, S.B., *"Turbo Coding,"* Kluwer Academic, Boston, MA, 1999.

[ISO/IEC 1993] ISO/IEC 3309:1993, *"Information technology — Telecommunications and information exchange between systems — High-level data link control (HDLC) procedures — Frame structure."*

[Lin 1983] Lin, S. and Costello, Jr. D., *"Error Control Coding,"* Prentice-Hall, Inc., Englewood Cliffs, NJ, 1983.

[MacWilliams 1977] MacWilliams, F.J. and Sloane N.J.A., *"The Theory of Error Correcting Codes,"* North-Holland, New York, 1977.

[McEliece 1986] McEliece, R.J. and Swanson, L. *"On the Decoder Error Probability for Reed — Solomon Codes,"* IEEE Trans. on Info. Theory, September, 1986, IT 32, pp. 701–703.

[McEliece 1997] McEliece, R.J., *"Gus' last theorem? Fast multiplication in GF (2^m),"* '10 Years PACRIM 1987–1997 — Networking the Pacific Rim.' 1997 IEEE Pacific Rim Conference on Communications, Volume: 2, pp. 736–738, 20–22 Aug 1997.

[Michelson 1985] Michelson, A.M. and Levesque, A.H., *"Error Control Techniques for Digital Communication,"* John Wiley, New York, 1985.

[Morelos-Zaragoza 2002] Morelos-Zaragoza, R.H., *"The Art of Error Correcting Coding,"* John Wiley, 2002.

[Odenwalder 1972] Odenwalder, J.P., et. al. *"Hybrid Coding Systems Study,"* Final Report, Contract NAS2-6722, Linkabit Corporation, La Jolla, CA, September 1972.

[Peterson 1972] Peterson, W.W. and Weldon, Jr., E.J., *"Error-Correcting Codes Second Edition,"* Colonial Press, 1972.

[Peterson 1961] Peterson, W. W. and Brown, D.T., *"Cyclic Codes for Error Detection,"* Proceedings of the IRE, January 1961, pp. 228–235.

[Pless 1998] Pless, V., *"Introduction to The Theory of Error-Correcting Codes Third Edition,"* John Wiley, New York, 1998.

[Ramsey 1970] Ramsey, J.L., *"Realization of Optimum Interleavers,"* IEEE Trans Info Theory, Vol. IT-16, No. 3, May 1970, pp. 338–345.

[Stallings 1997] Stallings, W., *"Data and Computer Communications, Fifth Edition,"* Prentice-Hall, Upper Saddle River, NJ, 1997.

[Starr 1999] Starr, T., Cioffi, J.M., and Silverman, P., *"Understanding Digital Subscriber Line Technology,"* Prentice-Hall, Upper Saddle River, NJ, 1999.

[Tong 1998] Tong, P., *"Efficient address generation for convolutional interleaving using a minimal amount of memory,"* United States Patent 5,764,649, June 9, 1998.

[Toti Rigatelli 1996] Toti Rigatelli, L., translated by John Denton, *"Evariste Galois, 1811–1832",* Birkhäuser, Boston, 1996.

[Toumpakaris 2003a] Toumpakaris, D., Yu, W., Cioffi, J.M., Gardan, D., and Ouzzif, M., *"A Byte-Erasure Method for Improved Impulse Immunity in DSL Systems using Soft Information from an Inner Code",* in ICC'2003.

[Toumpakaris 2003b] Toumpakaris, D., Yu, W., Cioffi, J.M., Gardan, D., and Ouzzif, M., *"A Simple Byte Erasure Method for Improved Impulse Noise Immunity in DSL,"* in ICC 2003.

[T1.413 1998] T1.413 Issue 2, *"Asymmetric Digital Subscriber Line (ADSL) Metallic Interface,"* ANSI T.413 — 1998.

[Wei 1987] Wei, L.-F., *"Trellis coded modulation with multi-dimensional constellations,"* IEEE Trans. on Info. Theory, 1987, IT-33, pp. 483–501.

[Zogakis 1994] Zogakis, T.N., Aslanis, J.T., Jr., and Cioffi, J.M., *"Analysis of a Concatenated Coding Scheme for a Discrete Multitone Modulation System,"* Military Communications Conference, MILCOM '94, October 2–5, 1994.

10

Advanced Coding Techniques for Digital Subscriber Lines

Evangelos Eleftheriou and Sedat Öl cer

CONTENTS

10.1 Introduction 272
10.2 Background on LDPC-Coding and Turbo-Coding Techniques 273
    10.2.1 LDPC Codes and Belief-Propagation (BP) Decoding 273
    10.2.2 The BP Decoding Algorithm in the Log Domain 275
        10.2.2.1 LLR-BP Decoding Based on the tanh Rule 276
        10.2.2.2 LLR-BP Decoding Based on Gallager's Approach 277
        10.2.2.3 LLR-BP Decoding Based on the Jacobian Approach 277
    10.2.3 Turbo Codes and Iterative BCJR Decoding 278
    10.2.4 The MAP Decoding Algorithm in the Log Domain 281
        10.2.4.1 The Log-MAP Decoding Algorithm 281
        10.2.4.2 The Max-Log-MAP Decoding Algorithm 283
10.3 Error-Correction Coding for Digital Subscriber Lines 284
10.4 Transmitter and Receiver Functions 286
10.5 Performance 291
    10.5.1 LDPC Coding 292
    10.5.2 Turbo Coding 292
    10.5.3 Latency 292
10.6 Complexity 295
10.7 Summary 296
References 296

ABSTRACT   The use of coding for error control is an integral part of the design of modern communication systems. Capacity-approaching codes such as turbo and LDPC codes, discovered or rediscovered in the past decade, offer near-Shannon-limit performance on the AWGN channel with rather low implementation complexity and are therefore increasingly being applied for error control in various fields of data communications. This chapter reviews the basic principles of low-density parity-check (LDPC) coding and turbo coding and provides a description of the main decoding algorithms used in this context. A generic multilevel modulation and coding scheme based on the use of turbo, turbo-like, or LDPC codes for DSL systems is described. It is shown that such codes provide significant gains in performance and allow an increase in data rate and loop reach that can be instrumental for the widespread deployment of future DSL services. Such techniques are also suitable for general multilevel modulation systems in other application areas.

10.1   Introduction

As described in Chapter 9, error-correcting codes have played an important role in achieving high data integrity in transmission systems. The T1.413 asymmetric digital subscriber line (ADSL) specification published by the American National Standards Institute (ANSI) in 1995 was the first DSL standard to incorporate error-correction coding. This ANSI document specifies the use of Reed–Solomon (RS) coding with code symbols from a Galois field (see Section 9.2.5) having 256 symbols, denoted as $GF(2^8)$, together with symbol-level convolutional interleaving (see Section 9.4), as a forward error-correction technique for ADSL systems that employ discrete multi-tone (DMT) modulation. Coding redundancy, number of DMT frames per RS codeword, and depth of interleaving are parameters that are selected from a pre-defined set of values in order to provide the best possible match to the transmission-channel characteristics and the application-specific constraints [Starr 1999]. As stated in Section 9.5, the ANSI document also includes the optional use of a 16-state, 4-dimensional trellis-coded modulation (TCM) scheme (see Chapter 8) as an inner coding mechanism to improve the communication reliability further. Subsequent ADSL specifications have retained this coding structure with some variations on the set of allowed parameter values. The current very-high-speed DSL (VDSL) specification [Standards Committee T1—Telecommunications: T1.424, European Telecommunications Standards Institute (ETSI): TS 101 270] has only included outer RS coding, whereas the Single-pair High-speed DSL (SHDSL) specification [International Telecommunication Union—Telecommunication Standardization Sector (ITU-T): G.991.2] has only included inner trellis coding without an outer error-correction code. Note that the G.992.3 ADSL2 Recommendation by ITU-T makes the use of inner TCM mandatory for upstream as well as downstream transmission.

It appears that DSL systems will still be evolving over many years under the auspices of several standards organizations. The solutions needed to meet the challenges of future systems will no doubt draw on the combined use of a number of advanced transmission techniques, including the use of selectable masks for the spectra of the transmitted signals [Ouyang 2003], dynamic spectrum management [Song 2002], and multi-input multi-output transmission techniques akin to those studied for wireless systems.

Yet another avenue is offered by the incorporation of more powerful, "near-Shannon-limit" coding techniques into DSL systems. These techniques were developed within the coding community, following the invention of turbo codes in 1993. They are now being adopted in many communication standards, in particular for wireless systems. It is shown in this chapter that the application of turbo [Berrou 1993] and low-density parity-check (LDPC) coding [Gallager 1962], [Gallager 1963] allows significant performance improvements in DSL without incurring a substantial increase in transceiver complexity. The results suggest that the inclusion of such coding techniques in DSL systems will facilitate the path toward higher service penetration, higher data rates, and more robust system operation.[1] The coding structure described herein is generic and can be used in conjunction with other capacity-approaching coding techniques (for example, block product-codes [Pyndiah 1998]) that are also decodable in an iterative fashion. The important question of whether these coding schemes should be part of a concatenated structure with outer RS coding will not be addressed here for space reasons but is a topic of intensive current research.

This chapter is organized as follows. In Section 10.2, the basic principles of LDPC coding and turbo-coding are reviewed, and a description of the main decoding algorithms used in this context is provided. The special issue of the *IEEE Communications Magazine*

---

[1] Such advanced coding techniques have been discussed to some extent in the ITU-T and Committee T1 for ADSL and VDSL.

[Spec. issue 2003] on capacity-approaching coding techniques provides an excellent introduction to this field and contains many useful references. A more in-depth description of the advances in this field can be found in the special issues of [Spec. issue 1998], [Spec. issue 2001a], [Spec. issue 2001b]. In Section 10.3, the performance improvements that can be expected from the introduction of advanced coding techniques into DSL systems, in terms of both loop reach and data rate, are discussed. In Section 10.4, specific LDPC coding and turbo-coding schemes are considered, and a generic structure for implementing coded modulation using these techniques is described. Examples of the performance achieved by the proposed approaches are given in Section 10.5 and comparisons with the ADSL TCM standard are provided.

10.2 Background on LDPC-Coding and Turbo-Coding Techniques

10.2.1 LDPC Codes and Belief-Propagation (BP) Decoding

Binary LDPC codes have been known since the early 1960s [Gallager 1962], [Gallager 1963] but their capacity-approaching performance has only been discovered in the past decade [MacKay 1999]. There is currently intensive research activity to explore deterministic constructions of such codes as well as to understand their theoretical limits. This section presents an introduction to LDPC codes and describes the most popular algorithms that are used for their decoding.

A binary (N, K) linear code is a set of binary vectors of length N, called codewords, that satisfy a set of $M \geq (N - K)$ parity-check equations. The length of the information vector is denoted by K. A parity-check equation can be conveniently represented as a binary vector of length N in which a "1" in position n, $1 \leq n \leq N$, implies that the nth codeword symbol participates in the parity check. The set of M parity checks can then be represented by an $M \times N$ parity-check matrix H. A "1" in the (m, n)th position in H indicates that the mth parity check involves the nth symbol of the codeword, or equivalently, that codeword symbol n participates in the mth parity check. This suggests a natural graphical representation of the parity-check matrix H as a "bipartite" graph with two kinds of nodes: N symbol or variable nodes that correspond to the codeword symbols, and M check nodes that correspond to the parity checks represented by the rows of matrix H [Tanner 1981], [Wiberg 1996]. The connectivity of the bipartite graph, known as the Tanner graph, is such that H is its incidence matrix; *i.e.*, for each "1" in the (m, n)th position, the graph has an edge connecting check node m with symbol node n.

A binary (N, K) LDPC code [Gallager 1962], [Gallager 1963], [MacKay 1999] is a linear block code described by a sparse $M \times N$ parity-check matrix H; *i.e.*, H has a low density of 1s. Each symbol is checked by a small number of parity checks, and each parity check involves a small number of symbols. An LDPC code is called (d_s, d_c)-regular if, in the corresponding bipartite graph, every symbol node is connected to d_s check nodes and every check node is connected to d_c symbol nodes; otherwise, it is called an irregular LDPC code. Figure 10.1 shows an example of the Tanner graph of a $(d_s = 3, d_c = 4)$-regular LDPC code of block length $N = 12$ having $M = 9$ parity-check equations.

The parity-check matrix of a regular LDPC code thus contains d_s ones in each column and d_c ones in each row. By contrast, the number of ones in each column, respectively, in each row, can vary widely for an irregular LDPC code. Although excellent performance can usually be achieved with regular LDPC codes, irregularity improves performance further and makes this class of codes capacity approaching.

The graphical representation of LDPC codes is attractive not only because it helps understand their parity-check structure but, more importantly, because it facilitates a powerful

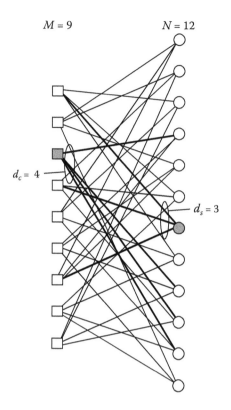

FIGURE 10.1
An example of a $(d_s = 3, d_c = 4)$-regular Tanner graph. Circles represent symbol nodes, and squares represent parity-check nodes.

decoding approach. The key steps in this decoding technique are the local application of the Bayes rule at each node and the exchange of the results, or messages, with "neighboring" nodes. At any given iteration, two types of messages are passed: probabilities from symbol nodes to check nodes, and probabilities from check nodes to symbol nodes. The message from symbol node n to check node m is the "belief" of the nth symbol node on the value of symbol n based on all parity checks involving n, except m. Similarly, the message from check node m to symbol node n is the "belief" of the mth check node on the value of symbol n based on all the symbols it checks, except n.

The message-passing decoding approach outlined above can be specified in an algorithmic form, leading to the so-called belief-propagation (BP) algorithm. Using a notation similar to that in [MacKay 1999], [Fossorier 1999], let $\mathbf{M}(n)$ denote the set of check nodes connected to symbol node n, *i.e.*, the positions of 1s in the nth column of the parity-check matrix H. Let $\mathbf{N}(m)$ denote the set of symbol nodes that participate in the mth parity-check equation, *i.e.*, the positions of 1s in the mth row of H. Furthermore, $\mathbf{N}(m)\backslash n$ represents the set $\mathbf{N}(m)$, excluding the nth symbol node, and similarly, $\mathbf{M}(n)\backslash m$ represents the set $\mathbf{M}(n)$, excluding the mth check node.

In addition, let $q_{n \to m}(0)$ and $q_{n \to m}(1)$ denote the message that symbol node n sends to check node m, indicating the probability of symbol n being 0 or 1, respectively, based on all the checks involving n except m. Similarly, $r_{m \to n}(0)$ and $r_{m \to n}(1)$ denote the message that the mth check node sends to the nth symbol node indicating the probability of symbol n being 0 or 1, respectively, based on all the symbols checked by m except n. Finally, $\mathbf{y} = [y_1, y_2, \ldots, y_N]$ denotes the received word corresponding to the transmitted codeword $\mathbf{x} = [x_1, x_2, \ldots, x_N]$.

In the probability domain, the inputs to the BP decoding algorithm are the a posteriori probabilities (APPs) $P(x_n = 0|y_n)$ and $P(x_n = 1|y_n)$, which are computed based on the channel statistics. The BP decoding algorithm is then summarized as follows.

Initialization: Each symbol node n is assigned APPs $q_{n \to m}(0) = P(x_n = 0|y_n)$ and $q_{n \to m}(1) = P(x_n = 1|y_n)$, which are sent to check node m as the initial message, for all m.

Step (i) (check-node update): For each m, for each $n \in \mathbf{N}(n)$, compute for $i = 0, 1$:

$$r_{m \to n}(i) = \sum_{x_{\{n': n' \in \mathbf{N}(m)\backslash n, \, \Sigma \oplus x_{n'} = i\}}} \prod_{n' \in \mathbf{N}(m)\backslash n} q_{n' \to m}(x_{n'}),$$

where $\Sigma \oplus x_{n'}$ indicates exclusive-OR addition of the summands $x_{n'}$.

Step (ii) (symbol-node update): For each n, for each $m \in \mathbf{M}(n)$, compute for $i = 0, 1$:

$$q_{n \to m}(i) = \mu_{n \to m} P(x_n = i|y_n) \prod_{m \in \mathbf{M}(n)} r_{m \to n}(i),$$

where the normalization constant $\mu_{n \to m}$ is chosen such that $q_n(0) + q_n(1) = 1$. For each n, compute the "pseudo-posterior" probabilities for $i = 0, 1$:

$$q_n(i) = \mu_n P(x_n = i|y_n) \prod_{m \in \mathbf{M}(n)} r_{m \to n}(i),$$

where the constant μ_n is chosen such that $q_{n \to m}(0) + q_{n \to m}(1) = 1$.

Step (iii) (decision): Quantize $\hat{x} = [\hat{x}_1, \hat{x}_2, \ldots, \hat{x}_N]$ such that $\hat{x}_n = 0$ if $q_n(0) \geq 0.5$ and $\hat{x}_n = 1$ if $q_n(0) < 0.5$. If $\hat{x}H^T = 0$, then halt the algorithm with $\hat{x}$ as the decoder output; otherwise go to *Step (i)*. If the algorithm does not halt within some maximum number of iterations, then declare a decoder failure.

Because the check-node update in *Step (i)* requires the computation of sums of products of probabilities, the BP algorithm is also called the "sum-product" algorithm.

10.2.2 The BP Decoding Algorithm in the Log Domain

In practical systems, using log-likelihood ratios (LLRs) as messages offers implementation advantages over using probabilities or likelihood ratios because multiplications are replaced by additions and the normalization step is eliminated. BP decoding of LDPC codes can be achieved in several different ways, all using LLRs as messages. This section first explains the motivations behind the most popular approaches [Gallager 1962], [Gallager 1963], [MacKay 1999], [Richardson 2001] and then describes alternative ways to perform BP decoding that are easy to implement and can reduce the decoding delay.

The LLR of a binary-valued random variable u is defined as

$$L(u) = \log \frac{P(u = 0)}{P(u = 1)},$$

where $P(u = 0)$ ($P(u = 1)$) denotes the probability that the random variable u takes on the value 0 (1). Noting that

$$P(u = 0) = e^{L(u)}/\left(1 + e^{L(u)}\right) \tag{10.1}$$

and

$$P(u = 1) = 1/\left(1 + e^{L(u)}\right) \tag{10.2}$$

yields $P(u = 0) - P(u = 1) = \tanh(L(u)/2)$. It can be shown (see [Hagenauer 1996] and [Richardson 2001]) that for two statistically independent binary random variables u and v, the so-called "tanh rule" is obtained, given by

$$L(u \oplus v) = 2\tanh^{-1}\left(\tanh\left(\frac{L(u)}{2}\right)\tanh\left(\frac{L(v)}{2}\right)\right), \qquad (10.3)$$

where the symbol $\oplus$ is used to indicate addition modulo 2.

A practical simplification follows from the fact that the functions $\tanh(x)$ and $\tanh^{-1}(x)$ are monotonically increasing and have odd symmetry. This implies that $\tanh(x) = \text{sign}(x)$ $\tanh(|x|)$ and $\tanh^{-1}(x) = \text{sign}(x)\tanh^{-1}(|x|)$. Therefore, the sign and the magnitude of $L(u \oplus v)$ are separable in the sense that the sign of $L(u \oplus v)$ depends only on the signs of $L(u)$ and $L(v)$, and the magnitude $|L(u \oplus v)|$ only on the magnitudes $|L(u)|$ and $|L(v)|$. Hence, in practice, the computation of the signs and the magnitudes of the outgoing LLRs can be separated, and the tanh rule is equivalently given by

$$L(u \oplus v) = \text{sign}(L(u))\text{sign}(L(v))2\tanh^{-1}\left(\tanh\left(\frac{|L(u)|}{2}\right)\tanh\left(\frac{|L(v)|}{2}\right)\right). \qquad (10.4)$$

This expression is readily generalized to the case where more than two independent binary random variables are involved.

In the following, the name LLR-BP decoding is used for the algorithm that uses LLRs as messages.

10.2.2.1   *LLR-BP Decoding Based on the* tanh *Rule*

The LLRs can be defined as:

$$Z_{n \to m}(x_n) = \log(q_{n \to m}(0)/q_{n \to m}(1))$$

and

$$L_{m \to n}(x_n) = \log(r_{m \to n}(0)/r_{m \to n}(1)).$$

Following the tanh rule, the LLR-BP is summarized as follows.

> **Initialization:**   Each symbol node n is assigned an a posteriori LLR $L(x_n|y_n) = \log(P(x_n = 0|y_n)/P(x_n = 1|y_n))$. In the case of equiprobable inputs on an additive white Gaussian noise (AWGN) channel, $L(x_n|y_n) = 2y_n/\sigma^2$, where σ^2 is the noise variance. For every position (m,n) such that $H_{n,m} = 1$,
>
> $$Z_{n \to m}(x_n) = L(x_n|y_n),$$
> $$L_{m \to n}(x_n) = 0.$$

Step (i) (check-node update):   For each m, and for each $n \in \mathbf{N}(m)$, compute

$$L_{m \to n}(x_n) = \left(\prod_{n' \in \mathbf{N}(m) \backslash n} \text{sign}(Z_{n' \to m}(x_{n'}))\right) 2\tanh^{-1}\left(\prod_{n' \in \mathbf{N}(m) \backslash n} \tanh\left(\frac{|Z_{n' \to m}(x_{n'})|}{2}\right)\right).$$

$$(10.5)$$

Step (ii) (symbol-node update):   For each n, and for each $m \in \mathbf{M}(n)$, compute

$$Z_{n \to m}(x_n) = L(x_n|y_n) + \sum_{m' \in \mathbf{M}(n) \backslash m} L_{m' \to n}(u_n).$$

For each n, compute

$$Z_n(x_n) = L(x_n|y_n) + \sum_{m \in \mathbf{M}(n)} L_{m \to n}(x_n).$$

Step (iii) (decision): Quantize $\hat{\mathbf{x}} = [\hat{x}_1, \hat{x}_2, \ldots, \hat{x}_N]$ such that $\hat{x}_n = 0$ if $Z_n(x_n) \geq 0$ and $\hat{x}_n = 1$ if $Z_n(x_n) < 0$. If $\hat{\mathbf{x}}H^T = 0$, then halt the algorithm with $\hat{\mathbf{x}}$ as the decoder output; otherwise go to *Step (i)*. If the algorithm does not halt within some maximum number of iterations, then declare a decoder failure.

The check-node updates are computationally the most complex part of the LLR-BP algorithm. Two issues influence their complexity: i) the topology used in computing the messages that a particular check node sends to the symbol nodes associated with it, and ii) the implementation of the core operation needed for computing these messages. For example, the core operation of the check-node update computation in *Step (i)* above is the hyperbolic tangent function, which is known to be difficult to implement in hardware. In a brute-force implementation of the check node update, $d_c(d_c - 1)$ multiplications are necessary per check node, with all multiplicands requiring the evaluation of the hyperbolic tangent core operation. Furthermore, it is necessary to compute the overall sign. Clearly, the higher the rate of the code, the higher the row degree d_c, thus leading to a larger number of multiplications. Therefore the brute-force topology and its corresponding core operation are not suited for high-speed digital applications.

10.2.2.2 LLR-BP Decoding Based on Gallager's Approach

In Gallager's original work [Gallager 1962], [Gallager 1963], it was proposed to compute the LLR messages in a check node based on the following identity

$$L(u \oplus v) = \text{sign}(L(u))\text{sign}(L(v))f(f(|L(u)|) + f(|L(v)|)). \tag{10.6}$$

The function

$$f(x) = \log\frac{e^x + 1}{e^x - 1} = -\log\left[\tanh\left(\frac{x}{2}\right)\right] \tag{10.7}$$

is an involution transform; *i.e.*, it has the property $f(f(x)) = x$. Clearly, Equation 10.6 can be obtained from Equation 10.4 by introducing the definition of $f(x)$. The check-node update in LLR-BP based on this approach is given by

$$L_{m \to n}(x_n) = \left(\prod_{n' \in \mathbf{N}(m)\backslash n} \text{sign}(Z_{n' \to m}(x_{n'}))\right) f\left(\sum_{n' \in \mathbf{N}(m)\backslash n} f\left(|Z_{n' \to m}(x_{n'})|\right)\right). \tag{10.8}$$

The second product term in Equation 10.5 has now been replaced by additions involving the $f(x)$ function. All other steps in the BP decoding algorithm remain the same. The implementation of Equation 10.8 is straightforward, provided that the involution transform can be implemented properly. This is easily and elegantly achieved in software where essentially infinite precision representation can be realized. The transform defined by Equation 10.7 is first done on all incoming LLR messages. Then all the terms are summed, and individual terms are subtracted to obtain the individual "pseudo-messages" from which the outgoing messages are obtained by means of the involution transform. The entire check-node update equation requires first the evaluation of Equation 10.7 and $2d_c$ additions. This approach is simple and amenable to a parallel implementation, and therefore looks promising for use in extremely high-speed applications. However, this decoding scheme is very sensitive to finite-precision implementation, requiring a large number of bits of precision [Chen 2003].

10.2.2.3 LLR-BP Decoding Based on the Jacobian Approach

The "tanh rule" of Equation 10.3 can alternatively be represented by [Hagenauer 1996]

$$L(U \oplus V) = \log\frac{1 + e^{L(U)+L(V)}}{e^{L(U)} + e^{L(V)}}.$$

Using the Jacobian logarithm twice yields

$$
\begin{aligned}
L(U \oplus V) &= \max\left(0, L(U) + L(V)\right) + \log\left(1 + e^{-|(L(U)+L(V)|}\right) \\
&\quad - \max\left(L(U), L(V)\right) - \log\left(1 + e^{-|(L(U)-L(V)|}\right) \\
&= \operatorname{sign}\left(L(u)\right)\operatorname{sign}(L(v))\min\left(|L(U)|, |L(V)|\right) \\
&\quad + \log\left(1 + e^{-|(L(U)+L(V)|}\right) - \log\left(1 + e^{-|(L(U)-L(V)|}\right).
\end{aligned}
\tag{10.9}
$$

An efficient way to obtain each of the d_c outgoing messages is the following. Consider a particular check node m with d_c connections from symbol nodes in $\mathbf{N}(m) = (n_1, n_2, \ldots, n_{d_c})$. The incoming messages are then $Z_{n_1 \to m}(x_{n_1})$, $Z_{n_2 \to m}(x_{n_2})$, $\ldots$, $Z_{n_{d_c} \to m}(x_{n_{d_c}})$. Let us define two sets of auxiliary binary random variables $f_1 = x_{n_1}$, $f_2 = f_1 \oplus x_{n_2}$, $f_3 = f_2 \oplus x_{n_3}, \ldots$, $f_{d_c} = f_{d_c-1} \oplus x_{n_{d_c}}$, and $b_{d_c} = x_{n_{d_c}}$, $b_{d_c-1} = b_{d_c} \oplus x_{n_{d_c-1}}, \ldots$, $b_1 = b_2 \oplus x_{n_1}$. Using Equation 10.9 repeatedly, one can obtain $L(f_1), L(f_2), \ldots, L(f_{d_c})$ and $L(b_1), L(b_2), \ldots, L(b_{d_c})$ in a recursive manner based on the knowledge of $Z_{n_1 \to m}(x_{n_1})$, $Z_{n_2 \to m}(x_{n_2})$, $\ldots$, $Z_{n_{d_c} \to m}(x_{n_{d_c}})$. Using the parity-check node constraint $(x_{n_1} \oplus x_{n_2} \oplus \cdots \oplus x_{n_{d_c}}) = 0$, one obtains $x_{n_i} = (f_{i-1} \oplus b_{i+1})$ for every $i \in \{2, 3, \ldots, d_c - 1\}$. Therefore, the outgoing message from check node m can simply be expressed as

$$
L_{m \to n_i}(x_{n_i}) = \begin{cases} L(b_2) & i = 1 \\ L(f_{i-1} \oplus b_{i+1}) & i = 2, 3, \ldots, d_c - 1 \;. \\ L(f_{d_c-1}) & i = d_c \end{cases}
$$

Clearly, this approach is essentially the forward-backward algorithm [Bahl 1974] applied to the trellis of a single parity-check code, requiring $3(d_c - 2)$ computations of the core operation $L(U \oplus V)$ in Equation 10.7 per check-node update. The underlying function $g(x) = \log(1 + e^{-|x|})$ in Equation 10.9 can be implemented using a look-up table or a piecewise linear function (see [Eleftheriou 2001] for examples).

There are many approaches that lead to reduced complexity LLR-BP while maintaining near optimum performance. All these approaches focus primarily on simplifying the more complex check-node update. The reader is referred to [Eleftheriou 2001], [Chen 2003], and the references therein for details on the topic.

10.2.3 Turbo Codes and Iterative BCJR Decoding

Various coding and decoding approaches according to the turbo-coding principle introduced in [Berrou 1993] have been proposed in the literature. In particular, parallel or serially concatenated convolutional codes, repeat accumulate codes, and various combinations thereof have been shown to approach the capacity of the AWGN channel. The common features of all these approaches are the interleaving and deinterleaving functions as well as the soft-input soft-output APP decoder usually implemented by the so-called BCJR algorithm. This section presents an introduction to turbo codes and describes two popular algorithms that are used for their decoding.

Turbo codes, as originally described in [Berrou 1993], are obtained by simultaneously encoding the information sequence to be transmitted by two convolutional encoders arranged in a parallel structure as shown in Figure 10.2.

The Convolutional Encoder 1 operates directly on the information sequence, however, the Convolutional Encoder 2 is provided with an interleaved version of the information sequence. The interleaver plays a key role in the performance of a turbo-coding scheme as both its size and type affect performance. Another important aspect is the fact that the convolutional encoders need to be in a recursive form. In this case, the performance

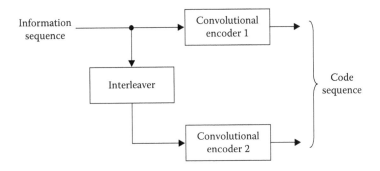

FIGURE 10.2
Parallel concatenation of two convolutional codes as defined in classical turbo coding.

increases with interleaver size K, a property known as "interleaver gain" [Benedetto 1996]. Figure 10.3 shows the block diagram of a transmission system that incorporates encoding and decoding for a typical turbo-coding scheme. Here, convolutional encoding with rate-1/2 codes has been assumed for simplicity; however, other code rates can in general be used, and the two encoders need not be identical. Both recursive convolutional encoders are in systematic form. For each systematic bit x_n^s, expressed in bipolar form $x_n^s = 2d_n - 1$, where d_n denotes the information bit to be encoded at time n, two parity bits $x_{1,n}^p$ and $x_{2,n}^p$ are generated by the encoder. If the systematic bit is transmitted along with both parity bits, a rate-1/3 code is obtained. In Figure 10.3, it is assumed that puncturing is applied to the sequences of parity bits so that the rate of the code can be increased (the transmitted parity bit x_n^p is also expressed in bipolar form). For example, if every other bit is punctured alternately in each parity sequence, a rate-1/2 code is obtained.

At the receiver, the channel output signals are denoted by $y_n^s = x_n^s + v_n^s$ and $y_n^p = x_n^p + v_n^p$, where v_n^s and v_n^p represent samples of a zero-mean white Gaussian noise process with variance σ^2. The turbo decoder processes the channel output signals and delivers estimates $\hat{d}_n$ of the transmitted bits. Because the puncturing mechanism is known at the receiver, decoding can be achieved without knowledge on the punctured parity bits.

It is possible to develop graphical representations for turbo codes similar to those introduced above for the LDPC codes. For example, the turbo code generated by the encoder in

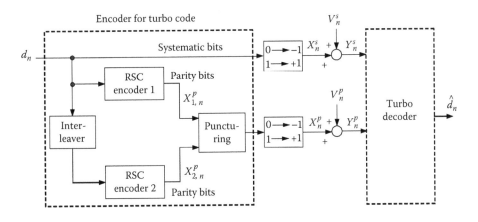

FIGURE 10.3
Transmission system comprising a parallel concatenation of two recursive systematic convolutional (RSC) codes, an additive white Gaussian noise channel, and a turbo decoder.

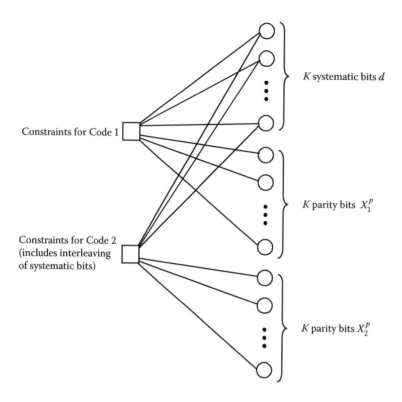

FIGURE 10.4
A Tanner graph representation of the turbo code generated by the encoder in Figure 10.3. The puncturing of parity bits is not shown.

Figure 10.3 can be represented as depicted in Figure 10.4. In fact, both LDPC and turbo codes can be regarded as particular forms of "codes defined on graphs," a powerful framework that has emerged in coding theory over the recent years.

Maximum-likelihood (ML) decoding of turbo codes would require considering all possible encoder states as well as all possible 2^K interleaver configurations simultaneously, an exceedingly large number in view of the interleaver-size values of K needed in practice. Instead, the turbo decoder resorts to a soft decoding algorithm that is applied alternately for convolutional Code 1 and convolutional Code 2, whereby decoding for Code 1 generates soft information that is used for the subsequent decoding for Code 2, and vice versa. The soft information that is exchanged between the two decoders is referred to as "extrinsic information" and represents new information about information bits that is made available by one decoder to the other. This iterative exchange of information, through which each decoder is able to improve its performance, is reminiscent of the efficient use of feedback in turbo engines, an analogy that motivated the name of turbo codes for the class of codes discussed in this section.[2] Therefore, instead of pursuing a "globally" optimum decoding strategy that would be far too complex to implement, such as ML decoding, the turbo decoder operates optimally only "locally" by performing maximum a posteriori (MAP) decoding alternately for Codes 1 and 2. (Similar local processing at symbol and check nodes has been discussed earlier for LDPC codes.) The algorithm described by Bahl, Cocke, Jelinek, and Raviv (BCJR)

---

[2] Although the qualifier "turbo" refers to the decoding process, it has become common practice to refer to this class of codes as "turbo codes." The more accurate terminology of "parallel concatenated convolutional codes" is also employed.

in [Bahl 1974] is a computationally efficient procedure for MAP decoding and was used in the original development of turbo codes. Together with its simplified versions, the BCJR algorithm represents the most popular approach for turbo decoding.

In the following, the BCJR algorithm is described in a form that involves LLRs rather than likelihood ratios as was originally described in [Bahl 1974]. The advantage of this "log-MAP" algorithm is that the multiplications needed in the MAP algorithm are replaced by additions. Recall that the same advantage was pointed out earlier when developing the LLR-BP decoding algorithm. Following the description of the log-MAP algorithm, the "max-log-MAP" algorithm is discussed. The "max-log-MAP" algorithm is suboptimum but offers a lower implementation complexity than the log-MAP algorithm does.

10.2.4 The MAP Decoding Algorithm in the Log Domain

10.2.4.1 The Log-MAP Decoding Algorithm

The principle of log-MAP decoding is to compute the a posteriori LLR of each information bit d_n,

$$L\left(d_n | y_1^K\right) = \log \frac{P\left(d_n = 0 | y_1^K\right)}{P\left(d_n = 1 | y_1^K\right)}, \quad n = 1, \ldots, K,$$

given the observation of the entire signal sequence $\mathbf{y} = y_1^K = \{y_1, y_2, \ldots, y_K\}$, where $y_n = \left(y_n^s, y_n^p\right)$ for $n = 1, \ldots, K$. Maximum a posteriori decisions $\hat{d}_n$ are generated by the decoder according to

$$\hat{d}_n = \begin{cases} 0 & \text{if } L\left(d_n | y_1^K\right) \geq 0 \\ 1 & \text{if } L\left(d_n | y_1^K\right) < 0 \end{cases}.$$

An efficient procedure for computing $L(d_n | y_1^K), n = 1, \ldots, K$ is obtained as follows. Let the state of the recursive systematic convolutional Encoder 1 or 2 at time n be denoted by $S_n \in \{0, 1, \ldots, M-1\}$, such that d_n represents the information bit associated with the transition from state S_{n-1} to state S_n. There are M possible state values at each time instant. The BCJR algorithm involves two types of recursions: forward and backward recursions. The forward recursions consist in the iterative computation of the quantities

$$\alpha_n(S_n = m) = \log\left\{P\left(S_n = m | y_1^n\right)\right\}, \quad m \in \{0, 1, \ldots, M-1\},$$

for an increasing time index n, and the backward recursions consist in the iterative computation of the quantities

$$\beta_n(S_n = m) = \log\left\{\frac{p\left(y_{n+1}^K | S_n = m\right)}{p\left(y_{n+1}^K | y_1^n\right)}\right\}$$

for a decreasing time index n. These recursions are obtained as [Robertson 1995]

$$\alpha_n(S_n = m) = \log \frac{\sum_{m'} \sum_{i=0}^{1} e^{\gamma_i(y_n, S_{n-1}=m', S_n=m) + \alpha_{n-1}(S_{n-1}=m')}}{\sum_m \sum_{m'} \sum_{i=0}^{1} e^{\gamma_i(y_n, S_{n-1}=m', S_n=m) + \alpha_{n-1}(S_{n-1}=m')}}, \quad n = 1, \ldots, K,$$

with the initial conditions $\alpha_0(S_0 = 0) = 1$, and $\alpha_0(S_0 = m) = 0$, for $m \neq 0$, and

$$\beta_n(S_n = m) = \log \frac{\sum_{m'} \sum_{i=0}^{1} e^{\gamma_i(y_{n+1}, S_n=m, S_{n+1}=m') + \beta_{n+1}(S_{n+1}=m')}}{\sum_m \sum_{m'} \sum_{i=0}^{1} e^{\gamma_i(y_n, S_n=m, S_{n+1}=m') \alpha_n(S_n=m)}}, \quad n = K-1, \ldots, 1,$$

with the initial conditions $\beta_K(S_K = m) = 1/M, \forall m$, and with $m, m' \in \{0, 1, \ldots, M-1\}$. Note that these conditions are chosen because the initial states for both encoders are, in this case, assumed to be the zero state and the final states are left unspecified. In the above expressions for $\alpha_n(S_n = m)$ and $\beta_n(S_n = m)$, the summation terms that involve consecutive state pairs are only meaningful if a transition between such pairs is defined on the encoder state machine: $S_{n-1} \rightarrow S_n$ or $S_n \leftarrow S_{n+1}$. For such "allowed" state pairs and state transitions, the quantity, $\gamma_i(y_n, S_{n-1}, S_n)$, which is defined for bit values $i = 0$ and 1 as

$$\gamma_i(y_n, S_{n-1}, S_n) = \log\{P(d_n = i, S_n, y_n | S_{n-1})\},$$

is computed by

$$\gamma_i(y_n, S_{n-1}, S_n) = \log\{p(y_n^s | d_n = i)\} + \log\{p(y_n^p | d_n = i, S_{n-1}, S_n)\}$$
$$+ \log\{P(S_n | S_{n-1})\}, i = 0, 1.$$

In this equation, the first term is easily evaluated because the noise process is assumed to have a Gaussian probability distribution:

$$p(y_n^s | d_n = i) = \frac{1}{\sqrt{2\pi}\sigma} e^{-(y_n^s - (2i-1))^2 / 2\sigma^2}.$$

The second term is likewise evaluated for the parity bit generated during the state transition $S_{n-1} \rightarrow S_n$ obtained for $d_n = i$. The third term $\log\{P(S_n | S_{n-1})\}$ plays an important role because it is through this term that the information delivered by the other decoder is included into the decoding process. Clearly, because the state transition $S_{n-1} \rightarrow S_n$ corresponds either to the input bit value $i = 0$ or to the input bit value $i = 1$, $P(S_n | S_{n-1})$ directly represents the a priori information of d_n being a "0" or a "1." A useful relationship between $P(S_n | S_{n-1})$ and the a priori LLR $L(d_n) = \log\{P(d_n = 0)/P(d_n = 1)\}$ is obtained by using the equation $P(d_n = 0) = 1 - P(d_n = 1)$. Thus, for a state pair corresponding to a transition with bit value $i = 0$,

$$\log\{P(S_n | S_{n-1})\} = L(d_n) - \log\{1 + e^{L(d_n)}\}, \tag{10.10}$$

and for a state pair corresponding to a transition with bit value $i = 1$,

$$\log\{P(S_n | S_{n-1})\} = -\log\{1 + e^{L(d_n)}\}, \tag{10.11}$$

similar to the expressions in Equations 10.1 and 10.2.

At this point, all the quantities $\alpha_n(S_n)$, $\beta_n(S_n)$, and $\gamma_i(y_n, S_{n-1}, S_n)$ for all state values and state transitions can be computed for all time indices $n = 1, \ldots, K$. Based on these computations, the LLRs $L(d_n | y_1^K)$ are obtained as

$$L(d_n | y_1^K) = \log \frac{\sum_m \sum_{m'} e^{\gamma_0(y_n, S_{n-1}=m', S_n=m) + \alpha_{n-1}(S_{n-1}=m') + \beta_n(S_n=m)}}{\sum_m \sum_{m'} e^{\gamma_1(y_n, S_{n-1}=m', S_n=m) + \alpha_{n-1}(S_{n-1}=m') + \beta_n(S_n=m)}}, \quad n = 1, \ldots, K.$$

To explain further how the exchange of soft information occurs between the two decoders during the turbo-decoding process, let $L^{(1)}(d_n)$ and $L^{(2)}(d_n)$ denote the a priori LLRs on bit d_n available to the log-MAP decoder 1 and the log-MAP decoder 2, respectively. At the start of the decoding process, both values of $d_n = 0$ and 1 are equally likely for Decoder 1, hence $L^{(1)}(d_n) = 0$ is used in Equation 10.10 or 10.11 for this decoder. After decoding for Code 1, Decoder 1 generates extrinsic information (see below) on d_n, denoted by $L_e^{(1)}(d_n)$, which is used by Decoder 2 as a priori information on d_n; i.e., Decoder 2 sets $L^{(2)}(d_n) = L_e^{(1)}(d_n)$ and uses it in evaluating Equation 10.10 or 10.11. After decoding for Code 2, the first "decoding round" is completed. For the second decoding round, Decoder 1 can use the extrinsic information $L_e^{(2)}(d_n)$ generated by Decoder 2 during the first decoding round and proceed with a new decoding step. The procedure is then repeated for all subsequent steps.

To complete the description of the turbo-decoding procedure, it is still necessary to define the generation of extrinsic information by the decoder. For this, the above expression for $L(d_n|y_1^K)$ can be written as

$$L(d_n|y_1^K) = \log\left\{\frac{p(y_n^s|d_n=0)}{p(y_n^s|d_n=1)}\right\}$$

$$+\log\left\{\frac{P(d_n=0)}{P(d_n=1)}\right\}$$

$$+\log\frac{\sum_m \sum_{m'} e^{p(y_n^p|d_n=0,\,S_{n-1}=m',\,S_n=m)+\alpha_{n-1}(S_{n-1}=m')+\beta_n(S_n=m)}}{\sum_m \sum_{m'} e^{p(y_n^p|d_n=1,\,S_{n-1}=m',\,S_n=m)+\alpha_{n-1}(S_{n-1}=m')+\beta_n(S_n=m)}},$$

where the double summation in the numerator (denominator) is over all state pairs (S_{n-1}, S_n) that are defined for an input bit value of 0 (1). The first term on the right-hand side of this expression is the systematic LLR $L_s(d_n)$, the second term the a priori LLR $L(d_n)$ discussed above, and the last term is the extrinsic LLR $L_e(d_n)$ generated by the decoder. Note that in the expression for $L_e(d_n)$, the terms in α include a dependency on $(y_1^s, y_1^p), \ldots, (y_{n-1}^s, y_{n-1}^p)$, and those in β, a dependency on $(y_{n+1}^s, y_{n+1}^p), \ldots, (y_K^s, y_K^p)$. The conditional probability $p(y_n^p|\cdot)$ terms include a dependency on the parity bit y_n^p at time n. Hence, $L_e(d_n)$ is independent of y_n^s (the dependency on this quantity has been moved to the term $L_s(d_n)$) and can thus be used by the other decoder as new additional information on bit d_n.

Finally, note that if parity information at time n is missing due to puncturing during the encoding process, then the corresponding probability term can be set to 0.5 for the computations.

10.2.4.2 The Max-Log-MAP Decoding Algorithm

It is possible to further simplify the log-MAP algorithm by introducing approximations in the computations of the forward and backward recursions. Note first that $\alpha_n(S_n = m)$ is equivalently written as [Robertson 1995]

$$\alpha_n(S_n = m) = \log\left\{\sum_{m'}\sum_{i=0}^{1} e^{\gamma_i(y_n,\,S_{n-1}=m',\,S_n=m)+\alpha_{n-1}(S_{n-1}=m')}\right\}$$

$$-\log\left\{\sum_m \sum_{m'}\sum_{i=0}^{1} e^{\gamma_i(y_n,\,S_{n-1}=m',\,S_n=m)+\alpha_{n-1}(S_{n-1}=m')}\right\}.$$

Using the approximation

$$\log\sum_i e^{u_i} \cong \max_i u_i,$$

the expression becomes

$$\alpha_n(S_n = m) \cong \max_{m',i}\{\gamma_i(y_n,\,S_{n-1}=m',\,S_n=m)+\alpha_{n-1}(S_{n-1}=m')\}$$

$$-\max_{m,m',i}\{\gamma_i(y_n,\,S_{n-1}=m',\,S_n=m)+\alpha_{n-1}(S_{n-1}=m')\}.$$

For $\beta_n(S_n = m)$,

$$\beta_n(S_n = m) \cong \max_{m',i}\{\gamma_i(y_{n+1},\,S_n=m,\,S_{n+1}=m')+\beta_{n+1}(S_{n+1}=m')\}$$

$$-\max_{m,m',i}\{\gamma_i(y_n,\,S_n=m,\,S_{n+1}=m')\,\alpha_n(S_n=m)\},$$

and for $L(d_n|y_1^K)$:

$$L(d_n|y_1^K) \cong \max_{m,m'} \{\gamma_0(y_n, S_{n-1} = m', S_n = m) + \alpha_{n-1}(S_{n-1} = m') + \beta_n(S_n = m)\}$$
$$- \max_{m,m'} \{\gamma_1(y_n, S_{n-1} = m', S_n = m) + \alpha_{n-1}(S_{n-1} = m') + \beta_n(S_n = m)\}.$$

Note that because $\alpha_{n-1}(S_{n-1} = m')$ appears in both maximizations in $L(d_n|y_1^K)$, the second $\max_{m,m',i}\{\cdot\}$ term in the expression for $\alpha_n(S_n = m)$ can be omitted without affecting the value of $L(d_n|y_1^K)$. Similarly, the second $\max_{m,m',i}\{\cdot\}$ term in the expression for $\beta_n(S_n = m)$ can be omitted. In this case, referring to the forward and backward recursion variables as $\bar{\alpha}_n(S_n = m)$ and $\bar{\beta}_n(S_n = m)$, respectively,

$$\bar{\alpha}_n(S_n = m) = \max_{m',i} \{\gamma_i(y_n, S_{n-1} = m', S_n = m) + \bar{\alpha}_{n-1}(S_{n-1} = m')\},$$
$$\bar{\beta}_n(S_n = m) = \max_{m',i} \{\gamma_i(y_{n+1}, S_n = m, S_{n+1} = m') + \bar{\beta}_{n+1}(S_{n+1} = m')\}.$$

It can readily be seen that the computations for $\bar{\alpha}_n$ correspond to the well-known Viterbi algorithm and the computations for $\bar{\beta}_n$ correspond to the Viterbi algorithm computed backward in time.

10.3   Error-Correction Coding for Digital Subscriber Lines

Before applying the coding techniques described in the previous section to DSL transmission, it is appropriate to illustrate the benefits of error-correction coding in DSL systems by means of a simple example. To this end, data transmission over an ordinary unshielded twisted-pair has been considered under the assumption that the received signal is disturbed by near-end crosstalk (NEXT) as well as AWGN. Furthermore, it is assumed that the NEXT disturbance is due to 49 ADSL downstream transmitters that share the same cable binder as the data-transmission system under consideration. The following questions are of interest: assuming operation at a fixed data rate, by how much can the cable length be increased for this system through error-correction coding (ECC)? Alternatively, by how much can the data rate be increased through ECC for a fixed cable length? It is possible to answer these questions by first computing the capacity of this communication channel for different data rates and cable lengths as discussed in Chapter 4. The results are summarized in Figure 10.5.

For example, Figure 10.5a shows that, for operation at 3 Mbit/s, the loop reach of this system is limited to 2355 m. If coding is employed with a net coding gain of 3 dB (6 dB), the loop reach can be extended by 315 m (645 m). As another example, Figure 10.5b shows that, for a loop reach of 2.0 km, an uncoded system can achieve a data rate of 3.5 Mbit/s and that a coding gain of 3 dB (6 dB) allows this data rate to be increased by 480 kbit/s (1 Mbit/s). This shows that coding can play an important role in increasing the performance of DSL systems in terms of data rate and loop reach. Clearly, similar results can be derived for a variety of channel and noise models, as discussed in Section 9.3.

Simulation studies indicate that coding gains of 6 dB and more can practically be achieved through the use of capacity-approaching coding techniques in DSL. Anticipating the results to be presented in Section 10.5, some examples are given in Figure 10.6. Here spectral-efficiency versus signal-to-noise ratio (SNR) representations, which are often used in digital communications to compare different modulation schemes, are shown in connection with 16, 256, and 4096 quadrature amplitude modulation (QAM).

As an example, one can consider 256-QAM. The "star" (A) indicates that uncoded transmission at a spectral efficiency of 8 bit/s/Hz requires an SNR of $E_b/N_0 = 24.4$ dB. This

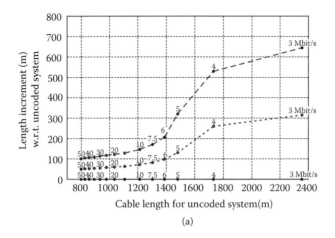

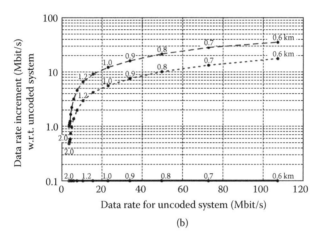

FIGURE 10.5
Examples of performance improvements resulting from coding for DSL transmission. (a) Cable-length increase achieved by coding. The numbers above the curves indicate achievable data rates in Mbit/s. (b) Data-rate increase achieved by coding. The numbers above the curves indicate cable length in km (the data-rate increment for the reference uncoded system is artificially displayed at the ordinate value of 0.1 instead of zero because of the logarithmic scale). The specific assumptions used to derive the results were a signal attenuation of 75 dB at 5 MHz for a cable length of 1 km; AWGN with a power spectral density of -140 dBm/Hz; NEXT due to 49 ADSL (ITU-T, Recommendation G.992.1) downstream transmitters that use all subchannels; transmit power of 15 dBm; SNR margin of 6 dB; gap to capacity for uncoded transmission of 9.95 dB, which implies operation at a symbol-error rate of 10^{-7}.

value corresponds to a bit-error rate (BER) of 10^{-7}. If an LDPC code mapping 2021 information bits to 2209 encoded bits is employed with the same 256-QAM constellation, the SNR value can be lowered to 17.8 dB while still ensuring operation at a BER of 10^{-7} (triangle, B). In that case, transmission occurs at a spectral efficiency of 7.49 bit/s/Hz because of the redundancy needed for coding. In other words, this LDPC-coded system achieves a coding gain of $\sim$6.6 dB at a BER of 10^{-7} with 256-QAM. Similar results can be read from this figure for the other constellation sizes, as well as for turbo-coding (diamonds) and TCM (circles). The dashed lines show the capacity of the band-limited AWGN channel with "discrete" QAM symbol constellations as a function of the SNR.

What are the important aspects that must be considered in designing coding schemes for DSL systems? First, high code rates are desirable to achieve high spectral efficiencies for bandwidth-constrained DSL transmission. Second, it should be possible to adapt

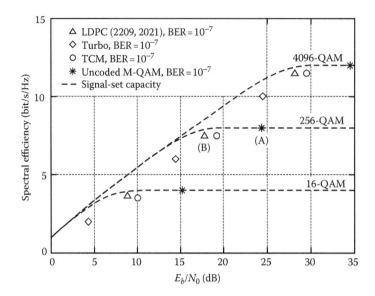

FIGURE 10.6
Comparison of several coded modulation schemes at 10^{-7} bit-error rate (BER).

the code parameters to given transmission-channel characteristics and application-specific constraints to achieve best performance. A simple solution would consist of allowing the receiver to select the most appropriate code from a small set of pre-defined codes after the channel-measurement phase during transceiver initialization. An approach that appears to be even more attractive, because of its greater flexibility, is one that would allow code construction "on the fly," provided that the processing effort needed to compute interleaver patterns for turbo codes, or parity-check matrices for LDPC codes, is small. A third aspect is linear-time encodability, meaning that encoding for a code of length N requires $O(N)$ operations. This very important property, which is natural in turbo coding, should also be achieved by LDPC codes.

In DSL transmission, overall delay, or latency, is another critical issue. "Voice" applications are known to demand rather low latency whereas other applications, such as video streaming, tolerate larger delays but need stronger error-correction capability. Thus, in studying new coding techniques for DSL, trade-offs between coding gain and latency have to be clearly established.

Finally, there is the implementation complexity associated with coding: complexity is a critical parameter, especially at central-office access multiplexors or at remote terminals, because it directly affects equipment cost and power consumption.

10.4   Transmitter and Receiver Functions

DMT-based DSL systems employ a flexible multi-carrier modulation method, whereby each subcarrier can be modulated by symbols taken from constellations of different sizes, such as BPSK, QPSK, 16, 32, 64-QAM, etc., ultimately up to a 2^{15}-symbol constellation. (See Chapter 7 and [Starr 1999] for more details.) The mapping of bit sequences to modulation symbols and the corresponding inverse mapping at the receiver thus represent an important functionality. An attractive generic realization of the modulator and demodulator in a DMT

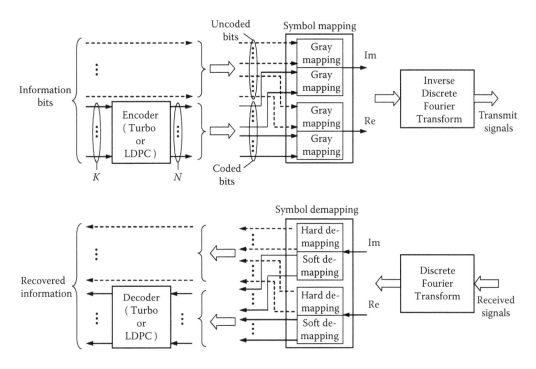

FIGURE 10.7
Encoding/symbol mapping and symbol demapping/decoding functions for DMT-based DSL transmission. Mapping and demapping functions are shown for the real (Re) and imaginary (Im) parts of a complex QAM symbol.

system that uses advanced coding schemes and iterative decoding is shown in the block diagram of Figure 10.7. The structure depicted is generic in that it can be employed to implement turbo-coded, LDPC-coded, or other capacity-approaching coded modulation schemes. Information bits representing data or control messages are encoded into a binary codeword of length N. Here, both the turbo and LDPC coding schemes are regarded as rate-K/N binary block codes. Therefore for turbo coding, the number of information bits K corresponds to the size of the interleaver. The symbol mapper collects groups of coded bits, possibly along with uncoded information bits as shown in the figure, and builds QAM symbols for frequency-domain modulation by an inverse discrete Fourier transform operation. In Figure 10.7, uncoded bits are mapped to the more significant bits of a QAM symbol, which are less prone to detection errors at the receiver and thus do not need the same level of protection as the less significant bits do. Furthermore, as explained below, a double Gray-code labelling is assumed for symbol mapping, wherein the less significant coded bits and the more significant uncoded bits are Gray-coded separately. As usual, the size of the QAM constellation used on each subchannel is determined through a "bit-loading" algorithm [Starr 1999]. (See Chapter 7.)

The symbol-mapping function is now explained in more detail with reference to the block diagram of Figure 10.8. If the constellations employed have square shape, soft demapping at the receiver is greatly simplified because the real and imaginary parts of the received noisy complex signals can be demapped independently. It can be assumed that transmit symbols are chosen from a 2^b-QAM symbol constellation, where $b = 1$ or $b = 2m$, with m a positive integer. When $b = 2m$, two binary m-tuples $v = (v_{m-1}, v_{m-2}, \ldots, v_1, v_0)$ and $w = (w_{m-1}, w_{m-2}, \ldots, w_1, w_0)$ independently select two L-ary real symbols, $L = 2^m$, representing the real and imaginary parts, respectively, of the complex QAM symbol to be

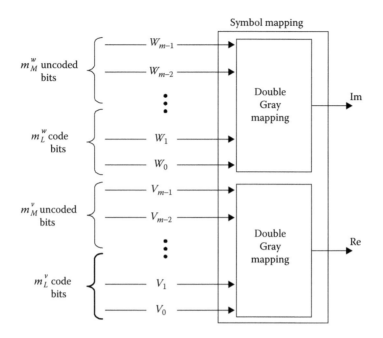

FIGURE 10.8

A more detailed representation of the multilevel encoding and symbol-mapping function shown in Figure 10.7.

transmitted. The L-ary symbols belong to the set

$$A = \{A_\ell = 2\ell - (L-1), \quad \ell = 0, 1, \ldots, L-1\}.$$

The real part of a QAM symbol conveys m_L^v code bits and m_M^v uncoded bits, with $m_L^v + m_M^v = m$. Similarly, the imaginary part conveys m_L^w code bits and m_M^w uncoded bits, with $m_L^w + m_M^w = m$. Symbol mapping relies on the partition of the set A into $2^{m_L^v}$ $[2^{m_L^w}]$ subsets such that the minimum Euclidean distance between the symbols within each subset is maximized. The m_L^v $[m_L^w]$ less-significant bits (LSBs) of v $[w]$ label the subsets of A following a Gray-coding rule. The remaining more-significant bits (MSBs) m_M^v $[m_M^w]$ label symbols within a subset following a separate Gray-coding rule. When $b = 1$, only the code bit v_0 is employed. This case corresponds to BPSK modulation.

The number of uncoded and coded bits per transmit symbol is a design parameter that involves a trade-off in terms of performance and decoding complexity. For full flexibility, this parameter can be specified by the receiver for each symbol constellation during initialization. Assuming, for example, that 1024-QAM is employed on each subchannel, with a total number of 200 subchannels, and that 3 coded bits are carried per dimension, a codeword of length $N = (3 + 3)$ bits $\times$ 200 = 1200 bits along with 4 bits $\times$ 200 = 800 uncoded bits can be mapped into a DMT frame. In this case, with a DMT frame rate of 4000 Hz, the line data rate is 10 bits $\times$ 200 $\times$ 4000 Hz = 8 Mbit/s.

At the receiver, the complex noisy symbols obtained at the output of the discrete Fourier transform are processed by a demapper (see Figure 10.7), whose function is to generate soft reliability information on the individual code bits and hard decisions on the uncoded information bits. Let y denote the real part of a noisy received signal:

$$y = A + v,$$

with $A \in A$ and v an AWGN sample with variance σ_n^2 (the imaginary part of the received signal is processed similarly). For soft demapping, the APP of code bit i conveyed by a symbol A being equal to $x = 0, 1$ is computed as

$$P(i = x|y) = \frac{\sum_{A_\ell \in \mathbf{A}_{i,x}} e^{-(y-A_\ell)^2/2\sigma_n^2}}{\sum_{A_\ell \in \mathbf{A}} e^{-(y-A_\ell)^2/2\sigma_n^2}}, \qquad i = 0, 1, \ldots, m_L^v - 1,$$

where $\mathbf{A}_{i,x}$ denotes the subset of all symbols $A \in \mathbf{A}$ with label value $x = 0, 1$ in position i.

The APPs generated in this manner are used for soft iterative LDPC or turbo decoding. Because the coded LSBs are Gray-coded, near-optimum extraction of soft information is possible. Likewise, because the uncoded MSBs can be recovered via simple threshold detection, the separate Gray coding used for them permits lowering the BER. The information bits recovered are finally output from the receiver.

Before illustrating the performance of LDPC and turbo-coded DSL systems employing the symbol-mapping scheme described above, let us show that double Gray-code labelling offers a good trade-off in terms of achievable performance and implementation complexity. To this end, the efficiency of symbol mapping based on this labelling technique is assessed by computing the capacity of the binary-valued-input and continuous-valued-output AWGN channel that also includes the bit-to-symbol mapping function. Ideal interleaving of the binary input sequence prior to the mapping of m consecutive bits into a multilevel symbol A is assumed. The capacity of this "bit-interleaved" channel is given by [Caire 1998]

$$C = m - \sum_{i=1}^{m} E_{x,y} \left\{ \log_2 \frac{\sum_{A \in \mathbf{A}} p(y|A)}{\sum_{A \in \mathbf{A}_{i,x}} p(y|A)} \right\},$$

where $E_{x,y}$ denotes expectation over x and y, the latter random variable representing the channel output signal.

Figures 10.9 to 10.12 show the capacity C in bit-per-dimension versus E_b/N_0 for the cases of 4-, 8-, 16-, and 32-level pulse-amplitude modulation (PAM), respectively. Several symbol-mapping functions are considered. The notation Gray(m_M, m_L) is used to indicate that the

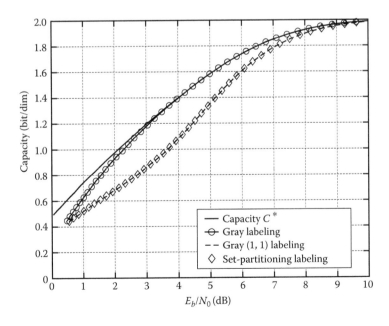

FIGURE 10.9
Capacity versus E_b/N_0 for transmission over the AWGN channel using different labelling schemes for a 4-PAM symbol mapping. The capacity C^* of the signal set is also shown.

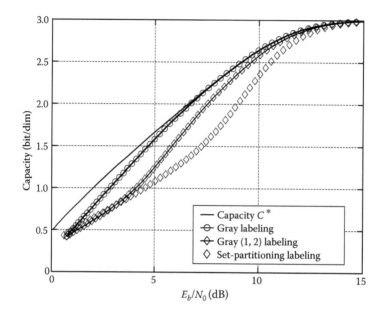

FIGURE 10.10
Capacity versus E_b/N_0 for transmission over the AWGN channel using different labelling schemes for an 8-PAM
symbol mapping. The capacity C^* of the signal set is also shown.

m_M MSBs and the m_L LSBs, with $m_M + m_L = m$, are separately Gray labelled. The figures
also show the capacity curves for (full) Gray labelling of all m bits, which is the optimum
labelling for bit-interleaved modulation, and for set-partitioning labelling. Furthermore,
the capacities C^* of the signal sets themselves [Ungerboeck 1982] are plotted as well. Note
that, in general, $C \leq C^*$.

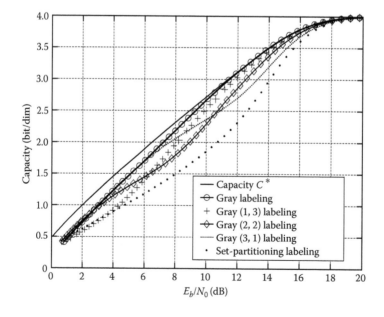

FIGURE 10.11
Capacity versus E_b/N_0 for transmission over the AWGN channel using different labelling schemes for a 16-PAM
symbol mapping. The capacity C^* of the signal set is also shown.

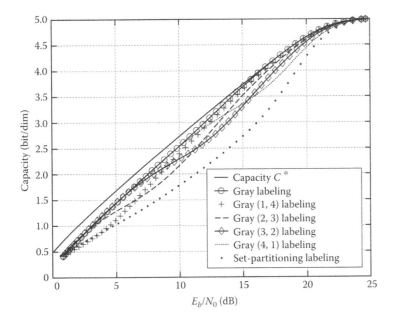

FIGURE 10.12
Capacity versus E_b/N_0 for transmission over the AWGN channel using different labelling schemes for a 32-PAM symbol mapping. The capacity C^* of the signal set is also shown.

The capacity computations show that full Gray labelling is preferable for quaternary (and binary) modulation. For 8-level modulation and higher, it is sufficient to employ $m_L = 2$ or 3 code bits per symbol and to leave the remaining m_M bits uncoded, provided that high-rate codes are employed. As double Gray labelling results in a smaller decoding complexity than full Gray labelling, Gray(m_M, 2) or Gray(m_M, 3) turn out to be a good trade-off in terms of capacity versus implementation complexity.

10.5 Performance

In this section, the typical performance that can be achieved by the turbo- and LDPC-coding schemes described above is illustrated by means of simulations. The telephone-twisted-pair channel introduces frequency-dependent signal distortion as well as several other forms of disturbances, of which crosstalk is the most important (see Chapter 3). In the following, disturbance by AWGN only will be assumed. The reason for this is that if each DMT subchannel has a sufficiently narrow bandwidth, then each one independently approximates an AWGN channel with a particular SNR value. Clearly, impulse noise and narrowband interference of various origins, *e.g.*, AM radio signals, also affect the reliability of communications in DSLs, as described in detail in Sections 3.1. and 3.2. Performance should ultimately be assessed using actual test-loop conditions. Note, finally, that many ADSL systems today use far more than 6 dB margin in practice because of the possibility of unforeseen crosstalk and other types of noise that arise during operation. The potential gain that might result in that respect from the use of capacity-approaching codes would need to be investigated.

To evaluate performance, both uncoded and coded systems are represented in terms of the symbol-error rate (SER) versus the normalized signal-to-noise ratio, SNR_{norm}, which, for a modulation and coding scheme operating at a given rate of η (in bits per two-dimensional symbol), is defined as [Eyuboglu 1992]

$$\text{SNR}_{\text{norm}} = \frac{\text{SNR}}{2^\eta - 1} = \frac{\eta}{2^\eta - 1} \frac{E_b}{N_0}.$$

Note that in the case of uncoded M-QAM transmission, $\eta = \log_2 M$ and the SER can be expressed as

$$P_S(E) \approx 4 \, Q\left(\sqrt{3 \, \text{SNR}_{\text{norm}}}\right), \text{ with } Q\left(\xi\right) = \frac{1}{\sqrt{2\pi}} \int_\xi^\infty e^{-z^2/2} \, dz,$$

which is nearly independent of the constellation size, provided the latter is sufficiently large. Note also that for a capacity-achieving scheme that transmits C bits/symbol, $\eta = C$ and $\text{SNR} = 2^C - 1$, which implies that $\text{SNR}_{\text{norm}} = 1$ (0 dB). Hence the curve of $P_S(E)$ versus SNR_{norm} also indicates the "gap to capacity" at a given SER.

10.5.1 LDPC Coding

It appears that high-rate LDPC codes with medium block length, whose parity-check matrices are constructed similarly to those of array codes [Fan 2000], exhibit as good a performance as random LDPC codes do. Array-code-based LDPC coding was shown in [Eleftheriou 2002] to offer a number of advantages for DSL transmission. The LPDC parity-check matrix is specified, in that case, by a small set of parameters and constructed deterministically without requiring "preprocessing" operations. Furthermore, array LDPC codes are amenable to linear-time encoding.

Figure 10.13 shows the SER performance of two array-code-based LDPC codes of block lengths $N = 529$ and 4489 bits, and code rates 0.870 and 0.940, respectively. For 16-QAM symbol mapping, all bits are coded. For 256- and 4096-QAM, there are 3 coded bits along each dimension, and the remaining bits are uncoded.

10.5.2 Turbo Coding

For turbo coding, a scheme with two component codes is used, similar to the architecture introduced in [Berrou 1993]. Figure 10.14 shows its SER performance with two different interleaver lengths and 16-, 256-, and 4096-QAM [Eleftheriou 2004]. For QAM symbol mapping, one systematic bit and one parity bit are used in each dimension, and the remaining bits are uncoded. The unused parity bits are "punctured." The resulting code rates for 16-, 256-, and 4096-QAM are 0.5, 0.75, and 0.833, respectively. These differences in code rates lead to different SER performance of the turbo codes depending on the constellation size. Note that it is also possible to encode all information bits; *i.e.*, no uncoded bits are used for symbol mapping. The advantage of this technique [Sadjadpour 1999] is its immunity to impulse noise, but it is computationally more complex.

10.5.3 Latency

As mentioned, latency is an important parameter in DSL systems. Generally speaking, if higher latencies can be tolerated, then longer—and hence more powerful codes—can be employed. Conversely, lower coding gains are imposed by small latencies.

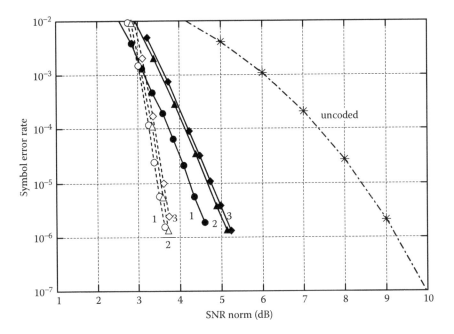

FIGURE 10.13
Performance of two LDPC codes for transmission over the AWGN channel. Solid (dashed) lines denote the case with a block length N of 529 (4489) bits. Curves 1: 16-QAM; curves 2: 256-QAM; and curves 3: 4096-QAM. For decoding, the sum-product algorithm is employed, with the number of iterations limited to 20.

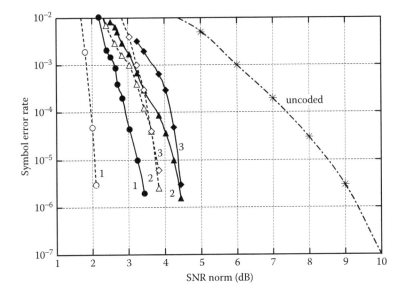

FIGURE 10.14
Performance of two turbo codes for transmission over the AWGN channel. Solid (dashed) lines denote the case with an interleaver length K of 462 (4224) bits. Curves 1: 16-QAM; curves 2: 256-QAM; and curves 3: 4096-QAM. Each component code is generated by an 8-state recursive systematic convolutional encoder with feedback and feedforward polynomials equal to (15)oct. and (17)oct., respectively. Semi-random interleaving is used. For decoding, a total of 20 iterations of the log-MAP algorithm are performed.

Recall from Chapter 7 that DMT symbols (frames) in ADSL systems are generated at the rate of 4000 Hz. Therefore, if one information block is encoded into one DMT frame, the encoding and decoding functions introduce a latency of 250 μs each, resulting in a total latency of 0.5 ms. When a codeword spans more than a single frame, latency increases accordingly.

Figure 10.15 shows the net coding gain as a function of latency for some turbo-coding and LDPC-coding schemes. To obtain the results, a simplified DMT system was assumed in which each DMT frame is transmitted via 100 or 200 subchannels. Furthermore, either

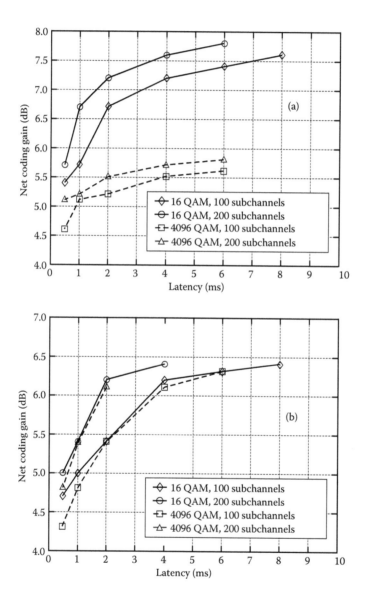

FIGURE 10.15

Net coding gains as a function of latency achieved at a SER of 10^{-7} by (a) turbo coding and (b) LDPC coding for 16- and 4096-QAM and a total number of 100 or 200 subchannels. No outer RS coding is included. Some of the points in the plots are omitted because computation times tend to become prohibitive for these cases. For turbo coding, the maximum interleaver size is 4800 bits, and the rates are 2 and 10 bit/symbol with 16- and 4096-QAM, respectively. For LDPC coding, the code rates are chosen in the range of 0.82 to 0.95, so that the rates vary in the range of 3.31 to 11.65 bit/symbol. The maximum code length is 7200 bits.

16- or 4096-QAM is used over all subchannels. Clearly, the number of subchannels, the constellation size, the number of uncoded bits, and the latency determine the codeword length. For example, 200 subchannels together with 16-QAM implies that 200×4 bits $=$ 800 bits are carried per DMT frame. If all the bits correspond to code bits, then a codeword of length $N = 800$ is carried with minimum latency (0.5 ms). If the latency is doubled, a codeword can be carried by two consecutive DMT frames, and the code length can be doubled.

It is seen from Figure 10.15 that good coding gains are achieved, even for very tight latency constraints. For a latency of 0.5 ms, the simulated turbo-coding scheme provides a coding gain of up to 5.7 and 5.1 dB for 16-QAM and 4096-QAM, respectively. Once the latency restriction is relaxed, higher coding gains can be achieved. The gain for turbo coding can be further improved by increasing the interleaver length or the number of states for each component code (from the eight states presented here to 16 or 32 states). Likewise, for a latency of 0.5 ms, LDPC coding provides a coding gain of up to 5.0 and 4.8 dB for 16-QAM and 4096-QAM, respectively. By increasing the codeword length to encompass more than one DMT frame, additional coding gains are realized.

Note that the TCM scheme defined in the ADSL standards achieves a net coding gain of 4.4 and 4.2 dB for 16-QAM and 4096-QAM, respectively, at a SER of 10^{-7}. The average rate is equal to 3.5 bits per two-dimensional symbol for 16-QAM and 11.5 bits per two-dimensional symbol for 4096-QAM. The encoding and decoding operations extend over one frame period each, hence latency amounts to 0.5 ms.

A word of caution is needed at this point. The objective of Figure 10.15 (or Figures 10.13 and 10.14) is not to compare turbo codes and LDPC codes, because they mostly operate at different spectral efficiencies, a fact that is apparent from Figure 10.6. A one-to-one performance comparison would require several parameters, such as code rate, latency, and implementation complexity, to be kept equal. It is only noted here that code rate and coding gain can often be traded off for DSL through the bit-loading process: for a specified minimum operating margin, a particular data rate that can be achieved with a specific code-rate/coding-gain combination can also be achieved, for example, by increasing the code rate and decreasing the coding gain, or vice versa.

10.6 Complexity

For encoding, it can be assumed that the computational complexity is essentially identical for turbo coding and for TCM. For LDPC coding, if the code word is obtained by multiplying the information block with the generator matrix of the code, encoding requires $O(N^2)$ operations, where N is the length of the code. However, the family of LDPC codes proposed in [Eleftheriou 2002] and used in the preceding section enjoys the desirable property of linear-time encodability, according to which encoding requires $O(N)$ operations. Indeed, it was shown for this case that the complexity of LDPC encoding typically amounts to three times the encoding complexity of TCM.

As mentioned, there are two basic algorithms for decoding turbo and LDPC codes: the BCJR algorithm and the sum-product or belief propagation algorithm, respectively. The complexity of these algorithms is linear in the code length. To minimize the number of multiplications in a practical implementation, it is advantageous for both algorithms to compute and propagate messages that represent log-likelihood ratios. In both cases, simplified algorithms exist that aim at lowering the implementation complexity at the price of some loss in performance. A comparison of the complexity of the various decoding algorithms would, however, exceed the scope of this chapter. Nevertheless, a generally accepted fact is that LDPC decoding by the sum-product algorithm is computationally less complex than turbo decoding by the BCJR algorithm.

Finally, it should be mentioned that turbo and LDPC coding techniques may have stringent memory requirements, especially for long codes. This is an important aspect in the design of DSL transceivers.

10.7   Summary

Capacity-approaching coding techniques can provide additional coding gains compared with the coding schemes used in current DSL standards. It was shown that this coding gain is a valuable resource for increasing the data rate and/or loop reach, which can be instrumental for an optimum usage of the local loop and a widespread deployment of DSL services. Two practical approaches based on turbo and LDPC coding were presented. It was not attempted to provide a one-to-one comparison of these two approaches because, in general, code parameters, encoding and decoding complexity, as well as other factors are different. The main conclusion is that both techniques appear to be practical for implementation with only a reasonable increase in transmitter/receiver complexity. It is expected that capacity-approaching coding techniques, such as those described in this chapter, will soon find their way into future generations of DSL modems and cable-transmission systems in general.

References

[Bahl 1974] L. Bahl, J. Cocke, F. Jelinek, and J. Raviv, *Optimal decoding of linear codes for minimizing symbol error rate*, IEEE Trans. Inform. Theory, Vol. IT-20, No. 2, pp. 284–287, Mar. 1974.

[Benedetto 1996] S. Benedetto and G. Montorsi, *Design of parallel concatenated convolutional codes*, IEEE Trans. Commun., Vol. 44, No. 5, pp. 591–600, May 1996.

[Berrou 1993] C. Berrou, A. Glavieux, and P. Thitimajshima, *Near Shannon limit error correcting coding and decoding: Turbo codes*, In *Proc. ICC'93*, Geneva, Switzerland, pp. 1064–1070, May 1993.

[Caire 1998] G. Caire, G. Taricco, and E. Biglieri, *Bit-interleaved coded modulation*, IEEE Trans. Inform. Theory, Vol. 44, No. 3, pp. 927–946, May 1998.

[Chen 2003] J. Chen, A. Dholakia, E. Eleftheriou, M. Fossorier, and X.Y. Hu, *Reduced-Complexity Decoding of LDPC Codes*, IBM Research Report RZ 3498, June 2003.

[Eleftheriou 2001] E. Eleftheriou, T. Mittelholzer, and A. Dholakia, *Reduced-complexity decoding algorithm for low-density parity-check codes*, IEE Electron. Lett., Vol. 37, pp. 102–104, Jan. 2001.

[Eleftheriou 2002] E. Eleftheriou and S. Öl cer, *Low-density parity-check codes for digital subscriber lines*. In *Proc. ICC2002*, New York, paper D21-3, Apr.–May 2002.

[Eleftheriou 2004] E. Eleftheriou, S. Öl cer, and H. Sadjadpour, *Application of capacity-approaching techniques to digital subscriber lines*, IEEE Communications Magazine, Vol. 42, No. 4, pp. 88–94, Apr. 2004.

[European Telecommunications Standards Institute (ETSI): TS 101 270] ETSI Technical Specification 101 270-2, V 1.2.1. *Very-high-speed digital subscriber line (VDSL). Part 2: Transceiver specification.*

[Eyuboglu 1992] M.V. Eyuboglu and G.D. Forney, *Trellis precoding: combined coding, precoding and shaping for intersymbol interference channels*, IEEE Trans. Inform. Theory, Vol. 38, No. 2, pp. 301–314, Mar. 1992.

[Fan 2000] J.L. Fan, *Array codes as low-density parity-check codes*. In *Proc. 2nd Int. Symposium on Turbo Codes and Related Topics*, Brest, France, pp. 543–546, Sept. 2000.

[Fossorier 1999] M.P.C. Fossorier, M. Mihaljevic, and H. Imai, *Reduced complexity iterative decoding of low density parity check codes based on belief propagation*, IEEE Trans. Commun., Vol. 47, pp. 673–680, May 1999.

[Gallager 1962] R.G. Gallager, *Low-density parity-check codes*, IRE Trans. Info. Theory, Vol. IT-8, pp. 21–28, Jan. 1962.

[Gallager 1963] R.G. Gallager, *Low Density Parity Check Codes*, Cambridge, MA: MIT Press, 1963.

[Hagenauer 1996] J. Hagenauer, E. Offer, and L. Papke, *Iterative decoding of binary block and convolutional codes*, IEEE Trans. Inform. Theory, Vol. 42, pp. 429–445, Mar. 1996.

[International Telecommunication Union - Telecommunication Standardization Sector (ITU-T): G.991.2] ITU-T Recommendation G.991.2 (2003) *Single-pair high-speed digital subscriber line (SHDSL) transceivers*.

[MacKay 1999] D. J. C. MacKay, *Good error-correcting codes based on very sparse matrices*, IEEE Trans. Inform. Theory, Vol. 45, pp. 399–431, Mar. 1999.

[Ouyang 2003] F. Ouyang, P. Duvaut, O. Moreno, and L. Pierrugues, *The first step of long-reach ADSL: Smart DSL technology, READSL*, IEEE Commun. Magazine, Vol. 41, No. 9, pp. 124–131, Sept. 2003.

[Pyndiah 1998] R. Pyndiah, *Near optimum decoding of product codes: Block turbo codes*, IEEE Trans. Commun., Vol. 46, No. 8, Aug. 1998, pp. 1003–1010.

[Richardson 2001] T.J. Richardson and R.L. Urbanke, *The capacity of low-density parity-check codes under message-passing decoding*, IEEE Trans. Inform. Theory, Vol. 47, pp. 599–618, Feb. 2001.

[Robertson 1995] P. Robertson, E. Villebrun, and P. Hoeher, *A comparison of optimal and sub-optimal MAP decoding algorithms operating in the log domain*. In *Proc. IEEE Int. Conf. on Communications*, ICC'95, Seattle, WA, pp. 1009–1013, June 1995.

[Sadjadpour 1999] H. Sadjadpour, *Application of parallel concatenated trellis coded modulation for discrete multi-tone modulation schemes*. In *Proc. Int. Conf. on Communications, ICC1999*, Vancouver, Canada, pp. 1022–1027.

[Song 2002] K.B. Song, S.T. Chung, G. Ginis, and J. Cioffi, *Dynamic spectrum management for next-generation DSL systems*, IEEE Commun. Magazine, Vol. 40, No. 10, pp. 101–109, Oct. 2002.

[Spec. issue 1998] Special issue on: *Concatenated coding techniques and iterative decoding: Sailing toward channel capacity*, IEEE J. Selected Areas in Commun., Vol. 16, No. 2, Feb. 1998.

[Spec. issue 2001a] Special issue on: *Codes and graphs and iterative algorithms*, IEEE Trans. on Inform. Theory, Vol. 47, No. 2, Feb. 2001.

[Spec. issue 2001b] Special issue on: *The turbo principle: From theory to practice*, IEEE J. Selected Areas in Commun., Vol. 19, No. 5, May 2001 and Vol. 19, No. 9, Sept. 2001.

[Spec. issue 2003] Special issue on: *Capacity approaching codes, iterative decoding algorithms, and their applications*, IEEE Commun. Mag., Aug. 2003.

[Starr 1999] T. Starr, J.M. Cioffi, and P.J. Silverman, *Digital Subscriber Line Technology*, Upper Saddle River, NJ: Prentice Hall, 1999.

[Tanner 1981] R.M. Tanner, *A recursive approach to low complexity codes*, IEEE Trans. Inform. Theory, Vol. 27, pp. 533–548, Sep. 1981.

[Ungerboeck 1982] G. Ungerboeck, *Channel coding with multilevel/phase signals*, IEEE Trans. Inform. Theory, Vol. IT-28, No. 1, pp. 56–67, Jan. 1982.

[Wiberg 1996] N. Wiberg, *Codes and Decoding on General Graphs*, Ph.D. Thesis, Linköping University, Linköping, Sweden, 1996.

11

DSL Channel Equalization

Ragnar Hlynur Jonsson

CONTENTS

11.1 Introduction 300
11.2 Background Theory 300
    11.2.1 Signal Representation 300
    11.2.2 Channel Model 301
    11.2.3 Inter-Symbol Interference (ISI) 303
    11.2.4 Causality and Signal Delay 303
11.3 Equalization Optimization Criteria 304
    11.3.1 Matched Filters 304
    11.3.2 Zero-Forcing Equalization (ZFE) 306
    11.3.3 MMSE Equalization 308
    11.3.4 MAP and ML Detection 310
11.4 Equalizer Structures 311
    11.4.1 Linear Equalization 312
    11.4.2 Decision Feedback Equalization 312
    11.4.3 Noise Prediction Filters 316
    11.4.4 Tomlinson–Harashima Precoding 318
    11.4.5 Maximum Likelihood Detection 319
    11.4.6 Frequency Domain Equalization 321
    11.4.7 Impulse Shortening Equalization 322
11.5 Closed-Form Equalizer Design 327
    11.5.1 MMSE Equalizer Design Algorithms 327
    11.5.2 Channel Probing 329
11.6 Adaptive Equalization 331
    11.6.1 The LMS Algorithm 331
    11.6.2 Equalizer Training 335
    11.6.3 Blind Equalizer Training 336
    11.6.4 Example Equalizer Training Procedure 338
11.7 Examples and Practical Design Issues 338
    11.7.1 Linear Versus Decision Feedback Equalizers 338
    11.7.2 Zero-Forcing Versus MMSE Equalization 339
    11.7.3 Fractionally Spaced Versus T-Spaced Equalizers 342
    11.7.4 Impulse Shortening Equalization 345
    11.7.5 Summary 348
References 349

ABSTRACT This chapter examines equalization in the context of DSL systems. Different optimization criteria are examined, in particular matched filters (MF), zero-forcing equalizers (ZFE), minimum mean-square error (MMSE) equalizers, and maximum likelihood (ML) equalization. Several different equalizer structures are discussed, in particular linear equalizers (LE), decision feedback equalization (DFE), noise prediction, precoding, maximum likelihood sequence estimation (MLSE), frequency domain equalization (FEQ), and impulse shortening equalization (ISE), which is also known as time domain (TEQ) in DMT systems. A generic algorithm is presented which can be used to design equalizers based on the MF, the ZFE, or the MMSE criteria given the symbol response of the channel and the noise characteristics. This generic algorithm is applicable to LE, DFE, precoding, and ISE equalizer structures. Adaptive equalization is discussed and the LMS algorithm is addressed. Different adaptation configurations are addressed, including adaptation based on decision errors, adaptation based on using training sequences, and adaptation based on blind training. Finally, several examples are presented to demonstrate different equalizer properties and to illustrate some practical equalizer design issues.

11.1 Introduction

Any DSL communication system can be modelled by a transmitter that sends a signal over a communication channel to a receiver on the other end. The channel may distort the transmitted signal and may introduce random noise into the signal. There is a fundamental difference between the channel distortion and the noise in that the distortion can be viewed as a deterministic operation on the signal, whereas noise introduces random changes to the signal that are not predictable beforehand. To reconstruct the transmitted signal, the receiver must compensate for the channel distortion and minimize the impact of the channel noise. Such compensation is called equalization and is at the heart of most DSL transceivers. This chapter addresses various forms of equalization and how they are used in DSL systems.

In DSL systems, the channel distortion is dominated by linear time-invariant (LTI) distortion, which can be modelled by convolution of the transmit signal with the impulse response of the channel. The characteristics of the distortions introduced by twisted pair loops are discussed in more detail in Chapter 2. The noise in DSL systems is usually dominated by the crosstalk noise from DSL systems on adjacent wire pairs. This noise can be modelled by additive colored Gaussian noise. The characteristics of the crosstalk noise are discussed in more detail in Chapter 3. Both the channel response and the noise can change with time, but for DSL systems these changes are usually very slow compared with some other systems such as mobile communication systems.

11.2 Background Theory

11.2.1 Signal Representation

In this chapter, the processing of analog signals, discrete signals, and digital signals will be at the heart of much of the discussion. In the context of the following discussion, analog signals can be viewed as signals that are continuous in time and in amplitude. Discrete signals can be viewed as signals that are discrete in time but may be continuous in amplitude. Digital signals are discrete in both time and amplitude, and are a special case of discrete signals.

A discrete signal, $x[n]$, can be converted to an analog signal, $x(t)$, according to

$$x(t) = \sum_n x[n]p(t - nT),$$

(11.1)

where $p(t)$ is the impulse response of the discrete-to-analog conversion and T is the sampling interval. If $p(t)$ and T are appropriately chosen, then $x[n]$ can be perfectly reconstructed from $x(t)$. (See [Oppenheim 1989] for details.)

An analog signal, $y(t)$, can be converted to a discrete signal, $y[n]$, according to

$$y[n] = \int y(t)q(nT - t)dt,$$

(11.2)

where $q(t)$ is the impulse response of the analog to discrete conversion and T is the sampling interval. If $q(t)$ and T are appropriately chosen and $y(t)$ is band limited, then $y(t)$ can be perfectly reconstructed from $y[n]$ (see [Oppenheim 1989] for details).

The signals that are transmitted over the twisted pair wires are analog signals. Most digital communication devices, on the other hand, generate and process discrete signals. The discrete signals in the transmitter are converted to analog signals using digital-to-analog conversion, transmitted over the wire as analog signals and then converted back into discrete signals by analog-to-digital conversion in the receiver. Unless otherwise stated, in this chapter it will be assumed that the digital-to analog and analog-to-digital conversions are done appropriately such that the discrete and analog signals can be considered to be equivalent.

11.2.2 Channel Model

The DSL channel can be modelled by an LTI filter and additive noise. Figure 11.1 shows a model of the transmission path that includes the transmitter, the transmit filter $P(f)$, the channel distortion filter $H(f)$, the additive noise $v'(t)$, the receive filter $Q(f)$, and the receiver, labelled Rx.

This model will be used in the remainder of this chapter as a reference when discussing various equalizer configurations. In the model, the transmitter, denoted Tx, generates a discrete signal $s[n]$, which is converted into an analog signal $x(t)$ according to

$$x(t) = \sum_n s[n]p(t - nT).$$

(11.3)

The analog transmit signal $x(t)$ then passes though the channel and undergoes amplitude and phase distortion due to the LTI filtering

$$y(t) = x(t) \star h(t) = \int h(\tau)x(t - \tau)d\tau.$$

(11.4)

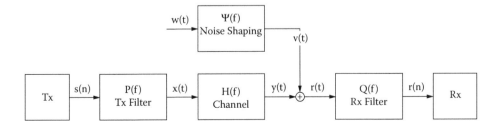

FIGURE 11.1
Channel model including transmitter and receiver.

The additive noise $v'(t)$ is added to $y(t)$ to generate the received analog signal

$$r(t) = y(t) + v'(t)(t). \tag{11.5}$$

The received analog signal undergoes analog filtering and is sampled to generate the receiver signal

$$r[n] = \int q(t)r(nT - t)dt. \tag{11.6}$$

The additive colored noise can be modelled by white Gaussian noise, $w(t)$, which has been filtered with a noise shaping filter $\Psi(f)$

$$v'(t) = \int w(\tau)\psi(t - \tau)d\tau. \tag{11.7}$$

The transfer function from the transmit symbols, $s[n]$, to the received discrete signal, $r[n]$, is given by

$$r[n] = \sum_k g(kT)s[n - k] + v[n], \tag{11.8}$$

where

$$g(t) = p(t) \star h(t) \star q(t), \tag{11.9}$$

is the aggregate channel transfer function referred to as the symbol transfer function. The discrete noise term $v[n]$ in Equation 11.8 is given by

$$v[n] = \sum_k \gamma[k]w((n - k)T), \tag{11.10}$$

where $\gamma[n]$ is the combined impulse response of the noise coloring filter and the receive filter

$$\gamma[n] = \gamma(nT) = \int q(\tau)\psi(nT - \tau)d\tau. \tag{11.11}$$

Figure 11.2 shows a discrete channel model, where the transmit and receive filters have been combined with the channel response according to Equation 11.9. The system models in Figures 11.1 and 11.2 will be used as references in the following discussion about the various equalizer configurations.

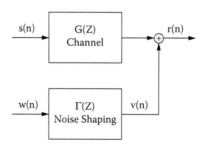

FIGURE 11.2
Discrete channel model.

11.2.3 Inter-Symbol Interference (ISI)

If the noise component in Equation 11.8 is ignored, then the relationship between the transmitted signal symbols $s[n]$ and the received signal $r[n]$ becomes

$$r[n] = \sum_k g(kT)s[n-k], \tag{11.12}$$

where $g(t)$ is the symbol transfer function given by Equation 11.9. The sum in Equation 11.12 can be broken into two parts with the desired signal term and the undesired inter-symbol interference (ISI) term as

$$r[n] = g(0)s[n] + \underbrace{\sum_{k \neq 0} g(kT)s[n-k]}_{ISI}. \tag{11.13}$$

It is clear from Equation 11.13 that the received signal $r[n]$ will be a perfect reconstruction of the transmit signal $s[n]$ if

$$g(nT) = \delta[n] = \begin{cases} 1 & \text{for } n = 0 \\ 0 & \text{else} \end{cases}. \tag{11.14}$$

By taking the Fourier transform of $g(nT)$ and observing that discrete time signals are periodic in frequency, it is straightforward to show that the requirement in Equation 11.14 is equivalent to (see [Lathi 1983])

$$\frac{1}{T} \sum_{n=-\infty}^{\infty} G\left(f - \frac{n}{T}\right) = 1, \tag{11.15}$$

where $G(f)$ is the Fourier transform of $g(t)$ and $\frac{1}{T}$ is the symbol rate. Transfer functions satisfying Equation 11.14 result in perfect reconstruction of the transmit signal, and the received signal $r[n]$ is free of ISI. The criteria in Equations 11.14 and 11.15 are equivalent and therefore signals that satisfy Equation 11.15 are also free of ISI. The requirement to satisfy Equation 11.15 is referred to as the "Nyquist criterion" or more precisely as the "Nyquist first criterion for zero ISI" [Lathi 1983].

11.2.4 Causality and Signal Delay

Real channels and real-life equipment introduce delays in the signal transmission path, such that a signal can only be observed at the receiver some time after it was transmitted. Therefore, it is, strictly speaking, not possible to satisfy $r[n] = s[n]$ without assuming that the receive filter $q(t)$ is noncausal. In reality, $q(t)$ has to be causal, and therefore it is impossible to reconstruct the transmit signal such that $r[n] = s[n]$. However, by introducing time delay n_0, it is possible to reconstruct the transmit signal such that $r[n] = s[n - n_0]$. Equivalently, it is possible to shift the time reference for the received signal such that $r[n'] = s[n]$, where $n' = n + n_0$. By introducing this delay, it is now possible to do "noncausal" filtering in the receiver (the filtering is noncausal relative to time index n', but is still causal relative to the "real" time index n).

The delayed reference point in the receiver is often referred to as the "cursor." Signal components that are received before the cursor are referred to as pre-cursor or noncausal parts of the signal. Likewise, the signal components received after the cursor are referred to as post-cursor or causal parts of the signal. This nomenclature can sometimes be a little confusing, but it can also sometimes provide better insight into what is going on in the equalization process.

11.3   Equalization Optimization Criteria

Good equalization should provide the optimal reconstruction of the original transmitted signal in the receiver. The meaning of optimal reconstruction, however, depends on the criterion for which the equalizer is optimized. The most commonly used criteria in equalizer design are the matched filter, zero-forcing, minimum mean-square error, and maximum likelihood criteria. Whenever discussing optimality, it is important to consider the criteria for the optimality and why these criteria are important given the problem at hand.

Matched filters (MF) are optimized to maximize the SNR in the context of a single transmitted symbol on a channel with additive white noise. The matched filter criterion does not address ISI, and ISI is not taken into consideration in the SNR calculations. Because the matched filters do not address ISI, they are often not considered to be equalizers in the strictest sense of the word. But when used, the matched filters are an integral part of the equalization process and, as such, belong in a discussion about equalization.

Zero-forcing equalizers (ZFE), on the other hand, are optimized to minimize ISI but do not consider channel noise. Because the ZFE criterion ignores the noise on the channel, the ZFE filters can end up amplifying the noise.

The minimum mean-square error (MMSE) criterion aims to minimize the power of the slicer error (the decision error), considering both additive noise on the channel and the ISI. In a way, the MMSE criterion combines the noise consideration of the matched filter criterion and the ISI consideration of the ZFE criterion. For most applications in modern communication systems, the MMSE criterion gives better performance than the matched filter or the ZFE criteria when used alone.

The maximum likelihood (ML) criterion aims to maximize the probability of correct slicer decisions. However, the ML criterion is not necessarily equivalent to minimizing the error probability, unless the transmitted symbols are independent and identically distributed (iid). For additive Gaussian noise, the ML criterion can be replaced by the MMSE criterion. This is because the mean-square error is proportional to the logarithm of the likelihood function, and maximizing the likelihood function becomes equivalent to minimizing the mean-square error.

In the following section, each of the above criteria is discussed in more depth. For the sake of simplicity, this discussion will be in the context of linear equalizers, but most of the criteria can easily be extended to other equalizer structures. Linear equalizers and alternative equalization structures are discussed in more detail in Section 11.4.

11.3.1   Matched Filters

Matched filters are optimal equalizers in the sense of maximizing the SNR of the received signals when the impulse response is known, and the noise is additive white noise. However, matched filters do not address ISI and therefore do not seek to minimize it.

The matched filter can be derived by observing the transmission of a single symbol S_0 over a channel with symbol response $g(t)$. Then the received analog signal $r(t)$ is given by

$$r(t) = S_0 g(t) + w(t), \tag{11.16}$$

where $w(t)$ is white additive noise. The matched filter maximizes the SNR by enhancing the received signal more than it enhances the additive noise. The output of the matched filter is given by

$$r_2(t) = S_0 \int_{-\infty}^{\infty} g(\tau) f(t - \tau) d\tau + \int_{-\infty}^{\infty} w(\tau) f(t - \tau) d\tau. \tag{11.17}$$

The first term of Equation 11.17 is due to the signal and the second term is due to the noise. At decision time, $t = 0$, the signal power is given by

$$\sigma_S^2 = S_0^2 \left(\int_{-\infty}^{\infty} g(\tau) f(-\tau) d\tau \right)^2. \tag{11.18}$$

The noise power is given by

$$\sigma_N^2 = E \left\{ \left(\int_{-\infty}^{\infty} w(\tau) f(-\tau) d\tau \right)^2 \right\} = \sigma_w^2 \int_{-\infty}^{\infty} f^2(-\tau) d\tau. \tag{11.19}$$

The SNR for the output of the matched filter is then

$$SNR = \frac{\sigma_S^2}{\sigma_N^2} = \frac{S_0^2 \left(\int_{-\infty}^{\infty} g(\tau) f(-\tau) d\tau \right)^2}{\sigma_w^2 \int_{-\infty}^{\infty} f^2(-\tau) d\tau}. \tag{11.20}$$

The well known Schwartz inequality [Papoulis 2002] states that

$$\left| \int_{-\infty}^{\infty} a(x) b(x) dx \right|^2 \leq \int_{-\infty}^{\infty} |a(x)|^2 dx \int_{-\infty}^{\infty} |b(x)|^2 dx, \tag{11.21}$$

with equality holding if and only if $a(x) = kb^*(x)$, where k is an arbitrary constant. Applying the Schwartz inequality to Equation 11.20 gives

$$SNR \leq \frac{S_0^2 \int_{-\infty}^{\infty} |g(\tau)|^2 d\tau \int_{-\infty}^{\infty} |f(-\tau)|^2 d\tau}{\sigma_w^2 \int_{-\infty}^{\infty} |f(-\tau)|^2 d\tau} = \frac{S_0^2}{\sigma_w^2} \int_{-\infty}^{\infty} |g(\tau)|^2 d\tau, \tag{11.22}$$

where the equality holds if and only if $f(t) = kg^*(-t)$. In other words, the matched filter, which maximizes the SNR, is given by

$$f(t) = kg^*(-t), \tag{11.23}$$

where k is an arbitrary gain factor. The frequency domain representation of matched filters is obviously

$$F(f) = kG^*(f). \tag{11.24}$$

It is worth noting that the matched filter is derived under the assumption of additive white noise (not necessarily Gaussian). For DSL systems, the noise is dominated by a highly colored cross-talk noise, so the matched filters are, in most cases, not directly applicable to DSL systems. However, the issue of colored noise can be resolved by first filtering the input signal with an input filter that whitens the noise, and then filtering the result with the matched filter, which is then matched to the combined channel and noise-whitening filter. Matched filters that are preceded by such noise-whitening filters are referred to as whitened matched filters (WMF) [Lee 1994].

Also note that the matched filter is optimized under the assumption that there is only one symbol transmitted over the channel. This implies that ISI is not taken into account in the derivation of the matched filter, but it is possible to choose transmit filters and matched receive filters such that ISI is minimized or eliminated. However, such considerations are in no way inherent to the concept of matched filtering.

One example of filters that consider both the matched filter criterion and ISI are the square root raised-cosine filters [Lee 1994]. The square root raised-cosine filters are interesting from a theoretical point of view and can be useful on channels that have virtually flat transfer functions, such as satellite channels. For DSL channels, on the other hand, square root raised-cosine filters are of little or no practical interest because the channel response for DSL channels is usually far from flat.

Another way to address ISI in matched filters is to do matched filtering in the analog domain and deal with ISI using digital processing such as digital equalizers or maximum-likelihood sequence estimation (MLSE) [Forney 1972].

Matched filters are most likely to be of value when used as analog filters before sampling the analog signal, *i.e.*, before analog-to-digital conversion (ADC). The problem is that the impulse response of the DSL channel depends heavily on the length and topology of the DSL loop. It is therefore not feasible to construct a single matched filter that fits some majority of DSL loops. Furthermore, the noise characteristics vary widely from one loop to another, and the cross-talk noise is likely to be very different from loop to loop and, for each loop, may even change with time, making noise whitening difficult. Therefore, matched filters are usually not directly applicable for DSL systems, but the basic considerations behind the derivation of matched filters can be valuable input in the design and implementation of analog front ends (AFE) for DSL systems.

Digital domain implementations of matched filters are usually of little or no value, except when the digital signal representation is sampled at a rate that is much higher than the symbol rate. In this case, matched digital filters may be useful, especially before down-sampling or at the input to a slicer. However, even in this case an adaptive equalizer based on the minimum mean-square error criterion is likely to give better performance.

In the context of DSL systems, matched filters are mainly of theoretical interest, but have limited direct practical value. Theoretically, matched filters are a very interesting concept, but their optimality and practical value tends to be overrated in the communication theory literature, especially in the context of modern communication systems.

11.3.2 Zero-Forcing Equalization (ZFE)

Zero-forcing equalizers minimize the ISI but ignore any impact that noise may have on the system. In other words, a ZFE corrects for distortion due to the ISI term in Equation 11.12 but ignores the effects of the additive noise component $v'(t)$. The ideal ZFE is the inverse of the symbol transfer function (presumably delayed in time by n_0 samples)

$$F(f) = \frac{e^{-j2\pi fTn_0}}{G(f)}. \tag{11.25}$$

Finding the ideal ZFE according to Equation 11.25 looks simple. However, due to problems such as filter stability and numerical accuracy, it is usually not practical to directly construct a ZFE according to Equation 11.25. Also, in real systems the symbol transfer function $G(f)$ is usually not precisely known, but rather is usually just an FIR filter approximation of the symbol transfer function. Unless this approximation is minimum-phase (with all zeros inside the unit circle), a solution according to Equation 11.25 will be unstable and is useless as an equalizer.

The problem of finding an optimal ZFE can be formulated as finding $f[n]$ such that

$$f[n] \star g[n] = \sum_m f[m]g[n-m] = \delta[n-n_0], \tag{11.26}$$

where n_0 is the delay introduced by the channel and the equalizer. If $g[n]$ is IIR, then there exists an FIR $f[n]$ that satisfies Equation 11.26, and then Equation 11.26 is satisfied for $n_0 = 0$ if

$$f[0] = \frac{1}{g[0]}$$

$$f[n] = \frac{-1}{g[0]} \sum_{m=0}^{n-1} f[m]g[n-m]. \tag{11.27}$$

For example, if the symbol transfer function is given by

$$g[n] = a^n, \quad \text{for } n \geq 0, \tag{11.28}$$

then ideal ZFE can be derived according to Equation 11.27 as follows.

$$f[0] = \frac{1}{g[0]} = 1$$

$$f[1] = \frac{-1}{g[0]} f[0]g[1] = -a$$

$$f[2] = \frac{-1}{g[0]} (f[0]g[2] + f[1]g[1]) = -(a^{-2} - a^{-2}) \tag{11.29}$$

$$f[n] = \frac{-1}{g[0]} \sum_{m=0}^{n-1} f[m]g[n-m] = 0, \quad \forall n \geq 2.$$

In this example, the ZFE only needed two taps to equalize an IIR symbol response.

For FIR symbol transfer functions, on the other hand, the solution of Equation 11.27 will result in an infinite-length impulse response that cannot be implemented with a finite-length FIR equalizer filter. Also, if $g[0]$ is relatively small compared to some other values of $g[n]$, then some of the values of $f[n]$ can become very large. This can cause significant amplification of the noise term $v[n]$ and can lead to various numerical problems. Because of these limitations, the approach in Equation 11.27 is usually not very useful in real systems.

In real systems, the length of the equalizer is limited and in many cases may not be sufficiently long to totally cancel all ISI. In this case, no $f[n]$ will perfectly satisfy Equation 11.26. But if it is not possible to eliminate ISI, at least the ISI can be minimized. Which metric to use when minimizing the ISI may depend on the application, but for many applications it makes sense to minimize the mean-square error.

For a received signal $r[n]$ given by Equation 11.12, the equalized signal is given by

$$\hat{s}[n] = \sum_m f[m]r[n-m] = \sum_m \sum_k f[m]s[k]g[n-m-k], \tag{11.30}$$

where $f[n]$ are the equalizer coefficients. As before, $g[n]$ represents the channel symbol response and $s[n]$ are the transmitted symbols. For this case the mean-square reconstruction error due to ISI is given by

$$E\left\{e_{ISI}^2[n]\right\} = E\left\{(s[n-n_0] - \hat{s}[n])^2\right\} = E\left\{\left(\sum_k \varepsilon[k]s[n-k]\right)^2\right\}, \tag{11.31}$$

where

$$\varepsilon[n] = \delta[n-n_0] - \sum_m f[m]g[n-m]. \tag{11.32}$$

Expanding the square term in Equation 11.31 gives

$$
\begin{aligned}
E\left\{e_{ISI}^2[n]\right\} &= E\left\{\sum_k \sum_i \varepsilon[n-k]s[k]\varepsilon[n-i]s[i]\right\} \\
&= \sum_k \sum_i \varepsilon[n-k]\varepsilon[n-i]E\left\{s[k]s[i]\right\}.
\end{aligned}
\tag{11.33}
$$

If the samples of $s[n]$ are not correlated, then

$$
E\left\{s[k]s[i]\right\} = \sigma_s^2 \delta[k-i],
\tag{11.34}
$$

and Equation 11.33 simplifies to

$$
E\left\{e_{ISI}^2[n]\right\} = \sigma_s^2 \sum_k \varepsilon^2[k].
\tag{11.35}
$$

This quadratic equation can be minimized with respect to each equalizer coefficient, $f[m]$, by finding the coefficients such that

$$
\frac{\partial E\left\{e_{ISI}^2[n]\right\}}{\partial f[n]} = 0.
\tag{11.36}
$$

It is easy to show that this implies

$$
\begin{aligned}
\frac{\partial E\left\{e_{ISI}^2[n]\right\}}{\partial f[n]} &= -2\sigma_s^2 \sum_k \varepsilon[k]g[k-n] \\
&= 2\sigma_s^2 \left(-g[n_0-n] + \sum_{m=0}^{M-1} f[m]\sum_k g[k-m]g[k-n]\right) = 0.
\end{aligned}
\tag{11.37}
$$

This in turn translates into M linear equations

$$
\sum_{m=0}^{M-1} f[m]R_g[n-m] = \sigma_s^2 g[n_0-n],
\tag{11.38}
$$

for $n = 0, 1, \ldots, M-1$, where

$$
R_g[n] = \sigma_s^2 \sum_k g[k]g[n+k].
\tag{11.39}
$$

The equations in Equation 11.38 can be solved for the M unknown equalizer taps, $f[m]$, using linear algebra. An algorithm for finding equalizer taps based on the above approach is discussed in more detail in Section 11.5.1.

11.3.3 MMSE Equalization

The zero-forcing equalization discussed above compensates for the ISI but ignores any effect that the noise may have on the signal. As a result, the ZFE may enhance additive noise as it suppresses the ISI. In minimum mean-square error equalization, both the ISI and the additive noise are considered.

For a received signal $r[n]$ given by Equation 11.8, the equalized signal is given by

$$
\hat{s}[n] = \sum_m f[m]r[n-m] = \sum_m \sum_k f[m]\left(g[k]s[n-m-k] + \gamma[k]w[n-m-k]\right),
\tag{11.40}
$$

where $w[n]$ is a white noise and $\gamma[n]$ is the noise coloring filter given by Equation 11.11. As before, $f[n]$ are the equalizer coefficients, $g[n]$ represents the channel symbol response, and $s[n]$ are the transmitted symbols.

The error metric for MMSE equalization is given by

$$E\left\{e^2[n]\right\} = E\left\{(s[n-n_0] - \hat{s}[n])^2\right\}. \tag{11.41}$$

This can be expressed as

$$E\left\{e^2[n]\right\} = E\left\{\left(\sum_k \varepsilon[k]s[n-k] + \sum_k \xi[k]w[n-k]\right)^2\right\}, \tag{11.42}$$

where

$$\xi[n] = \sum_m f[m]\gamma[n-m] \tag{11.43}$$

and $\varepsilon[n]$ is again given by Equation 11.32. This can also be expressed as

$$\begin{aligned} E\left\{e^2[n]\right\} = {} & \sum_k \sum_l \varepsilon[k]\varepsilon[l]E\left\{s[n-k]s[n-l]\right\} \\ & + 2\sum_k \sum_l \varepsilon[k]\xi[l]E\left\{s[n-k]w[n-l]\right\} \\ & + \sum_k \sum_l \xi[k]\xi[l]E\left\{w[n-k]w[n-l]\right\}. \end{aligned} \tag{11.44}$$

Assuming that $s[n]$ and $w[n]$ are not correlated with each other and both signals are uncorrelated (white) signals, this simplifies to

$$E\left\{e^2[n]\right\} = \sigma_s^2 \sum_k \varepsilon^2[k] + \sigma_w^2 \sum_k \xi^2[k]. \tag{11.45}$$

This is again a quadratic equation that can be minimized with respect to each equalizer coefficient, $f[m]$, by finding the coefficients satisfying Equation 11.36. It can be shown that this implies

$$-\sigma_s^2 \sum_k \varepsilon[k]g[k-n] + \sigma_w^2 \sum_k \xi[k]\gamma[k-n] = 0 \tag{11.46}$$

or

$$-\sigma_s^2 g[n_0-n] + \sum_{m=0}^{M-1} f[m]R_g[n-m] + \sum_{m=0}^{M-1} f[m]R_\gamma[n-m] = 0, \tag{11.47}$$

where

$$R_\gamma[n] = \sigma_w^2 \sum_k \gamma[k]\gamma[n+k]. \tag{11.48}$$

This again translates into M linear equations

$$\sum_{m=0}^{M-1} f[m]\left(R_g[n-m] + R_\gamma[n-m]\right) = \sigma_s^2 g[n_0-n], \tag{11.49}$$

for $n = 0, 1, \ldots, M-1$. The equations in Equation 11.49 can be solved for the M unknown equalizer taps, $f[m]$, using linear algebra.

If the channel response $g[n]$ and the noise coloring filter $\gamma[n]$ are known, then the MMSE equalizer can be constructed using the linear equalizations in Equation 11.49. However, in most cases the channel characteristics are not known a priori. In this case either the channel characteristics have to be obtained using channel probing (see Section 11.5.2), or the equalizer needs to be constructed directly from the received signal using adaptive algorithms such as the LMS algorithm (LMS) (see Section 11.6.1).

Both MMSE equalizers and ZFE are optimum solutions for their respective criteria. However, in real DSL systems the MMSE equalizers normally give much better performance than ZFE equalizers. This is because in real systems the MMSE criterion is closer to minimizing the probability of transmission errors in the communication channel. Actually, if both ISI noise and the additive noise are Gaussian noise sources, then the MMSE criterion is the criterion that minimizes the probability of transmission errors. In line with the central limit theorem, both the ISI and the additive noise tend to be almost Gaussian in actual DSL systems, so in most cases the MMSE criterion is close to minimizing the probability of decision error. However, as the combined ISI and additive noise is not exactly Gaussian, the MMSE criterion introduces a very small bias in the equalized symbols [Lee 1994]. It is possible to construct an unbiased mean-square error (MSE) estimate and by minimizing this unbiased MSE, it should be possible to get a minor improvement in error probability relative to the MMSE solution [Cioffi 1995][Lee 1994].

Because minimizing MSE is almost equivalent to minimizing the probability of errors and because of how easy it is to work with and analyze the MMSE, equalization in most real DSL systems is based on the MMSE criterion in one form or another.

11.3.4 MAP and ML Detection

Communication systems are usually required to operate virtually error free, and the allowable error probability is usually limited. Therefore, the error probability under given operating conditions is usually the relevant performance metric in most communication systems. However, the criteria discussed above for matched filters, ZFE, and MMSE equalization do not directly address the issue of minimizing the error probability. Under some conditions, the MMSE criterion may approximate the criterion for minimizing the error probability. Likewise, a ZFE tends to reduce the error probability. However, truly minimizing the error probability requires consideration of the joint or conditional probabilities of the transmitted symbols and the received signal.

The probability of correct detection can be maximized by estimating the transmit symbol as the symbol with the highest conditional probability given the received signal, r_0. The transmit symbol estimate, s_0, then satisfies

$$P_{s|r}(s_0|r_0) \geq P_{s|r}(s|r_0) \quad \forall s. \tag{11.50}$$

This is known as the maximum a posteriori (MAP) detection or as Bayes detection. Applying Bayes rule gives

$$P_{s|r}(s|r) = \frac{P_s(s)P_{r|s}(r|s)}{P_r(r)}. \tag{11.51}$$

Because for any observed receive signal r_0, the probability $P_r(r_0)$ is constant, it is trivial to show that MAP detection according to Equation 11.50 can equivalently be expressed in terms of the joint probability as

$$P_{r,s}(r_0, s_0) \geq P_{r,s}(r_0, s) \quad \forall s. \tag{11.52}$$

The MAP decision rule is the optimum decision rule if it is only desirable to minimize the error probability and all errors have the same weight. However if decision errors have different weight, then the MAP detection rule can be generalized to consider the "cost" of each false detection. The generalized Bayes decision rule then becomes

$$\sum_j C(s_j, s_0) P_{s|r}(s_j|r_0) \leq \sum_j C(s_j, s) P_{s|r}(s_j|r_0) \quad \forall s, \tag{11.53}$$

where $C(s_j, s)$ is the cost of estimating s_j when s was actually transmitted.

If all the possible transmit symbols have the same probability (*i.e.*, $P_s(s)$ is constant), then the MAP decision rule simplifies to

$$P_{r|s}(r_0|s_0) \geq P_{r|s}(r_0|s) \quad \forall s. \tag{11.54}$$

This is known as the maximum likelihood detection rule and is used in maximum likelihood sequence estimation. Maximum likelihood estimation is widely used for parameter and sequence estimation. This is because the ML criterion can often lead to elegant simplifications of problems, especially in the case of additive Gaussian noise. It is important, however, to keep in mind that the ML estimation is only optimal under the assumption of all transmit symbols being equally likely (*i.e.*, $P_s(s)$ is constant).

The MAP and ML decision rules are basically sequence estimation methods rather than conventional equalization methods. However, because sequence estimation based on these methods can replace conventional equalization, it is appropriate to consider them in this discussion about equalization.

11.4 Equalizer Structures

Equalization can be implemented using several different equalization structures. Each of these structures has its own properties and characteristics. This section will address some of the most commonly used equalization structures applicable to DSL systems.

The simplest equalizers use only linear time-invariant filters for the equalization. Such equalizers are often referred to as linear equalizers. Linear equalizers can be implemented as analog filters, but today it is far more common to implement linear equalizers using digital filters.

By introducing nonlinear elements in the equalizer structures, it is possible to enhance the quality and the flexibility of the equalization. The most common nonlinear equalization structures in DSL systems are the decision feedback equalization (DFE) and the related Tomlinson–Harashima precoding. Noise predictors (NP) are also closely related to DFE systems, and NP structures can have equivalent DFE structure representation. Another commonly used nonlinear equalization structure is MLSE.

Discrete multi-tone (DMT) systems use frequency-domain equalizers (FEQ). The frequency domain equalization is usually linear by nature but is, strictly speaking, not time invariant. Also associated with the DMT modulation are impulse shortening equalizers, which in DMT systems are referred to as time-domain equalizers (TEQ). The purpose of the TEQ is to shorten the channel impulse response such that, after equalization, the channel response is no longer than the cyclic prefix used by the DMT system.

The following sections discuss the various equalization structures mentioned above in more detail. The properties of each equalization structure are discussed, and the equalization structures are compared to one another.

11.4.1   Linear Equalization

Linear equalizers are based on simple LTI filters and do not possess any of the nonlinear elements that some other equalizer structures may incorporate. This makes it relatively simple to analyze linear equalizers using conventional signal processing theory.

Linear equalizers can be implemented in either the analog or digital domain. The analog version of a linear equalizer can be represented as

$$\hat{s}(t) = f(t) \star r(t) = \int f(\tau) r(t - \tau) d\tau, \tag{11.55}$$

where $f(t)$ is the impulse response of the linear equalizer and $r(t)$ is the received signal. The corresponding discrete linear equalizer is

$$\hat{s}[n] = f[n] \star r[v] = \sum_{m=0}^{M-1} f[m] r[n - m], \tag{11.56}$$

where $f[n]$ is the impulse response of the discrete linear equalizer and $r[n]$ is the discrete representation of the received signal.

Because of their simplicity, linear filters were assumed in the discussion about the matched filter, zero-forcing equalizers, and the minimum mean-square error equalizers in Section 11.4 above. The optimal matched filter is given by Equation 11.23 in Section 11.3.1. The optimal zero-forcing linear equalizer (ZF-LE) can be constructed based on Equation 11.38 in Section 11.3.2. The optimal minimum mean-square error linear equalizer (MMSE-LE) can be constructed based on Equation 11.49 in Section 11.3.3. A generic equalizer design algorithm, which, among other things, can be used to design optimal MF, ZF-LE, and MMSE-LE, is discussed in Section 11.5.1.

In discrete equalizers, the input signal, $r[n]$, may be over-sampled (sampled at a sampling rate higher than the symbol rate). Equalizers with over-sampled input signals are referred to as fractionally spaced equalizers. The fractionally spaced equalizers can still be implemented as regular LTI filters, but the output of the equalizer needs to be down-sampled before the decision process. If the equalizer input signal is sampled at Q times the symbol rate, then a fractionally spaced linear equalizer can be implemented as

$$\hat{s}[n] = \sum_{m=0}^{M-1} f[m] r[nQ - m]. \tag{11.57}$$

If the over-sampling ratio, Q, is equal to one (input sampling rate is the same as the symbol rate), then Equation 11.57 reduces to Equation 11.56. Equalizers with an over-sampling ratio equal to one are referred to as T-spaced equalizers. Equalizers with an over-sampling ratio higher than one are referred to as fractionally spaced or T/Q-spaced equalizers. The pros and cons of using fractionally spaced equalizers are discussed in Section 11.7.3.

11.4.2   Decision Feedback Equalization

Equalizer performance can be significantly improved by introducing nonlinear elements into the equalizer structures. This can be done without a significant increase in complexity. One of the best-known examples of nonlinear equalizer structures is the decision feedback equalizer [Belfiore 1979]. As shown in Figure 11.3, the DFE feeds the slicer decisions into a feedback loop as part of the equalization structure, hence the name "decision feedback."

The basic assumption for the DFE structure is that almost all the slicer decisions are correct, and therefore the symbols out of the slicer are (almost all) correct reconstructions of

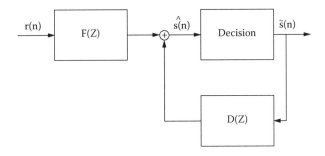

FIGURE 11.3
Structure of decision feedback equalizer (DFE) system.

the original transmitted symbols. This is equivalent to assuming that the original transmit symbols are available as input into the equalizer structure. Because the slicer only provides access to symbols that have already been received, the feedback part of the DFE can only use past symbols as its input and is therefore strictly causal.

The equalized samples in the DFE structure are given by

$$\hat{s}[n - n_0] = \sum_{m=0}^{M-1} f[m]r[n - m] - \sum_{k=1}^{K-1} d[k]\tilde{s}[n - n_0 - k], \tag{11.58}$$

where the symbol values $\tilde{s}[n]$ are the output of the slicer and n_0 is the delay through the communication channel, including the delay introduced by the linear equalizer $f[n]$. In DFE systems, the linear filter $f[n]$ is usually referred to as the feedforward filter (FFF) and the feedback filter $d[n]$ is referred to as the decision feedback filter (DFF).

To better understand the DFE, it is interesting to examine the structure in Figure 11.4. If there are no decision errors in the slicer (a reasonable assumption for this analysis), then the structure in Figure 11.4 is equivalent to the DFE structure in Figure 11.3. If the channel transfer function is given by

$$G(Z) = \frac{B(Z)}{A(Z)}, \tag{11.59}$$

then an ideal zero-forcing equalizer (ignoring casualty, stability, etc.) for the structure in Figure 11.4 would be

$$F(Z) = A(Z), \tag{11.60}$$

and

$$D(Z) = 1 - B(Z). \tag{11.61}$$

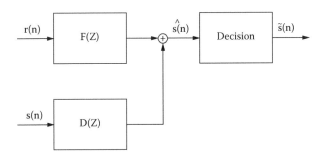

FIGURE 11.4
Equivalent structure to DFE.

This implies that, ideally, the feedforward filter addresses the recursive portion $A(Z)$ of the channel response, and the decision feedback filter addresses the FIR portion $B(Z)$ of the channel response. In real systems, things are not quite this simple because of causality constraints, stability of filters, and other such practical issues. Further, when noise is also considered as in the MMSE criterion, then a simple channel inverse as above can boost the noise, and the optimal equalizer is no longer just the channel inverse. However, the observation above can provide useful insight into the behavior of DFE systems when designing or analyzing DFE systems.

Considering that both decision feedback filters and stable recursive filters are strictly causal, it is tempting (but incorrect) to conclude that the decision feedback filter is ideally suited for cancelling a recursive channel response. For example, if a channel transfer function is equivalent to

$$y[n] = a \cdot y[n-1] + x[n], \tag{11.62}$$

where $0 < a < 1$, then an ideal decision feedback filter would be

$$d[k] = a^k, \quad \text{for} \quad k > 0. \tag{11.63}$$

If the decision feedback filter were infinitely long, then the equalizer would be a perfect zero-forcing equalizer. The problem is that real filters have finite length, so the decision feedback filter must be truncated, which in turn introduces ISI. For the case of the simple recursive filter in Equation 11.62, the ISI introduced by truncating the decision feedback filter can be addressed by adding a recursive term at the end of the decision feedback filter [Crespo 1991]. Although this approach may work, it is far more complicated than simply using a two-tap feedforward filter with $f[0] = 1$ and $f[1] = -a$, such that

$$\hat{x}[n] = y[n] - a \cdot y[n-1] = (a-a) \cdot y[n-1] + x[n] = x[n]. \tag{11.64}$$

This simple example demonstrates again that the feedforward filter is usually better suited than the decision feedback filter for cancelling any recursive parts of the channel response.

The discussion above may provide insight into the behavior of DFE systems, but it does not provide a generic approach to designing DFE filters. In most DSL systems, the equalizers are optimized under the MMSE criterion. The optimum linear equalizers under the MMSE criterion were derived in Section 11.3.3. This derivation can be extended to DFE systems by also incorporating the decision feedback filter. Substituting Equation 11.8 into Equation 11.58, the equalizer output is

$$\hat{s}[n] = \sum_m \sum_k f[m] \left(g[k]s[n-m-k] + \gamma[k]w[n-m-k] \right) - \sum_{k=1}^{K-1} d[k]s[n-n_0-k]. \tag{11.65}$$

Following the steps in Section 11.3.3 gives the mean-square error for the DFE case as

$$E\{e^2[n]\} = \sigma_s^2 \sum_k \varepsilon^2[k] + \sigma_w^2 \sum_k \xi^2[k] - \sigma_s^2 \sum_{k=1}^{K-1} d[k] \left(d[k] - 2\varepsilon[k+n_0] \right), \tag{11.66}$$

where, as before, $\varepsilon[n]$ and $\xi[n]$ are given by Equations 11.32 and 11.43, respectively.

The error surface in Equation 11.66 is quadratic, so the optimum decision feedback filter coefficients must satisfy

$$\frac{\partial E\{e^2[n]\}}{\partial d[n]} = -\sigma_s^2 \left(2d[k] - 2\varepsilon[k+n_0] \right) = 0. \tag{11.67}$$

Therefore, the decision feedback filter coefficients that minimize Equation 11.66 are given by

$$d[k] = \varepsilon[k + n_0]. \tag{11.68}$$

Substituting these optimal decision feedback coefficients into Equation 11.66 gives

$$E\{e^2[n]\} = \sigma_s^2 \sum_k \varepsilon^2[k] + \sigma_w^2 \sum_k \xi^2[k] - \sigma_s^2 \sum_{k=1}^{K-1} \varepsilon^2[k + n_0]. \tag{11.69}$$

Again following the steps in Section 11.3.3 gives

$$\sum_{m=0}^{M-1} f[m](R_g[n-m] + R_y[n-m] - r_g[n-n_0, m-n_0]) = \sigma_s^2 g[n_0 - n] \tag{11.70}$$

for $n = 0, 1, \ldots, M-1$, where $R_g[n]$ and $R_y[n]$ are given by Equations 11.39 and 11.48, respectively, and

$$r_g[n, m] = \sigma_s^2 \sum_{k=1}^{K-1} g[k-n]g[k-m]. \tag{11.71}$$

The derivation above was derived to gain some insight into how the DFE works. But this derivation also provides the foundations for a generic algorithm for designing MMSE-DFE. The M equations in Equation 11.70 can be solved for the M unknown feedforward equalizer taps, $f[m]$, using linear algebra according to

$$\mathbf{f} = \mathbf{R}^{-1}\mathbf{g}, \tag{11.72}$$

where

$$\mathbf{R}_{m,n} = R_g[n-m] + R_y[n-m] - r_g[n-n_0, m-n_0], \tag{11.73}$$

$$\mathbf{g} = \sigma_s^2 [g[n_0], g[n_0 - 1], \ldots, g[n_0 - M + 1]]^T, \tag{11.74}$$

and

$$\mathbf{f} = [f[0], f[1], \ldots, f[M-1]]^T. \tag{11.75}$$

Once the feedforward filter coefficients have been determined, the feedback coefficients can be calculated according to

$$d[k] = \varepsilon[k + n_0] = -\sum_{m=0}^{M-1} f[m]g[k + n_0 - m], \tag{11.76}$$

for $k = 1, 2, \ldots, K-1$. Equations 11.72 through 11.76 provide an algorithm for determining the optimal MMSE filter taps for a DFE structure [Thormundsson 2001]. By setting the DFF length to zero ($K = 0$), this same algorithm can be used to find optimal MMSE filter taps for linear equalizers. By ignoring the noise component and setting the noise variance $\sigma_w^2 = 0$, this algorithm can be used to find optimal ZFE filter taps. This algorithm is generalized in Section 11.5.1 to address complex valued bandpass signals and also to address fractionally spaced equalizers.

It is interesting to compare the error estimate in Equation 11.69 for a DFE with the error estimate in Equation 11.45 for linear equalizers. The two error estimates are identical except for the subtraction of the last term in Equation 11.69. Considering that each squared error term is positive, it is clear from this comparison that optimal DFE systems will perform better than a linear equalizer system (with the same number of feedforward filter coefficients).

It is also interesting to examine the error estimate in Equation 11.69 in terms of understanding DFE behavior. The last (negative) term is identical to the first term, except for the range of the summation. In essence, the last term implies that any ISI contributed by the previous $K - 1$ symbols should be ignored when determining the optimal feedforward filter coefficients. The explanation for this can be found in Equation 11.76 where the optimal decision feedback filter coefficients are selected to exactly cancel any ISI from the previous $K - 1$ symbols that remains after the feedforward equalization. One way to interpret this is to say that the decision feedback filter gives the feedforward filter increased flexibility by allowing it to ignore the ISI from previous $K - 1$ symbols. Another interpretation is to say that the feedforward filter acts as a noise whitening and impulse shortening filter, such that the remaining equalization can be done by the decision feedback filter. It is interesting to note that the feedforward filter in a DFE system plays a similar role to an impulse shortening equalizer in DMT systems (see Section 11.4.7).

One of the benefits of the DFE structure is that as long as the correct decisions are made in the slicer, the DFE feedback loop remains stable even if the same DFE coefficients in the form of an equivalent IIR filter would be unstable. The problem is that if an error occurs, then the DFE starts generating poor equalization signals, which in turn can cause more slicer errors, causing worse equalization signals, etc. In other words, a single slicer error can cause a long string of errors. This propagation of errors implies that errors in DFE systems tend to come in error bursts.

It can be shown [Lee 1994] that the theoretically achievable SNR at the output of an infinitely long MMSE-DFE is the geometric mean

$$SNR_{MMSE-DFE} = \exp\left(\frac{1}{w}\int_w \ln\left(1 + \frac{S_s(f)\,|G(f)|^2}{S_v(f)}\right) df\right),\qquad(11.77)$$

where $S_s(f)$ is the power spectral density (PSD) of the transmit signal $s(t)$ and $S_v(f)$ is the PSD of the additive noise $v(t)$. This is the maximum SNR that can theoretically be achieved, meaning that an infinitely long MMSE-DFE can achieve the theoretically minimum mean-square error and is therefore an ideal MMSE equalizer. Of more practical significance, real DFE systems of reasonable size can come strikingly close to this theoretical performance.

11.4.3 Noise Prediction Filters

If the slicer error (noise) is correlated (colored), then the mean-square error at the slicer can be reduced by removing the correlation from the error. Noise predictor schemes like the one in Figure 11.5 can be used to remove the correlation (coloring) from the error signal.

If the slicer error signal $e[n]$ has autocorrelation

$$r_e[n] = E\{e^*[m]e[m + n]\},\qquad(11.78)$$

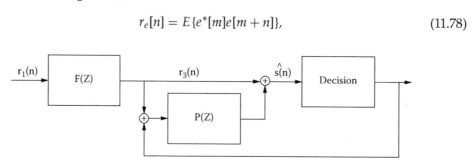

FIGURE 11.5
Noise predictor system.

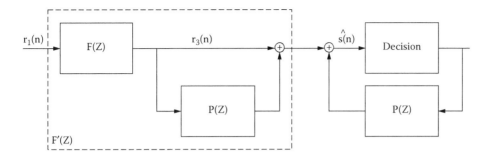

FIGURE 11.6
Noise predictor equivalent to DFE.

then the power of the slicer error can be minimized by removing correlation (color) of the slicer error signal. The Z-transform $R_e(Z)$ of the autocorrelation function can be factored into a constant A times two factors, where one is a monic loosely minimum-phase filter and the other is its reflected transfer function [Lee 1994]

$$R_e(Z) = AR_{min}(Z)R^*_{min}(1/Z^*). \tag{11.79}$$

The inverse of the monic minimum-phase filter can now be used to filter the noise such that the filtered error signal has an autocorrelation function satisfying

$$R'_e(Z) = R^{-1}_{min}(Z)R_e(Z)R^{*-1}_{min}(1/Z^*) = A. \tag{11.80}$$

In other words, the filtered signal is white with variance A. The constant A can be computed as the geometric mean of the error signal power spectral density

$$A = \exp\left(\frac{1}{2\pi} \int_{-\pi}^{\pi} \ln\left(R_e(e^{j\omega})\right) d\omega\right). \tag{11.81}$$

The noise predictor structure shown in Figure 11.5 has much in common with the DFE structure itself. Actually the NP can be split into feedforward and feedback portions as shown in Figure 11.6.

The DFE plus NP system shown in Figure 11.7 can be described by

$$\hat{s}[n] = r_1[n] \star f[n] + s[n] \star (d[n] + p[n]) + r_3[n] \star p[n]. \tag{11.82}$$

This can easily be rewritten as

$$\hat{s}[n] = r_1[n] \star (f[n] \star (\delta[n] + p[n])) + s[n] \star ((\delta[n] + d[n]) \star (\delta[n] + p[n]) - \delta[n]). \tag{11.83}$$

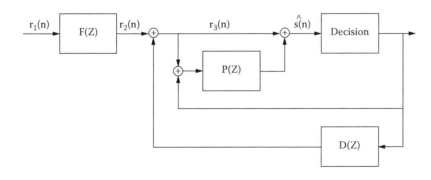

FIGURE 11.7
DFE structure with noise predictor (NP).

This is equivalent to a DFE system with feedforward filter

$$f_2[n] = (f[n] \star (\delta[n] + p[n]))$$ (11.84)

and decision feedback filter

$$d_2[n] = ((\delta[n] + d[n]) \star (\delta[n] + p[n]) - \delta[n]).$$ (11.85)

In other words, any DFE and NP system as shown in Figure 11.7 has an equivalent DFE-only representation. This means that theoretically, an NP system does not do anything that cannot be done equivalently using only a DFE structure. This would imply that NPs are relatively useless additions to DFE systems. However, the introduction of an NP helps resolve practical issues in real systems, in particular when adaptive algorithms are used to update the filter coefficients of the DFE and NP structures. The NP structure can also help resolve issues related to limited arithmetic accuracy.

11.4.4 Tomlinson–Harashima Precoding

Decision feedback equalizers have two main drawbacks: one is propagation of errors, and the other is that DFE systems do not work well in systems that have long decoding delay, such as trellis coded systems. One way to address both these problems is to use precoding methods such as the Tomlinson–Harashima precoding (THP).

Most common precoding schemes, including THP, can be modelled by the system in Figure 11.8. Ideally the precoder filter, $D(z)$, and the receiver linear equalizer, $F(z)$, are chosen such that

$$D(Z) = 1 - G(Z)F(Z).$$ (11.86)

In this case, the transfer function from $s[n]$ to $r_2[n]$ is given by

$$\frac{F(Z)G(Z)}{1 - D(Z)} = \frac{F(Z)G(Z)}{1 - 1 + G(Z)F(Z)} = \frac{F(Z)G(Z)}{F(Z)G(Z)} = 1.$$ (11.87)

This means that the channel is fully equalized and the signal is received undistorted. However, it may not be possible to choose $D(z)$ such that $1/(1 - D(z))$ is stable (*i.e.*, $1 - D(z)$ has all its roots inside the unit circle). If $1/(1 - D(z))$ is unstable, then the signal $x_2[n]$ out of the precoding filter may be unbounded even for a bounded input signal, $s[n]$. This means that an IIR prefilter like this is not a practical solution unless special measures are taken to limit the output of the prefilter. This can be countered by adding a side signal $\mu_1[n]$ to the output of the prefilter as shown in Figure 11.8. By appropriately choosing $\mu_1[n]$, it is possible to limit the value of $x_2[n]$ and make the precoder output bounded. This simple trick takes care of the instability problem, but introduces a new problem. Because $\mu_1[n]$ has now been added to the transmitted signal $x_2[n]$, this signal must be "removed" again to

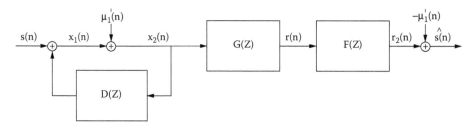

FIGURE 11.8
Precoding structure.

reconstruct $s[n]$. This problem can be solved by observing that the transfer function from $s[n]$ to $r_2[n]$ is $H(z) = 1$, so

$$r_2[n] = (s[n] + \mu_1[n]) \star \delta[n] = s[n] + \mu_1[n]. \tag{11.88}$$

Simply subtracting $\mu_1[n]$ from $r_2[n]$ reconstructs $s[n]$.

The problem now becomes how to reconstruct in the receiver the $\mu_1[n]$ used in the transmitter. One solution to this problem is to construct the signal $\mu_1[n]$ from a set that is "orthogonal" to the signal set for $s[n]$. The two signal sets are orthogonal in the sense that adding $\mu_1[n]$ to any valid value of $s[n]$ can never result in another valid value for $s[n]$. In THP, this is done by constructing $\mu_1[n]$ using a modulo operator. If the transmit signal is a PAM signal with symbols in the range from $-A$ to A such that

$$s[n] \in [-A; A) \quad \forall n, \tag{11.89}$$

then all values of $\mu_1[n]$ are multiples of $2A$ such that

$$-A \leq x_2[n] < A. \tag{11.90}$$

The output of a THP is inherently uniformly distributed over the range from $-A$ to A. As a result, the THP signals are slightly less energy efficient than the original PAM signal. It is possible, however, to make a relatively simple modification to the THP to improve the energy efficiency of the transmit signal. This can be done by choosing the $\mu_1[n]$ values such that they are still multiples of $2A$, but they are chosen based on knowledge about the transmit filter to minimize the power out of the transmit filter. Such methods have been reported to provide shaping gain of 0.75 dB [Orckit 1999] (compared the theoretical maximum shaping gain of 1.53 dB).

It is possible to do precoding using a more general structure than the THP. For example, the precoder used in V.92 upstream transmission is based on selecting transmit levels for each symbol from a set of "equivalent" levels, so-called equivalence classes. There are numerous other methods possible to choose the signal $\mu_1[n]$ based on the transmitted symbols and the precoder filter output. But most, if not all, of these can be viewed as relatively simple generalizations of the structure in Figure 11.8.

11.4.5 Maximum Likelihood Detection

The equalization structures discussed above are all based on processing the received signal to minimize ISI and noise before the signal detection process. In these methods, the equalization and the detection are done separately. It is possible, however, to do the equalization as an integral part of the detection process. The most commonly used method for joint equalization and decision is maximum likelihood sequence estimation [Forney 1972], usually implemented with the Viterbi algorithm [Viterbi 1971].

Suppose the received signal, $r[n]$, is given by

$$r[n] = \sum_{m=0}^{N-1} g[m]s[n-m] + v[n], \tag{11.91}$$

where $g[n]$ is the symbol response of the channel, $s[n]$ is the transmitted symbol, and $v[n]$ is the additive noise. Using vector representations for $g[n]$ and $s[n]$, this can be expressed as

$$r[n] = \mathbf{g}^T \mathbf{s}_{(N)}[n] + v[n], \tag{11.92}$$

where $\mathbf{g}$ is a vector of length N and

$$\mathbf{s}_{(N)}[n] = [s[n], s[n-1], \ldots, s[n-N+1]]^T. \tag{11.93}$$

For a given sequence, $s_{(N)}[n]$, of transmit symbols, the receive signal is known apart from the additive noise component, $v[n]$. It is obvious from Equation 11.92 that

$$v[n] = r[n] - \mathbf{g}^T \mathbf{s}_{(N)}[n]. \tag{11.94}$$

Therefore, the probability of receiving an observed value $r_0[n]$ at sample time nT given a known transmit sequence $s_{(M)}[n]$ is obviously

$$P_{r|s}\left(r_0[n]|\mathbf{s}_{(N)}[n]\right) = P_v\left(r[n] - \mathbf{g}^T \mathbf{s}_{(N)}[n]\right), \tag{11.95}$$

where $P_v(v)$ is the probability distribution of the noise, $v[n]$. If the noise is white, then the noise samples are not correlated, and the probability of receiving the noise sequence $v_{(K)}[n]$ is given by

$$P_v\left(v[n], v[n-1], \ldots, v[n-K+1]\right) = \prod_{k=0}^{K-1} P_v\left(v[n-k]\right). \tag{11.96}$$

In this case, the probability of receiving a sequence $\mathbf{r}_{(K)}[n]$, given the transmit symbol sequence $\mathbf{s}_{(N+K)}[n]$ is given by

$$P_{r|s}\left(r_{(K)}[n]|\mathbf{s}_{(N+K)}[n]\right) = \prod_{k=0}^{K-1} P_v\left(r[n-k] - \mathbf{g}^T \mathbf{s}_{(N)}[n-k]\right). \tag{11.97}$$

This can also be expressed as a sum of logarithms according to

$$P_{r|s}\left(r_{(K)}[n]|\mathbf{s}_{(N+K)}[n]\right) = \exp\left(\sum_{k=0}^{K-1} \ln\left(P_v\left(r[n-k] - \mathbf{g}^T \mathbf{s}_{(N)}[n-k]\right)\right)\right). \tag{11.98}$$

This implies that for additive white Gaussian noise, the probability of receiving a sequence $\mathbf{r}_{(K)}[n]$, given the transmit symbol sequence $\mathbf{s}_{(N+K)}[n]$ is given by

$$P_{r|s}\left(r_{(K)}[n]|\mathbf{s}_{(N+K)}[n]\right) = \frac{-1}{\sigma_v \sqrt{2\pi}} \exp\left(\frac{1}{2\sigma_v^2} \sum_{k=0}^{K-1} \left(r[n-k] - \mathbf{g}^T \mathbf{s}_{(N)}[n-k]\right)^2\right). \tag{11.99}$$

The ML criterion (see Section 11.3.4) states that given an observed receive signal, $\mathbf{r}_{(K)}[n]$, the most likely transmit sequence, $s_{0(M+K)}[n]$, must satisfy

$$P_{r|s}\left(r_{(K)}[n]|\mathbf{s}_{0(N+K)}[n]\right) \geq P_{r|s}\left(r_{(K)}[n]|\mathbf{s}_{(N+K)}[n]\right) \quad \forall \mathbf{s}_{(N+K)}[n]. \tag{11.100}$$

For additive white Gaussian noise, substitution of Equation 11.99 into Equation 11.100 gives

$$\sum_{k=0}^{K-1} \left(r[n-k] - \mathbf{g}^T \mathbf{s}_{0(N)}[n-k]\right)^2 \leq \sum_{k=0}^{K-1} \left(r[n-k] - \mathbf{g}^T \mathbf{s}_{(N)}[n-k]\right)^2 \quad \forall \mathbf{s}_{(N+K)}[n]. \tag{11.101}$$

This implies that for additive white Gaussian noise, the ML criterion is equivalent to minimizing the mean-square error

$$e\left\{\mathbf{s}_{(N+K)}[n]\right\} = \frac{1}{K} \sum_{k=0}^{K-1} \left(r[n-k] - \mathbf{g}^T \mathbf{s}_{(N)}[n-k]\right)^2. \tag{11.102}$$

The maximum likelihood sequence estimation can now be done by searching for the sequence that gives the lowest mean-square error. If each symbol, $s[n]$, can take on L distinct values, then the vector $\mathbf{s}_{(M+K)}[n]$ can take on $L^{(N+K)}$ distinct values. This makes exhaustive search for the best ML sequence impractical for large M and K values. It is possible, however, to simplify the task of finding the best ML sequence by applying the well-known Viterbi algorithm [Viterbi 1971].

In DSL systems the typical channel impulse response has a very long tail. As a result, the symbol response $g[n]$ is typically very long and N in Equation 11.102 is very large. Therefore, the MLSE becomes computationally intensive even if the efficient Viterbi algorithm is used. This makes MLSE impractical for DSL systems unless special measures are taken to reduce the complexity of the calculations in Equation 11.102.

One way to reduce the computational complexity of the MLSE is to use impulse shortening filters (see Section 11.4.7 below) to reduce M by making $g[n]$ shorter. An interesting and a very practical way to do this is to use the MMSE-DFE system to construct the impulse shortening equalizer and to generate an estimate of the (shortened) symbol response. The FFF from the DFE system serves as an impulse shortening and noise whitening filter (see discussion in Section 11.4.2), and the DFF coefficients are used directly to generate the channel response vector. In this case, the error calculation in Equation 11.102 is replaced by

$$e\left\{\mathbf{s}_{(N+K)}[n]\right\} = \frac{1}{K}\sum_{k=0}^{K-1}\left(r'[n-k] - \mathbf{d}^T\mathbf{s}_{(N)}[n-k]\right)^2, \tag{11.103}$$

where

$$r'[n] = \sum_{m=0}^{M-1} f[m]r[n-m] \tag{11.104}$$

and

$$\mathbf{d} = [d[0], d[1], \ldots, d[N-1]]^T. \tag{11.105}$$

The value of N can now be selected such that the calculations in Equation 11.103 have reasonable complexity and the only constraint is that the DFF filter in the DFE system should also be of length N.

The computational complexity associated with calculating error estimates in Equation 11.102 or Equation 11.103 makes the MLSE far more computationally intensive than the DFE for the THP systems discussed in previous sections. Therefore, MLSE systems need to have clear performance advantage if they are to be a viable alternative to the DFE and THP systems. One setting where MLSE has advantage over DFE systems is when trellis coding is used (see Chapter 8). Because of delays introduced by the decoding of trellis codes, it is not practical to use the decision feedback values from the trellis decoding. It might be tempting to use a simple slicer to generate the decision feedback symbols, but this would result in too high an error rate and propagation of errors. Therefore, it is not desirable to use DFE systems in combination with trellis coding. In this case, it can sometimes be feasible to use MLSE, especially because trellis codes are typically decoded using the Viterbi algorithm, so the MLSE and the trellis decoding can be combined in a single process. However, THP can also be used with trellis coding and, in most cases, the THP has lower computational complexity than MLSE.

11.4.6 Frequency Domain Equalization

As discussed above, the channel distortion of the transmit signal is best modelled as a convolution of the channel impulse response with the transmitted signal (see Equation 11.30). Convolution in the time domain is equivalent to multiplication (or windowing) in the frequency domain. The LTI channel distortion can be modelled in the frequency domain as multiplication of the channel transfer function, $H(f)$, and the transmit signal, $X(f)$. This suggests that zero-forcing equalization can very easily be done in the frequency domain as a multiplication of the received signal and the inverse of the channel transfer function.

A well-known method for implementing convolution in the time domain is to transform the time domain signals into the frequency domain, multiply the frequency domain signal with the frequency domain representation of the filter, and then transform the resulting signal back into the time domain. Because of how efficiently discrete Fourier transforms (DFT) can be implemented using the fast Fourier transform (FFT) algorithm [Oppenheim 1989], this approach is in many cases more computationally efficient than implementing convolution directly in the time domain. The problem is that because of the circular nature of DFT, multiplication of two signals in the DFT domain translates into circular convolution in the time domain (see [Oppenheim 1989]). However, in processing time domain signals such as communication signals, it is usually desirable to implement linear convolution, not circular convolution. This issue can be solved by using techniques such as the overlap-save or overlap-add methods (see [Oppenheim 1989]) to implement linear convolution using circular convolution based on FFT. This could potentially lead to efficient equalizer implementations, especially if the FFT used is much longer than the length of the equalizer impulse response.

A more interesting approach, however, is to introduce circular characteristics into the transmit signal so that equalization can be done in the frequency domain without special methods such as overlap-add or overlap-save. This is the idea behind introducing the cyclic prefix in DMT modulation (see Section 7.3). The signal is segmented into blocks, where each block is of suitable size for a DFT (in DMT, each block is generated by an inverse DFT). Each block is then extended by replicating the v last samples of the block in a prefix in front of each block. If the impulse response of the channel (minus one sample) is no longer than the length of the cyclic prefix, then all the samples in each received block are completely independent of the samples in all the other blocks. Furthermore, because the cyclic prefix is taken from the last samples of each block, the linear convolution of the channel distortion looks like a circular convolution within each block. This in turn means that equalization can be done by taking the DFT of each signal block and multiplying the frequency domain representation of the signal block by the inverse of the channel response.

If the transmit signal is modulated using single-carrier modulation techniques like PAM, QAM, or CAP, then the equalized signal must be transformed back to the time domain using an IDFT (see [Falconer 2002]). On the other hand, if DMT modulation is used, then the demodulation is done directly in the frequency domain. Chapter 7 discusses in more detail how frequency domain equalization can be implemented and utilized for DMT modulation.

11.4.7   Impulse Shortening Equalization

A key issue in the use of the cyclic prefix to make signals circular within each block is the assumption that the channel impulse response is shorter than the length of the cyclic prefix. If the channel response is more than one sample longer than the cyclic prefix, then special measures must be taken to "shorten" the impulse response so that the equalization of each block can again be done using circular convolution. The impulse response of the channel can be shortened by filtering the received signal with an impulse shortening filter (*i.e.*, an impulse shortening equalizer). If the channel has impulse response $h[n]$, then the impulse shortening equalizer should have impulse response $f[n]$ such that the combined impulse response

$$h_{eq}[n] = f[n] \star h[n] = \sum_m f[m]h[n-m] \qquad (11.106)$$

is no longer than the cyclic prefix. In other words, if the cyclic prefix is of length ν, then the combined impulse response should ideally satisfy

$$h_{eq}[n] = 0 \quad \forall n \notin Z_1, \tag{11.107}$$

where

$$Z_1 = \{n_0, n_0 - 1, \ldots, n_0 - \nu + 1\}. \tag{11.108}$$

From the perspective of impulse shortening, the equalized impulse response $h_{eq}[n]$ can take on any value in the interval Z_1, but must have at least one nonzero value.

The convolution in Equation 11.106 can be expressed in matrix form as

$$\mathbf{h}_{eq} = \mathbf{H}_c \mathbf{f}, \tag{11.109}$$

where $\mathbf{H}_c$ is the convolution matrix

$$\mathbf{H}_c = \begin{bmatrix} h[0] & 0 & \cdots & & 0 \\ h[1] & h[0] & \cdots & & 0 \\ \vdots & \vdots & \vdots & & \vdots \\ 0 & \cdots & h[N_h - 1] & h[N_h - 2] \\ 0 & \cdots & 0 & h[N_h - 1] \end{bmatrix}. \tag{11.110}$$

The condition in Equation 11.107 can now be expressed as

$$\mathbf{H}_{c0}\mathbf{f} = \mathbf{0}, \tag{11.111}$$

where

$$\mathbf{H}_{c0} = \begin{bmatrix} h[0] & 0 & \cdots & & 0 \\ h[1] & h[0] & \cdots & & 0 \\ \vdots & \vdots & \vdots & & \vdots \\ h[n_0 - 1] & h[n_0 - 2] & \cdots & & h[n_0 - M] \\ h[n_0 + \nu] & h[n_0 + \nu - 1] & \cdots & & h[n_0 + \nu - M + 1] \\ \vdots & \vdots & \vdots & & \vdots \\ 0 & \cdots & h[N_h - 1] & h[N_h - 2] \\ 0 & \cdots & 0 & h[N_h - 1] \end{bmatrix}. \tag{11.112}$$

The only solution for Equation 11.111 is the trivial solution $\mathbf{f} = \mathbf{0}$. This is obvious from the observation that if $h[0]$ is nonzero, then the product of $\mathbf{f}$ and the first row of $\mathbf{H}_{c0}$ can only be zero if $f[0]$ is zero. By iterating this thought process for each row of $\mathbf{H}_{c0}$ and each element of $\mathbf{f}$, it becomes apparent that the only $\mathbf{f}$ that satisfies Equation 11.111 is the trivial solution $\mathbf{f} = \mathbf{0}$.

This means that it is not possible to perfectly satisfy Equation 11.111 if T-spaced impulse shortening equalization is used. It is possible, however, to construct fractionally spaced impulse-shortening equalizers such that the impulse response is truly zero for all values n not in Z_0. For a Q times over-sampled fractionally spaced equalizer, the impulse shortening criterion becomes

$$h_{eq}[nQ] = 0 \quad \forall n \notin Z_1. \tag{11.113}$$

Unlike the criterion for T-spaced impulse shortening equalizers, this criterion can be satisfied.

Even though T-spaced impulse shortening equalizers cannot perfectly satisfy the impulse shortening criterion in Equation 11.111, they can come arbitrarily close to satisfying this criterion if the impulse shortening filters are made long enough. For impulse shortening filters of limited length, the impulse shortening filter can be optimized by minimizing the error

$$\mathbf{e} = \mathbf{H}_{c0}\mathbf{f}, \tag{11.114}$$

under the additional constraint about f not being the trivial solution $f = \mathbf{0}$.

This additional constraint can be specified in several different ways including the following.

$$\sum_{n \in Z_1} h_{eq}^2[n] = 1, \tag{11.115}$$

$$h_{eq}[n_0] = 1, \tag{11.116}$$

or

$$\mathbf{f}^T\mathbf{f} = 1. \tag{11.117}$$

These three criteria are not equivalent but have very similar implications. Which one makes the best additional constraint depends on the details of the implementation of the impulse shortening equalization.

It is possible to use many different metrics for the error in Equation 11.114, but it is most common to use the MMSE metric. For the MMSE metric, the error function to be minimized is

$$E = \mathbf{e}^T\mathbf{e} = (\mathbf{H}_{c0}\mathbf{f})^T \mathbf{H}_{c0}\mathbf{f} = \mathbf{f}^T\mathbf{R}_0\mathbf{f}, \tag{11.118}$$

where

$$\mathbf{R}_0 = \mathbf{H}_{c0}^T\mathbf{H}_{c0}. \tag{11.119}$$

The minimization of Equation 11.118 under the constraint in Equation 11.117 is trivial. It is well known from elementary matrix theory (see [Apostol 1969][Melsa 1996]) that the vector f that minimizes Equation 11.118 such that $f^T f = 1$ is the eigenvector of $\mathbf{R}_0$ corresponding to the smallest eigenvalue of $\mathbf{R}_0$. Therefore, the optimum impulse shortening equalizer according to the criterion in Equation 11.118 that satisfies the constraint in Equation 11.117 is e_{min}, the eigenvector of $\mathbf{R}_0$ corresponding to the smallest eigenvalue of $\mathbf{R}_0$. Furthermore, the error E as defined in Equation 11.118 is equal to the value of the smallest eigenvalue of $\mathbf{R}_0$.

A similar (but slightly more involved) approach is proposed in [Melsa 1996]. In this approach, the error in Equation 11.118 is minimized, satisfying the constraint in Equation 11.115. For this case, the constraint can be expressed as

$$\mathbf{f}^T\mathbf{H}_{c1}^T\mathbf{H}_{c1}\mathbf{f} = \mathbf{f}^T\mathbf{R}_1\mathbf{f} = 1, \tag{11.120}$$

where

$$\mathbf{H}_{c1} = \begin{bmatrix} h[n_0] & h[n_0 - 1] & \cdots & h[n_0 - M + 1] \\ h[n_0 + 1] & h[n_0] & \cdots & h[n_0 - M] \\ \vdots & \vdots & \vdots & \vdots \\ h[n_0 + v - 1] & h[n_0 + v - 2] & \cdots & h[n_0 + v - M] \end{bmatrix}, \tag{11.121}$$

and

$$\mathbf{R}_1 = \mathbf{H}_{c1}^T\mathbf{H}_{c1}. \tag{11.122}$$

If $\mathbf{R}_1$ is full rank, then $\mathbf{R}_1$ is invertible and it is possible to use Cholesky factorization to construct invertible $M \times M$ matrix $\mathbf{D}$ that is the square root of $\mathbf{R}_1$, such that

$$\mathbf{R}_1 = \mathbf{D}^T\mathbf{D}. \tag{11.123}$$

The condition in Equation 11.120 can now be expressed as

$$\mathbf{f}^T \mathbf{R}_1 \mathbf{f} = \mathbf{f}^T \mathbf{D}^T \mathbf{D} \mathbf{f} = \mathbf{f}_1^T \mathbf{f}_1 = 1, \tag{11.124}$$

where

$$\mathbf{f}_1 = \mathbf{D} \mathbf{f} \tag{11.125}$$

and

$$\mathbf{f} = \mathbf{D}^{-1} \mathbf{f}_1. \tag{11.126}$$

The error criteria in Equation 11.118 can then be written as

$$E = \mathbf{f}^T \mathbf{R}_0 \mathbf{f} = \left(\mathbf{D}^{-1} \mathbf{f}_1 \right)^T \mathbf{R}_0 \mathbf{D}^{-1} \mathbf{f}_1 = \mathbf{f}_1^T \mathbf{R}_0' \mathbf{f}_1, \tag{11.127}$$

where

$$\mathbf{R}_0' = \mathbf{D}^{-T} \mathbf{R}_0 \mathbf{D}^{-1}. \tag{11.128}$$

Now the optimum solution for Equation 11.118 that satisfies Equation 11.115 is $\mathbf{e}_{min}'$, the eigenvector of $\mathbf{R}_0'$ that corresponds to the minimum eigenvalue of $\mathbf{R}_0'$, and the optimum impulse shortening equalizer becomes

$$\mathbf{f} = \mathbf{D}^{-1} \mathbf{e}_{min}'. \tag{11.129}$$

As before, the error E is equal to the smallest eigenvalue of $\mathbf{R}_0'$.

Both of the methods discussed above can have issues with the numerical accuracy of the calculations, especially if the matrices $\mathbf{R}_0$ and $\mathbf{R}_1$ are singular. An appendix to [Melsa 1996] describes a method for constructing the impulse shortening equalizers if $\mathbf{R}_1$ is singular. This basic approach can be used for both methods described above. But if this method is used, special care must always be taken with numerical accuracy when dealing with singular or almost singular matrices.

The minimization of Equation 11.118 using the constraint in Equation 11.115 maximizes the shortening SNR (SSNR), which is given by

$$SSNR = 10 \log_{10} \left(\frac{\sum_{n \in Z_1} h_{eq}^2 [n]}{\sum_{n \notin Z_1} h_{eq}^2 [n]} \right) = 10 \log_{10} \left(\frac{\mathbf{f}^T \mathbf{H}_{c1}^T \mathbf{H}_{c1} \mathbf{f}}{\mathbf{f}^T \mathbf{H}_{c0}^T \mathbf{H}_{c0} \mathbf{f}} \right). \tag{11.130}$$

In this respect, it is better to use the constraint in Equation 11.115 than the one in Equation 11.117. However, the SSNR criterion is defined in an ad hoc way and does not guarantee best performance of the overall communication system.

One alternative approach to look at impulse shortening equalizers is to compare them to the feedforward filters in the DFE structure. In the DFE discussion in Section 11.4.2, it was observed that, according to Equation 11.68, the optimal feedback filters, are $d[n] = \varepsilon[n + n_0]$. As a result, the values of $\varepsilon[n]$ in the range from $n_0 + 1$ to $n_0 + K - 1$ become "don't care" values in the optimization of the feedforward filter coefficients. A similar approach can be used to account for the "don't care" values in the combined impulse response of the channel and the impulse shortening equalizer.

For any given impulse shortening equalizer, $f[n]$, the equalized impulse response, $h_{eq}[n]$, is given by Equation 11.106. A hypothetical decision feedback filter, $d[n]$, can be specified as

$$d[n] = -\delta[n - n_0] + h_{eq}[n - n_0], \quad \text{for} \quad n = 0, 1, \dots, K - 1, \tag{11.131}$$

where $K = v$. This imaginary decision feedback filter is not really implemented, but only used to account for the "don't care" values in the equalized impulse response, The optimum impulse shortening equalizer can now be derived using the same basic approach as was used to find optimal feedforward filters for the DFE structure. Assuming there is no noise in the signal, a hypothetical equalized signal (including the hypothetical decision feedback filter) is given by

$$\hat{x}[n] = \sum_m f[m]r[n-m] - \sum_{k=1}^{K-1} d[k]x[n-n_0-k] = \sum_{m \notin Z_1} h_{eq}[m]x[n-m] + x[n]. \quad (11.132)$$

Therefore, minimizing the impulse shortening criterion becomes equivalent to minimizing the error

$$e[n] = x[n] - \hat{x}[n]. \quad (11.133)$$

Under the MMSE criterion, this becomes the minimization of

$$E\{e^2[n]\} = E\{(x[n] - \hat{x}[n])^2\}. \quad (11.134)$$

Using the approach from Section 11.3.2 , it is trivial to show that the optimum impulse shortening equalizer under the MMSE criterion must satisfy

$$\sum_{m=0}^{M-1} f[m]R_h[m,n] = h[n_0-n], \quad (11.135)$$

where

$$R_h[m,n] = \sum_k h[k-m]h[k-n] - \sum_{k=0}^{K-1} h[k-m]h[k-n] = \sum_{k \notin Z_1} h[k-m]h[k-n]. \quad (11.136)$$

The coefficients for the optimum impulse shortening equalizer, $f[n]$, can then be computed according to Equation 11.135 using linear algebra.

It is interesting to note that the impulse shortening equalizer criteria given in Equation 11.135 is a "zero-forcing" criterion that does not account for noise. Just as for regular zero-forcing equalizers, it is possible that the "zero-forcing" impulse shortening equalizer will amplify any noise that may be present in the signal. This noise amplification can be addressed by including the noise term, $v[n]$, in the derivation of the optimal impulse shortening filter as given by Equation 11.135. This is done by using the autocorrelation function

$$R_h'[m,n] = R_h[m,n] + R_y[m-n], \quad (11.137)$$

where, as before, $R_y[m,n]$ is given by Equation 11.48.

For DMT systems, a criterion that is more closely related to overall performance would be to maximize the maximum theoretically achievable bit rate C for the equalized signal. The maximum achievable bit rate is given by

$$C = \int_0^{F_0} \log\left(1 + \frac{S_1(f)}{S_0(f) + S_n(f)}\right) df, \quad (11.138)$$

where $S_1(f)$ is the PSD of the signal component due to H_{c1} (the desired signal), $S_0(f)$ is the PSD of the signal due to H_{c0} (intersymbol interference), and $S_n(f)$ is the noise PSD. Minimizing E as given by Equation 11.118 under any of the constraints in Equation 11.115, Equation 11.116, or Equation 11.117 are all good first approximations of optimizing for the criterion in Equation 11.138. A better approximation could be to maximize the geometric mean SNR given by

$$SNR_{GM} = \frac{\exp\left\{\int_0^{F_0} \ln(S_1(f))df\right\}}{\exp\left\{\int_0^{F_0} \ln(S_0(f) + S_n(f))df\right\}} = \exp\left\{\int_0^{F_0} \ln\left(\frac{S_1(f)}{S_0(f) + S_n(f)}\right)df\right\}. \quad (11.139)$$

A method for maximizing the geometric mean SNR can be found in [Al-Dhahir 1996]. This method maximizes the geometric mean SNR under the constraint in Equation 11.117 using Lagrange multipliers. This results in a nonlinear optimization problem that does not have an analytic closed-form solution. Although this approach is one step closer to optimizing the overall performance of a DMT system, it does not take into account practical DMT implementation issues such as bit loading and windowing due to finite length DFTs. Deriving the optimal impulse shortening equalizer for a criterion that takes all the practical DMT issues into account is beyond the scope of this chapter, but one interesting method to optimize the SNR for each tone is the "per-tone" equalization proposed in [Van Acker 2001].

It is interesting to observe the similarities between the impulse shortening equalizer and the feedforward filter in a DFE structure. They are usually implemented using the same basic structure, and they basically provide the same functionality of shortening the impulse response such that the rest of the system can more easily process the signal. This is just one one of many examples where the processing needed for single-carrier modulation (PAM, QAM, and CAP) is strikingly similar to the processing needed for multi-carrier modulation (DMT).

11.5   Closed-Form Equalizer Design

In the theoretical discussion above, it was always assumed that the channel impulse response and noise spectrum were known beforehand. In the real world, this is usually not the case. However, it is often possible to use "channel probing" to estimate the channel properties, including the channel impulse response and the noise spectrum. If the channel impulse response and noise spectrum can be obtained through "channel probing" or by other means, then it is possible to construct optimal equalizers based on this knowledge.

11.5.1   MMSE Equalizer Design Algorithms

In Section 11.4.2 a method was derived for finding optimal feedforward and decision feedback filter coefficients for the DFE structure. This method gives rise to the general MMSE-DFE design algorithm presented below. The algorithm has been extended to address fractionally spaced equalizers and support complex valued passband signals. This algorithm can address as special cases most of the other equalization methods discussed above. By setting the decision feedback filter to zero (*i.e.*, $K = 0$), this algorithm reduces to the MMSE algorithm for a linear equalizer. By setting the noise term $R_y[n]$ to zero, the algorithm reduces to the ZFE algorithm discussed in Section 11.3.2. A matched filter can be derived by setting $R_g[n, m]$ to zero, ignoring ISI and $R_{yg}[n] = \delta[n]$ (white noise), and a whitened matched filter can be derived by setting $R_g[n, m]$ to zero but including the $R_y[n]$ term. The algorithm can be used directly (or with minor modifications) to design impulse shortening equalizer coefficients, as discussed in Section 11.4.7. Of all the equalizer structures discussed in Section 11.4, the only two cases where the general MMSE-DFE algorithms are not directly applicable are the MLSE and the frequency domain equalization. But even for these cases, the DFF $d[n]$ are directly applicable if the MLSE or the FEQ is preceded by an impulse shortening filter designed by the general MMSE-DFE algorithm. The basic equalizer design method for the MMSE-DFE as given in Section 11.4.2 can be extended to address complex valued passband signals and also to address fractionally spaced equalizers. For fractionally spaced equalizers, with Q times over sampling, the linear equations for the feedforward filter coefficients $f[n]$ are given by

$$\sum_{m=0}^{M-1} f[m](R_g[m,n] + R_\gamma[m-n] - r_{g0}[m,n]) = \sigma_s^2 g^*[n_0 Q - n],$$ (11.140)

where

$$R_g[m,n] = \sigma_s^2 \sum_k g^*[kQ - m]g[kQ - n],$$ (11.141)

$$R_\gamma[m-n] = \sigma_w^2 \sum_k \gamma^*[kQ - m]\gamma[kQ - n],$$ (11.142)

and

$$r_{g0}[m,n] = \sigma_s^2 \sum_{k=1}^{K-1} g^*[kQ + n_0 Q - m]g[kQ + n_0 Q - n].$$ (11.143)

The corresponding optimal decision feedback filter coefficients are given by

$$d[k] = -\sum_{m=1}^{M-1} f[m]g[kQ + n_0 Q - m] \quad \text{for } k = 1, 2, \ldots, K - 1.$$ (11.144)

The M equations in Equation 11.140 can be expressed in matrix form as

$$\mathbf{g}_0^* = \mathbf{R}\mathbf{f},$$ (11.145)

where

$$\mathbf{R}_{m,n} = R_g[m,n] + R_\gamma[m-n] - r_{g0}[m,n],$$ (11.146)

is an $M \times M$ matrix and

$$\mathbf{g}_0 = \sigma_s^2 [g[n_0 Q], g[n_0 Q - 1], \ldots, g[n_0 Q - M + 1]]^T.$$ (11.147)

If the matrix $\mathbf{R}$ is full rank, then Equation 11.145 can be solved according to

$$\mathbf{f} = \mathbf{R}^{-1}\mathbf{g}_0^*.$$ (11.148)

Once the feedforward filter coefficients have been calculated according to Equation 11.148, the decision feedback filter coefficients can be calculated according to Equation 11.144.

Stated more formally as an algorithm, the MMSE-DFE coefficients can be computed as follows. Given $g[n]$, σ_s^2 and $R_\gamma[n]$:

1. Construct $\mathbf{g}_0$ according to Equation 11.147:

$$\mathbf{g}_0 = \sigma_s^2 [g[n_0 Q], g[n_0 Q - 1], \ldots, g[n_0 Q - M + 1]]^T.$$

2. Compute correlation matrices according to Equation 11.141, Equation 11.143, and Equation 11.146:

$$R_g[m,n] = \sigma_s^2 \sum_k g^*[kQ - m]g[kQ - n],$$

$$r_{g0}[m,n] = \sigma_s^2 \sum_{k=1}^{K-1} g^*[kQ + n_0 Q - m]g[kQ + n_0 Q - n],$$

$$r_{g0}[m,n] = \sigma_s^2 \sum_{k=1}^{K-1} g^*[kQ + n_0 Q - m]g[kQ + n_0 Q - n].$$

3. Find FFF coefficients according to Equation 11.148:

$$\mathbf{f} = \mathbf{R}^{-1}\mathbf{g}_0^*.$$

4. Find DFF coefficients according to Equation 11.144:

$$d[k] = -\sum_{m=1}^{M-1} f[m]g[kQ + n_0Q - m] \quad \text{for} \quad k = 1, 2, \ldots, K-1.$$

The computation in Equation 11.148 requires a matrix inversion, which can be computationally intensive for large matrices. In many cases, however, it is possible to utilize the structure of the matrix R and its submatrices to reduce the computational complexity of inverting R, but the details of efficient computation of Equation 11.148 are beyond the scope of this chapter. It is also possible to find the optimal f using iterative algorithms such as steepest descent [Alexander 1986][Haykin 1996] or the conjugate gradient algorithm [Golub 1989]. In this case, Step 3 in the algorithm above is replaced by the iterative algorithm, but the other steps remain the same.

An alternative algorithm for computing optimum MMSE–DFE coefficients can be found in [Al-Dhahir 1995a] and [Al-Dhahir 1995b]. The derivation of this algorithm is along similar lines to the derivation in Section 11.4.2, but the resulting algorithm is slightly different. The algorithm in [Al-Dhahir 1995a] requires the channel symbol response, $g[n]$, to be of finite length, N_g, and requires the inversion of an $(M + N_g) \times (M + N_g)$ matrix. This can be computationally intensive for DSL channels that are typically characterized by very long impulse response (in reality, infinitely long). However, the complexity of the matrix inversion is reduced by an efficient algorithm presented in [Al-Dhahir 1995b], which makes use of the special structure of the matrices that need to be inverted.

11.5.2 Channel Probing

In the MMSE discussion above, it was assumed that the channel impulse response was known. In DSL systems, the channel impulse response usually varies significantly from one loop (channel) to the next. Therefore, if the above algorithms are to be used, the channel impulse response must be obtained for each channel.

The simplest way to obtain the channel impulse response is to send a single pulse (approximately an impulse) over the channel and observe the resulting received signal. If there is no noise on the channel, this could give a good estimate of the channel impulse response. In reality, however, there is always some noise on the channel. Therefore, in reality an estimate of the channel impulse response that is based on a single transmitted impulse will always be noisy. It is possible to reduce the effect of the noise by making many measurements of the impulse response and taking the average of the measured impulse responses. The transmitted impulses must be spaced sufficiently far apart so that the resulting channel impulse responses do not overlap. It can therefore take some time to send enough impulses to build up sufficiently good channel estimates.

An alternative approach to sending a single impulse or a sequence of impulses is to send probing sequences with auto- or cross-correlation properties that make it easy to extract the channel impulse response from the received signal. Good examples of such sequences are the maximum length sequences or m-sequences [Ziemer 1985]. The m-sequences can be constructed in a simple manner using the same basic structure as used for self-synchronizing scramblers.

What makes the m-sequences particularly good signals for channel probing are their auto-correlation properties. If an m-sequence is mapped to a 2-PAM signal, $s_m[n]$, with levels 1 and -1, then the auto-correlation function of an m-sequence of length (period) M is

$$R_m[n] = \sum_{m=0}^{N-1} s_m[m]s_m[(m-n)_{\bmod N}] = \begin{cases} N, & (m-n)_{\bmod N} = 1 \\ -1, & \text{otherwise} \end{cases}. \tag{11.149}$$

This property can be utilized for channel probing as follows. The channel probing transmit signal, $s_m[n]$, is generated by mapping a periodic m-sequence with period N to a 2-PAM constellation. The signal is distorted by the channel, resulting in the received signal

$$y[n] = s_m[n_{\bmod N}] \star h[n] + v[n]. \tag{11.150}$$

The received signal is then passed through a filter with coefficients consisting of one period of the m-sequence (time reversed). The resulting filter output is given by

$$\begin{aligned} h_p[n] &= \sum_{m=0}^{N-1} s_m[N-1-m]y[n-m] \\ &= \sum_{m=0}^{N-1} s_m[N-1-m]\left(\sum_k h[k]s_m[(n-m-k)_{\bmod N}] + v[n-m]\right) \\ &= \sum_k h[k]R_m[n-k] + \sum_{m=0}^{N-1} s_m[N-1-m]v[n-m] \\ &= N \cdot h[n_{\bmod N}] - h_{mean} + \sum_{m=0}^{N-1} s_m[N-1-m]v[n-m], \end{aligned} \tag{11.151}$$

where h_{mean} is the mean of $h[n]$ over all n. For noise with mean m_v and variance σ_v^2, the noise term after filtering will have the same mean and variance σ_v^2/N. Ignoring the noise term, the filter output is the desired impulse response minus an offset. The offset can be determined from the correlation filter output signal by observing that

$$mean\,\{h_p[n]\} = mean\,\{N \cdot h[n]\} - h_{mean} + mean\,\{v[n]\} = (N-1)h_{mean} + v_{mean}. \tag{11.152}$$

The channel impulse response $h[n]$ can be approximated based on the filtered probe signal, $h_p[n]$, according to

$$\tilde{h}[n] = \frac{1}{N}\left(h_p[n] - \frac{1}{N-1}(mean\,\{h_p[n]\} - v_{mean})\right). \tag{11.153}$$

Therefore, the channel impulse response can be derived directly from the output of the correlation filter. If the channel impulse response is shorter than the period of the m-sequence, then Equation 11.153 provides an accurate reconstruction of the channel impulse response. If the channel impulse response is longer than one period of the m-sequence, then the impulse response estimate is distorted by the tail of the impulse response that overlaps and adds to the signal according to

$$\tilde{h}[n] = \sum_k h[n+kN]. \tag{11.154}$$

The channel impulse response for a DSL channel is usually characterized by a decaying exponential and is, therefore, IIR in nature. The m-sequence needs to be sufficiently long so that the "folding" of the impulse response tail, according to Equation 11.154, does not cause significant degradation in the impulse response estimate.

The m-sequence channel probing described above is based on using PAM signals. It is possible, however, to do the same kind of probing for passband signals. There exist complex signals, such as constant amplitude zero auto-correlation (CAZAC) sequences (Milewski sequences) [Milewski 1983], [Chevillat 1987], which, like m-sequences, have auto-correlation properties that make them well suited for channel probing. It is also possible to map m-sequences to complex constellations and do probing that way. For example, probing signals used in ADSL (for example, in ITU Recommendation G.992.1 [ITU-T G.992.1]) are based on modulating a pseudo-random sequence generated by the scrambler polynomial $1 + x^{18} + x^{23}$. This pseudo-random sequence is actually an m-sequence. The ADSL probing sequence is actually much shorter than the $2^{23} - 1$ bits, which is the period of the 23rd-order m-sequence, so only part of the sequence is used. However, the ADSL probing signals are made periodic by repeating them over and over again for each symbol. This generates a "white" periodic probing signal with nice properties for channel probing.

11.6 Adaptive Equalization

The methods in Section 11.5 for obtaining the equalizer coefficients were based on solving closed-form equations for a given channel response. It was observed that the channel impulse response is usually not known beforehand but can be obtained by using channel probing. However, it is frequently desirable to obtain the equalizer coefficients directly based on the received signal, without the intermediate step of determining the channel impulse response through channel probing. Adaptive equalization derives the best equalizer taps directly from the received signal [Qureshi 1985] without the intermediate step of determining the channel impulse response. In DSL systems, it is probably far more common to obtain the equalizer coefficients through adaptive algorithms than it is to use closed-form algorithms. There are several reasons why adaptive algorithms are more common than closed-form algorithms for obtaining equalizer coefficients in DSL systems. The main reason for the popular use of the adaptive algorithms is their proven track record. Adaptive algorithms have been used for a long time, and they have proved to be robust and give good performance. Another reason for using adaptive algorithms is that they tend to fit well into the structure of equalizer hardware, whereas closed-form algorithms tend to use some processing that is "different" from the equalizer hardware structure and may therefore need special hardware. As DSL technology is becoming more and more based on software implementations on digital signal processors, this is becoming less of an issue. Another benefit of adaptive algorithms is that they can be updated during normal data mode (Showtime) operation of the DSL modem, allowing the modem to track and adjust to changes in channel conditions. The main advantage of closed-form algorithms is that channel probing and closed-form equalizer coefficient design can usually be done in much less time than is needed for adaptive equalizers to converge.

11.6.1 The LMS Algorithm

The most common adaptive algorithm for equalizers is the least mean-square algorithm [Qureshi 1985][Alexander 1986][Haykin 1996]. The LMS algorithm is very robust and almost always gives good performance. This along with the simplicity of its implementation has made the LMS algorithm extremely popular for adaptive filtering of all kinds, including adaptive equalization. Even though there exist other well-known adaptive algorithms, such as the recursive least squares (RLS) algorithms [Alexander 1986][Cioffi 1984], which can have faster convergence and sometimes better performance, the LMS algorithm and its variants continue to be the most commonly used algorithms for adaptive equalization.

The update formula for the LMS algorithm is very simple. Assume that $\mathbf{w}$ is a coefficient vector that is used to generate output signal $y[n]$, based on input vector $\mathbf{x}[\mathbf{n}]$ according to

$$y[n] = \mathbf{w}^T \mathbf{x}[n]. \tag{11.155}$$

Also assume that it is desired to minimize the mean-square of the error

$$e[n] = d[n] - y[n]. \tag{11.156}$$

Then the LMS update formula is given by

$$\mathbf{w}[n+1] = \mathbf{w}[n] + \alpha\, e^*[n]\mathbf{x}[n], \tag{11.157}$$

where α is the step-size coefficient that controls the convergence rate of the algorithm.

To understand how and why the LMS algorithm works, it is useful to look at it as a simplification of the steepest descent algorithm. The mean-square error is given by

$$E = E\{e^2[n]\} = E\{(d[n] - \mathbf{w}^T \mathbf{x}[n])^2\}, \tag{11.158}$$

which can be rewritten as

$$E = E\{d^2[n]\} - 2\mathbf{w}^T E\{d[n]\mathbf{x}[n]\} + \mathbf{w}^T E\{\mathbf{x}[n]\mathbf{x}^T[n]\}\mathbf{w}. \tag{11.159}$$

The aim is to find $\mathbf{w}_*$ that minimizes the mean-square error. The steepest descent algorithm [Alexander 1986] is based on successive approximation of the optimum $\mathbf{w}_*$, where in each step the approximation is improved by moving $\mathbf{w}$ in the direction of steepest descent in the error surface. In other words, $\mathbf{w}$ is updated in each iteration by moving it one step toward the minimum error. The direction toward the minimum is given by the opposite direction to the gradient of the error surface, and the update formula for the steepest descent algorithm is given by

$$\mathbf{w}[n+1] = \mathbf{w}[n] - \mu \nabla_w E, \tag{11.160}$$

where $\nabla_w E$ is the error surface gradient and μ is the update step size. In the steepest descent algorithm, the gradient is usually computed according to [Alexander 1986][Haykin 1996]

$$\nabla_w E = -2E\{d[n]\mathbf{x}[n]\} + 2E\{\mathbf{x}[n]\mathbf{x}^T[n]\}\mathbf{w}. \tag{11.161}$$

An alternative expression for the gradient is

$$\nabla_w E = \frac{\partial}{\partial \mathbf{w}^*} E\{e^2[n]\} = E\left\{2e[n]\frac{\partial}{\partial \mathbf{w}^*}e[n]\right\} = -2E\{e[n]\mathbf{x}[n]\}. \tag{11.162}$$

When the steepest descent algorithm is used, the expected values in Equation 11.161 or Equation 11.162 must be known beforehand. The problem is that these expected values are not always known and must therefore be approximated. Based on Equation 11.162, one obvious approximation for the gradient is

$$\nabla_w E = -2E\{e[n]\mathbf{x}[n]\} \approx -2e[n]\mathbf{x}[n]. \tag{11.163}$$

This is not a very good approximation for each individual value of $e[n]\mathbf{x}[n]$, but the expected value of $e[n]\mathbf{x}[n]$ is the correct gradient. Therefore, on average, $e[n]\mathbf{x}[n]$ is a good approximation of the gradient. Now if this approximation of the gradient is substituted into the steepest descent update formula Equation 11.160, then the result is the LMS update

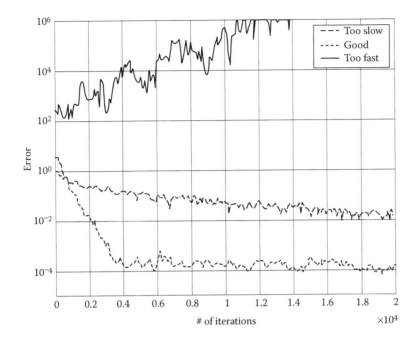

FIGURE 11.9
Convergence of the LMS algorithm with different update coefficients.

formula Equation 11.157. In other words, the LMS update formula is an approximation of the steepest descent update formula. The difference is that unlike steepest descent, where the expected values must be known beforehand, the LMS update is based on observed values each time.

Because the LMS update is based on somewhat random values, and not the actual gradient, the update step-size α must be chosen with care. The value for α must be sufficiently small to ensure that the LMS algorithm converges instead of randomly jumping around or even diverging. But α must also be sufficiently large so that the LMS algorithm converges in reasonable time. Figure 11.9 shows how the "convergence" of the LMS algorithm depends on α. If α is chosen too small, the LMS algorithm converges very slowly, but if α is chosen too large, the LMS algorithm may diverge instead of converging. If α is chosen just right, then the LMS algorithm converges fairly quickly to its "noise floor" but does not improve much beyond that.

The convergence rate of the LMS algorithm can be analyzed based on the statistical behavior of $x[n]$ and $d[n]$, but such analysis tends to become rather complicated. In real applications, it is very common to base the selection of the update step-size α on empirical results and experience. As a result, the tuning of LMS algorithms is often much more an art than a science.

The tuning of the step-size can become especially difficult if there is much variation in the $x[n]$ values or if $x[n]$ has high variance. In this case, the update steps become inherently large when $x[n]$ has large values and small when $x[n]$ has small values. Such variations in the $x[n]$ values can be compensated for by normalizing the update error by $x^T[n]x[n]$. In this case, the update formula for the normalized LSM becomes

$$\mathbf{w}[n+1] = \mathbf{w}[n] + \alpha \frac{e^*[n]\mathbf{x}[n]}{\mathbf{x}^T[n]\mathbf{x}[n]}. \tag{11.164}$$

This normalized LMS has much more predictable behavior and it is much easier to predict its convergence behavior.

In most communication systems, it is common to use automatic gain control (AGC) to scale the input signal before equalization is done. The ideal AGC adjusts the signal level such that the average signal level after the AGC is constant and independent of the attenuation of the transmission channel. The constant average signal level allows better utilization of the dynamic range of the digital representation. This is especially important if fixed point arithmetic is used. An additional benefit of the AGC is that the input signal **x[n]** into the equalizer becomes normalized, and there is no need for the normalization in Equation 11.164. Therefore, not only does the AGC improve the utilization of the dynamic range, but it also normalizes the LMS update.

The convergence characteristics of the LMS algorithm have been studied extensively [Haykin 1996]. Yet in real communication systems, the tuning of the LMS algorithm tends to be more of an art then a science. This is because much of the theoretical analysis is based on assumed knowledge about correlation matrices and other characteristics of the signals. In real life, these characteristics are usually not known at design time, so rules based on specific signal characteristics are often of limited direct value, even though they can give good insight into the expected behavior of the LMS algorithm.

One of the most interesting observations about LMS convergence characteristics states that the convergence rate for normalized LMS is approximately [Schultheiss 1988]

$$E_n = E\{e^2[n]\} = E_0 \cdot \left(1 - \frac{\alpha(2-\alpha)}{N}(1-b^2)\right)^n, \tag{11.165}$$

where α is the update step-size, N is the length of **w**, and b is a coefficient dependent on how colored **x[n]** is (with $b = 0$ for white signals). This equation was derived for an autoregressive (AR) signal, but it is a good approximation of the normalized LMS behavior for many other types of signals. Examination of Equation 11.165 provides interesting insight into the convergence characteristics of the normalized LMS algorithm. One thing that can be concluded from Equation 11.165 is that $\alpha = 1$ is the update rate that gives the fastest convergence for the normalized LMS and that for $\alpha > 2$, the normalized LMS algorithm can become unstable.

The error in Equation 11.165 can be expressed in dB as

$$E_n = 10\log_{10}(E_0) + n \cdot 10\log_{10}\left(1 - \frac{\alpha(2-\alpha)}{N}(1-b^2)\right) \text{ [dB]}. \tag{11.166}$$

Applying the series expansion

$$\ln(1+a) = a - \frac{1}{2}a^2 + \frac{1}{3}a^3 - \dots, \tag{11.167}$$

to Equation 11.166 gives the following approximation for the error estimate (valid for reasonably large N):

$$E_n \approx 10\log_{10}(E_0) - n\frac{\alpha(2-\alpha)}{N}(1-b^2) \cdot 4.34 \text{ dB}. \tag{11.168}$$

This equation gives a simple rule of thumb for the convergence rate of the normalized LMS algorithm. In particular, for white signals ($b = 0$) and $\alpha = 1$, the error goes down by approximately 4 dB for every N iterations. Another thing to observe is that the convergence time increases linearly with increasing filter length.

The error estimate in Equation 11.165 only holds while the error is considerably higher than the noise floor. Once the error approaches the noise floor, the convergence slows down.

For white signals and $\alpha = 1$, the normalized LMS will converge to error level approximately 3 dB above the noise level. By decreasing α, the LMS algorithm will converge closer to the noise floor. In LMS training it is common to start out with fast adaptation ($\alpha = 1$ for normalized LMS) and then gradually lower the update rate α as the error approaches the error floor.

11.6.2 Equalizer Training

The LMS algorithm can be used to obtain equalizer coefficients in equalizer training. Figure 11.10(a) shows an adaptive linear equalizer configuration. The coefficients of the

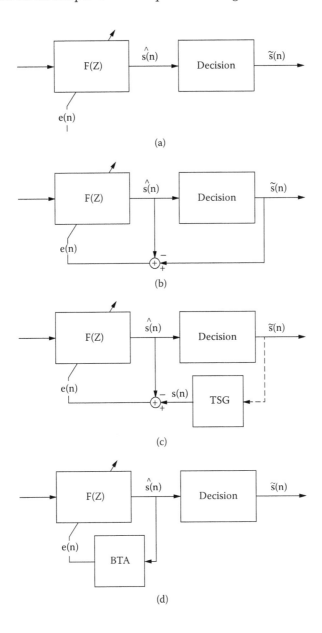

(a)

(b)

(c)

(d)

FIGURE 11.10
Adaptive equalizer configurations: (a) generic adaptive equalizer structure; (b) adaptation based on decision error (slicer error); (c) adaptation using training sequence generator (TSG); (d) adaptation using blind training adaptation.

feedforward filter $f[n]$ are updated based on the error signal $e[n]$. This update would typically be done using the LMS algorithm but could be done using other adaptive algorithms. The error signal $e[n]$ may be constructed in several different ways. How the error signal is constructed depends on how the equalizer training is configured.

Ideally, the update error signal would be constructed as

$$e[n] = s[n] - \hat{s}[n]. \qquad (11.169)$$

However, this assumes that the transmit symbols $s[n]$ are available in the receiver, which may not be the case. Therefore, the error cannot usually be constructed directly according to Equation 11.169, and additional steps must be taken to construct/estimate the update error signal. There are three main methods that are used to construct the error signal: one is to use decision error (slicer error), another is to transmit deterministic training signals (for example, a pseudo-random signal) that can be reconstructed in the receiver, and the third method is to use known signal properties to generate an update error signal that may be wrong at any given instance, but is correct on the average (blind training). Figure 11.10 shows the reference configuration for each of these three methods.

Figure 11.10(b) shows a configuration for updating the equalizer coefficients based on the decision error (slicer error). If the transmit signal, $s[n]$, can be accurately regenerated in the receiver, then this is equivalent to generating the error signal directly according to Equation 11.169, and this configuration would essentially be equivalent to forced training using a training sequence generator (TSG) as illustrated in Figure 11.10(c). On the other hand, if the error probability is high, then at any given instance the error signal may not be a correct estimate of the error signal in Equation 11.169, but on the average the error estimate will be correct. In this case, the configuration in Figure 11.10(b) behaves more like the blind training configuration in Figure 11.10(d).

In normal data mode operation (Showtime), the equalizers have usually been trained reasonably well, and the decision values become a good estimate of the transmit symbols. This makes the configuration in Figure 11.10(b) well suited for updating the equalizer coefficients to track changing loop conditions. During initial start-up, on the other hand, the error probability in the decision output is usually very high and the configuration in Figure 11.10(b) behaves more like blind equalizer training.

If the transmitted training sequence has a known generating structure and the state of the generating structure can be determined in the receiver, then the training sequence can be regenerated in the receiver and used for forced equalizer training. Figure 11.10(c) shows an equalizer training configuration where a TSG is used to reconstruct the training sequence $s[n]$. Examples of training sequences that could be reconstructed relatively easily in the receiver are periodic sequences and pseudo-random sequences. If the training sequence is a periodic sequence, then the receiver needs to synchronize to the period of the transmit sequence and can then perfectly reconstruct the transmitted training sequence internally. If a pseudo-random training sequence is used, then the receiver needs to synchronize to the "seed" of the pseudo-random generator. This is particularly easy if the pseudo-random generator is based on self-synchronizing scramblers.

11.6.3 Blind Equalizer Training

Blind equalization is usually based on using some known statistical or structural properties of the transmit signal $s[n]$ to update the equalizer coefficients. One of the oldest and best-known blind equalization algorithms is the Sato algorithm [Sato 1975][Ding 2001]. The update formula for the Sato algorithm is

$$\mathbf{f}[n+1] = \mathbf{f}[n] + \alpha\,\hat{e}[n]\mathbf{r}[n], \qquad (11.170)$$

where

$$\hat{e}[n] = \hat{s}[n] - S_1 \cdot sign(\hat{s}[n]), \tag{11.171}$$

and

$$S_1 = \frac{E\{s^2[n]\}}{E\{|s[n]|\}}. \tag{11.172}$$

The Sato algorithm can be considered to be a special version of the LMS algorithm. Instead of assuming knowledge about each transmitted symbol, the update is only based on the equalized signal and knowledge about the statistical properties of the transmit signal.

The BGR algorithm [Benveniste 1980][Ding 2001] is an extension of the Sato algorithm such that

$$\hat{e}[n] = \psi(\hat{s}[n]) - S_\psi \cdot sign(\hat{s}[n]), \tag{11.173}$$

where

$$S_\psi = \frac{E\{\psi(s[n])s[n]\}}{E\{|s[n]|\}} \tag{11.174}$$

and $\psi(.)$ is an odd twice differentiable function such that

$$\frac{d^2\psi(x)}{dx^2} \geq 0, \quad \forall x \geq 0. \tag{11.175}$$

The Sato algorithm is clearly a special case of the BGR algorithm with $\psi(x) = x$.

The Sato and BGR algorithms as described above are specific to real (PAM) signals, but can easily be extended to complex signals by applying the error function to the real and imaginary terms separately according to

$$\hat{e}[n] = (\psi(\Re\{\hat{s}[n]\}) - S_r \cdot sign(\Re\{\hat{s}[n]\})) + j(\psi(\Im\{\hat{s}[n]\}) - S_i \cdot sign(\Im\{\hat{s}[n]\})). \tag{11.176}$$

Another extension of the Sato algorithm is the so-called Godard or constant modulus algorithms [Godard 1980][Ding 2001]. In the Godard algorithm, the coefficients are again updated according to Equation 11.170, but the error estimate is given by

$$\hat{e}[n] = \left(|\hat{s}[n]|^q - S_q\right) \cdot |\hat{s}[n]|^{q-2} \hat{s}[n], \tag{11.177}$$

with

$$S_q = \frac{E\{s^{2q}[n]\}}{E\{|s[n]|^q\}}. \tag{11.178}$$

For $q = 1$, the error estimate again becomes the Sato error estimate. For $q = 2$, the error estimate becomes

$$\hat{e}[n] = \left(|\hat{s}[n]|^2 - S_2\right) \cdot \hat{s}[n], \tag{11.179}$$

which clearly becomes zero when the modulus of $\hat{s}[n]$ is equal to the constant S_2. Therefore, this algorithm is sometimes referred to as a constant modulus algorithm.

There exist a number of other blind equalization algorithms, but addressing all of them is beyond the scope of this chapter. For an excellent treatment of blind equalizer algorithms, see [Ding 2001].

11.6.4   Example Equalizer Training Procedure

In most (if not all) DSL systems, part of the "activation sequence" is intended for equalizer training. In PAM-based DSL systems such as HDSL [ITU-T G.991.1], HDSL2 [T1.418-2000], SDSL, and SHDSL [ITU-T G.991.2] [ETSI TS 101 524], part of the startup sequence is a transmission of pseudo-random 2-PAM signals that are intended (among other things) for equalizer training. The pseudo-random sequence is generated by feeding constant signal (0 or 1) into a self-synchronizing scrambler to generate a "random" sequence. Because the 2-PAM sequence only has two levels, it is relatively easy to do initial blind training on the sequence. For the 2-PAM case, a (blind) decision-directed equalizer training is equivalent to doing blind training using the Sato algorithm (see Section 11.6.3). In this case the blind decision-directed training is usually sufficient to obtain the initial equalizer coefficients, and there is no need for more advanced blind equalization algorithms.

Once the equalizer coefficients have been trained sufficiently well, the probability of detection error becomes fairly low. Then the self-synchronizing property of the scrambler can be utilized to lock the descrambler to the received signal. Once the descrambler has been locked to the transmit sequence, the descrambler can be used to perfectly reconstruct in the receiver the transmitted pseudo-random sequence. The reconstructed pseudo-random sequence can now be used to do forced equalization training using the configuration in Figure 11.10(c).

The approach described above leads to a relatively simple startup procedure, where existing system components are utilized to speed up the convergence and enhance the robustness of the equalizer training. It is possible to use a similar approach for QAM and CAP systems, using 4-QAM constellations and self-synchronizing scramblers.

11.7   Examples and Practical Design Issues

It is possible to use the methods discussed in Chapters 2, 3, and 4 to construct simulation models for DSL channels. These models can be very useful in evaluating how various equalization configurations will perform for real DSL systems.

Figure 11.11 shows the impulse response and the frequency response of a typical DSL loop. This is a 2 km (6 kft) long 0.4 mm (26 AWG) loop with signal transformers at either end (loop parameters are taken from [ETSI TS 101 388]). In the following examples, this loop is used as a benchmark to compare various equalization configurations. These examples show typical behavior of real DSL systems and can help improve the understanding of how various equalizer configurations will behave in real systems. However, due to the diversity of real-life conditions, these examples only show how the equalization configurations might behave on a subset of real-life DSL loops. Real DSL designs require considerably more evaluation than just these simple examples.

All the examples below assume a baseband signal with sampling rate of 2 MHz (bandwidth of 1 MHz) and a transmit power of -40 dBm/Hz across the 1 MHz bandwidth. This does not correspond to any real DSL system, but the symbol response is representative of real DSL symbol responses. In the examples, the background noise is assumed to be white -110 dBm/Hz noise. This is not very representative for the highly colored cross-talk noise but was chosen to keep the examples simpler.

11.7.1   Linear Versus Decision Feedback Equalizers

As discussed in Section 11.4.2, decision feedback equalizers provide a significant performance improvement over linear equalizers. Figure 11.12 shows a comparison of the

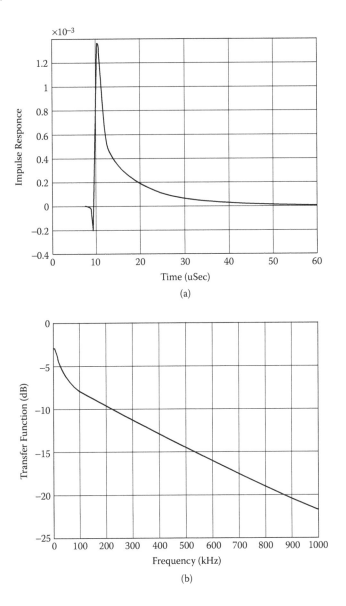

FIGURE 11.11
Example impulse response and frequency response of 2 km long 0.4 mm (26 AWG) wire.

performance of LE and DFE systems. For this example, the DFE has 6–7 dB better per-formance than an LE of comparable complexity. It is interesting to observe that increasing the number of LE taps beyond 30 taps does not improve the LE performance significantly, and the SNR for the LE is limited to approximately 15.5 dB. For comparison, the SNR for the DFE is limited to about 22.5 dB. This example clearly demonstrates the performance improvement that the DFE gives over the LE. Similar improvements can be expected for precoded systems (see Section 11.4.4) and for MLSE (see Section 11.4.5).

11.7.2 Zero-Forcing Versus MMSE Equalization

To better understand how equalizers work, it is interesting to look at how different equalizer configurations perform their two main tasks: to eliminate ISI and to minimize the effect

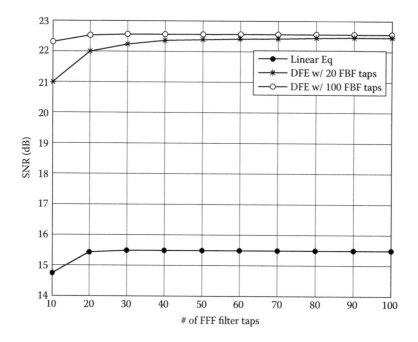

FIGURE 11.12
Comparison of performance of linear equalizers and DFE.

of noise. Figure 11.13 compares how well four different equalizer configurations eliminate ISI. The SNR values for these examples are listed in the "no noise" column in Table 11.1. To better illustrate the various deficiencies of the different equalizer structures, the equalizers were deliberately chosen such that they did not perform very well. To this end, the equalizers were chosen to have relatively few taps. The linear equalizers in Figure 11.13(a) and (c) have 30 FFF taps (and obviously no DFF taps). The DFEs in Figure 11.13(b) and (d) have 20 FFF taps and 10 DFF taps. Ideally, all samples should be zero except for $n = 20$, because any fluctuations from zero correspond to ISI. As can be seen in Figure 11.13(a), the ZF-LE does a good job of inverting the channel response and eliminating ISI. The signal to "noise" ratio (the noise is actually ISI) for the ZF-LE in this case is a respectable 62.4 dB. The equalization of a ZF-DFE is shown in Figure 11.13(b). Again, there is very little ISI in the equalized signal and, as expected, the DFE does even better than the LE and has SNR of 72.7 dB.

Figure 11.13(c) shows the result of equalizing the channel with a 30 tap MMSE-LE. The MMSE-LE does not eliminate the ISI as well as the ZF equalizers. This is to be expected, because unlike the ZF criterion, the MMSE criterion does not only consider the minimization of ISI, but also the effects of additive noise. Figure 11.13(d) shows equalization with MMSE-DFE. The MMSE-DFE does a fairly good job of eliminating the ISI, but with 53.4 dB SNR it still has 9 dB more ISI noise than the ZF-LE and almost 20 dB more ISI noise than the ZF-DFE.

Figure 11.14 shows the performance of equalizers in the presence of noise. This figure was generated using exactly the same equalizers as were used in Figure 11.13. The only difference is that this time the signal is noisy. The SNR values for these examples are given in the "with noise" column in Table 11.1. Comparison of the plots for the ZF-LE and the MMSE-LE shows that the MMSE-LE suppresses the noise slightly better than the ZF-LE and has almost 1 dB better SNR in the presence of noise. This is because unlike the ZF criterion, the MMSE criterion considers the effect of the additive noise on the SNR.

Figure 11.14(b) shows a signal equalized by ZF-DFE. It is clear that in this figure the ZF-DFE fails to equalize the signal. The reason for this is error propagation in the decision

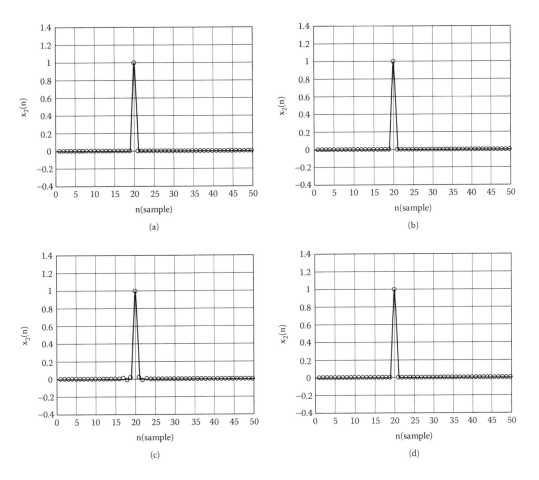

FIGURE 11.13
Comparison of different equalizer configurations in the absence of noise: (a) ZF-LE; (b) ZF-DFE; (c) MMSE-LE; (d) MMSE-DFE.

feedback path. At sample 7, a decision error occurs, where the −1 is detected instead of 0. As a result, a wrong value is used in the DFF, which in turn causes the following sample to be wrong. This again causes the sample after that to be wrong, and so on. The single error at sample 7 propagates throughout the rest of the signal. This kind of error propagation is probably the biggest weakness of the DFE structure, because it causes the DFE to fail altogether until it somehow reaches an error-free state again.

Figure 11.14(d), on the other hand, shows the strength of the decision feedback equalization structure. The MMSE-DFE clearly performs much better than any of the other equalizers

TABLE 11.1

Comparison of Different Equalizer Configurations

Equalizer Configuration	SNR, in dB (no noise)	SNR, in dB (with noise)
ZF-LE	62.4	15.2
ZF-DFE	72.7	—
MMSE-LE	42.7	16.1
MMSE-DFE	53.4	21.6

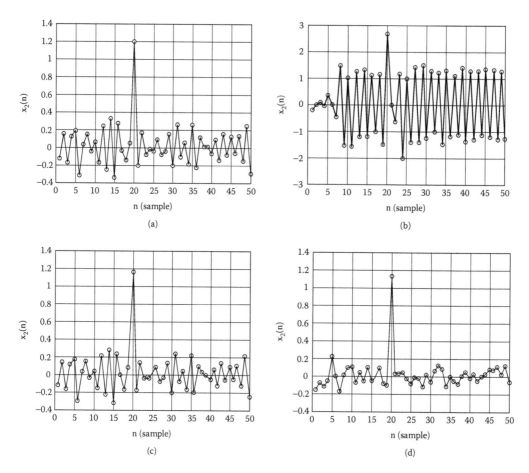

FIGURE 11.14
Comparison of different equalizer configurations in the presence of noise: (a) ZF-LE; (b) ZF-DFE; (c) MMSE-LE; (d) MMSE-DFE.

in Figure 11.14. The SNR for the MMSE-DFE is 21.6 dB compared to 16.1 dB for the MMSE-LE, which is a difference of 5.5 dB.

11.7.3   Fractionally Spaced Versus T-Spaced Equalizers

As discussed in Section 11.4.7, the sampling rate into the feedforward filter can be the same as the symbol rate, or it can be higher than the symbol rate. Equalizers with sampling rate equal to the symbol rate are referred to as T-spaced equalizers. Equalizers with sampling rate higher than the symbol rate are referred to as fractionally spaced equalizers. The main disadvantage with using fractionally spaced equalizers is that all processing before the equalization, including the ADC, must be done at a higher rate than would be needed for T-spaced equalizers. This can increase the cost and power consumption of systems based on fractionally spaced equalizers. However, fractionally spaced equalizers have several advantages that often can justify this increased complexity.

One advantage of fractionally spaced equalizers over T-spaced equalizers is relative immunity to sampling phase. In this context, sampling phase refers to the time offset of the sampling instance relative to the symbol clock. For more detailed discussion about sampling phase and timing recovery, see Chapter 12.

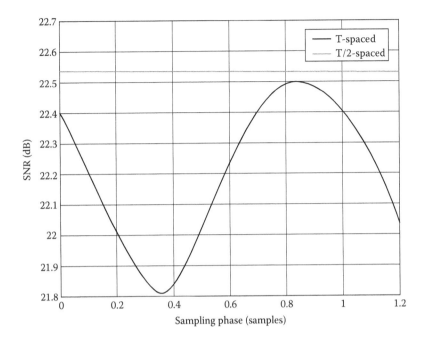

FIGURE 11.15

Comparison of the affect of sampling phase on the performance of T-spaced versus fractionally spaced equalizers.

Figure 11.15 shows how the SNR of the equalized signal changes with sampling phase. For this particular example, the SNR for T-spaced equalizers changes by approximately 0.7 dB from worst to the best sampling phase. For many channel responses, this difference can be even bigger, especially for equalizers with relatively few equalizer taps. The SNR for the T/2-spaced equalizer, on the other hand, hardly changes at all with different sampling phase. This immunity to relative sampling phase can simplify the timing recovery for fractionally spaced equalizers.

It is a common misconception that T/2-spaced equalizers need twice the number of FFF taps needed for T-spaced equalizers to achieve the same performance. This misconception is based on the observation that because T/2-spaced equalizers have two times the sampling rate, the channel response is two times longer in terms of samples, and therefore the FFF must be twice as long. The flaw in this reasoning is that equalization is all about channel inversion, and the FFF taps correspond to the inverse of the channel response and not to the channel response itself. Therefore, a longer channel response does not necessarily imply a longer FFF.

Figure 11.16 shows a comparison of the performance of T-spaced and T/2-spaced equalizers with different numbers of equalizer taps. It is interesting to observe that when only 20 DFF taps are used, the T-spaced equalizers have better performance given the same number of FFF taps. There are even cases where the T/2-spaced equalizer needs almost twice the number of taps needed for T-spaced equalizers. For example, the T/2-spaced equalizer needs almost 40 FFF taps to achieve the same performance as a T-spaced equalizer with 20 FFF taps, so for this particular case, the misconception discussed above is actually true. However, as the number of equalizer taps (and performance) increases, the difference between T-spaced and T/2-spaced equalizers disappears. With 20 DFF taps and 50 or more FFF taps, T-spaced and T/2-spaced equalizers have virtually the same performance. With 100 DFF taps, the T/2-spaced equalizers perform better than the T-spaced equalizers with the same number of FFF taps.

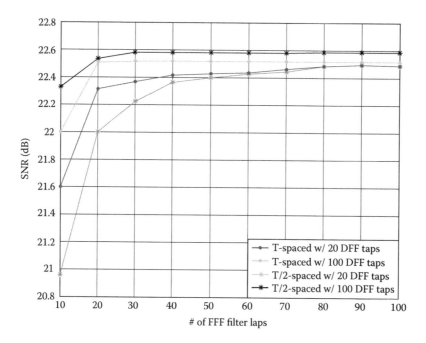

FIGURE 11.16
Comparison of performance of T-spaced versus fractionally spaced equalizers.

The results in Figure 11.16 are specific to the loop and noise characteristics used in this example, but Figure 11.16 demonstrates a simple rule of thumb about the relative complexity of T-spaced and T/2-spaced equalizers. For high performance equalizers, the T/2-spaced equalizers perform the same or better than T-spaced equalizers with the same number of taps, whereas for moderate-performance equalizers, the T/2-spaced equalizers need about 20 to 50 percent more FFF taps than T-spaced equalizers to archive the same performance.

It is interesting to observe in Figure 11.16 that for the 100 tap DFF cases, the performance of both the T-spaced and the T/2-spaced equalizers has saturated and does not increase with increasing number of FFF taps. This is because the equalizers are approaching the theoretically maximum performance. Another interesting observation is that the T/2-spaced equalizers perform slightly better than the T-spaced equalizers (there is about 0.065 dB difference for the example in Figure 11.16). This difference is because the receive filter is not perfect, and there is some signal energy at frequencies above the Nyquist frequency (half the sampling rate). For T-spaced equalizers, this "excess bandwidth" is folded into the baseband signal and causes an aliasing effect that can slightly degrade the signal quality. The T/2-spaced equalizers, on the other hand, are over-sampled, so there is no aliasing at sampling time. The T/2-spaced equalizers can even make use of the aliasing by aligning the phase of the aliasing signal so that it adds to the "baseband" signal in a constructive way, slightly enhancing the overall SNR.

Matched filters can also be used to make constructive use of signal aliasing. The impulse response of the MF is the time reversed channel response (see Equation 11.23), so the overall response of the channel and the MF is inherently linear phase. This implies that all frequencies are delayed equally, which in turn means that the aliasing adds to the "baseband" signal in a constructive way. Therefore, T-spaced equalizers with analog MF before sampling should be able to achieve approximately the same performance as T/2-spaced equalizers.

Figure 11.17 compares the performance of T-spaced equalizers with an analog MF with the performance of T-spaced and T/2-spaced equalizers with a reasonably good Butterworth

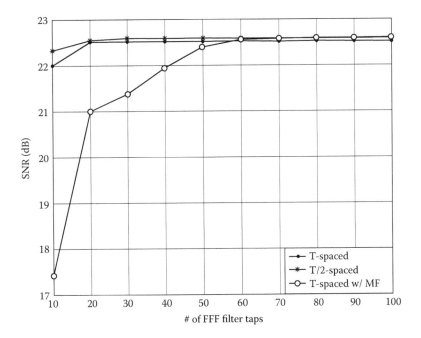

FIGURE 11.17
The effect of using matched filter for T-space equalizers (assuming 100 DFF taps).

anti-aliasing filter. The introduction of the MF lengthens the impulse response and makes the equalization more difficult. Therefore, the T-spaced equalizer with an analog MF performs considerably worse when relatively short FFFs are used. However, for long FFF (more than 60 taps), the MF improves the performance of the T-spaced equalizer, and its performance becomes virtually as good as for T/2-spaced equalizers.

This experiment demonstrates three things. First, it shows that it is possible to improve the performance of T-spaced equalizers by using an analog MF before sampling the signal. Second, it shows that this improvement is very minor and that any reasonably good anti-aliasing filters can do virtually as well as matched filters. Third, it demonstrates that fractionally spaced equalizers do not "need" MF, because fractionally spaced equalizers can themselves provide directly any performance gain that MF may provide for T-spaced equalizers.

11.7.4 Impulse Shortening Equalization

As discussed in Section 11.4.7, impulse shortening equalizers (ISE), also known as time domain equalizers (TEQ) in DMT systems, "shorten" the impulse response of the channel. Ideally, the shortened impulse response is zero outside a given window, but inside the window the shortened impulse response can take on any value (except all zeros).

Table 11.2 and Figures 11.18 through 11.21 show the performance of example impulse shortening equalizers. The equalizer in Figure 11.18 is designed using the maximum SSNR method based on Equations 11.120 through 11.129. In this discussion about example ISE designs, this method will be referred to as Method 1. The equalizers demonstrated in Figures 11.19 through 11.21 are designed using the "DFE-like" approach described in Equations 11.131 through 11.137. In the following discussion, this method will be referred to as Method 2. The equalizer in Figure 11.19 was designed using the ZF criterion, where the effects of noise are ignored, but the equalizer in Figures 11.20 and 11.21 were designed using

TABLE 11.2

Comparison of Different Impulse Shortening Equalizer Design Methods

Equalizer Design Method	SSNR [dB]	SNR [dB]	SNR$_{GM}$ [dB]	Bit Rate [bits/Hz]
Method 1 (ZF)	44.9	19.3	19.7	6.64
Method 2 – ZF	99.9	17.1	19.9	6.69
Method 2 – MMSE	44.4	26.2	20.4	6.86
Method 2 – MMSE (T/2)	49.2	26.3	21.7	7.26

the MMSE criterion. The equalizer in Figure 11.21 is T/2-spaced, and all the other equalizers are T-spaced.

As can be seen by comparing Figures 11.18(a) through 11.21(a), the equalized channel response is very different for the four equalizers. However, they all have in common that almost all the energy of the equalized channel response is concentrated in 32 samples (from $n = 51$ to $n = 82$). This demonstrates that all four equalizers are doing a good job of their primary objective, to shorten the impulse response to 32 samples. The SSNR column in Table 11.2 shows how well the equalizers suppress the channel response outside the desired 32 taps. The equalizer in Figure 11.19 (Method 2 – ZF) has SSNR of 99.9 dB, which is more than 50 dB better SSNR than for the other equalizers, which still have a respectable SSNR of about 45 dB. What is particularly interesting in this context is that the equalizer in Method 1 is designed to optimize SSNR, whereas Method 2 does not explicitly attempt to maximize SSNR; still Method 2 has much better SSNR than Method 1. The reasons for this are issues related to numerical accuracy and stability of calculations. Method 1 is based on calculations that tend to be numerically sensitive. These calculations involve "inverting" matrices that are typically not of full rank (some eigenvalues are zero). A clever method for dealing with this matrix "inversion" is described in [Melsa 1996], but these calculations are still fairly sensitive to errors due to numerical accuracy. This comparison of Method 1 and Method 2 – ZF clearly demonstrates one very important issue in practical equalizers design, which is that the theoretically optimum approach may not be the optimum solution in practice if that approach is sensitive to the imperfections of practical implementations.

Figures 11.18(b) through 11.21(b) show the equalized signals in the presence of noise. As is to be expected, the noise is stronger for the ZF equalizers in Figures 11.18 and 11.19,

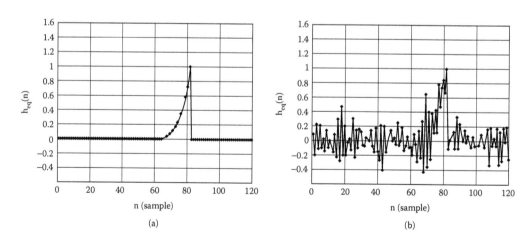

(a)   (b)

FIGURE 11.18

Impulse shortening equalization (TEQ) using equalizer design Method 1: (a) channel response after equalization; (b) channel response after equalization in the presence of noise.

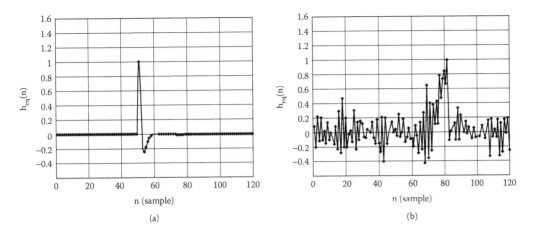

FIGURE 11.19
Impulse shortening equalization (TEQ) using equalizer design Method 2 with ZF criterion: (a) channel response after equalization; (b) channel response after equalization in the presence of noise.

than it is for the MMSE equalizers in Figures 11.20 and 11.21. This difference is clearly captured in the SNR column of Table 11.2, which shows that the MMSE equalizers have more than 6 dB better SNR than the ZF equalizers. However, the SNR out of the impulse shortening equalizer is not a direct indicator of how well DMT systems will perform. The geometric mean SNR (SNR_{GM}) is a much better indicator of the expected performance of DMT systems. The SNR_{GM} can be calculated according to Equation 11.139. Comparison of the SNR_{GM} shows that the MMSE equalizers perform a little better than the ZF equalizers and that the T/2-spaced MMSE equalizer performs slightly better than the T-spaced MMSE equalizer. The difference in performance between the four equalizers is as much as 2 dB, which is significant, but not as drastic as the difference in SSNR and "plain" SNR.

The Bit Rate column in Table 11.2 shows the maximum theoretical bit rate (the Shannon limit) given the signal and noise power spectral densities at the output of each impulse shortening equalizer. As is to be expected, these bit rate values show the same characteristics

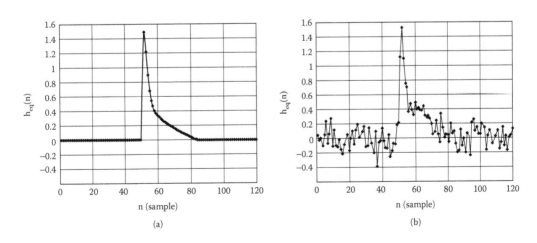

FIGURE 11.20
Impulse shortening equalization (TEQ) using equalizer design Method 2 with MMSE criterion: (a) channel response after equalization; (b) channel response after equalization in the presence of noise.

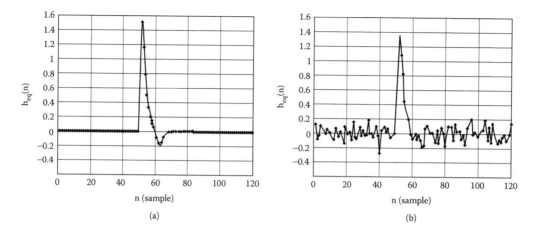

FIGURE 11.21

Impulse shortening equalization (TEQ) using equalizer design Method 2 with MMSE criterion and T/2-spaced equalizer: (a) channel response after equalization; (b) channel response after equalization in the presence of noise.

as the SNR_{GM}. Again, there is a noticeable difference between the different equalizers, with the MMSE equalizers showing better performance than the ZF equalizers, and the T/2-spaced equalizer showing better performance than the T-spaced equalizer. The same characteristics can be expected from real DMT systems, but it is important to keep in mind that the bit rate calculations in this example do not take into account the various practical limitations of real DMT systems, such as the Shannon gap, windowing effect of finite length DFT, bit-loading, etc.

The above experiments demonstrate that the SSNR and "plain" SNR of impulse shortening equalizer output are not necessarily good metrics of the quality of the ISE when used in the context of DMT systems. It is also interesting to note, as a comparison of SSNR for Method 1 and Method 2 ZF shows, that methods that should be optimal in theory may not be optimal in practice.

11.7.5 Summary

This chapter examined equalization in the context of DSL systems. Different optimization criteria lead to different equalizer designs. Among possible equalizer designs are matched filters, zero-forcing equalizers, minimum mean-square error equalizers, and maximum likelihood equalization. The MF is usually of little direct interest in DSL systems. The ZFE is interesting, but in real DSL systems MMSE or ML equalization should normally be preferred. In DSL systems, the noise is usually dominated by additive Gaussian noise, so the MMSE and the ML criteria become equivalent.

Several different equalizer structures were discussed, in particular, linear equalizers, decision feedback equalization, noise prediction, precoding, maximum likelihood sequence estimation, frequency domain equalization, and impulse shortening equalization, which is also known as time domain equalization in DMT systems. Systems based on linear equalizers do not have as good performance as DFE systems, but DFE systems can come very close to being optimal MMSE equalizers. The noise predictor systems can be shown to be equivalent to DFE systems. Precoded systems have much in common with DFE systems, but precoded systems are better suited for systems that have long decision delays (such as trellis coded systems). The MLSE can combine the equalization process with the decision/decoding process, which is particularly interesting for trellis coded systems, but

MLSE tends to be more computationally intensive than DFE or precoding. Frequency-domain equalization can provide computational savings, especially in the context of DMT modulation where frequency domain equalization can also become very flexible. The ISE is mainly used in the context of DMT systems, but ISE can also be interesting to reduce the complexity of MLSE. It is interesting to observe that ISE has much in common with the feedforward filters in DFE and precoded systems.

The chapter also presented generic algorithms that can be used to design MF, ZFE, MMSE equalizers, and ISE, given the symbol response of the channel and the noise characteristics. It is possible to use these or similar algorithms, in combination with line probing, to design equalizers on the fly in DSL modems. In real systems, however, it is far more common to use adaptive algorithms such as LMS to train the equalizers. The training can be done using blind-training, forced-training, or a combination of blind- and forced-training.

References

[Al-Dhahir 1995a] N. Al-Dhahir and J.M. Cioffi, *MMSE decision-feedback equalizers: finite-length results*, IEEE Transactions on Information Theory, Volume: 41, Issue: 4, pp. 961–975, July 1995.

[Al-Dhahir 1995b] N. Al-Dhahir and J.M. Cioffi, *Fast computation of channel-estimate based equalizers in packet data transmission*, IEEE Transactions on Signal Processing, Volume: 43, Issue: 11, pp. 2462–2473, Nov 1995.

[Al-Dhahir 1996] N. Al-Dhahir and J.M. Cioffi, *Optimum finite-length equalization for multicarrier transceivers*, IEEE Transactions on Communications, Volume: 44, Issue: 1, pp. 56–64, Jan 1996.

[Alexander 1986] S.T. Alexander, *Adaptive Signal Processing*, New York: Springer-Verlag, 1986.

[Apostol 1969] T.M. Apostol, *Calculus – Volume II*, 2nd ed., New York: Wiley, 1969.

[Belfiore 1979] C.A. Belfiore and J.H. Park Jr., *Decision Feedback Equalization*, Proceedings of the IEEE, Volume: 67, No. 8, pp.1143–1156, Aug 1979.

[Benveniste 1980] A. Benveniste, M. Goursat, and G. Ruget, *Robust identification of a nonminimum phase system: Blind adjustment of a linear equalizer in data communications*, IEEE Transactions on Automatic Control, Volume: 25, Issue: 3, pp. 385–399, June 1980.

[Chevillat 1987] P. Chevillat, D. Maiwald, and G. Ungerboeck, *Rapid training of a voiceband data-modem receiver employing an equalizer with fractional-T spaced coefficients*, IEEE Transactions on Communications, Volume: 35, Issue: 9, pp. 869–876, Sept. 1987.

[Cioffi 1984] J. Cioffi and T. Kailath, *Fast, recursive-least-squares transversal filters for adaptive filtering*, IEEE Transactions on Acoustics, Speech, and Signal Processing, Volume: 32, Issue: 2, pp. 304–337, Apr 1984.

[Cioffi 1995] J.M. Cioffi, G.P. Dudevoir, M.V. Eyuboglu, and G. D. Forney Jr., *MMSE decision-feedback equalizers and coding. I. Equalization results*, IEEE Transactions on Communications, Volume: 43, Issue: 10, pp. 2582–2594, Oct 1995.

[Crespo 1991] P.M. Crespo, and M.L. Honig, *Pole-zero decision feedback equalization with a rapidly converging adaptive IIR algorithm*, IEEE Journal on Selected Areas in Communications, Volume: 9, Issue: 6, pp. 817–829, Aug 1991.

[Ding 2001] Z. Ding and Y. Li, *Blind Equalization and Identification*, New York: Marcel Dekker, 2001.

[ETSI TS 101 388] ETSI TS 101 388 V1.3.1 (2002–05), *Asymmetric Digital Subscriber Line(ADSL) –European specific requirements*, ETSI Technical Specification, 2002.

[ETSI TS 101 524] ETSI TS 101 524 V1.2.1 (2003–03), *Symmetric Single Pair High Bitrate Digital Subscriber Line (SDSL)*, ETSI Technical Specification, 2003.

[Falconer 2002] D. Falconer, S.L. Ariyavisitakul, A. Benyamin-Seeyar, and B. Eidson, *Frequency domain equalization for single-carrier broadband wireless systems*, IEEE Communications Magazine, Volume: 40, Issue: 4, pp. 58–66, April 2002.

[Forney 1972] G. Forney Jr., *Maximum-likelihood sequence estimation of digital sequences in the presence of intersymbol interference*, IEEE Transactions on Information Theory, Volume: 18, Issue: 3, pp. 363–378, May 1972.

[Godard 1980] D. Godard, *Self-Recovering Equalization and Carrier Tracking in Two-Dimensional Data Communication Systems*, IEEE Transactions on Communications, Volume: 28, Issue: 11, pp. 1867–1875, Nov 1980.

[Golub 1989] G.H. Golub and C.F. Van Loan, *Matrix Computations*, 2nd ed., Baltimore: Johns Hopkins University Press, 1989.

[Haykin 1996] S. Haykin, *Adaptive Filter Theory*, 3rd ed., Upper Saddle River, NJ: Prentice-Hall, 1996.

[ITU-T G.991.1] ITU-T G.991.1 (10/1998), *High bit rate Digital Subscriber Line (HDSL) transceivers*, ITU-T Recommendation, 1998.

[ITU-T G.991.2] ITU-T G.991.2 (02/2001), *Single-pair high-speed digital subscriber line(SHDSL) transceivers*, ITU-T Recommendation, 2001.

[ITU-T G.992.1] ITU-T G.992.1 (06/1999), *Asymmetric digital subscriber line (ADSL)transceivers*, ITU-T Recommendation, 1999.

[Lathi 1983] B.P. Lathi, *Modern Digital and Analog Communication Systems*, New York: Holt-Saunders, 1983.

[Lee 1994] E.A. Lee and D.G. Messerschmitt, *Digital Communication*, 2nd ed., Boston: Kluwer, 1994.

[Melsa 1996] P.J.W. Melsa, R.C. Younce, and C.E. Rohrs, *Impulse response shortening for discrete multitone transceivers*, IEEE Transactions on Communications, Volume: 44, Issue: 12, pp. 1662–1672, Dec. 1996.

[Milewski 1983] A. Milewski, *Periodic sequences with optimal properties for channel estimation and fast start-up equalization*, IBM J. Res. Develop., Volume: 27, pp. 426–431, Sep 1983.

[Oppenheim 1989] A.V. Oppenheim and R.W. Schafer, *Discrete-Time Signal Processing*, Englewoods Cliffs, NJ: Prentice-Hall, 1989.

[Orckit 1999] Orckit Communications Ltd., *Combined Constellation Shaping and Reduction of Peak-to-Average Ratio*, ETSI TM6 Contribution, 993t24a1, Edinburgh, Sep 1999.

[Papoulis 2002] A. Papoulis and S.U. Pillai, *Probability, Random Variables and Stochastic Processes*, 4th ed., New York: McGraw-Hill, 2002.

[Proakis 1989] J.G. Proakis, *Digital Communication*, 2nd ed., New York: McGraw-Hill, 1989.

[Qureshi 1985] S.U.H. Qureshi, *Adaptive Equalization*, Proceedings of the IEEE, Volume: 73, No. 9, 1349–1387, Sept 1985.

[Sato 1975] Y. Sato, *A method of self-recovering equalization for multilevel amplitude-modulation systems*, IEEE Transactions on Communications, Volume: 23, Issue: 6, pp. 679–682, June 1975.

[Schultheiss 1988] U. Schultheiss, *Über die Adaption eines Kompensators für akustische Echos*, Fortschr.-Ber. VDI, Reihe 10, Nr. 90, Düsseldorf: VDI-Verlag, 1988.

[T1.418-2000] ANSI T1.418, *High Bit Rate Digital Subscriber Line—2nd Generation (HDSL2)*, ANSI Standard, 2000.

[Thormundsson 2001] T. Thormundsson, *Fast Methods for Designing MMSE-DFE and ZF-DFE from an Imperfect Channel ID using Separate MMSE and Optimization of the DFE's FFF and FBF*, Conexant Systems internal communication, Aug 2001.

[Van Acker 2001] K. Van Acker, G. Leus, M. Moonen, O. van de Wiel, and T. Pollet, *Per tone equalization for DMT-based systems*, IEEE Transactions on Communications, Volume: 49, Issue: 1, pp. 109–119, Jan 2001.

[Viterbi 1971] A. Viterbi, *Convolutional codes and their performance in communication systems*, IEEE Transactions on Communications, Volume: 19, Issue: 5, pp. 751–772, Oct 1971.

[Ziemer 1985] R.E. Ziemer and R.L. Peterson, *Digital Communication and Spread Spectrum Systems*, New York: Macmillan, 1985.

Synchronization of DSL Modems

Sverrir Olafsson

CONTENTS

12.1 Overview 352
    12.1.1 Quality of Timing Error Signal Generation 353
    12.1.2 Examples 354
12.2 PAM, QAM, and CAP Timing Recovery 355
    12.2.1 Maximum-Likelihood Timing Recovery 355
    12.2.2 Band-Edge Energy Maximization 359
        12.2.2.1 The Band-Edge Energy 360
        12.2.2.2 BETR Timing Function for QAM 362
    12.2.3 Timing Signals Based on Nonlinearity-Induced
        Spectral Line 365
        12.2.3.1 Analysis of Squarer-Induced Spectral
                Line—Without Noise 366
    12.2.4 Timing Recovery Based on Equalizer Taps 371
    12.2.5 The Gardner Timing Function 375
    12.2.6 Mueller–Müller-Based Methods 376
    12.2.7 Gradient-Descent Timing Recovery 381
12.3 DMT Synchronization 386
    12.3.1 Synchronization to the DMT Symbol 388
    12.3.2 Effect of Timing Phase on a Demodulated
        DMT Symbol 389
    12.3.3 Pilot-Based Timing Acquisition and Tracking 391
    12.3.4 Nonpilot-Based Timing Acquisition and Tracking 392
        12.3.4.1 Effect of Timing Jitter on DMT Performance 393
References 394

ABSTRACT Synchronization methods are required in most communication systems to align the timing of receivers to the clock domain of their corresponding transmitter. In DSL systems, this typically involves regenerating the network clock in the customer premises modem and using it to decode the receive signal as well as to generate the signal transmitted back to the network.

    This chapter discusses timing recovery methods to regenerate signal clocking. For single-carrier modulation, maximum-likelihood, spectral line, and decision-directed methods are addressed. Methods for DMT modulation are also presented.

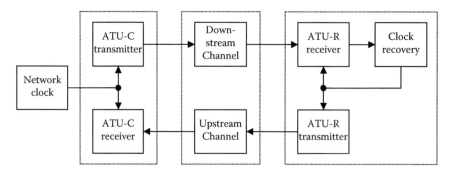

FIGURE 12.1
Typical DSL clocking configuration.

12.1 Overview

Figure 12.1 shows a typical DSL system. Most DSL systems use a clocking configuration referred to as slave timing. The network transmitter on the left operates synchronously with the network clock, generating transmit symbols at a fixed rate relative to the network timing. The customer premises equipment on the right, being driven from an independent clock source such as a crystal oscillator, must recover the symbol clock in order to demodulate the signal and deliver the received bitstream at the same rate it was transmitted. The customer equipment will then use the regenerated clock to shift in the upstream bits and generate its transmit signal. The network receiver then only needs to adapt to the sampling phase of its receive signal, as the sampling rate has been synchronized.

Figure 12.2 shows a typical timing recovery system in the customer premises transceiver (in this case, an ATU-R). The input signal is filtered and digitized, and then a timing error signal is extracted. That error signal is then filtered in a timing error loop filter before being directed to clock control circuitry. The clock control can take on different forms, but is typically a phase-locked loop (PLL) of some sort. The PLL controls the sampling phase of the input signal based on the filtered timing error. In the steady state, this closed loop system will effectively regenerate the transmit timing in the PLL, allowing digitization of samples that are synchronous with the transmit clock.

Alternatively, a fixed sample clock may be used followed by an interpolation filter to control the sampling phase. DSL systems are increasingly using interpolation filters for sampling control.

Figure 12.3 shows the slightly simpler network side (in this case, an ATU-C) clocking system. The same type of timing error signal needs to be generated, but now only to potentially

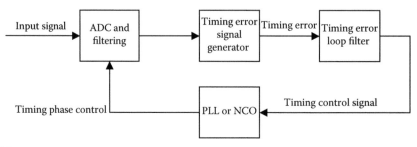

FIGURE 12.2
Typical ATU-R timing recovery system.

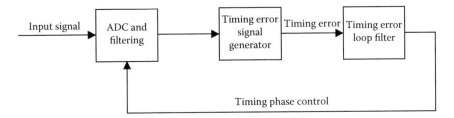

FIGURE 12.3
Typical ATU-C timing recovery system.

adjust the sampling phase to improve the quality of the digitized signal. A PLL configuration can still be used, but now it will not attempt to change the overall sampling rate. In systems with fractionally spaced equalization, no timing phase adaptation is required on the network side, as the timing phase can be adjusted by the receiver filtering adaptively.

The quality of the timing tracking depends on the following factors:

- The quality of the timing error signal generator
- The timing error signal filtering
- The clock generator (PLL) characteristics

The main focus in this section will be on the timing error signal generation. Several different methods for PAM, QAM, and CAP modulation are explained and analyzed, and options for how to generate a timing error signal for DMT modulation are considered. The discussion will inevitably be somewhat uneven, as timing recovery is generally a more difficult problem for single-carrier modulation but relatively straightforward for DMT. In the case of single-carrier modulation, the selection of which method to use is part of an overall system design question. The most critical criterion is the sampling rate, as some methods require a sampling rate higher than the symbol rate. The equalization strategy can also play a role, as some methods may interact with the equalization. While analyzing the different methods independent of the rest of the receiver, restrictions and limitations to other system components will be identified. Although the presentation attempts to isolate salient characteristics of the clock generation function, there are difficulties in presenting a full analysis of the whole timing loop in a general manner.

12.1.1 Quality of Timing Error Signal Generation

Ideally, if a signal is sampled at a sample time t_n and that particular sample time is found to have a timing error τ_n, the subsequent sample time t_{n+1} is found simply as:

$$t_{n+1} = t_n + T_s + \tau_n, \tag{12.1}$$

where $1/T_s$ is the nominal sampling rate. In most cases, however, it is only possible to calculate a stochastic, nonlinear timing error estimate function $z(\tau)$ with expected value $E[z(\tau)] = \xi(\tau)$ and variance $E[z^2(\tau)] - \xi^2(\tau) = \eta(\tau)$. If the timing update equation above is modified to be

$$t_{n+1} = t_n + T_s + \alpha z_n, \tag{12.2}$$

where z_n is the error function output corresponding to sample time t_n, and α is a suitably chosen update gain, it is often possible to eventually approach the desired sampling time and subsequently maintain it. Whether convergence can be guaranteed, how quickly convergence will occur, and how accurately the sampling phase can be maintained depends on the timing error function.

Convergence is normally guaranteed as long as the sign of the expected value of the error function matches the sign of the timing error in a large enough interval around $\tau = 0$; for example, if

$$sgn(\xi(\tau)) = sgn(\tau), \quad \text{for} \quad -T_s/2 < \tau < T_s/2. \tag{12.3}$$

How quickly convergence takes place depends on the effective SNR of the timing function in the convergence interval, namely, the ratio of the squared expected value $\xi^2(\tau)$ to the variance $\eta(\tau)$.

Normally, the tracking performance, *i.e.*, how accurately the desired sampling phase can be maintained with $\tau = 0$, is of interest. In particular, the magnitude of the error function for small timing errors in comparison with the variance is important: if there is a small timing error, how large an indication will the error function provide in comparison with the noise in the error function samples? This ratio will determine how well the correct sampling phase can be maintained and how quickly the system can correct for possible shifts (*e.g.*, due to micro-interruptions). The quality metric of a timing error estimator is, therefore, defined as:

$$\lambda = \frac{(\xi'(0))^2}{\eta(0)}, \tag{12.4}$$

where $\xi'(0)$ is the slope of the expected value of the timing error function around the tracking point $\tau = 0$. This quality metric will be evaluated where appropriate for the various estimates analyzed in this chapter.

It is often useful to separate the variance into two components:

$$\eta(\tau) = \eta_S(\tau) + \eta_N(\tau), \tag{12.5}$$

where the signal-related component is $\eta_S(t)$ and the noise-related component is $\eta_N(t)$. The signal-related component, which is self-noise arising from the statistical properties of the transmitted signal itself, can sometimes be removed completely by filtering and careful design of the error signal. The channel noise component depends on signal components as well as the noise and channel characteristics.

12.1.2 Examples

To give better understanding of some performance aspects, this chapter uses a simple example system based on a square-root raised cosine transmit filter, where the excess bandwidth can be varied in a controlled manner. This will provide a qualitative idea of how the timing error and variance may change as the signal bandwidth is varied. The filter impulse response is [Im 1995]:

$$h(t) = \frac{\sin(\pi t)}{\pi t} \frac{\cos(\pi Rt)}{1 - (2Rt)^2}, \tag{12.6}$$

which has an associated frequency response

$$H(f) = \begin{cases} T, & |f| < \dfrac{1-R}{2T} \\ 1 + \dfrac{T}{2}\cos\left(\dfrac{\pi T}{R}\left(|f| - \dfrac{1-R}{2T}\right)\right), & \dfrac{1-R}{2T} \leq |f| \leq \dfrac{1-R}{2T} \\ 0, & |f| > \dfrac{1+R}{2T}. \end{cases} \tag{12.7}$$

Values of $R = 0.1, 0.5, 1.0$ will be considered, with a finite-length approximation of Equation 12.6. This leads, in the case of PAM transmission, to signal spectra as shown in Figure 12.4.

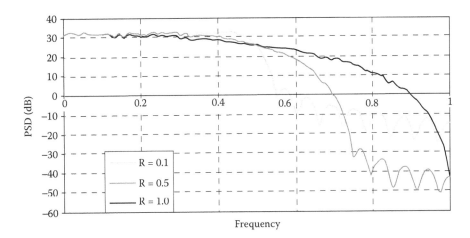

FIGURE 12.4
Frequency response of transmit signal.

12.2 PAM, QAM, and CAP Timing Recovery

Many of the timing error functions considered in this section apply to PAM, QAM, and CAP. In most cases, the focus will be on PAM methods, which can usually be extended to QAM and CAP in a straightforward manner.

There are two fundamental system design issues to keep in mind when selecting a timing recovery method for PAM, QAM, and CAP: the sampling rate of the signal and the sampling rate of the adaptive equalization. Often these are the same, either at the symbol rate or multiples thereof, but many systems first sample the receive signal at a multiple of the symbol rate and then decimate the sampling rate to the symbol rate before adaptive equalization. Some of the methods discussed later require a sampling rate higher than the symbol rate, independent of the equalizer sample rate. Some are particularly suited to select an optimal sampling phase when downsampling to the symbol rate. If fractionally spaced equalization is employed, the timing recovery only needs to maintain a consistent sampling phase, matching the sampling rate at the transmit side.

Another important criterion is whether the timing recovery method is decision-directed. During acquisition, the transmitted symbols may be unknown and difficult to estimate reliably, in which case a purely decision-directed approach may prove disastrous. Methods that do not require symbol decisions can acquire and track the timing phase independent of other receiver functions. Decision-directed methods, on the other hand, often have difficulty with equalizer interaction, where the adaptive equalizer may absorb time shifts and eventually set the equalizer phase on a random walk. Although remedies are straightforward, they need to be factored in when the overall cost is projected.

The analysis begins with a class of timing recovery algorithms that are motivated by maximum-likelihood analysis. Conventional maximum-likelihood analysis leads to systems with symbol-spaced equalization with decision-directed timing error functions.

12.2.1 Maximum-Likelihood Timing Recovery

Timing recovery can be viewed as a parameter estimation problem. The timing recovery parameters, the sampling phase, and the carrier phase can be estimated based on the

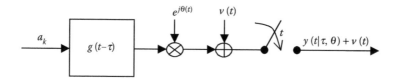

FIGURE 12.5
ML channel model.

observed receive signal. Ideally, the criterion for choosing the timing recovery parameters would be selected such that the overall performance of the receiver is optimized. However, deriving such criteria is in many cases not trivial, and applying the resulting criterion may become too complex to be practical. An alternative approach for optimizing the timing recovery parameters is to estimate the most likely timing recovery parameters given the received signal. This is the basic idea behind maximum likelihood (ML) timing recovery methods.

Suppose data symbols a_k are transmitted through a channel with impulse response $g(t)$ and delay τ followed by a phase rotation $\theta(t)$ and additive noise $v(t)$. Suppose further that the signal is sampled at time t, as shown in Figure 12.5. Then the received signal can be modelled as

$$r(t) = y(t|\tau, \theta) + v(t), \tag{12.8}$$

where

$$y(t|\tau, \theta) = \sum_k a_k g(t - kT - \tau) \exp(j\theta(t)). \tag{12.9}$$

If the channel symbol response $g(t)$ and the transmitted symbols a_k are known, then the signal term $y(t|\tau, \theta)$ is known for any given values of τ and θ. Because of the additive noise, statistical methods must be used to estimate their true values.

The most obvious probabilistic approach for estimating the timing recovery parameters is to choose the most likely parameters, $\hat{\tau}_0$ and $\hat{\theta}_0$, given the received signal, $r(t)$. This can be expressed as

$$P(\hat{\tau}_0, \hat{\theta}_0 | r(t)) \geq P(\tau, \theta | r(t)) \quad \forall \tau, \theta. \tag{12.10}$$

This is the maximum a posteriori (MAP) estimate of the timing recovery parameters.

An alternative approach is to choose the parameter estimates as the parameter values that are most likely to have produced the received signal. This is the maximum likelihood estimate, which can be expressed as

$$P(r(t) | \hat{\tau}_0, \hat{\theta}_0) \geq P(r(t) | \tau, \theta)) \quad \forall \tau, \theta, \tag{12.11}$$

where

$$P(r(t) | \tau, \theta) = P_v(v(t)) = P_v(r(t) - y(t|\tau, \theta)). \tag{12.12}$$

It is important to keep in mind the difference between the MAP criterion and the ML criterion. The MAP estimate provides the most likely timing parameters, given the observed received signal. The ML estimate, on the other hand, gives the timing parameters that find the observed received signal the most likely to occur. The two criteria are not equivalent unless the timing parameters τ and θ are uniformly distributed. It is normally a reasonable assumption that the timing parameters τ and θ are uniformly distributed, but in the case of a timing tracking loop, the knowledge of past history renders that assumption questionable. However, it is reasonable to assume the estimator ignores the value of previous estimates, and in the following it is shown how the ML analysis leads to a well-known receiver structure, namely, the matched filter DFE receiver.

For additive Gaussian noise, the probability density function for $v(t)$ is given by

$$p(x) = \frac{1}{\sigma\sqrt{2\pi}} \exp\left(-\frac{1}{2\sigma^2}(x-m)^2\right),$$ (12.13)

where σ^2 is the variance and m is the mean of the noise. In the reminder of this section, it will be assumed that $v(t)$ is zero mean; *i.e.*, $m = 0$. For additive Gaussian noise, the conditional probability in Equation 12.12 is given by

$$P(r(t)|\tau, \theta) = \frac{1}{\sigma\sqrt{2\pi}} \exp\left(-\frac{1}{2\sigma^2}(r(t) - y(t|\tau, \theta))^2\right).$$ (12.14)

This implies that the ML estimate can equivalently be expressed as

$$(r(t) - y(t|\hat{\tau}_0, \hat{\theta}_0))^2 \leq (r(t) - y(t|\tau, \theta))^2 \quad \forall \tau, \theta.$$ (12.15)

That is, the ML criterion is equivalent to minimizing the square error

$$\varepsilon(\tau, \theta) = (r(t) - y(t|\tau, \theta))^2.$$ (12.16)

Assuming $v(t)$ is a white signal, the probability of observing given values $v_0(t_m)$ at given times $\{t_0, t_1, t_2, \ldots t_{M-1}\}$ is given by

$$P(v_0(t_0), v_0(t_1), \ldots, v_0(t_{M-1})) = \prod_{m=0}^{M-1} P(v_0(t_m)) = \exp\left(\sum_{m=0}^{M-1} \ln(P(v_0(t_m)))\right).$$ (12.17)

After some rationalizing, one can conclude that the probability of observing a specific noise signal, $v_0(t)$, over the interval $[t_0; t_1]$ is proportional to

$$p(v_0(t); t_0 \leq t \leq t_1) = \exp\left(\int_{t0}^{t1} \ln(P_v(v_0(t))dt)\right).$$ (12.18)

For additive white Gaussian noise, this becomes

$$p(v_0(t); t_0 \leq t \leq t_1) = \frac{1}{\sigma\sqrt{2\pi}} \exp\left(-\frac{1}{2\sigma^2}\int_{t0}^{t1} v_0^2(t)dt\right).$$ (12.19)

Substituting this into Equation 12.12 gives the probability of observing the received signal $r(t)$ over the time interval $[t_0; t_1]$, given timing parameters τ and θ, as being proportional to

$$p(r(t); t_0 \leq t \leq t_1|\tau, \theta) = \frac{1}{\sigma\sqrt{2\pi}} \exp\left(-\frac{1}{2\sigma^2}\int_{t0}^{t1}(r(t) - y(t|\tau, \theta))^2 dt\right).$$ (12.20)

Therefore, the ML parameter estimate for τ and θ, given an observed receive signal $r(t)$ over the interval $[t_0; t_1]$, are the τ and θ that minimize

$$\varepsilon(\tau, \theta) = \int_{t0}^{t1}(r(t) - y(t|\tau, \theta))^2 dt.$$ (12.21)

The error function $\varepsilon(\tau, \theta)$ is a convex error surface and can be minimized using the steepest descent algorithm. In the steepest descent algorithm, the parameter estimate is updated according to

$$\gamma_{n+1} = \gamma_n + \alpha \nabla_\gamma \varepsilon,$$ (12.22)

where $\nabla_\gamma \varepsilon$ is the gradient of the error surface with respect to the parameter γ that is being estimated.

At this point,

$$\frac{\partial}{\partial \tau} \varepsilon(\tau, \theta) = \int_{t0}^{t1} \frac{\partial}{\partial \tau} ((r(t) - y(t|\tau, \theta))^*(r(t) - y(t|\tau, \theta)))dt$$

$$= \int_{t0}^{t1} 2\text{Re}\{(r(t) - y(t|\tau, \theta))^* \frac{\partial}{\partial \tau} y(t|\tau, \theta)\}dt$$

$$= -2\text{Re}\left\{ \int_{t0}^{t1} (r(t) - y(t|\tau, \theta))^* \sum_k a_k g'(t - kT - \tau) \exp(j\theta(t))dt \right\} \quad (12.23)$$

$$= -2\text{Re}\left\{ \sum_k a_k \int_{t0}^{t1} (r(t) - y(t|\tau, \theta))^* g'(t - kT - \tau) \exp(j\theta(t))dt \right\},$$

and by substituting for $y(t|\tau, \theta)$, this can also be expressed as

$$\frac{\partial}{\partial \tau} \varepsilon(\tau, \theta) = -2\text{Re}\left\{ \sum_k a_k \int_{t0}^{t1} r^*(t)g'(t - kT - \tau) \exp(j\theta(t))dt \right\}$$

$$-2\text{Re}\left\{ \sum_{k,m} a_k a_m \int_{t0}^{t1} g^*(t - mT - \tau)g'(t - kT - \tau)dt \right\}. \quad (12.24)$$

Similarly,

$$\frac{\partial}{\partial \theta} \varepsilon(\tau, \theta) = \int_{t0}^{t1} \frac{\partial}{\partial \theta} ((r(t) - y(t|\tau, \theta))^*(r(t) - y(t|\tau, \theta)))dt$$

$$= \int_{t0}^{t1} 2\text{Im}\{(r(t) - y(t|\tau, \theta))^* y(t|\tau, \theta)\}dt$$

$$= -2\text{Im}\left\{ \int_{t0}^{t1} (r(t) - y(t|\tau, \theta))^* \sum_k a_k g(t - kT - \tau) \exp(j\theta(t))dt \right\} \quad (12.25)$$

$$= -2\text{Im}\left\{ \sum_k a_k \int_{t0}^{t1} (r(t) - y(t|\tau, \theta))^* g(t - kT - \tau) \exp(j\theta(t))dt \right\},$$

which again can be expressed as

$$\frac{\partial}{\partial \theta} \varepsilon(\tau, \theta) = -2\text{Im}\left\{ \sum_k a_k \int_{t0}^{t1} r^*(t)g(t - kT - \tau) \exp(j\theta(t))dt \right\}$$

$$-2\text{Im}\left\{ \sum_{k,m} a_k a_m \int_{t0}^{t1} g^*(t - mT - \tau)g(t - kT - \tau)dt \right\}. \quad (12.26)$$

This implies that the ML estimates for τ_0 and θ_0 can be obtained by using the iterations

$$\tau_{n+1} = \tau_n + \alpha \frac{\partial}{\partial \tau} \varepsilon \quad (12.27)$$

and

$$\theta_{n+1} = \theta_n + \beta \frac{\partial}{\partial \theta} \varepsilon, \quad (12.28)$$

where $\frac{\partial}{\partial \tau}\varepsilon$ and $\frac{\partial}{\partial \theta}\varepsilon$ are given by given by Equations 12.24 and 12.26, respectively. These iterations will result in the true ML parameter estimate, assuming that α and β are chosen such that Equations 12.27 and 12.28 converge.

It is possible to approximate the update in Equation 12.27 by dropping the second term in the error estimate. This seems plausible, as its expected value is independent of $r(t)$. In this case, the iteration becomes

$$\tau_{n+1} = \tau_n - \alpha\, Re\left\{\sum_k a_k \int_{t0}^{t1} r^*(t)g'(t - kT - \tau)\exp(j\theta(t))dt\right\}. \tag{12.29}$$

Likewise, it is possible to approximate the update in Equation 12.28 by

$$\theta_{n+1} = \theta_n - \beta\, Im\left\{\sum_k a_k \int_{t0}^{t1} r^*(t)g(t - kT - \tau)\exp(j\theta(t))dt\right\}. \tag{12.30}$$

The approximations in Equations 12.29 and 12.30 are very similar to some of the heuristic decision-directed timing recovery algorithms that will be introduced later in this chapter. These equations indicate that the timing adjustment is derived from a decision feedback filter, where the taps are convolutions of the received signal with segments of the derivative of the channel impulse response. In the case of carrier recovery, the adjustment uses taps derived from a convolution with the channel impulse response, similar to a matched filtering operation.

The discussion above assumed that the transmitted symbols a_k were known and could be used as part of the timing recovery process. Such timing recovery methods are referred to as decision-directed or data-aided timing recovery. However, if the transmit symbols cannot be reliably obtained in the receiver, then it is necessary to use nondecision-directed algorithms. In [Franks 1980], it is shown how taking the expected value over the symbol distribution leads to timing recovery estimators that resemble nonlinearly induced spectral line (NISL) methods that will be explored later in the chapter.

Analysis of timing recovery methods based on ML and MAP parameter estimation can provide very useful insight into the timing recovery problem. For example, ML timing recovery can address both decision-directed and nondecision-directed timing recovery using the same basic approach. However, one should take care to not overrate the optimality of the ML and MAP approaches. The ML and MAP criteria do provide the "most likely" parameter estimates, but these estimates are not necessarily the optimal parameter estimates in terms of overall system performance. In particular, the ML and MAP criteria fail to take into account the quality of the error recovery signals in terms of variance, stability, rate of convergence, etc. Furthermore, ML-based analysis of the timing recovery problem tends to involve a fair amount of imprecise assumptions and approximations that may or may not be applicable to real communication systems. So, although ML analysis may suggest specific timing recovery methods, these methods should be subject to the same scrutiny as any other in terms of acquisition characteristics, stability, and tracking performance.

12.2.2 Band-Edge Energy Maximization

Band-edge timing recovery (BETR) represents a class of timing recovery methods used primarily with QAM modulated signals where symbol-spaced equalization is utilized. BETR methods aim for a sampling phase that results in constructive, rather than destructive, aliasing effects.

Because of aliasing, the power of the sampled signal can vary with the sampling phase when sampled at the symbol rate. That is easily seen in the case of a sinusoid with frequency equal to half the sample rate, where the power of the sampled signal can vary from 0 to twice the pre-sampled power. The sinusoid vanishes because of destructive aliasing, where in effect the phase at $f = +1/2T$ is opposite to the phase at $f = -1/2T$. A similar effect can occur with data signals if the phase of the channel response at one band-edge is opposite to the phase at the other. At a certain sampling phase, the effective channel response can go to zero. Correspondingly, at another sampling phase, the effective channel response is maximized. Band-edge timing recovery seeks to find the phase that creates that maximum response.

Generally speaking, increasing the overall energy of the sampled signal will enhance SNR, as the noise energy will typically not be dependent on the sample phase. Although a sample phase that maximizes the band-edge energy is not guaranteed to maximize the energy of the sampled signal, the two criteria will usually find a similar optimal sample phase, and often they will coincide.

12.2.2.1 The Band-Edge Energy

First, a baseband channel is considered, as shown in Figure 12.6. The baseband channel may be either a real-valued PAM channel or a complex baseband-analytic channel in the case of QAM or CAP. It is assumed the channel $G(f)$ includes transmit filtering, the channel itself, and receive filtering. Suppose data symbols a_k are transmitted across that channel. The signal that should be sampled at the symbol rate $\frac{1}{T}$ is

$$y(t) = \sum_{m=-\infty}^{\infty} a_m g(t - mT). \tag{12.31}$$

Suppose this signal is sampled at times $\tau + kT$ to obtain:

$$y_k(\tau) = \sum_{m=-\infty}^{\infty} a_m g(\tau + (k - m)T). \tag{12.32}$$

The symbol-rate sampling will create a new channel that is dependent on the sampling phase τ. This new channel is composed of symbol-rate-wide segments of the actual channel response $G(f)$, shifted to the baseband. The phase of the segments will depend on the sampling phase, and hence the overall sum. This procedure has effectively created a sample-time-dependent symbol-rate channel $G_T(f, \tau)$:

$$G_T(f, \tau) = \sum_{n=-\infty}^{\infty} G\left(f - \frac{n}{T}\right) e^{j2\pi(f-n/T)\tau}. \tag{12.33}$$

Considering only narrowband signals where $G(f) = 0$ for $|f| > 1/T$, most terms of the expression can be omitted if only the baseband symbol-period, $|f| < 1/2T$ (noting that $G_T(f + n/T, \tau) = G_T(f, \tau)$), is considered:

$$G_T(f, \tau) = G\left(f + \frac{1}{T}\right) e^{j2\pi(f+1/T)\tau} + G(f)e^{j2\pi f\tau} + G\left(f - \frac{1}{T}\right) e^{j2\pi(f-1/T)\tau}. \tag{12.34}$$

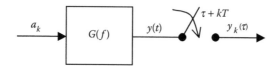

FIGURE 12.6
Baseband channel model.

It is now possible to examine how the band-edge energy varies with the sampling phase τ. Denoting the band-edge response in polar form,

$$G\left(\frac{1}{2T}\right) = G_+ . e^{j\theta_+}$$

$$G\left(-\frac{1}{2T}\right) = G_- . e^{j\theta_-}$$

(12.35)

it is clear that

$$G_T\left(\frac{1}{2T}, \tau\right) = G_+ . e^{j(\theta_+ + \pi\tau/T)} + G_- . e^{j(\theta_- - \pi\tau/T)},$$

(12.36)

and the energy is

$$\left\|G_T\left(\frac{1}{2T}, \tau\right)\right\|^2 = \left\|G_+ . e^{j(\theta_+ + \pi\tau/T)} + G_- . e^{j(\theta_- - \pi\tau/T)}\right\|^2$$

$$= G_+^2 + G_-^2 + 2G_+ G_- \cos(\theta_+ - \theta_- + 2\pi\tau/T).$$

(12.37)

In the case of PAM, the channel $G(f)$ is real, and therefore $G_+ = G_-$, and $\theta_+ = -\theta_-$, resulting in

$$\left\|G_T\left(\frac{1}{2T}, \tau\right)\right\|^2 = 2G_+^2\left(1 + \cos\left(2\theta_+ + \frac{2\pi\tau}{T}\right)\right).$$

(12.38)

This will vanish if the sample time is selected as

$$\tau_{\text{min,PAM}} = \left(k + \frac{1}{2} - \frac{\theta_+}{\pi}\right)T$$

(12.39)

for any integer k. Correspondingly, the channel response of the symbol-rate channel at the band edge is maximized if the sampling time is

$$\tau_{\text{max,PAM}} = \left(k - \frac{\theta_+}{\pi}\right)T.$$

(12.40)

In the case of QAM, the signal will not vanish at the band-edge unless $G_+ = G_-$, which is generally not the case. The maximum band-edge energy is clearly obtained when the cosine term is 1, namely when the timing phase is

$$\tau_{\text{max,QAM}} = \left(k - \frac{\theta_+ - \theta_-}{2\pi}\right)T.$$

(12.41)

To obtain this phase, it is necessary to create a timing function $f(\tau)$ that has a positive-going zero-crossing at the sample times that result in the maximum energy. One obvious example is a sinusoid. For PAM,

$$f_{\text{PAM}}(\tau) = \sin\left(2\theta_+ + \frac{\tau}{T}\right),$$

(12.42)

and for QAM:

$$f_{QAM}(\tau) = \sin\left(\theta_+ - \theta_- + \frac{2\pi\,\tau}{T}\right). \tag{12.43}$$

Next, methods of calculating these timing functions are investigated.

12.2.2.2 BETR Timing Function for QAM

The timing function proposed above suggests that if somehow the product of the upper band-edge signal component and the conjugate of the lower band-edge component is created, a useful timing function may be obtained. Consider the timing function

$$\gamma(t) = y_U(t)y_L^*(t), \tag{12.44}$$

where components $y_U(t)$ and $y_L(t)$ are created using bandpass filters $H_L(f)$ and $H_U(f)$ with impulse responses $h_L(t)$ and $h_U(t)$, as illustrated in Figure 12.7. Of interest is the expected value of the product $\gamma(t)$, including the effects of additive noise $v(t)$. It is assumed the noise is independent of the signal, and the noise components at the output of the upper and lower bandpass filters are denoted as $v_U(t)$ and $v_L(t)$. Evaluating the expected value over independent symbol sequences $\{a_k\}$ and the noise at a sample time τ yields

$$E[\gamma(\tau)] = E[(y_U(\tau + kT) + v_U(\tau + kT))(y_L^*(\tau + kT) + v_L^*(\tau + kT))]$$

$$= \sigma_a^2 \sum_{m=-\infty}^{\infty} g_U(\tau + mT)g_L^*(\tau + mT) + \int_{-\infty}^{\infty}\int_{-\infty}^{\infty} R_v(t - s)\,h_U(t)h_L^*(s)\,dt\,ds, \tag{12.45}$$

where $E[a_n a_n^*] = \sigma_a^2$, $R_v(t - s) = E[v(t)v^*(s)]$ and the channel response and bandpass filters in $G_L(f)$ and $G_U(f)$ have been combined. Expanding the first term with the Poisson sum formula, the result becomes

$$\sigma_a^2 \sum_{m=-\infty}^{\infty} g_U(\tau + mT)g_L^*(\tau + mT) = \frac{\sigma_a^2}{T} \sum_{n=-\infty}^{\infty} A_n e^{j2\pi nt/T}, \tag{12.46}$$

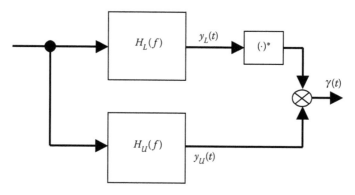

FIGURE 12.7
QAM timing function.

where

$$A_n = \int_{-\infty}^{\infty} g_U(\tau) g_L^*(\tau) e^{-j2\pi n\tau/T} d\tau$$

$$= \int_{-\infty}^{\infty} H_U\left(\frac{n}{T} + f\right) H_L^*(f) G\left(\frac{n}{T} + f\right) G^*(f) df. \tag{12.47}$$

Also,

$$\int_{-\infty}^{\infty}\int_{-\infty}^{\infty} R_v(t-s) h_U(t) h_L^*(s) dt\, ds = \int_{-\infty}^{\infty}\int_{-\infty}^{\infty}\int_{-\infty}^{\infty}\int_{-\infty}^{\infty} R_v\,(t-s)\, H_U(u) H_L^*(r) e^{j2\pi(ut-rs)} du\, dr\, dt\, ds$$

$$= \int_{-\infty}^{\infty}\int_{-\infty}^{\infty} H_U(u) H_L^*(r) \int_{-\infty}^{\infty}\int_{-\infty}^{\infty} R_v(t-s) e^{j2\pi(ut-rs)} dt\, ds\, du\, dr. \tag{12.48}$$

If it is assumed the filters have a bandwidth less than $1/T$, *i.e.*,

$$H_U(f) = 0, \quad f \le 0 \text{ and } f \ge 1/T$$
$$H_L(f) = 0, \quad f \le -1/T \text{ and } f \ge 0, \tag{12.49}$$

then $A_n = 0$, $n \ne 1$, and the noise term vanishes as $H_U(f) H_L^*(f) = 0$ in Equation 12.48. In that case,

$$E\left[\gamma\left(\tau\right)\right] = \sigma_a^2 \frac{1}{T} A_1 e^{j2\pi\tau/T}, \tag{12.50}$$

and a complex exponential indicates the timing phase. It will be sufficient to allow phase tracking at an arbitrary phase, simply by taking the imaginary part of the timing function:

$$f(\tau) = \mathrm{Im}[E[\gamma(\tau)]] = \frac{\sigma_a^2}{T} \|A_1\| \sin\left(\arg\left(A_1\right) + \frac{2\pi\tau}{T}\right). \tag{12.51}$$

However, optimality of the sampling phase has not been guaranteed, because $\arg\left(A_1\right)$ is still dependent on the bandpass filters. That dependency can be removed by letting them have the same shape, for example, as frequency translated versions of a low-pass response $H_0(f)$:

$$H_L(f) = H_0(f + 1/2T), \quad h_L(t) = h_0(t) e^{-j\pi t/T}$$
$$H_U(f) = H_0(f - 1/2T), \quad h_U(t) = h_0(t) e^{j\pi t/T}. \tag{12.52}$$

Then,

$$A_1 = \int_{-\infty}^{\infty} \left\| H_0\left(f - f_c + \frac{1}{2T}\right) \right\|^2 G\left(\frac{1}{T} + f\right) G^*(f) df$$

and $\arg\left(A_1\right)$ now only depends on $G(f)$. Under the further simplifying assumption that the bandpass filters are very narrow, *i.e.*,

$$H_0(f) = \begin{cases} 1, & |f| < \Delta f \\ 0, & |f| \ge \Delta f \end{cases} \tag{12.53}$$

it is clear that

$$A_1 \approx 2\Delta f G_+ G_- e^{j(\theta_+ - \theta_-)}, \tag{12.54}$$

where the notation in Equation 12.35 has been used. The timing function is now

$$E\left[\gamma(\tau)\right] = \frac{\sigma_a^2}{T} 2\Delta f G_+ G_- e^{j(2\pi\tau/T + \theta_+ - \theta_-)}, \tag{12.55}$$

and taking the imaginary part provides the desired result:

$$f(\tau) = \mathrm{Im}\left[E\left[\gamma(\tau)\right]\right] = \frac{\sigma_a^2}{T} 2\Delta f G_+ G_- \sin\left(\frac{2\pi\tau}{T} + \theta_+ - \theta_-\right). \tag{12.56}$$

Thus, as the filters become narrower, the desired function of Equation 12.43 is approached. If, on the other hand, overlapping filters are used, more terms of the Poisson sum are active, in particular A_0:

$$A_0 = \int_{-\infty}^{\infty} H_U(f) H_L^*(f) \|G(f)\|^2 df. \tag{12.57}$$

A constant offset also results due to the noise term. This combined offset will shift the timing phase away from its optimum value. [Ungerboeck 1990] suggests a simple way of removing this offset, simply by using the difference $\eta(t) = \gamma(t) - \gamma(t - T/2t)$. Then the offset term and the noise term cancel out, and

$$E[\eta(t)] = \frac{\sigma_a^2}{T}\left(A_1 e^{j2\pi t/T} - A_1 e^{j2\pi(t - T/2)/T}\right)$$

$$= \frac{2\sigma_a^2}{T} A_1 e^{j2\pi t/T}. \tag{12.58}$$

Taking the imaginary part of $\eta(t)$ yields the same timing function as before, apart from a gain factor of 2.

Variance of $\gamma(t)$ If it is assumed that the band-pass filters do not overlap, *i.e.,* $H_U(f)H_L(f) = 0$, it can be shown that the variance is

$$E[\|\gamma(t) - E[\gamma(t)]\|^2] = \left(E\left[\|a_k\|^4\right] - 2\sigma_a^4\right)\left(\Lambda_{UL}(0) + 2Re\left[\Lambda_{UL}\left(\frac{2\pi}{T}\right)e^{j2\pi/Tt}\right]\right)$$

$$+ \left(\sigma_a^2 \Lambda_{0L} + \sigma_{vL}^2\right)\left(\Lambda_{0U} + \sigma_{vU}^2\right), \tag{12.59}$$

where

$$\Lambda_{UL}(0) = \int_{-\infty}^{\infty} \|g_L(t)\|^2 \|g_U(t)\|^2 dt$$

$$\Lambda_{UL}\left(\frac{2\pi}{T}\right) = \int_{-\infty}^{\infty} \|g_L(t)\|^2 \|g_U(t)\|^2 e^{-j2\pi t/T} dt$$

$$\tag{12.60}$$

$$\Lambda_{0L} = \int_{-\infty}^{\infty} \|g_L(t)\|^2 dt$$

$$\Lambda_{0U} = \int_{-\infty}^{\infty} \|g_U(t)\|^2 dt$$

and

$$
\sigma_{\upsilon U}^2 = \int_{-\infty}^{\infty} \|N(f)\|^2 \, \|H_U(f)\|^2 \, df
$$

(12.61)

$$
\sigma_{\upsilon L}^2 = \int_{-\infty}^{\infty} \|N(f)\|^2 \, \|H_L(f)\|^2 \, df.
$$

The variance for the narrow bandpass filters satisfying Equation 12.53 is now evaluated. In that case,

$$
E[\|\gamma(t) - E[\gamma(t)]\|^2] = \left(E\left[\|a_k\|^4\right] - 2\sigma_a^4 \right) \frac{16}{3} G_+^2 G_-^2 \, (\Delta f)^3
$$
$$
+ 4 \, (\Delta f)^2 \left(\sigma_a^2 G_-^2 + N_-^2 \right) \left(\sigma_a^2 G_+^2 + N_+^2 \right),
$$

(12.62)

where N_- and N_+ represent the noise spectral density at the lower and upper band-edges, respectively. Under the assumption that the variance of $\mathrm{Im}[\gamma(t)]$ is half the variance of $\gamma(t)$, the quality metric is

$$
\lambda = \frac{16\pi^2}{T^4 \left(\frac{4}{3} \Delta f \left(\frac{E[\|a_k\|^4]}{\sigma_a^4} - 2 \right) + \left(1 + \frac{N_-^2}{\sigma_a^2 G_+^2} \right) \left(1 + \frac{N_+^2}{\sigma_a^2 G_-^2} \right) \right)}.
$$

(12.63)

Although the derivation above was made on the assumption that the signal is baseband, it is easily extended to the passband. All that is involved is a frequency translation.

Using the Complex Exponential as Timing Function Although using the imaginary part of $\gamma(\tau)$ or $\eta(\tau)$ is perfectly acceptable, [Ungerboeck 1990], [Jablon 1992], and [Jablon 1988] suggest using the complex exponential itself. [Ungerboeck 1990] suggests using $\eta(\tau)$ to compare against a reference, using the phase difference between $\eta(\tau)$ and the reference as a timing error function. This allows tracking to an arbitrary phase, which may be convenient when using this method with fractionally spaced equalizers, where the equalizer is initialized with an arbitrary phase and the receiver needs to maintain that phase. In [Jablon 1988], the complex exponential is used to obtain a fast estimate of the timing frequency offset, allowing the receiver to very quickly adapt the sampling rate itself to the received signal.

12.2.3 Timing Signals Based on Nonlinearity-Induced Spectral Line

One of the more traditional methods of timing recovery is based on creating a spectral line by passing the receive signal through a nonlinearity of some kind. Many types of nonlinearities have been used, most commonly squarers, absolute value functions, and fourth-power nonlinearities. Maximum-likelihood analysis has motivated some logarithmic-based nonlinear functions [Panayirci 1996]. The spectral line is created by the correlation of signal components spaced $1/T$ apart in the frequency domain. As the spectrum of the original symbol sequence will be periodic with period $1/T$, the presence of periodic components, even after amplitude and phase changes, will ensure the creation of a spectral line at the frequency of $1/T$, or multiples thereof.

A typical NISL system is shown in Figure 12.8. After some optional prefiltering, the signal is passed through the nonlinearity and then the spectral line is cleaned up with a narrowband filter. The filter removes the inevitable DC component of the signal after the nonlinearity, as well as other harmonics apart from reducing the overall variance. The

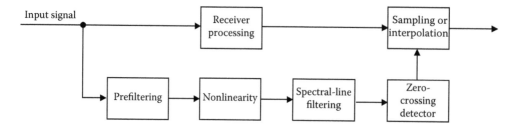

FIGURE 12.8
Typical timing recovery based on nonlinearity-induced spectral line.

output of the spectral line filter will typically be a jittery sine wave. The frequency of the expected value of that sine wave will be exactly equal to an integer multiple of the transmitter symbol rate. By locking the receiver timing to zero crossings spaced by the symbol time T, the receiver will be synchronized. In a properly designed system, the positions of the zero crossings will stay constant and only the amplitude will vary.

This section concentrates on the squarer nonlinearity, as its analysis is less difficult than for others. It has not been shown that one type of nonlinearity is universally better than others, and the analysis presented here for the squarer should give a reasonable indication of the performance of other types of nonlinearities.

12.2.3.1   Analysis of Squarer-Induced Spectral Line — Without Noise

Suppose the continuous-time received signal component, after suitable receiver processing, is given by:

$$x(t) = \sum_{k=-\infty}^{\infty} a_k g(t - kT),\qquad(12.64)$$

where a_k are the independent transmit symbols and $g(\cdot)$ is the combined symbol response of the transmitter, channel, and receiver processing (including pre-filtering). Then the expected value of the signal squared, taken with respect to the data sequence $\{a_k\}$, is given by

$$\xi_0(t) = E[x^2(t)] = \sigma_a^2 \sum_{k=-\infty}^{\infty} g^2(t - kT),\qquad(12.65)$$

where $\sigma_a^2 = E[a_k^2]$. Clearly, $\xi_0(t)$ is periodic with period T, so it can be represented by a Fourier series, or the Poisson sum formula can be used to obtain

$$\xi_0(t) = \frac{\sigma_a^2}{T} \sum_{m=-\infty}^{\infty} A_m e^{j2\pi mt/T},\qquad(12.66)$$

where

$$A_m = \int_{-\infty}^{\infty} g^2(t) e^{-j2\pi mt/T} dt\qquad(12.67)$$

is the Fourier transform of $g^2(t)$ at $f = m/T$, which can also be expressed in terms of the Fourier transform $G(f)$ of $g(t)$ as

$$A_m = \int_{-\infty}^{\infty} G\left(\frac{m}{T} - f\right) G(f) df.\qquad(12.68)$$

The function $\xi_0(t)$ can thus be decomposed into a sum of sinusoids of frequency m/T, where the components depend on how quickly the combined frequency response dies off.

Indeed, if it is assumed that $G(f) = 0$ for $|f| \leq 1/T$, there are only three nonzero elements in the sum, resulting in

$$\xi_0(t) = \frac{\sigma_a^2}{T} \left(\|G\|^2 + 2\|A_1\| \cos\left(\arg(A_1) + 2\pi \frac{t}{T}\right)\right) \tag{12.69}$$

by using the relation $A_{-m} = A_m^*$. Thus, the expected value of the square of the signal contains a DC component and a sinusoid periodic in $1/T$. By removing the DC component, a useful timing function can be created.

The traditional method is to remove the DC component by filtering to produce a sinusoidal timing error function, periodic in $1/T$:

$$\xi_1(t) = \frac{2\sigma_a^2 \|A_1\|}{T} \cos\left(\arg(A_1) + 2\pi \frac{t}{T}\right). \tag{12.70}$$

It is also possible to effectively remove the DC component by sampling the sinusoid at twice the symbol rate and calculating the difference. Denoting the sampling time by τ,

$$\xi_2(\tau) = \gamma(\tau) - \gamma(\tau - T/2)$$
$$= \frac{2\sigma_a^2 \|A_1\|}{T} \left(\cos\left(\arg(A_1) + 2\pi \frac{\tau}{T}\right) - \cos\left(\arg(A_1) + 2\pi \frac{\tau - T/2}{T}\right)\right) \tag{12.71}$$
$$= \frac{4\sigma_a^2 \|A_1\|}{T} \cos\left(\arg(A_1) + 2\pi \frac{\tau}{T}\right).$$

Equations 12.69 and 12.72 indicate that the strength of the timing signal depends, through A_1, on the excess bandwidth of the received signal. Thus, if the transmit filter allows no transmit energy at and above $f = 1/2T$, the timing signal will be zero. Further, if the energy around the band-edge is attenuated by the channel, the performance of a square-law-based timing recovery will change correspondingly.

Variance As noted previously, it is often appropriate to distinguish between signal-induced and channel noise-induced variance in the timing error estimate. This subsection first focuses on the signal-induced component to find how careful system design can greatly reduce it.

To evaluate the variance, the spectral line filter must be taken into account. Although the expected value of the timing function only depends on the filter response at the spectral line at $f = 1/T$, the variance will depend on the overall response of the filter. Denote the impulse response of the spectral line filter by $h(t)$ and its output by $y(t)$. Then

$$y(t) = h \otimes x^2(t) = \int_{-\infty}^{\infty} h(t - s)x^2(s)\,ds$$
$$= \int_{-\infty}^{\infty} h(t - s) \sum_{k=-\infty}^{\infty} \sum_{m=-\infty}^{\infty} a_k a_m g(s - kT)g(s - mT)\,ds$$
$$= \sum_{k=-\infty}^{\infty} \sum_{m=-\infty}^{\infty} a_k a_{k+m} \int_{-\infty}^{\infty} h(t - kT - s)g(s)g(s - mT)\,ds \tag{12.72}$$
$$= \sum_{k=-\infty}^{\infty} \sum_{m=-\infty}^{\infty} a_k a_{k+m} q_m(t - kT),$$

where

$$q_m(t) = \int_{-\infty}^{\infty} h(t - s)g(s)g(s - mT)\,ds, \tag{12.73}$$

or, equivalently,

$$Q_m(f) = H(f) \int_{-\infty}^{\infty} G(f - \lambda)G(\lambda)e^{-j2\pi mT\lambda}d\lambda. \tag{12.74}$$

If it is assumed the frequency response at the symbol rate is $H(1/T) = 1$, then $E[y(t)] = \tilde{\xi}(t)$. The self-noise component of the variance is found to be

$$E[y^2(t)] - \tilde{\xi}^2(t) = E\left[a_k^4\right] \sum_{k=-\infty}^{\infty} q_0^2(t - kT)$$

$$+ E\left[\sum_{k=-\infty}^{\infty} \sum_{\substack{l=-\infty \\ l \neq k}}^{\infty} a_k^2 a_l^2 q_0(t - kT)q_0(t - lT)\right]$$

$$+ E\left[\sum_{k=-\infty}^{\infty} \sum_{\substack{m=-\infty \\ m \neq 0}}^{\infty} a_k^2 a_{k+m}^2 q_m^2(t - kT)\right] \tag{12.75}$$

$$+ E\left[\sum_{k=-\infty}^{\infty} \sum_{m=-\infty}^{\infty} a_k a_{k+m} a_{k+m} a_k q_m(t - kT)q_{-m}(t - (k + m)T)\right] - \tilde{\xi}^2(t)$$

$$= \left(E\left[a_k^4\right] - 3\sigma_a^4\right) \sum_{k=-\infty}^{\infty} q_0^2(t - kT) + 2\sigma_a^4 \sum_{k=-\infty}^{\infty} \sum_{m=-\infty}^{\infty} q_m^2(t - kT).$$

As the variance is periodic in T, Equation 12.75 can be expanded using the Poisson sum formula to get a Fourier series with components at multiples of $1/T$:

$$E[y^2(t)] - \tilde{\xi}^2(t) = \left(E\left[a_k^4\right] - 3\sigma_a^4\right)\frac{1}{T} \sum_{n=-\infty}^{\infty} Q_0^{(2)}\left(\frac{n}{T}\right)e^{j2\pi nt/T}$$

$$+ 2\sigma_a^4 \frac{1}{T} \sum_{m=-\infty}^{\infty} \sum_{n=-\infty}^{\infty} Q_m^{(2)}\left(\frac{n}{T}\right)e^{j2\pi nt/T}, \tag{12.76}$$

where

$$Q_m^{(2)}\left(\frac{n}{T}\right) = \int_{-\infty}^{\infty} q_m^2(t)e^{-j2\pi nt/T}\,dt. \tag{12.77}$$

Expanding again,

$$Q_m^{(2)}\left(\frac{n}{T}\right) = \int_{-\infty}^{\infty} \int_{-\infty}^{\infty} \int_{-\infty}^{\infty} h(t-r)g(r)g(r-mT)h(t-s)g(s)g(s-mT)dr\,ds\,e^{-j2\pi nt/T}dt$$

$$= \int_{-\infty}^{\infty} \int_{-\infty}^{\infty} H_{r-s}^{(2)}\left(\frac{n}{T}\right)g(r)g(r-mT)g(s)g(s-mT)e^{-j2\pi nr/T}ds\,dr, \tag{12.78}$$

where

$$H_{r-s}^{(2)}\left(\frac{n}{T}\right) = \int_{-\infty}^{\infty} h(t)h(t+r-s)e^{-j2\pi nt/T}\,dt$$

$$= \int_{-\infty}^{\infty} H\left(\frac{n}{T} - f\right)H(f)e^{j2\pi f(r-s)}\,df. \tag{12.79}$$

If the condition

$$H(f) = 0; \quad ||f| - 1/T| \le 1/2T \tag{12.80}$$

is imposed, namely, that the spectral line filter only passes through frequencies in a band of width less than the symbol rate around the spectral line, then $H_{r-s}^{(2)}(n/T) = 0$ for $n \notin \{-2, 0, 2\}$. Hence, the variance is of the form

$$E[y^2(t)] - \tilde{\xi}^2(t) = A + B \cos\left(\frac{4\pi t}{T} + \theta\right). \tag{12.81}$$

This function has two minima within the period of T. The actual level and phase are dependent on both $H(f)$ and $G(f)$. If $G(f)$ is made to be bandlimited, few of the terms in the double-sum in Equation 12.76 remain nonzero. In fact, it can be shown [Franks 1974] that under certain conditions, the self-noise component of the variance vanishes at the zero crossings of the timing function $\tilde{\xi}(t)$. These conditions are the following:

$$a_k = \pm 1$$

$$G(f) = \begin{cases} 0, & |f| < \dfrac{1}{4T} \\ G^*\left(\dfrac{1}{T} - f\right), & \dfrac{1}{4T} \le |f| < \dfrac{3}{4T} \\ 0, & |f| \ge \dfrac{3}{4T} \end{cases} \tag{12.82}$$

$$H(f) = \begin{cases} 0, & |f| < \dfrac{1}{2T} \\ H^*\left(\dfrac{2}{T} - f\right), & \dfrac{1}{2T} \le |f| < \dfrac{3}{2T} \\ 0, & |f| \ge \dfrac{3}{2T} \end{cases}.$$

The conditions are equivalent to stating the data is bipolar, the combined channel response conjugate is symmetric around $1/2T$ and less than $1/2T$ wide, and the spectral line filter conjugate-symmetric around the spectral line and bandlimited to $1/2T$ on either side. The condition on $H(f)$ is straightforward to fulfill, but the condition on $G(f)$ is difficult in the case of unknown channels. It involves having to filter the incoming signal with a response that will result in a combined response that is symmetric around the band-edge.

Analysis of Squarer-Induced Spectral Line — With Noise   Adding a noise component $v(t)$ with variance σ_n^2, the input to the squarer becomes

$$x(t) = \sum_{k=-\infty}^{\infty} a_k g(t - kT - \tau) + v(t). \tag{12.83}$$

Taking the expectation now over both the data sequence and the noise,

$$\xi(t) = E[x^2(t)] = \sigma_a^2 \sum_{k=-\infty}^{\infty} g^2(t - kT - \tau) + \sigma_n^2. \tag{12.84}$$

After removing the DC component, the expected value is the same as in the noiseless case. It can be shown that with the symmetry conditions in Equation 12.82, the additional variance caused by the noise term is independent of t [Franks 1974].

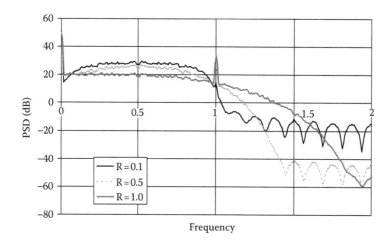

FIGURE 12.9
Frequency response of transmit signal squared.

Example   Suppose the following PAM signal is transmitted:

$$x_1(t) = \sum_{k=-\infty}^{\infty} a_k h_R(t - kT),\tag{12.85}$$

where $a_k = \pm 1$ and $h_R(t)$ is the raised cosine response defined in Equation 12.6. Without receiver filtering, the squared signal will result in the spectrum shown in Figure 12.9.

With the low excess bandwidth, the symbol-rate spectral line is almost buried in the spectrum. To obtain a self-noise-free timing estimator at the zero crossings of the timing function, the signal must be filtered with the raised-cosine response shifted by $1/T$. This will significantly enhance the spectral lines as shown in Figure 12.10. Filtering now with a conjugate-symmetric spectral-line filter with no DC component, the timing function shown in Figure 12.11 is obtained. As expected, the greater excess bandwidth results in a stronger timing function.

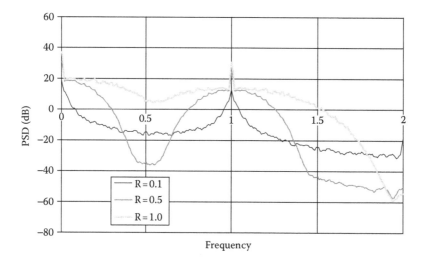

FIGURE 12.10
Frequency response of prefiltered signal squared.

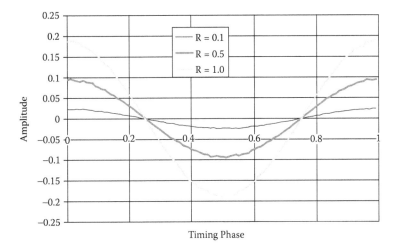

FIGURE 12.11
Timing function.

Finally, evaluating the variance shows that at the zero crossings of the timing function, the self-noise component of the variance vanishes, as shown in Figure 12.12.

12.2.4 Timing Recovery Based on Equalizer Taps

As pointed out previously, an adaptive equalizer, as long as it has an adaptive feedforward component, can absorb a timing offset. In the case of fractionally spaced feedforward equalization, absorption of the timing offset will not necessarily cause any degradation of performance, whereas a symbol-spaced equalizer will perform optimally only at a specific timing phase. In either case, a timing error signal can be derived from the adaptive taps to control a timing recovery loop.

Advantages of equalizer-tap-based timing recovery are primarily the following:

- Ability to jointly lock timing and optimize equalizer positioning
- Possibility to generate timing function based on a relatively clean signal

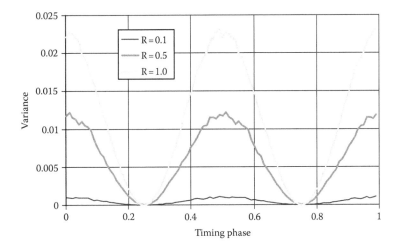

FIGURE 12.12
Variance of timing function.

The former advantage may be important when equalizer resources are scarce and channel characteristics difficult to predict. The latter can prove significant at low SNR or, in particular, when a high level of echo is only subtracted after equalization. The averaging performed in the equalizer update serves to lower the effects of noise and echo. However, the disadvantages of this method can cause difficulties. First, the equalizer must be in tracking mode for timing recovery to be active, and the update gain must be high enough to allow fast enough tracking. This condition can place difficult restraints on the equalization strategy, where freezing equalizer taps for preservation or simply to save resources makes clock tracking impossible. Further, the update rate required to allow enough tracking gain may result in unacceptably high tap noise. A second disadvantage is that fractionally spaced equalizers may give misleading transient indications, which are caused by the redundant degrees of freedom in fractionally spaced equalizers. Tap leakage algorithms may circumvent this problem, but it has been observed that adjustment of sample phase in one direction can cause positive feedback that renders the timing loop unstable. A third disadvantage of this method is that the timing function may be difficult to predict because of the dependence of equalizer taps on sampling phase as well as the channel. If the algorithm is not designed carefully, the timing function may have multiple local minima in certain conditions.

In any case, these methods depend heavily on the equalization strategy and update algorithms. As a result, performance is difficult to quantify independently. Therefore, no performance analysis is attempted here; only qualitative comments are provided.

Simple Equalizer Tap-Based Algorithms   The simplest form of an equalizer tap-based algorithm is to compare the energy of a subset of equalizer taps before and after an equalizer "center." If the response of the squared equalizer taps is guaranteed to be monotonic when filtered by a rectangular window of size M, a stable timing function will be created by calculating the difference in energy. However, the shape of the equalizer response will usually not be independent of the sampling phase, and a monotonic timing function is by no means guaranteed. Formally, the algorithm can be stated as follows. Given a sample-phase dependent equalizer response $\{f_n(\tau), 0 \leq n < N\}$, the timing error function $z(\tau)$ is

$$z(\tau) = \sum_{k=0}^{K-1} \| f_n(\tau) \|^2 - \sum_{k=N-K}^{N-1} \| f_n(\tau) \|^2, \tag{12.86}$$

where $K < N/2$ defines the number of taps on each end of the equalizer whose energy is compared. If the timing phase shifts in one direction, the energy in the K taps on one side will tend to grow, and the energy in the K taps on the other side will be reduced. If it can be ensured that the derivative with sample phase is always positive, *i.e.*,

$$\frac{\partial}{\partial \tau} z(\tau) > 0 \tag{12.87}$$

in a large enough region around where $z(\tau) = 0$, the timing function will be stable. As noted before, however, the redundant degrees of freedom in fractionally spaced equalizers may make fulfillment of this condition difficult to ensure. In the case of symbol-spaced equalizers, the nonoptimal sample phases may also result in unexpected timing functions.

A variant to this algorithm for T-spaced equalizers is to use alternating signs of the equalizer taps, summing before squaring:

$$z_1(\tau) = \left\| \sum_{k=0}^{K-1} (-1)^k f_n(\tau) \right\|^2 - \left\| \sum_{k=N-K}^{N-1} (-1)^k f_n(\tau) \right\|^2. \tag{12.88}$$

The same condition applies to the derivative as far as dynamic stability. This timing function makes use of the fact that band-edge components tend to appear more on one side of an equalizer response when timing lags and the other side when timing leads. This assumes the channels are predictable enough that this phenomenon can be relied on; when that assumption holds, not only will a stable timing phase be found, but this phase will also tend to reduce the band-edge gain of the equalizer along with maximizing the band-edge energy to approach an optimal timing phase.

In the case of DFE-equalization, the combined response of the feedforward and feedback sections may be used for the two sums. More simply, the DFE taps alone can be used to represent the postcursor band-edge energy, using the function

$$z_2(\tau) = \left\| \sum_{k=0}^{K_1-1} (-1)^k f_n(\tau) \right\|^2 - \left\| \sum_{k=N_2-K_2}^{N_2-1} (-1)^k g_n(\tau) \right\|^2, \tag{12.89}$$

where $g_n(\tau)$, $\tau = 0, 1, \ldots, N_2 - 1$ are the feedback filter taps, and K_1 and K_2 are now chosen to provide a balanced timing function. This choice will inevitably depend on anticipated channel conditions.

Calculation of the Effect of Fractional Delays on Symbol-Spaced Sampling An alternative to using the equalizer taps directly is to use the equalizer taps to estimate the effect of shifting the sampling time forward or backward, and then shift the phase based on the results. Unlike fractionally spaced sampling, the effect of alternate sample timing upon the received samples cannot be directly calculated in the case of symbol-spaced sampling. However, the effect on the equalizer can be calculated. Given an equalizer impulse response $\{h_k(t)\}$, which can either be a feedforward response or a combined feedforward and feedback response, by convolving with delay filters, the resulting response can be predicted. For example, using a filter $\{d_k(\tau)\}$ with constant amplitude response and small group delay, the equalizer response corresponding to a delayed sample time $t + \tau$ can be estimated as

$$h(t + \tau) = h(t) \star d(\tau). \tag{12.90}$$

Correspondingly, the equalizer response corresponding to a phase advance could be calculated with a filter with negative delay:

$$h(t - \tau) = h(t) \star d(-\tau). \tag{12.91}$$

Although such a filter is not realizable as such, the equivalent result could be achieved with two filters $\{d_k(nT + \tau)\}$ and $\{d_k(nT - \tau)\}$ if the delayed and advanced impulse responses $\{h_k(t + \tau)\}$ and $\{h_k(t - \tau)\}$ were calculated and evaluated against the unshifted response $\{h_k(t)\}$. Given an appropriate cost function, it is feasible to evaluate the three responses and shift the sample time based on the cost of each sampling phase. If the lowest cost is associated with the optimal sampling phase and monotonically increasing as the sample phase is moved away in either direction, a stable timing recovery loop can be constructed. One possibility is to simply evaluate the power of the impulse response, as the optimal sampling phase will closely correspond with the equalizer response with lowest power; *i.e.*,

$$c(\tau) = \sum_k \|h_k(\tau)\|^2. \tag{12.92}$$

A more sensitive measure may be achieved by omitting the center taps of the equalization response, as they tend to be less sensitive to timing phase than the taps further out. That leads to something that can be referred to as an energy compactness measure:

$$c(\tau) = \sum_{k<K_1, k>K_2} \|h_k(\tau)\|^2. \tag{12.93}$$

To be an efficient timing recovery method, the delay filter must provide for efficient calculation. [Jonsson 2002] suggests using filters of the form $d_k = r_k \pm s_k$, where $\{r_k\}$ is a delay of a fixed number of symbols:

$$r_k = \begin{cases} 1, & k = k_0 \\ 0, & k \neq k_0 \end{cases} \qquad (12.94)$$

and $\{s_k\}$ is an asymmetric response:

$$s_k = \begin{cases} 0, & k = k_0 \\ s_{2k_0-k}, & k \neq k_0 \end{cases}. \qquad (12.95)$$

The two delay filters are then $d_k = r_k + s_k$ and $d_k = r_k - s_k$, allowing a significant reduction in the complexity of calculating the difference in energy between the two nonzero delays:

$$\begin{aligned} c_k(t+\tau) - c_k(t-\tau) &= |h_k(t+\tau)|^2 - |h_k(t-\tau)|^2 \\ &= |(h(t) \star r)_k + (h(t) \star s)_k|^2 - |(h(t) \star r)_k - (h(t) \star s)_k|^2 \\ &= 4(h(t) \star r)_k (h(t) \star s)_k \\ &= 4h_{k+k_0}(t)(h(t) \star s)_k. \end{aligned} \qquad (12.96)$$

Although nontrivial, the calculation of the combined equalizer response and calculation of the cost difference can typically be performed at a rate substantially lower than the symbol rate, often using hardware components of the equalizer itself.

Relation to Gradient Descent Method As noted earlier, if there is a timing error, an adaptive equalizer will tend to absorb that error. As this will occur through the tap update, it may be possible to extract timing information there. Denote the equalizer taps by $c_k^{(n)}$, where n is a time index and k is a tap index. If the error is e_n and the equalizer input is x_n, the update equation is

$$c_k^{(n+1)} = c_k^{(n)} - \beta e_n x_{n-k}^*, \qquad (12.97)$$

where β is the update gain. If $c_k^{(n)}$ is being modified in the direction of $c_{k+1}^{(n)}$, the effect is to delay the timing phase of the equalizer. Hence, if the correction $e_n x_{n-k}^*$ correlates with $c_{k+1}^{(n)} - c_k^{(n)}$, the timing phase should be advanced. Similarly, if the correction lines up with $c_{k-1}^{(n)} - c_k^{(n)}$, the timing phase should be delayed. Therefore, a usable timing function is

$$\begin{aligned} \phi_k^{(n)} &= \mathrm{Re}\left[\left(c_{k-1}^{(n)} - c_k^{(n)} \right) (e_n x_{n-k}^*)^* - \left(c_{k+1}^{(n)} - c_k^{(n)} \right) (e_n x_{n-k}^*)^* \right] \\ &= \mathrm{Re}\left[\left(c_{k-1}^{(n)} - c_{k+1}^{(n)} \right) (e_n^* x_{n-k}) \right]. \end{aligned} \qquad (12.98)$$

Summing over k yields

$$\begin{aligned} \phi^{(n)} &= \sum_{k=0}^{K-1} \phi_k^{(n)} \\ &= \mathrm{Re}\left[\sum_{k=0}^{K-1} \left(c_{k-1}^{(n)} - c_{k+1}^{(n)} \right) e_n^* x_{n-k} \right] \\ &= \mathrm{Re}\left[e_n^* \left(\sum_{k=1}^{K-1} c_k^{(n)} x_{n+1-k} - \sum_{k=0}^{K-2} c_k^{(n)} x_{n=1-k} \right) \right] \\ &\approx \mathrm{Re}\left[e_n^* (y_{n+1} - y_{n-1}) \right], \end{aligned} \qquad (12.99)$$

which, incidentally, is the same timing function as will be derived later in Equation 12.132.

12.2.5 The Gardner Timing Function

Gardner [Gardner 1986] proposed a timing error function using $T/2$-spaced samples as

$$z_k^{(mg-2)}(\tau) = \frac{1}{2\sigma_a^2} y\left(\tau + \left(k - \frac{1}{2}\right)T\right)(y(\tau + kT) - y(\tau + (k-1)T)).$$ (12.100)

The motivation is that for bipolar signaling, the "in-between" sample would give a useful measure of timing error in the case of symbol transitions, with the sign taken care of by the symbol difference being approximated by the received symbols, and zeroed by the symbol difference otherwise. The basic principle would extend, albeit with increased variance, to multi-level signaling. Based on this reasoning, a modified $T/2$-spaced timing error function is also considered, using the quantized symbols instead of the received symbols:

$$z_k^{(mg-1)}(\tau) = \frac{1}{2\sigma_a^2} y\left(\tau + \left(k - \frac{1}{2}\right)T\right)(a_k - a_{k-1}).$$ (12.101)

For this timing function, the expected value is well defined as

$$
\begin{aligned}
E\left[z_k^{(mg-1)}(\tau)\right] &= \frac{1}{2\sigma_a^2} E\left[\left(y\left(\tau + \left(k - \frac{1}{2}\right)T\right) + v\left(\tau + \left(k - \frac{1}{2}\right)T\right)\right)(a_k - a_{k-1})\right] \\
&= \frac{1}{2\sigma_a^2} E\left[\sum_{m=-\infty}^{\infty} a_m g\left(\tau + \left(k - \frac{1}{2} - m\right)T\right)(a_k - a_{k-1})\right] \\
&\quad + \frac{1}{2\sigma_a^2} E\left[v\left(\tau + \left(k - \frac{1}{2}\right)T\right)(a_k - a_{k-1})\right] \\
&= \frac{1}{2}\left(g\left(\tau - \frac{1}{2}T\right) - g\left(\tau + \frac{1}{2}T\right)\right),
\end{aligned}
$$ (12.102)

where additive noise $v(t)$ has been included. The timing function will then converge to a point where the combined response is equal at half-symbol times to either side.

The original timing function is straightforward to evaluate, yet somewhat less intuitive:

$$
\begin{aligned}
&E\left[z_k^{(mg-2)}(\tau)\right] \\
&= \frac{1}{2\sigma_a^2} E\left[\begin{array}{c}\left(y\left(\tau + \left(k - \frac{1}{2}\right)T\right) + v\left(\tau + \left(k - \frac{1}{2}\right)T\right)\right) \\ (y(\tau + kT) + v(\tau + kT) - y(\tau + (k-1)T) - v(\tau + (k-1)T))\end{array}\right] \\
&= \frac{1}{2\sigma_a^2}\left(\begin{array}{c}E\left[\begin{array}{c}\sum_{m=-\infty}^{\infty} a_m g\left(\tau + \left(k - \frac{1}{2} - m\right)T\right)\sum_{n=-\infty}^{\infty} a_n g(\tau + (k-n)T) \\ -\sum_{m=-\infty}^{\infty} a_m g\left(\tau + \left(k - \frac{1}{2} - m\right)T\right)\sum_{n=-\infty}^{\infty} a_n g(\tau + (k-1-n)T)\end{array}\right] \\ +E\left[v\left(\tau + \left(k - \frac{1}{2}\right)T\right)v(\tau + kT)\right] - E\left[v\left(\tau + \left(k - \frac{1}{2}\right)T\right)v(\tau + (k-1)T)\right]\end{array}\right) \\
&= \frac{1}{2}\sum_{m=-\infty}^{\infty} g\left(\tau + \left(m - \frac{1}{2}\right)T\right)(g(\tau + mT) - g(\tau + (m-1)T)).
\end{aligned}
$$ (12.103)

In both cases there is a clear relation to the Mueller–Müller methods that will be considered in the next section, although in the original $z_k^{(mg-2)}(\tau)$ function, which is of course not decision-directed, convergence can be difficult to guarantee.

Variance   Starting with the modified timing function, the variance is

$$
\begin{aligned}
E\left[\left(z_k^{(mg-1)}(\tau)\right)^2\right] - E^2\left[z_k^{(mg-1)}(\tau)\right] &= \frac{1}{4\sigma_a^4} E\left[y^2\left(\tau + \left(k - \frac{1}{2}\right)T\right)(a_k - a_{k-1})^2\right] \\
&\quad - \frac{1}{4}\left(g\left(\tau - \frac{1}{2}T\right) - g\left(\tau + \frac{1}{2}T\right)\right)^2 \\
&= \frac{E\left[a_k^4\right] - \sigma_a^4}{4\sigma_a^4}\left(g^2\left(\tau - \frac{1}{2}T\right) + g^2\left(\tau + \frac{1}{2}T\right)\right) \quad (12.104) \\
&\quad - \frac{1}{4}\left(g\left(\tau - \frac{1}{2}T\right) - g\left(\tau + \frac{1}{2}T\right)\right)^2 \\
&\quad + \frac{1}{2}\sum_{m=-\infty}^{\infty} g^2\left(\tau + \left(m + \frac{1}{2}\right)T\right).
\end{aligned}
$$

Assuming the combined response is bandlimited, *i.e.*, $G(f) = 0$ for $|f| > 1/T$, using the Poisson sum formula, the summation term is sinusoidal in τ with a period of T:

$$
\begin{aligned}
E\left[\left(z_k^{(mg-1)}(\tau)\right)^2\right] - E^2\left[z_k^{(mg-1)}(\tau)\right] &= \frac{1}{4\sigma_a^4} E\left[y^2\left(\tau + \left(k - \frac{1}{2}\right)T\right)(a_k - a_{k-1})^2\right] \\
&\quad - \frac{1}{4}\left(g\left(\tau - \frac{1}{2}T\right) - g\left(\tau + \frac{1}{2}T\right)\right)^2 \\
&= \frac{E\left[a_k^4\right] - 2\sigma_a^4}{4\sigma_a^4}\left(g^2\left(\tau - \frac{1}{2}T\right) + g^2\left(\tau + \frac{1}{2}T\right)\right) \quad (12.105) \\
&\quad - \frac{1}{2}g\left(\tau - \frac{1}{2}T\right)g\left(\tau + \frac{1}{2}T\right) \\
&\quad + \frac{1}{2T}\left(\|G\|^2 - 2A_1\cos\left(\frac{2\pi\tau}{T} + \arg(A_1)\right)\right),
\end{aligned}
$$

where

$$
A_1 = \int_{-\infty}^{\infty} G(\lambda)G\left(\frac{1}{T} - \lambda\right)d\lambda. \quad (12.106)
$$

Example   Using a transmit signal shaped by the raised cosine filter as in Equation 12.6 and varying the bandwidth parameter R, Figure 12.13 illustrates that there is surprisingly little variation with the excess bandwidth in the expected value of the modified Gardner error function. However, the variance decreases with wider bandwidth, becoming negligible at $R = 1.0$, as shown in Figure 12.14. The expected value of the original Gardner function, on the other hand, has a high dependence on excess bandwidth, whereas the variance behaves similar to the modified function. Figures 12.15 and 12.16 illustrate.

12.2.6   Mueller–Müller-Based Methods

In their landmark paper of 1976 [Mueller 1976], Kurt H. Mueller and Markus Müller formalized a class of timing recovery methods where a timing function is defined in terms

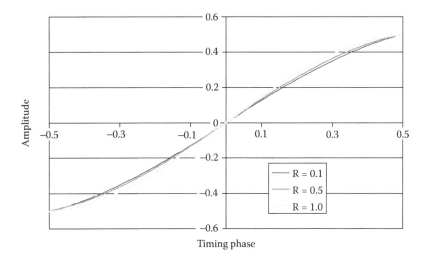

FIGURE 12.13
S-curve for different excess bandwidth — modified Gardner function.

of the channel impulse response (including receiver processing), and then an estimate is calculated using a weighted sum of products of decoded symbols and channel outputs. Meuller–Müller-based methods will converge to a sampling phase that depends on the channel response but is not necessarily optimal in any sense. Given an equalized channel response, a Meuller–Müller timing function may be derived that optimizes the sampling time (in some sense) relative to the channel response.

The primary advantage of Meuller–Müller-based methods is that they can operate on symbol-spaced signals, thus requiring only symbol-spaced sampling. If, however, a feed-forward adaptive equalizer is required to obtain meaningful channel outputs, the targeted sampling phase may drift as timing shifts are absorbed by the equalizer.

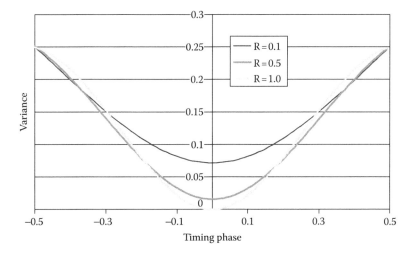

FIGURE 12.14
Variance for different excess bandwidth — modified Gardner function.

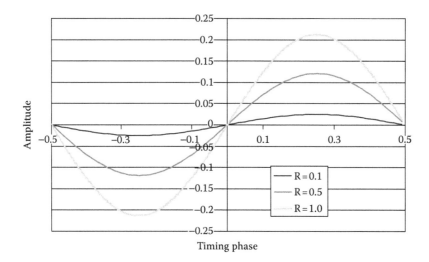

FIGURE 12.15
S-curve for different excess bandwidth — original Gardner function.

Timing Functions   Suppose a receive signal is given as

$$x(t) = \sum_{k=-\infty}^{\infty} a_k g(t - kT),$$ (12.107)

where a_k are the independent transmit symbols, and $g(\cdot)$ is the combined symbol response of the transmitter, channel, and receiver processing. Assume that the desired sampling instants are at even multiples of the symbol time T. If this were not already the case, $g(\cdot)$ could be redefined with a time shift. Define a timing function $f(\tau)$ as

$$f(\tau) = \sum_{n} u_n g(\tau + nT),$$ (12.108)

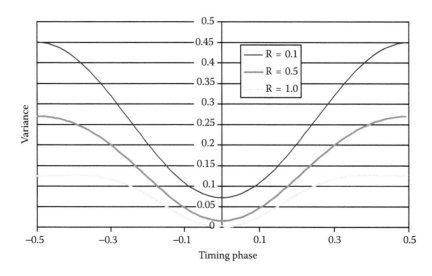

FIGURE 12.16
Variance for different excess bandwidth — original Gardner function.

where the constants u_i are selected such that $f(0) = 0$. If, in addition, it can be ensured that:

$$\frac{\partial f(\tau)}{\partial \tau} > 0 \qquad (12.109)$$

for all τ within a large enough interval around 0, an approximation of $f(\tau)$ will allow the timing phase to be tracked.

For symmetric transfer functions, it is always possible to find such a function that produces a sampling phase at the point of symmetry by selecting $u_0 = 0$ and $u_i = -u_{-i}$ for $i \neq 0$. In that case, clearly $f(\tau) = 0$, and for most practical cases condition Equation 12.109 is also satisfied.

A common example is to set $u_1 = -u_{-1} = \frac{1}{2}$ and other $u_i = 0$; *i.e.,*

$$f(\tau) = \frac{1}{2}(g(\tau + T) - g(\tau - T)). \qquad (12.110)$$

In this case, the slope of the impulse response one symbol period from the center in each direction will determine the slope, and thereby effectiveness, of the timing function.

Another important example, especially in the case of asymmetric responses, is simply to use a single zero crossing of the impulse response. Typically, the first precursor is used, namely $u_{-1} = 1$ (which is the only nonzero component), and:

$$f(\tau) = g(\tau - T). \qquad (12.111)$$

This timing function is common in conjunction with DFE-based equalization, in particular where the size of the feedforward section is limited, as in 2B1Q U-interface modems [Aboulnasr 1994].

Note that although the signal samples used are symbol-spaced, the timing function does depend on the excess bandwidth.

Approximating Timing Functions This subsection considers how to calculate estimates of the timing functions above. The treatment is limited to linear combinations of receive samples, forming an estimator $z_k(\tau)$ that has expected value $f(\tau)$:

$$z_k(\tau) = \sum_{m=0}^{M-1} \alpha_m x(\tau + (k - m)T)$$

$$E[z_k(\tau)] = f(\tau). \qquad (12.112)$$

Mueller and Müller [Mueller 1976] derive general expressions for the expected value and variance of $z_k(\tau)$, and provide guidelines for selecting a minimum variance estimator given a timing function. The reader is referred to their paper for the general case; the focus here will shift toward some examples.

Examples This subsection revisits the examples introduced earlier. First,

$$f^{1a}(\tau) = g(\tau - T). \qquad (12.113)$$

This timing function can be approximated by the function

$$z_k^{(1a)}(\tau) = x(\tau + (k - 1)T)a_k / \sigma_a^2. \qquad (12.114)$$

Clearly,

$$E[x(\tau + kT)a_m] = g(\tau + (k - m)T). \qquad (12.115)$$

Thus, the condition $E[z_k^{(1a)}(\tau)] = f(\tau) = g(\tau - T)$ is satisfied. The variance of this estimator, including a noise term $\upsilon(\tau + kT)$, can be evaluated as well:

$$E\left[\left(z_k^{(1a)}(\tau) - E\left[z_k^{(1a)}(\tau)\right]\right)^2\right] = \left(\frac{E\left[a_k^4\right]}{\sigma_a^4} - 2\right)g^2(\tau - T) + \sum_{m=-\infty}^{\infty} g^2(\tau + mT) + \frac{\sigma_\upsilon^2}{\sigma_a^2}.$$

(12.116)

The tracking figure of merit is thus

$$\lambda^{(MM-1a)} = \frac{(g'(-T))^2}{\sum_{m=-\infty}^{\infty} g^2(mT) + \frac{\sigma_\upsilon^2}{\sigma_a^2}}.$$

(12.117)

Note that in the steady state, when τ is small, $f(\tau) = g(\tau - T) = 0$, and the first term of the variance vanishes. The remaining term is representative of the channel energy and the noise-to-signal ratio. The former will prevent reaching an arbitrarily low variance as the noise vanishes. However, the estimator can be modified to remove the "main" term $g^2(0)$ without affecting the expected value. In the ISI-free case, the other terms of the summation are zero. The estimator becomes

$$z_k^{(1b)}(\tau) = (x(\tau + (k-1)T) - a_{k-1}g(0))a_k/\sigma_a^2.$$

(12.118)

The expected value is unchanged, because $E[a_{k-1}a_k] = 0$:

$$E\left[z_k^{(1b)}(\tau)\right] = g(\tau - T),$$

(12.119)

but the variance is now

$$E\left[\left(z_k^{(1b)}(\tau) - E\left[z_k^{(1b)}(\tau)\right]\right)^2\right] = \left(\frac{E\left[a_k^4\right]}{\sigma_a^4} - 2\right)g^2(\tau - T) + \sum_{m=-\infty}^{\infty} g^2(\tau + mT)$$

$$+(g(0) - 2g(\tau))g(0) + \frac{\sigma_\upsilon^2}{\sigma_a^2},$$

(12.120)

and the tracking figure of merit is

$$\lambda^{(MM-1b)} = \frac{(g'(-T))^2}{\sum_{m=-\infty, \, m\neq 0}^{\infty} g^2(mT) + \frac{\sigma_\upsilon^2}{\sigma_a^2}}.$$

(12.121)

Again, the first term will vanish in the steady state, and if ISI has been eliminated, the summation term will now vanish. Absent any noise, the variance will thus disappear.

Consider the more balanced timing function

$$f^{(2)}(\tau) = \frac{1}{2}(g(\tau + T) - g(\tau - T)).$$

(12.122)

This timing function can be approximated by the estimator

$$z_k^{(2)}(\tau) = (x(\tau + kT)a_{k-1} - x(\tau + (k-1)T)a_k)/2\sigma_a^2.$$

(12.123)

Clearly,

$$E\left[z_k^{(2)}(\tau)\right] = \frac{1}{2\sigma_a^2}(E[x(\tau + kT)a_{k-1}] - E[x(\tau + (k-1)T)a_k])$$

$$= \frac{1}{2}(g(\tau + T) - g(\tau - T)),$$

(12.124)

so the condition $E[z_k(\tau)] = f(\tau)$ is satisfied. The variance of this estimator can be evaluated as well as

$$E\left[\left(z_k^{(2)}(\tau) - E\left[z_k^{(2)}(\tau)\right]\right)^2\right] = \left(\frac{E\left[a_k^4\right] - 2\sigma_a^4}{4\sigma_a^4}\right)(g^2(\tau + T) + g^2(\tau - T))$$

$$+ \frac{1}{2}\sum_{m=-\infty,\ m\neq 0}^{\infty} g^2(\tau - mT) + \frac{\sigma_v^2}{2\sigma_a^2}. \tag{12.125}$$

The tracking figure of merit is thus

$$\lambda^{(MM-2)} = \frac{(g'(T) - g'(-T))^2}{2\sum\limits_{m=-\infty,\ m\neq 0.}^{\infty} g^2(mT) + (g^2(T) + g^2(-T))\left(\frac{E\left[a_k^4\right] - 2\sigma_a^4}{\sigma_a^4}\right) + 2\frac{\sigma_v^2}{\sigma_a^2}} \tag{12.126}$$

PAM Examples   This subsection investigates the response of the error functions to a PAM transmit signal. The signal is shaped by the raised cosine filter as in Equation 12.6, and the bandwidth parameter R is varied. Figures 12.17 and 12.19 illustrate that for small excess bandwidth, timing functions MM-1a and MM-1b become almost one-sided. The high dependence on excess bandwidth is, in fact, somewhat unexpected, because "baseband" signals are used to calculate the error signal.

Considering the variance, Figure 12.20 illustrates that the modification in MM-1b changes the variance from being highest at $\tau = 0$ for MM-1a (see Figure 12.18) to being lowest there.

Considering the more balanced timing function MM-2, Figure 12.21 shows, not surprisingly, that the expected value is anti-symmetric around $\tau = 0$ but is still heavily dependent on the excess bandwidth. Figure 12.22 shows the variance for timing function MM-2.

12.2.7   Gradient-Descent Timing Recovery

A class of methods closely related to the Meuller–Müller methods is based on symbol-decision-error gradient descent. These methods are also referred to as MMSE timing recovery [Sari 1986], [Daneshrad 1995].

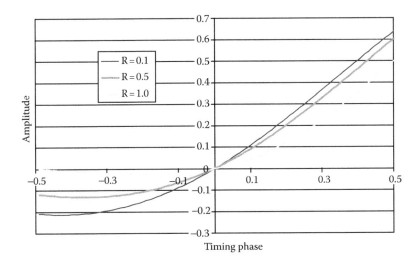

FIGURE 12.17
Expected value of timing function MM-1a.

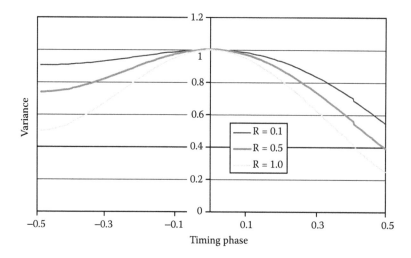

FIGURE 12.18
Variance of timing function MM-1a.

The gradient of the decision error squared as a function of time is given by

$$\frac{\partial e^2(t)}{\partial t} = 2e(t)\frac{\partial e(t)}{\partial t} = 2e(t)\frac{\partial(y(t) - a_k)}{\partial t}. \qquad (12.127)$$

As the transmit symbols a_k are effectively fixed,

$$\frac{\partial e^2(t)}{\partial t} = 2e(t)\frac{\partial y(t)}{\partial t}. \qquad (12.128)$$

This result is reminiscent of the maximum likelihood timing recovery functions, where the derivative of the channel response typically appears. In the case of symbol-spaced sampling, or where equalized symbols are only available at the symbol spacing (which is normally

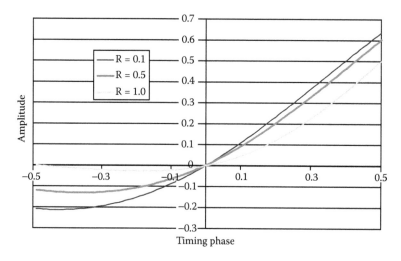

FIGURE 12.19
Expected value of timing function MM-1b.

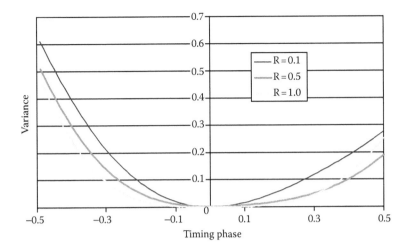

FIGURE 12.20
Variance of timing function MM-1b.

the case in DSL systems), the derivative is difficult to evaluate. However, it is possible to hypothesize and examine some potential approximations, such as

$$\frac{\partial y(t)}{\partial t} \approx \frac{y(t+T) - y(t-T)}{2T}. \tag{12.129}$$

In fact, this approximation only needs to be valid on average; *i.e.*, if

$$E\left[\frac{\partial y(t)}{\partial t} \cdot \frac{y(t+T) - y(t-T)}{2T}\right] > 0, \tag{12.130}$$

the timing function may be useful. Another alternative, considering that the equalized samples are trying to approximate the transmit symbols, would be to use the transmit

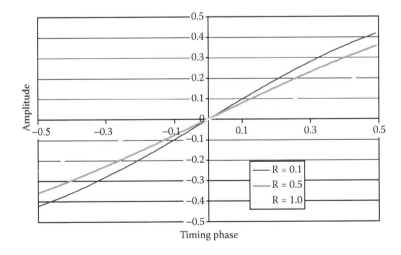

FIGURE 12.21
Expected value of timing function MM-2.

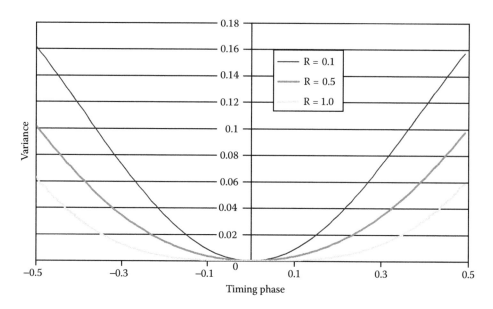

FIGURE 12.22
Variance of timing function MM-2.

symbols themselves:

$$\left.\frac{\partial y(t+kT)}{\partial t}\right|_{t=0} \approx \frac{a_{k+1} - a_{k-1}}{2T}. \tag{12.131}$$

Next, two timing functions are analyzed:

$$z_k^{(1)}(\tau) = \frac{1}{2\sigma_a^2} e_k(y(\tau + (k+1)T) - y(\tau + (k-1)T))$$

$$= \frac{1}{2\sigma_a^2}(y(\tau + kT) - a_k)(y(\tau + (k+1)T) - y(\tau + (k-1)T)) \tag{12.132}$$

and

$$z_k^{(2)}(\tau) = \frac{1}{2\sigma_a^2} e_k(a_{k+1} - a_{k-1})$$

$$= \frac{1}{2\sigma_a^2}(y(\tau + kT) - a_k)(a_{k+1} - a_{k-1}). \tag{12.133}$$

Starting with the second, the expected value is derived as

$$E\left[z_k^{(2)}(\tau)\right] = \frac{1}{2\sigma_a^2} E\left[(y(\tau + kT) - a_k)(a_{k+1} - a_{k-1})\right]$$

$$= \frac{1}{2\sigma_a^2} E\left[\left(\sum_{m=-\infty}^{\infty} a_m g(\tau + (k-m)T) - a_k\right)(a_{k+1} - a_{k-1})\right] \tag{12.134}$$

$$= \frac{1}{2}(g(\tau + T) - g(\tau - T)),$$

which is the same as the second Meuller–Müller timing function. On the other hand, the variance is actually larger:

$$
E\left[\left(z_k^{(2)}(t)\right)^2\right] - E^2\left[\left(z_k^{(2)}(t)\right)\right] = \frac{1}{4\sigma_a^4} E\left[(y(\tau + kT) + v(\tau + kT) - a_k)^2 (a_{k+1} - a_{k-1})^2\right]
$$

$$
-\frac{1}{4}(g(\tau + T) - g(\tau - T))^2
$$

$$
= \frac{E\left[a_k^4\right] - 2\sigma_a^4}{4\sigma_a^4}(g^2(\tau - T) + g^2(\tau + T))
$$

$$
+\frac{1}{2}\sum_{m=-\infty}^{\infty} g^2(\tau + mT) \tag{12.135}
$$

$$
+\frac{1}{2}(1 - g(\tau - T)g(\tau + T)) - g(\tau) + \frac{\sigma_v^2}{2\sigma_a^2}
$$

$$
= \frac{E\left[a_k^4\right] - 2\sigma_a^4}{4\sigma_a^4}(g^2(\tau - T) + g^2(\tau + T))
$$

$$
+\frac{1}{2}\sum_{m=-\infty,\ m\neq 0}^{\infty} g^2(\tau + mT)
$$

$$
+\frac{1}{2}\{(1 - g(\tau))^2 - g(\tau + T)g(\tau - T)\} + \frac{\sigma_v^2}{2\sigma_a^2}.
$$

The figure of merit is thus

$$
\lambda^{GD-2} = \frac{(g'(T) + g'(-T))^2}{\frac{E[a_k^4] - 2\sigma_a^4}{\sigma_a^4}(g^2(-T) + g^2(T)) + 2\sum_{m=-\infty,\ m\neq 0}^{\infty}\left(\begin{array}{c} g^2(mT) \\ +2((1 - g(0))^2 - g(-T)g(T)) \\ +2\frac{\sigma_v^2}{\sigma_a^2}\end{array}\right)}.
$$

$$\tag{12.136}$$

Returning to the first function, the expected value is again the same:

$$
E\left[z_k^{(1)}(\tau)\right]
$$

$$
= \frac{1}{2\sigma_a^2} E[(y(\tau + kT) - a_k)(y(\tau + (k+1)T) - y(\tau + (k-1)T))]
$$

$$
= \frac{1}{2\sigma_a^2} E\left[\left(\sum_{m=-\infty}^{\infty} a_m g(\tau + (k-m)T) - a_k\right)\left(\begin{array}{c}\sum_{m=-\infty}^{\infty} a_m g(\tau + (k+1-m)T) \\ -\sum_{m=-\infty}^{\infty} a_m g(\tau + (k-1-m)T)\end{array}\right)\right]
$$

$$
= \frac{1}{2}(g(\tau - T) + g(\tau + T)) \tag{12.137}
$$

or the same as $[z_k^{(2)}(t)]$. The variance, however, can be shown to be higher.

PAM Examples The response of the gradient descent error functions to a PAM transmit signal is now investigated. The signal is assumed to be shaped by the raised cosine filter as

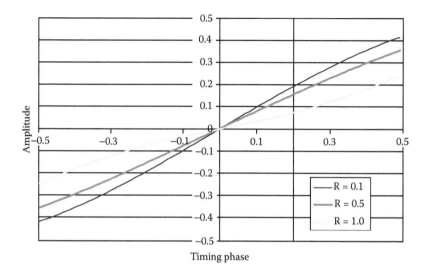

FIGURE 12.23
Expected value of timing function GD-1.

in Equation 12.6 and the bandwidth parameter R is varied. The response for both is similar to the MM-2 function, as Figures 12.23 through 12.26 illustrate.

12.3   DMT Synchronization

In the case of DMT or OFDM, acquiring synchronization is a two-step process that is followed by a clock-tracking phase. First, a receiver needs to synchronize to the DMT

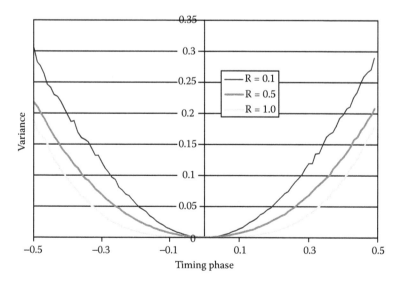

FIGURE 12.24
Variance of timing function GD-1.

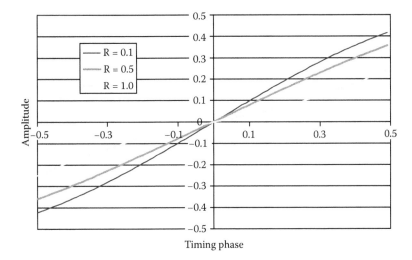

FIGURE 12.25
Expected value of timing function GD-2.

symbol in order to perform meaningful Fourier transforms. Second, the receiver needs to acquire a rough estimate of the timing frequency offset between the transmit and receive clocks to avoid distortion caused by the timing offset. Then, in the steady state, the receiver needs to perform clock tracking in a manner similar to other modulation methods such as PAM and QAM.

Early DMT systems, as used in ADSL (T1-413 [T1.413 1998], G.992.1 [G.992.1 1998]), incorporated a pilot signal to aid in timing acquisition. With this scheme, one of the subcarriers effectively carries no data to provide a fixed reference for timing purposes. Although potentially allowing simpler implementations, more recent designs have omitted the pilot, assuming timing can be extracted from the data signal without any redundancy. Generally, such schemes will be more robust than pilot-based schemes, in particular against

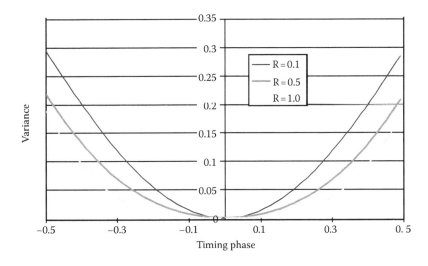

FIGURE 12.26
Variance of timing function GD-2.

impairments that happen to be located at frequencies around the pilot tone, such as radio interference.

Note that as in the single-carrier case, some of the timing phase can be absorbed by the time-domain and frequency-domain equalization. The design must therefore avoid interference between the two, in which a shift in the timing recovery algorithm one way is absorbed by an equalizer change in the other. This interference would eventually lead to an improper sampling window for the input signal and, as a result, intersymbol interference.

12.3.1 Synchronization to the DMT Symbol

Finding the beginning and end of a DMT symbol is not necessarily a simple task. Although the more coarse estimation can be done using correlation methods, the finer estimation ends up as part of the time-domain equalization problem. This section focuses on the coarse estimation.

In traditional ADSL systems, the training sequence begins with transmission of a sequence of repeated symbols without cyclic prefixes. In that case, an estimate of the DMT symbol position can be obtained by simple correlation, where the correlation can be improved by averaging to provide a reliable estimate. Alternatively, the repeated symbols can be used to create a channel estimate from which the symbol delay can be deduced.

If such a repetitive sequence is not available, it is possible to take advantage of the redundancy of the cyclic prefix to find the symbol position. Assume transmission of a symbol with $\overline{N}$ subchannels, resulting in N output samples (where $N = 2\overline{N}$ as in Chapter 7) in each DMT symbol and v samples in a cyclic prefix. Then, the signal is cyclo-stationary with period $N + v$, where the cyclic prefix samples are repeated:

$$x_{k(N+v)+n} = x_{k(N+v)+N+n}, \quad 0 \leq n < v \tag{12.138}$$

Without any channel distortion, an estimator

$$\rho_m = x_m \cdot x_{m-N}. \tag{12.139}$$

could be created, where, because of the prefix,

$$E[\rho_m] = \begin{cases} E\left[x_n^2\right] & k(N+v)+N \leq m < (k+1)(N+v), k \in Z \\ 0 & k(N+v) \leq m < k(N+v)+N, k \in Z \end{cases} . \tag{12.140}$$

Note that the fact that a timing offset will shift the periodicity has been ignored. For example, in a G.992.1 system, the period will shift by 5 percent of a sample time each DMT symbol in the case of a 0.01 percent timing offset.

Averaging over a small number of DMT symbols would quickly reveal a clear indication of where the cyclic prefix repeated itself. Alternatively, a moving average as suggested by maximum likelihood analysis [van de Beek 1997] could be used. On actual channels with attenuation and phase distortion and noise, the correlation will inevitably get smeared. Denoting the channel impulse response by $\{h_n\}$ and noise samples by η_m, the received signal is actually

$$y_m = \sum_{j=-\infty}^{\infty} x_{m-k}h_k + \eta_m, \tag{12.141}$$

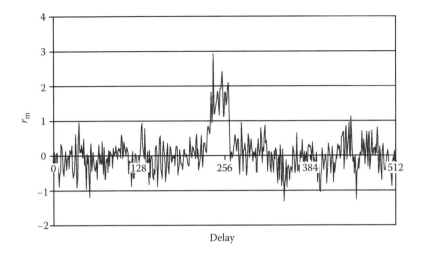

FIGURE 12.27
Correlator output for 0.4 mm 5 km loop at 20 dB SNR after 20 symbol-averages.

and the estimator becomes

$$E[\rho_m] = E[y_m \cdot y_{m-N}]$$

$$= \sum_{j=-\infty}^{\infty} \sum_{k=-\infty}^{\infty} E[x_{m-j} x_{m-N-k}] h_j h_k + E[\eta_m \eta_{m-N}]. \tag{12.142}$$

As the DMT symbol is typically much larger than the channel impulse response, it can be assumed that $h_j = 0$ for $j < 0$ and $j > N$. On the other hand, because the noise may be caused by crosstalk from a like system, the noise term cannot be assumed to be zero. As a result,

$$E[\rho_m] = \sigma_x^2 \sum_{(m-j)\bmod(N+\nu)\geq N}^{h_j^2} + E[\eta_m \eta_{m-N}]$$

$$= \sigma_x^2 \sum_{j=(m-N-\nu+1)\bmod(N+\nu)}^{(m-N)\bmod(N+\nu)} h_j^2 + E[\eta_m \eta_{m-N}], \tag{12.143}$$

where $E[x_k^2] = \sigma_x^2$. This is simply the channel impulse response squared and integrated over a moving and wrapping rectangular window. Figure 12.27 shows an example of the correlation output on a 0.4 mm loop that is 5 km in length at 20 dB SNR, where the output of 20 consecutive frames of correlation is averaged. Even for this extreme channel, a clear indication results. Figure 12.28 shows the result of applying a moving average with a window size equal to the cyclic prefix to the correlation output in Figure 12.27.

12.3.2 Effect of Timing Phase on a Demodulated DMT Symbol

As in the case of a regular Fourier transform, a timing shift in the time domain results in a phase change in the DFT of a signal. If the DFT of a signal $x(t)$, sampled at intervals of T/N, is:

$$X_k = \sum_{n=0}^{N-1} x\left(\frac{nT}{N}\right) e^{-j2\pi nk/N}, \tag{12.144}$$

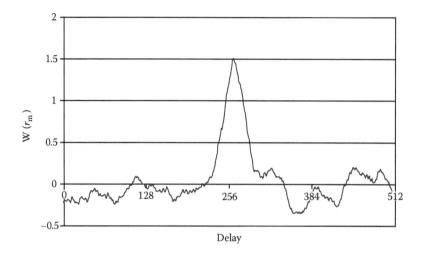

FIGURE 12.28

Correlator output for 0.4 mm 5 km loop at 20 dB SNR at 20 dB SNR after 20 symbol-averages and moving average.

then, if sampled at a timing offset of τ,

$$
\begin{aligned}
X_k(\tau) &= \sum_{n=0}^{N-1} x\left(\tau + \frac{nT}{N}\right) e^{-j2\pi nk/N} \\
&= \sum_{n=0}^{N-1} x\left(\tau + \frac{nT}{N}\right) e^{-j2\pi k(n/N+\tau/T)} e^{j2\pi k\tau/T} \\
&= e^{j2\pi k\tau/T} \sum_{n=0}^{N-1} x\left(T\left(\frac{n}{N}+\frac{\tau}{T}\right)\right) e^{-j2\pi k(n/N+\tau/T)} \\
&= e^{j2\pi k\tau/T} X_k,
\end{aligned}
\tag{12.145}
$$

which indicates the demodulated DMT symbol is rotated by a factor of $2\pi k\tau/T$. If the sampling rate is f_s in the transmitter and $(1+\alpha)f_s$ in the receiver, the phase will change by a factor

$$
\begin{aligned}
\phi &= \frac{2\pi k\tau}{T} \\
&= \frac{2\pi k}{T}\left(\frac{1}{1+\alpha}-1\right) f_s(N+v) \\
&= 2\pi k \frac{\alpha}{1+\alpha}\frac{N+v}{N}
\end{aligned}
\tag{12.146}
$$

after each DMT symbol. However, if the sampling rates are not equal, the effects of the loss of orthogonality will also appear. Taking the DFT at the wrong sampling rate of a transmitted symbol sequence

$$
x\left(\frac{n}{f_s}\right) = \sum_{k=0}^{N-1} X_k e^{j2\pi nk/N},
\tag{12.147}
$$

results in the decoded DMT symbols being

$$
\tilde{X}_k = \sum_{n=0}^{N-1} x\left(\frac{n}{(1+\alpha)\,f_s}\right) e^{-j2\pi nk/N}
$$

$$
= \sum_{n=0}^{N-1}\sum_{m=0}^{N-1} X_m e^{j(2\pi n/N)(m/(1+\alpha)-k)} \tag{12.148}
$$

$$
= \sum_{n=0}^{N-1}\sum_{m=0}^{N-1} X_m e^{j(2\pi n/N)(m-k)} e^{j(2\pi nm/N)(\alpha/1+\alpha)},
$$

and the subcarriers are no longer independent. The impact will be more prevalent at higher subcarriers. As a result, taking the receiver DFT at the wrong sampling rate in itself leads to severe distortion. Although not impossible, reducing that distortion is very costly. Hence, synchronization to the remote sampling rate should occur before the receiver applies a DFT. A second consequence is that during acquisition of timing synchronization, the timing indication derived from higher subcarriers is noisier than that of lower subcarriers. The higher subcarriers should therefore be weighted less or omitted from a calculation of timing phase error during acquisition.

12.3.3   Pilot-Based Timing Acquisition and Tracking

In legacy DMT systems, one of the subcarriers transmits the same subsymbol during each symbol period to allow simple acquisition and tracking of timing phase. This subcarrier is referred to as a pilot. Denoting the index of the pilot subcarrier by k_p, the objective of the timing recovery is to keep $\tilde{X}_{k_p}$ constant. Denoting the mth receive symbol by $\tilde{X}_{k_p,m}$, Equation 12.147 reveals that

$$
\arg(\tilde{X}_{k_p,m}) - \arg(\tilde{X}_{k_p,m-1}) = 2\pi k_p \frac{\alpha}{1+\alpha}\frac{N+\nu}{N}, \tag{12.149}
$$

and the sampling frequency error is

$$
\alpha \approx \frac{N}{2\pi k_p(N+\nu)}(\arg(\tilde{X}_{k_p,m}) - \arg(\tilde{X}_{k_p,m-1}))
$$

$$
= \frac{N\arg(\tilde{X}_{k_p,m}\tilde{X}^*_{k_p,m-1})}{2\pi k_p(N+\nu)} \tag{12.150}
$$

$$
\approx \frac{N}{2\pi k_p(N+\nu)}\frac{\mathrm{Im}[\tilde{X}_{k_p,m}\tilde{X}^*_{k_p,m-1}]}{\mathrm{Re}[\tilde{X}_{k_p,m}\tilde{X}^*_{k_p,m-1}]}.
$$

This relation can be used to obtain initial estimates of the timing frequency offset.
During tracking, the algorithm can be simplified by establishing a tracking objective of

$$
\mathrm{Im}[\tilde{X}_{k_p,m}(\tau)] = 0.
$$

Suppose the estimator given by

$$
z_m(\tau) = \frac{T}{2\pi k_p \|\tilde{X}_{k_p,m}(\tau)\|}\mathrm{Im}[\tilde{X}_{k_p,m}(\tau)] \tag{12.151}
$$

is used. From Equation 12.145, the expected value of the estimator is

$$E[z_m(\tau)] = \frac{T}{2\pi k_p} E\left[\frac{Im[\tilde{X}_{k_p,m}(\tau)]}{\|\tilde{X}_{k_p,m}(\tau)\|}\right]$$

$$\approx \tau, \tag{12.152}$$

which is linear in the sampling phase error as long as the phase error is small. This simple error signal, which is a scaled version of the imaginary part of the pilot subcarrier, can be used as input to a tracking loop.

However, this approach can have disastrous consequences if the pilot frequency happens to be in a frequency region of low SNR, which can (for example) be caused by radio-frequency interference or by the presence of a bridged tap in the subscriber loop [Sands 2002], [Pollet 1999]. Further, a less noisy timing phase error indicator can be obtained by calculating an error estimate for each frequency bin and forming a weighted average based on SNR at the individual subcarriers, thereby taking full advantage of the diversity of the signal.

12.3.4 Nonpilot-Based Timing Acquisition and Tracking

Without a pilot subcarrier (tone), timing acquisition can be done using a sequence of repeated DMT symbols to obtain an estimate of the sampling frequency offset. Then, $X_{k,m} = X_{k,m-1}$, and a timing phase error can be calculated for each subchannel during acquisition similar to the pilot-based case:

$$\arg(\tilde{X}_{k,m}) - \arg(\tilde{X}_{k,m-1}) = 2\pi k \frac{\alpha}{1+\alpha} \frac{N+\nu}{N}, \tag{12.153}$$

or, equivalently,

$$\arg(\tilde{X}_{k,m}\tilde{X}^*_{k,m-1}) = 2\pi k \frac{\alpha}{1+\alpha} \frac{N+\nu}{N} \tag{12.154}$$

for all signal-bearing subcarriers. An estimator of the frequency offset can then be calculated as

$$z_m(\alpha) = \frac{N}{2\pi(N+\nu)} \sum_k \frac{w_k}{k} \arg(\tilde{X}_{k,m}\tilde{X}^*_{k,m-1}), \tag{12.155}$$

where the weight factors w_k are chosen to weight individual estimates according to their reliability. Assuming the SNR at each subcarrier is known and equal to SNR_k, it can be shown [Sands 2002] that the optimal weighting is proportional to $k^2 SNR_k$, leading to an estimator

$$z_m^{opt}(\alpha) = \frac{N}{2\pi(N+\nu)\sum_k k^2 SNR_k} \sum_k k SNR_k \arg(\tilde{X}_{k,m}\tilde{X}^*_{k,m-1}), \tag{12.156}$$

where the effect of nonorthogonality should be included in the SNR estimate. However, the practical ramifications of estimating the SNR during acquisition are obvious. A practical system would typically use a predetermined set of weights based on expected SNR profiles.

Regarding tracking, a decision-directed approach can be used, where symbol decision estimates are used instead of the known pilot symbol. Denoting the symbol decisions as

$\hat{X}_{k,m}$, an appropriate estimator is

$$z_m(\tau) = \frac{T}{2\pi} \sum_k \frac{w_k}{k} \arg(\tilde{X}_{k,m}(\tau)\hat{X}^*_{k,m}), \tag{12.157}$$

which, using Equation 12.145, has expected value

$$E[z_m(\tau)] = \sum_k w_k \tau = \tau, \tag{12.158}$$

if it is assumed there are no symbol decision errors. In most cases, decision errors will not have a meaningful effect, as they will average out. Their effect can be further limited by increasing the weight on subcarriers with a high bit loading. Assuming the error probability is uniform over the subcarriers, the effect of a decision error on the phase estimate is lower as the constellation gets denser and the received symbol larger. So although the same weighting factors as above provide the optimal estimate in the case of no decision errors, a higher emphasis on higher SNR subcarriers may result in better estimates in practice [Sands 2002], [Pollet 1999].

12.3.4.1   *Effect of Timing Jitter on DMT Performance*

In DSL systems based on PAM and QAM modulation techniques, constellation sizes are typically small enough that timing recovery loops can be easily tuned to reduce timing jitter to a point where it has a negligible effect on performance. Performance of DMT systems, however, with as many as 15 bits per symbol or 32,768 constellation points, will degrade quickly with even small amounts of timing jitter. Assuming an energy E_m for an m-bit constellation with distance 2, the distance between decision points $d_{\min}$ can be estimated as a function of the signal variance σ and number of bits m:

$$d_{\min} = 2\frac{\sigma}{\sqrt{E_m}}. \tag{12.159}$$

Estimating the distance to the outermost signal points, which will be the most sensitive to phase errors, as

$$r_{\max} = (2^{m+1/2} - 1)\frac{\sigma}{\sqrt{E_m}}, \tag{12.160}$$

the amount of rotation that will lead to decision errors, in the absence of noise, is

$$\phi_{err} \approx \frac{1}{2}\frac{d_{\min}}{r_{\max}} = 2^{-(m+1)/2}. \tag{12.161}$$

Relating this back to a timing phase error, the minimum timing slip that will cause an error in the absence of noise is

$$\tau_{err} = \phi_{err}\frac{T}{2\pi k} = \frac{T}{\pi k 2^{\frac{m+3}{2}}}. \tag{12.162}$$

For a subcarrier at the upper end of the transmit band, where $k \approx N/2$, this corresponds to a timing shift of 0.1 percent of a sample. Of course, even smaller timing errors will increase the error probability in the presence of noise. This places extremely stringent requirements on the allowable jitter in the tracking loop.

References

References on Gradient Descent/MMSE Methods

[Daneshrad 1995] B. Daneshrad and H. Samueli, *"A 1.6 Mbps Digital-QAM System for DSL Transmission,"* IEEE Journal on Selected Areas in Communications, no. 9, December 1995, pp. 1600–1610.

[Sari 1986] H. Sari, L. Desperben, and S. Moridi, *"Minimum Mean-Square Error Timing Recovery Schemes for Digital Equalizers,"* IEEE Transactions on Communications, No. 7, July 1986, pp. 694–702.

References on Signal Interpolation Methods

[Farrow 1988] C.W. Farrow, *"A Continuously Variable Digital Delay Element,"* in Proc. IEEE Int. Symp. Circuits. (ISCAS-88), Vol. 3, pp. 2641–2645, Espoo, Finland, June 6–9, 1988.

[Laakso 1996] T.I. Laakso, V. Valimaki, M. Karjalainen, and U.K. Laine, *"Splitting the Unit Delay,"* IEEE Signal Processing Magazine, January 1996, pp. 30–60.

References on Maximum Likelihood Methods

[D'Andrea 1988] N.A. D'Andrea and U. Mengali, *"Tracking Performance of Synchronizers Driven by Trellis-Code Modulated Signals,"* Proceedings ICC, Philadelphia, 1988.

[Franks 1980] L.E. Franks, *"Carrier and Bit Synchronization in Data Communication — A Tutorial Review,"* IEEE Transactions on Communications, No. 8, August 1980, pp. 1107–1121.

[Hirosaki 1982] B. Hirosaki, T. Kato, and Y. Fujinobu, *"Suboptimal Maximum Likelihood Timing Estimator for a PCM Regenerative Repeater,"* IEEE Transactions on Communications, No. 10, October 1982, pp. 2376–2384.

[Kobayashi 1971] H. Kobayashi, *"Simultaneous Adaptive Estimation and Decision Algorithm for Carrier Modulated Data Transmission Systems,"* IEEE Transactions on Communication Technology, No. 3, June 1971, pp. 268–280.

[Meyr 1994] H. Meyr, M. Oerder, and A. Polydoros, *"On Sampling Rate, Analog Prefiltering, and Sufficient Statistics for Digital Receivers,"* IEEE Transactions on Communications, No. 12, December 1994, pp. 3208–3214.

[Meyers 1980] M.H. Meyers and L.E. Franks, *"Joint Carrier Phase and Symbol Timing Recovery for PAM Systems,"* IEEE Transactions on Communications, No. 8, August 1980, pp. 1121–1129.

[Moeneclaey 1980] M. Moeneclaey, *"Synchronization Problems in PAM Systems,"* IEEE Transactions on Communications, No. 8, August 1980, pp. 1130–1136.

[Morelli 1997] M. Morelli, A.N. D'Andrea, and U. Mengali, *"Feedforward ML-Based Timing Estimation with PSK Signals,"* IEEE Communications Letters, No. 3, May 1997, pp. 80–82.

[Ungerboeck 1974] G. Ungerboeck, *"Adaptive Maximum-Likelihood Receiver for Carrier-Modulated Data-Transmission Systems,"* IEEE Transactions on Communications, No. 5, May 1974, pp. 624–636.

References on BETR Methods

[Caron 1977] F.G. Caron, A.E. Desblache, D.N. Godard, and F.P. Maddens, US patent 4,039,748, *"Method and device for synchronizing the receiver clock in a data transmission."*

[Farhang-Boroujeny 1977] B. Farhang-Boroujeny, *"Near Optimum Timing Recovery for Digitally Implemented Data Receivers,"* IEEE Transactions on Communications, No. 9, September 1990, pp. 1333–1336.

[Godard 1978] D.N. Godard, *"Passband Timing Recovery in an All-Digital Modem Receiver,"* IEEE Transactions on Communications, No. 5, May 1978, pp. 517–523.

[Godard 1980] D. Godard, US patent 4,227,252, *"Method and device for acquiring the initial phase of the clock in a synchronous data receiver."*

[Godard 1982] D. Godard, US patent 4,309,770, *"Method and device for training an adaptive equalizer by means of an unknown data signal in a transmission system using double sideband-quadrature carrier modulation."*

[Jablon 1988] N.K. Jablon, C.W. Farrow, and S.-N. Chou, *"Timing Recovery for Blind Equalization,"* in 22nd Asilomar Conf. Signals, Syst., Comput. Rec. (Pacific Grove, CA), Oct 31–Nov 2, 1988, pp. 112–118.

[Jablon 1992] N.K. Jablon, *"Joint Blind Equalization, Carrier Recovery, and Timing Recovery for High-Order QAM Signal Constellations,"* IEEE Transactions on Signal Processing, No. 6, June 1992, pp. 1383–1398.

[Lyon 1975] D.L. Lyon, *"Timing Recovery in Synchronous Equalized Data Communication,"* IEEE Transactions on Communications, No. 2, February 1975, pp. 269–274.

[Qureshi 1982] S. Qureshi, US patent 4,344,176, *"Time recovery circuitry in a modem receiver."*

[Ungerboeck 1990] G. Ungerboeck, US patent 4,969,163, *"Timing control for Modem receivers."*

References on Mueller and Müller Methods

[Aboulnasr 1994] T. Aboulnasr, M. Hage, B. Sayar, and S. Aly, *"Characterization of a Symbol Rate Timing Recovery Technique for a 2B1Q Digital Receiver,"* IEEE Transactions on Communications, No. 2/3/4, Feb/Mar/Apr 1994, pp. 1409–1414.

[Bergmans 1995] J.W.M. Bergmans and H. Wong-Lam, *"A Class of Data-Aided Timing-Recovery Schemes,"* IEEE Transactions on Communications, No. 2/3/4, Feb/Mar/Apr 1995, pp. 1819–1827.

[Cowley 1994] W.G. Cowley and L.P. Sabel, *"The Performance of Two Symbol Timing Recovery Algorithms for PSK Demodulators,"* IEEE Transactions on Communications, No. 6, June 1994, pp. 2345–2355.

[Fertner a) 1997] A. Fertner and C. Sölve, *"Symbol-Rate Timing Recovery Comprising the Optimum Signal-to-Noise Ratio in a Digital Subscriber Loop,"* IEEE Transactions on Communications, No. 8, August 1997, pp. 925–936.

[Fertner b) 1997] A. Fertner and C. Sölve, *"Symbol-Rate Timing Recovery Comprising the Optimum Signal-to-Noise Ratio in a Digital Subscriber Loop,"* IEEE Transactions on Communications, No. 8, August 1997, pp. 925–936.

[Gysel 1998] P. Gysel and D. Gilg, *"Timing Recovery in High Bit-Rate Transmission Systems Over Copper Pairs,"* IEEE Transactions on Communications, No. 12, December 1998, pp. 1583–1586.

[Hwang 2002] I.-S. Hwang and Y.H. Lee, *"Optimization of baud-rate timing recovery for equalization,"* IEEE Transactions on Communications, No. 4, April 2002, pp. 550–552.

[Mueller 1976] K.H. Mueller, and M. Müller, *"Timing Recovery in Digital Synchronous Data Receivers,"* IEEE Transactions on Communications, No. 5, May 1976, pp. 516–531.

[Tzeng 1986] C.-P.J. Tzeng, D.A. Hodges, and D.G. Messerschmitt, *"Timing Recovery in Digital Subscriber Loops Using Baud-Rate Sampling,"* IEEE Journal on Selected Areas in Communications, No. 8, November 1986, pp. 1302–1311.

References on Equalizer-Based Methods

[Haar 2002] S. Haar, D. Daecke, R. Zukunft, and T. Magesacher, *"Equalizer-Based Symbol-Rate Timing Recovery for Digital Subscriber Line Systems,"* GLOBECOM 2002 — IEEE Global Telecommunications Conference, No. 1, November 2002, pp. 320–324.

[Jonsson 2002] R. Jonsson, S. Olafsson, and E. Bjarnason, US patent 6,414,990, *"Timing recovery for a high speed digital data communication system based on adaptive equalizer impulse response characteristics."*

[Qureshi 1977] S.U.H. Qureshi and G.D. Forney, Jr., US patent 4,004,226, *"QAM receiver having automatic adaptive equalizer."*

[Ungerboeck 1976] G. Ungerboeck, *"Fractional Tap-Spacing Equalizer and Consequences for Clock Recovery in Data Modems,"* IEEE Transactions on Communications, No. 8, August 1976, pp. 856–864.

References on Gardner Methods

[D'Andrea 1993] N.A. D'Andrea and M. Luise, *"Design and Analysis of a Jitter-Free Clock Recovery Scheme for QAM Systems,"* IEEE Transactions on Communications, No. 9, September 1993, pp. 1296–1299.

[D'Andrea 1996] N.A. D'Andrea and M. Luise, *"Optimization of Symbol Timing Recovery for QAM Data Demodulators,"* IEEE Transactions on Communications, No. 3, March 1996, pp. 399–406.

[Gardner 1986] F.M. Gardner, *"A BPSK/QPSK Timing-Error Detector for Sampled Receivers,"* IEEE Transactions on Communications, No. 5, May 1986, pp. 423–429.

[Knutson 1999] P.G. Knutson, US Patent 5,987,073, *"Symbol Timing Recovery Network for a Carrierless Amplitude Phase (CAP) signal."*

[Oerder 1987] M. Oerder, *"Derivation of Gardner's Timing-Error Detector from the Maximum Likelihood Principle,"* IEEE Transactions on Communications, No. 6, June 1987, pp. 684–685.

References on DMT

[Barbarossa 2002] S. Barbarossa, M. Pompili, and G.B. Giannakis, *"Channel-Independent Synchronization of Orthogonal Frequency Division Multiple Access Systems,"* IEEE Journal on Selected Areas in Communications, No. 2, Feburary 2002, pp. 474–486.

[Coulson a) 2001] A.J. Coulson, *"Maximum Likelihood Synchronization for OFDM Using a Pilot Symbol: Algorithms,"* IEEE Journal on Selected Areas in Communications, No. 12, December 2001, pp. 2486–2494.

[Coulson b) 2001] A.J. Coulson, *"Maximum Likelihood Synchronization for OFDM Using a Pilot Symbol: Analysis,"* IEEE Journal on Selected Areas in Communications, No. 12, December 2001, pp. 2495–2503.

[G.992.1 1998] ITU Recommendation G.992.1, *"Asymmetrical Digital Subscriber Line (ADSL) transceivers,"* 1998.

[Keller 2001] T. Keller, L. Piazzo, P. Mandarini, and L. Hanzo, *"Orthogonal Frequency Division Multiplex Synchronization Techniques for Frequency-Selective Fading Channels,"* IEEE Journal on Selected Areas in Communications, No. 6, June 2001, pp. 999–1008.

[Larsson 2001] E.G. Larsson, G. Liu, J. Li, and G.B. Giannakis, *"Joint Symbol Timing and Channel Estimation for OFDM Based WLANs,"* IEEE Communications Letters, No. 8, August 2001, pp. 325–327.

[Lashkarian 2000] N. Lashkarian and S. Kiaei, *"Globally Optimum ML Estimation of Timing and Frequency Offset in OFDM Systems,"* ICC 2000 — IEEE International Conference on Communications, No. 1, June 2000, pp. 1044–1048.

[NG-027 1999] I. Reuven, Temporary Document NG-027, SG15 Nuremberg, 2–6 August 1999.

[Park 2002] B. Park, H. Cheon, C. Kang, and D. Hong, *"A Novel Timing Estimation Method for OFDM Systems,"* GLOBECOM 2002 — IEEE Global Telecommunications Conference, No. 1, November 2002, pp. 277–280.

[Pollet 1999] T. Pollet and M. Peeters, *"Synchronization with DMT Modulation,"* IEEE Communications Magazine, No. 4, April 1999, pp. 80–86.

[Pollet 2000] T. Pollet and M. Peeters, *"A New Digital Timing Correction Scheme for DMT Systems Combining Temporal and Frequential Signal Properties,"* ICC 2000 — IEEE International Conference on Communications, No. 1, June 2000, pp. 1805–1808.

[Sands 2002] N.P. Sands and K.S. Jacobsen, *"Pilotless Timing Recovery for Baseband Multicarrier Modulation,"* IEEE Journal on Selected Areas in Communications, No. 5, June 2002, pp. 1047–1054.

[Sathananthan 2001] K. Sathananthan and C. Tellambura, *"Probability of Error Calculation of OFDM Systems with Frequency Offset,"* IEEE Transactions on Communications, No. 11, November 2001, pp. 1884–1888.

[Schmidl 1997] T.M. Schmidl and D.C. Cox, *"Robust Frequency and Timing Synchronization for OFDM,"* IEEE Transactions on Communications, No. 12, December 1997, pp. 1613–1621.

[Sliskovic 2001] M. Sliskovic, *"Carrier and Sampling Frequency Offset Estimation and Correction in Multicarrier Systems,"* GLOBECOM 2001 — IEEE Global Telecommunications Conference, No. 1, November 2001, pp. 285–289.

[T1.413 1998] ANSI, *"Network and Customer Installation Interfaces, Asymmetric Digital Subscriber Line (ADSL), Metallic Interface,"* T1.413-1998, 1998.

[van de Beek 1997] J.-J. van de Beek, M. Sandell, and P.O. Borjesson, *"ML Estimation of Time and Frequency Offset in OFDM systems,"* IEEE Trans. Sig. Processing, Vol. 45, No. 7, July 1997, pp. 1800–5.

[Wiese 2002] B. Wiese, K. Jacobsen, N.P. Sands, and J. Chow, US patent 6,434,119, *"Initializing communications in systems using multi-carrier modulation."*

[Yang 1999] B. Yang, K.B. Letaief, R.S. Cheng, and Z. Cao, *"An Improved Combined Symbol and Sampling Clock Synchronization Method for OFDM Systems,"* WCNC 1999 — IEEE Wireless Communications and Networking Conference, No. 1, September 1999, pp. 1153–1157.

[Yang 2000] B. Yang, K.B. Letaief, R.S. Cheng, and Z. Cao, *"Timing Recovery for OFDM Transmission,"* IEEE Journal on Selected Areas in Communications, No. 11, November 2000, pp. 2278–2291.

[Zhao 2001] Y. Zhao and S.-G. Häggman, *"Intercarrier Interference Self-Cancellation Scheme for OFDM Mobile Communication Systems,"* IEEE Transactions on Communications, No. 7, July 2001, pp. 1185–1191.

References on NISL

[Brophy 1986] S.G. Brophy and D.D. Falconer, *"Investigation of Synchronization Parameters in a Digital Subscriber Loop Transmission System,"* IEEE Journal on Selected Areas in Communications, No. 8, November 1986, pp. 1312–1316.

[Franks 1974] L.E. Franks and J.P. Bubrouski, *"Statistical Properties of Timing Jitter in a PAM Timing Recovery Scheme,"* IEEE Transactions on Communications, No. 7, July 1974, pp. 913–920.

[Franks 1980] L.E. Franks, *"Carrier and Bit Synchronization in Data Communication–A Tutorial Review,"* IEEE Transactions on Communications, No. 8, August 1980, pp. 1107–1121.

[Im 1995] G.-H. Im, D.D. Harman, G. Huang, A.V. Mandzik, M.-H. Nguyen, and J.-J. Werner, *"51.84 Mb/s 16-CAP ATM LAN Standard,"* IEEE Journal on Selected Areas in Communications, No. 4, May 1995, pp. 620–632.

[Meyers 1980] M.H. Meyers and L.E. Franks, *"Joint Carrier Phase and Symbol Timing Recovery for PAM Systems,"* IEEE Transactions on Communications, No. 8, August 1980, pp. 1121–1129.

[Order 1988] M. Oerder and H. Meyr, *"Digital Filter and Square Timing Recovery,"* IEEE Transactions on Communications, No. 5, May 1988, pp. 605–612.

[Panayirci 1996] E. Panayirci and E.Y. Bar-Ness, *"A New Approach for Evaluating the Performance of a Symbol Timing Recovery System Employing a General Type of Nonlinearity,"* IEEE Transactions on Communications, No. 1, January 1996, pp. 29–33.

[Panayirci 1997] E. Panayirci, *"Timing Recovery for DSL Transceivers in the Presence of Residual Echo and Impulsive Noise,"* IEEE Transactions on Communications, No. 8, August 1997, pp. 917–920.

13

Radio-Frequency Interference Suppression in DSL

Rickard Nilsson, Thomas Magesacher, Steffen Trautmann, and Tomas Nordström

CONTENTS

13.1 Introduction 400
    13.1.1 Short RFI Ingress Analysis 401
    13.1.2 What Are the Main Sources of RFI? 402
        13.1.2.1 AM Radio 402
        13.1.2.2 Amateur ("HAM") Radio 404
    13.1.3 Other Sources of Narrowband Interference 406
13.2 Suppression Strategies—An Overview 406
13.3 Analog Suppression Techniques 407
    13.3.1 Common-Mode Choke 407
    13.3.2 Analog Filtering 408
    13.3.3 Active Analog Reference-Based Cancellation 408
        13.3.3.1 Coupling Mechanism and Model 409
        13.3.3.2 An Example: Mixed-Signal RFI Canceller 411
13.4 Digital Suppression Techniques 415
13.5 Passive Digital Suppression Techniques 416
    13.5.1 Digital Receiver Filtering 416
    13.5.2 Adaptive Digital Notch Filters 417
    13.5.3 Receiver Windowing 417
13.6 Active Digital RFI Cancellation 420
    13.6.1 Model-Based Digital RFI Cancellation 421
    13.6.2 Deterministic RF Signal Model 421
        13.6.2.1 Appendix: Taylor Parameterization 423
    13.6.3 Stochastic RF Signal Model 425
        13.6.3.1 Narrowband Signal Model 426
        13.6.3.2 DFT Processing 426
        13.6.3.3 LMMSE Estimator 427
        13.6.3.4 Optimal Low-Rank Approximation 429
        13.6.3.5 Rank of the RFI 431
        13.6.3.6 Partial RFI Cancellation 432
        13.6.3.7 Cancellation Complexity 433
    13.6.4 Nonmodel-Based RFI Cancellation 433
        13.6.4.1 Convergence Speed 434
        13.6.4.2 Canceller Performance 436
    13.6.5 Frequency Invariance 437
13.7 Alternative Methods to Suppress RFI 438
13.8 Evaluation of Digital Suppression Methods 439

13.8.1 Suppression Performance 439
        13.8.1.1 Passive RFI Suppression 439
        13.8.1.2 Active RFI Cancellation Methods for DMT 442
13.8.2 Complexity 445
        13.8.2.1 Passive methods 446
        13.8.2.2 Active Methods for DMT 446
13.9 Summary 448
References 448

ABSTRACT This chapter focuses on techniques for suppressing radio-frequency interference (RFI) in DSL. Amateur radio and AM radio are identified as the main RFI sources capable of severely degrading DSL performance. The chapter describes effective methods for suppressing RFI in both the analog and the digital part of the DSL modem, evaluates their performance, compares their complexity, and gives advice on the selection of canceller parameters.

13.1   Introduction

As discussed in Chapter 1, most of today's copper network was already installed long before new broadband DSL technologies such as ADSL, SDSL, and VDSL became popular. Most existing telephone cables were therefore not originally designed to transport broadband signals with frequencies up to several megahertz. Just as with any metallic conductor, with increasing frequency of the transmitted signal, wires in the cable act more and more as an antenna, emitting electromagnetic waves and therefore causing so-called radio-frequency interference (RFI) egress. Conversely, external radio-frequency (RF) sources that overlap with the spectra of the transmitted signal may induce interfering signals in the wire. This type of noise in xDSL transmission systems is referred to as RFI ingress.

Fortunately, Graham Bell showed great technological foresight when he invented shielded differential-mode signalling over twisted wire pairs as early as 1876 (see Chapter 2). Twisting of the wire pairs helps immensely to improve their egress and ingress properties, at least up to frequencies of several hundred kHz. However, with increasing frequency, imbalances in the cable become more and more prominent, making the transmission more and more vulnerable to both ingress and egress.

Typically, RFI noise can be regarded as narrowband interference compared to the overall bandwidth of DSL systems. One would therefore expect that, in particular, DSL systems based on discrete multi-tone (DMT), such as ADSL and VDSL, which split the broadband channel into many narrowband subchannels, would be perfectly suited to deal with narrowband interference. However, the blockwise processing of the DMT system involves rectangular windowing of the received time-domain signal, which can be regarded as a convolution with the $\sin x/x$ function in the frequency domain. Thus, only for the pathological case when the frequency of a single tone interferer is placed exactly on the frequency grid of the DMT system, interference is limited to one particular DMT tone. Otherwise, the energy of the interferer leaks with $\sin x/x$ characteristics to a large number of neighboring DMT tones, potentially disturbing the whole DMT symbol. This effect is most severe if the interferer is located exactly in the middle of two DMT tones.

Therefore, practically all present broadband DSL technologies have to face the RFI ingress problem. This chapter gives an extensive overview of state of the art RFI suppression.

13.1.1 Short RFI Ingress Analysis

Before concentrating on the two main sources of RFI noise, a short overview of RFI ingress calculation is given in this subsection.

A simple ingress model can be derived if the transmitter is assumed to be an omnidirectional point source [Foster 1995]. The field strength is given by

$$E = \sqrt{\frac{P_t \cdot Z_0}{4 \cdot \pi \cdot d^2}}, \tag{13.1}$$

where P_t is the the transmit power, d is the distance from the point source, and Z_0 is the characteristic impedance of free space. This equation only holds true for the far field, where the electromagnetic waves can be regarded as uniform and planar, yielding a homogeneous electric field. Nevertheless, as long as $(2 \cdot \pi \cdot d)/\gamma > 10$ holds, where γ is the wavelength of the RF signal, Equation 13.1 provides a useful approximation with an error of less than 1 dB [ITU-R 368-7].

According to elementary antenna theory, the differential-mode voltage that is induced in a twisted-pair link exposed to a homogeneous electric field of strength E over a certain length L can be determined from the reciprocal egress problem [Stolle b) 2000] as

$$V_d = \frac{E}{b_i} \cdot \frac{Z_g}{Z_g + Z_d} \cdot \frac{\pi \cdot L^2}{\gamma}, \tag{13.2}$$

where Z_g specifies the input impedance of the modem (for example, $135\,\Omega$), Z_d is the characteristic impedance of the cable in differential mode, and b_i is the current balance of the cable. It is assumed that the part of the cable which is exposed to the field is located at the end of the cable; *i.e.*, the common-mode termination is considered open circuited. If this is not the case, then some additional differential-mode attenuation has to be considered. Note also that Equation 13.2 is slightly modified compared to [Stolle b) 2000] in order to preserve consistency with the definition of the cable balance. In this chapter, current balance of the cable is also defined as $b_i = I_{cm}/I_d$, but in the ingress sense, *i.e.*, for the common-mode-to-differential-mode conversion. Thus, provided that the system under consideration shows reciprocal behavior, the resulting value for b_i will be reciprocal to the definition given in [Stolle a) 2000]. The current balance in the dB scale, also known as longitudinal conversion loss (LCL), is then given by

$$\text{LCL} = 20 \cdot \log_{10}(b_i) \text{ [dB]}, \tag{13.3}$$

provided that b_i equals its counterpart voltage balance $b_v = V_{cm}/V_d$. But this is only true if the characteristic impedances of the common mode and differential mode are identical. However, because the characteristic impedance of the common mode is usually much higher than that of the differential mode, a more accurate relation between b_v and b_i reads [Stolle a) 2000]

$$b_v = b_i \cdot \frac{Z_{cm}}{Z_d}. \tag{13.4}$$

Similarly, the exact definition of LCL for nonequal characteristic impedances becomes

$$\text{LCL} = 10 \cdot \log_{10}\left(b_i^2 \cdot Z_{cm}/Z_d\right)$$
$$= 10 \cdot \log_{10}\left(b_v^2 \cdot Z_d/Z_{cm}\right). \tag{13.5}$$

According to Equation 13.2, the induced voltage is dependent on the length L of the exposed part of the cable. However, measurements at BT Laboratories have indicated that cables will, at worst, receive an induced common mode voltage equal to the incident field strength [Foster 1995]. Thus, with $Z_0 = 377\,\Omega$, a simple expression for the induced common-mode voltage can be derived from Equation 13.1:

$$V_{cm} = \frac{5.48 \cdot \sqrt{P_t \cdot V/A}}{d/m}. \tag{13.6}$$

Taking the voltage balance of the cable b_v into account, the differential interfering voltage V_d could then be as high as

$$V_d = b_v \cdot \frac{5.48 \cdot \sqrt{P_t \cdot V/A}}{d/m}. \tag{13.7}$$

Thus, the balance of the cable has a strong impact on the amount of RFI noise leaking into the differential-mode signal.

In general, the cable balance or LCL can be considered frequency-dependent. Although below 1 MHz, the LCL of typical TP cables is 50 dB or better, it may drop to 30–40 dB in the 10 megahertz range [Daecke b) 2000].

The amount of RFI ingress noise depends not only on the balance of the cable, but also on the symmetry of the DSL modem's analog front-end (AFE) against ground potential, and other influences such as splitters, in-house and POTS wiring, as well as connected devices. All influences together form the so-called system balance, which on the dB scale is defined as

$$LCL_{sys} = -20 \cdot \log_{10}\left(10^{-LCL_{cable}/20} + 10^{-LCL_{xDSL}/20} + 10^{-LCL_{other}/20}\right). \tag{13.8}$$

LCL_{sys} is clearly dominated by the smallest, *i.e.*, worst, number among the individual balances. But taking the above numbers for cable balance, especially below 1 MHz, into account, the cable may not necessarily be the worst performer. For example, current DSL standards only require a modem balance of 40 dB or better. Furthermore, due to the drop wire and in-house wiring, RFI ingress is typically worse at the CPE side than at the CO side. Worst-case RFI levels at the CO side may be 20 dB lower than at the CPE side, especially when no central splitter is installed. RFI noise caused by the POTS wiring can be very strong, nonstationary, and unpredictable, especially when the in-house wiring is very unbalanced.

13.1.2 What Are the Main Sources of RFI?

In the frequencies used by DSL (*i.e.*, up to 30 MHz), the two main sources of RFI that influence performance are AM radio and amateur (known as "HAM") radio. Although practically all current DSL flavors have to deal with AM radio generated RFI noise, HAM radio ingress will only affect systems with bandwidths larger than 1.8 MHz, such as VDSL or ADSL2plus. In general, HAM radio is more harmful than AM radio, not only in terms of the noise level, but also due to its nonstationary, unpredictable characteristics.

13.1.2.1 AM Radio

In Europe, two separate frequency bands are used for AM radio: the long wave (LW) band, ranging from 148.5 to 283.5 kHz, and the medium wave (MW) band, ranging from 526.5 to 1606.5 kHz. Stations are spaced at 9 kHz, except for two LW stations at 177 kHz and 183 kHz. This limits the bandwidth of the modulating audio signal to 4.5 kHz.

The maximum allowed transmission power is 2000 kW, but only a few strong LW stations transmit with 2000 kW. For MW AM carriers, the maximum power is typically limited to about 600 kW. In general, LW stations transmit with higher power than MW stations. Although 80 percent of all LW stations transmit with more than 200 kW, only 20 percent of all MW stations exceed this power. Sometimes several transmitters send the same signal with strict phase lock. This is common practice, for example, in Great Britain.

In Europe, with such strong transmitters, the number of AM transmitters close to major cities is typically very small. In general, it is not possible to be close to more than one LW

AM radio station. MW stations are more widespread but mostly transmit at lower power. Therefore, only one or, at most, two strong interferers are to be expected in the near vicinity (within 10 km) of any location in Europe [Daecke a) 2000, Reusens a) 2002].

The worst case is likely to be found in Great Britain, because Great Britain is one of the few countries where both LW and MW stations are located close together. Overhead telephone lines that are more susceptible to RFI are also common in Great Britain. Furthermore, Great Britain has a very high total effectively radiated power (ERP) relative to its size. One example of the RFI severity in Great Britain is Droitwich, close to Birmingham, which has four strong AM stations (one LW station at 198 kHz, two MW stations within the ADSL band at 693 and 1053 kHz, and one MW station above the ADSL band at 1215 kHz).

The situation in the United States is very different. First of all, the lower LW band is not used. AM stations in the United States use frequencies assigned in 10 kHz increments from 540 to 1700 kHz. The channel bandwidth is standardized to 10 kHz, which limits the highest audio frequency to 5 kHz. However, in practice, an audio bandwidth of 10 kHz is also possible, which results in AM signals of 20 kHz bandwidth. Therefore, in order to prevent overlapping, strong local AM stations in the United States are typically more than 30 kHz apart in frequency. The maximum power licensed in the United States is only 50 kW, which leads to a completely different distribution of AM stations compared to Europe. Due to the limited transmit power, typically many MW transmitters surround a single urban area. Carrier frequencies are categorized according to the intended coverage area into clear channel, regional channel, or local channel frequencies.

To provide a short analysis of AM modulation, some basics are addressed first. Modulation in general means that an analog or digital baseband signal $m(t)$ (the modulating signal) is encoded into a bandpass signal $s(t)$ (the modulated signal), which is given by [Couch 2001]

$$s(t) = \text{Re}\{g(t)e^{j\omega_c t}\}, \tag{13.9}$$

where $\omega_c = 2\pi f_c$ with f_c as the carrier frequency. The function $g(t) = g[m(t)]$ is called the complex envelope and performs a mapping operation on the modulating signal $m(t)$. For AM modulation, the complex envelope is defined as

$$g(t) = A_c \cdot [1 + m(t)], \tag{13.10}$$

where A_c specifies a certain amplitude and thus a certain power level. The modulated signal becomes

$$s(t) = A_c \cdot [1 + m(t)] \cdot \cos \omega_c t. \tag{13.11}$$

The spectrum of an AM modulated signal consists of two components, the main carrier and two sidebands, one to the left and one to right of the main carrier.

Several characteristic parameters of AM modulation can be derived. The modulation percentage or modulation index [Couch 2001], for example, is determined by:

$$\% \text{ modulation} = \frac{A_{max} - A_{min}}{2 A_c} \times 100 = \frac{\max[m(t)] - \min[m(t)]}{2} \times 100. \tag{13.12}$$

If $m(t)$ has a positive peak value of $+1$ and a negative peak value of -1, the AM signal is said to be 100 percent modulated. Another characteristic parameter is the so-called modulation efficiency:

$$E = \frac{\mathcal{M}}{1 + \mathcal{M}} \times 100 \; [\%], \tag{13.13}$$

which determines the percentage of the total power of the modulated signal that actually carries information [Couch 2001]. $\mathcal{M}$ stands for the time average of $m^2(t)$. The highest efficiency for a 100 percent modulated AM signal is 50 percent.

Although the main carrier of an AM signal as a single sine tone is quite stable in amplitude and frequency, the sidebands can be rapidly changing. The power spectral density (PSD) of a typical AM modulating signal is definitely not white. It usually has a peak at around 500 Hz, and most energy lies within 500–1000 Hz distance from the carrier. This is because typical music signals contain dominant low frequencies.

For the unmodulated carrier, the crest factor, *i.e.*, the ratio between peak amplitude value and root mean square (RMS) value, would be exactly $\sqrt{2}$ (3 dB). The crest factor of the modulating signal $m(t)$ depends on the type of signal. Speech typically has a crest factor of about 12 dB, whereas music can have crest factors in the range of 16 to 20 dB, depending on the nature of the music. On the other hand, highly compressed music signals such as modern pop and rock music may have crest factors of only about 4 dB.

The typical crest factor of the modulated signal is always less than 9 dB, because the RMS value of the modulated carrier is typically larger than the RMS value of the unmodulated carrier [Reusens b) 2002]. This is independent from the crest factor of the modulating signal, even if the peak amplitude value of the modulated carrier is twice as large as the peak amplitude of the unmodulated carrier, which corresponds to the maximum modulation index of 100 percent. Furthermore, the modulation index for AM stations in the field is usually kept to about 80 percent in order to prevent over-modulation.

Interestingly, if the carrier frequency of an AM signal aligns closely to a DMT tone frequency in the 4.3125 kHz grid of subcarriers, the interference caused by the carrier itself is more or less restricted to the closest DMT tone only. Then, interference to the other DMT tones is dominated by the sidebands, which results in a larger variation of the interfering signal, *i.e.*, a higher crest factor.

Another observation can be made at night: nonstationarity of the AM RFI noise becomes more prominent because of the better MW propagation. Nevertheless, thanks to its constant carrier part, AM radio noise is easier to track than HAM radio noise (see next section), and therefore perfectly suited to adaptive cancellation suppression strategies.

RFI noise caused by AM radio stations is also much smaller than HAM radio interference. Levels of less than -90 dBm [Daecke c) 2000] are typical, and worst case levels are around -40 dBm.

13.1.2.2 Amateur ("HAM") Radio

Amateur ("HAM") radio is confined to a number of relatively small frequency bands, most of which are standardized by international agreement. As can be seen from Table 13.1, there is no HAM band in the ADSL and SDSL frequency range, and only the first HAM band between 1.81 and 1.85 MHz overlaps with the extended bandwidth of the new ADSL2plus

TABLE 13.1

Allocated Amateur Radio Bands for ITU Region 1 (Europe, Middle East, Africa, and North Asia)

Start Frequency (MHz)	End Frequency (MHz)
1.810	1.850 (in some countries 2.000)
3.500	3.800
7.000	7.100
10.100	10.150
14.000	14.350
18.068	18.168
21.000	21.450
24.890	24.990
28.000	29.700

standard. Thus, with a total of four frequency bands in the VDSL frequency range, amateur radio is mainly an issue for VDSL.

Radio amateurs typically transmit at the maximum permitted power. For example, in Great Britain, transmit power is limited to 400 W ERP, whereas in the United States up to 1 kW is allowed. The transmit signal is radiated in all directions equally from an omnidirectional antenna. HAM signals may occasionally change frequency but typically stay within a single HAM band.

The most typical modulation type for HAM radio is single-sideband suppressed carrier (SSB-SC), which can be derived in two steps from normal AM modulation. In a first step, the carrier is eliminated, which leads to double-sideband suppressed carrier (DSB-SC) modulation. The complex envelope becomes

$$g(t) = A_c \cdot m(t), \tag{13.14}$$

where $m(t)$ is assumed to have no DC component. Then, the modulated signal simplifies to

$$s(t) = A_c \cdot m(t) \cdot \cos \omega_c t. \tag{13.15}$$

The modulation index for DSB-SC is infinite, because a carrier spectral line component does not exist. Modulation efficiency becomes 100 percent, because no power is wasted for the carrier.

A further reduction in power/bandwidth can be obtained if the DSB signal is appropriately combined with the Hilbert transform $\hat{m}(t)$ of the modulating signal $m(t)$, such that one of the sidebands is suppressed:

$$g(t) = A_c \cdot [m(t) \pm j\hat{m}(t)]. \tag{13.16}$$

Accordingly, the modulated signal changes to

$$s(t) = A_c \cdot [m(t) \cos \omega_c t \pm \hat{m}(t) \sin \omega_c t]. \tag{13.17}$$

Typically, SSB-SC modulation occupies a bandwidth of 2.5 to 4 kHz.

Despite the smaller bandwidth, there are other significant differences that distinguish a HAM radio signal from normal broadcast transmission. For AM signals, more than half of the transmit power is put into the carrier. Therefore, an AM signal contains a dominant fixed spectral component that is always active and therefore behaves a lot more predictably than a HAM signal. On the other hand, amateur radio transmission usually follows a half-duplex scheme where RF power is not continuously emitted. A signal is only transmitted if the radio amateur is talking, whereas for all other periods the transmitter is quiet with little or no RF power radiated. HAM radio has the same talk-spurt characteristic of speech and also exhibits a similar crest factor of up to 12 dB. Due to its very adverse statistics, interference from SSB transmissions cannot be adaptively notched out [Foster 1995]. This has to be taken into account if one develops a method for suppressing HAM radio signals.

If the radio amateur is located in the direct neighborhood, induced voltages due to RFI from the HAM transmitter can be quite high and may overload the DSL modem's ADC [Foster 1995]. Measurements show that on a typical access dropwire at a distance of 10 m from the amateur radio antenna, induced RMS voltages in the region of 0.2–0.3 V may occur anywhere in the amateur bands [Foster 1995].

In [Daecke a) 2000], some interesting statements are made concerning the distribution of radio amateurs in Great Britain: for a typical English household, there is only a 50 percent likelihood that there is no amateur within 800 m, which drops to 10 percent likelihood if the radius is increased to 1400 m. In combination with Equation 13.7, the probability of worst-case RFI levels due to HAM radio may be computed, as well. Depending on the overall system balance, there is a 50 percent probability that an overhead drop wire can pick up RFI levels of −40 dBm and more. With a more realistic model for neighborhood housing density, one obtains a 1 percent probability of −10 dBm interference which directly translates to the typical 99 percent worst case used for test purposes. There is even a 0.1 percent probability of 0 dBm interference.

13.1.3 Other Sources of Narrowband Interference

Radio services other than AM radio and HAM radio, such as fixed radio, radio navigation, maritime, and aeronautical radio [Daecke a) 2000], are beyond the scope of this chapter. When these other sources of RFI are narrowband in nature, they have good prerequisites to be effectively handled by some of the suppression methods in this chapter. However, if they have a more wideband nature (as wideband crosstalk), they can be treated either as constant additional background noise or impulse noise, depending on whether they are stationary.

13.2 Suppression Strategies—An Overview

Suppression strategies can be classified according to certain criteria. In general, a single optimal strategy that fits all possible scenarios does not exist. Whether a certain suppression method can be applied depends on specific conditions such as strength of the induced RFI noise, characteristics of the RFI noise, complexity limitations, and others.

Next, this chapter provides a short overview of the different classes of RFI suppression strategies and their purposes. As pointed out previously, the RFI noise from HAM radio may be significantly larger than the received signal itself. Thus, the disturbing signal can drive the DSL ADC into saturation, which turns the linear RFI noise component into a nonlinear one. It is not possible by any means to compensate for this nonlinear distortion later on in the digital domain. The only solution would be an ADC with increased dynamic range that has enough margin to cope with the worst-case RFI noise while at the same time keeping an acceptable level of resolution for the actual received signal. This, of course, would not be very efficient in terms of power consumption and cost. Thus, there is a clear need for methods that suppress RFI before the receiver's ADC. These methods are referred to as analog suppression techniques. However, because operating in the analog domain is more costly, less flexible, and often dependent on the nonideal behavior of analog circuitry, one would restrict the effort to be put into those analog RFI suppression methods such that the remaining signal does not exceed the maximum amplitude of the ADC. The logical consequence is a second suppression step in the digital domain, which reduces the RFI to the final desired level. Methods that suppress RFI after the receiver's ADC and thus operate exclusively in the digital domain are referred to as digital suppression techniques.

Optimal results can only be achieved with suppression methods that allow active tracking of the disturbing signal. For those active methods, the compensator parameters are adjusted according to the present RFI noise at least once at start-up, if the interference is assumed to be stationary in time, or continuous during steady-state operation. However, sometimes the implementation of an active method is too costly for the desired application, or the nonpredictive, time-varying characteristics of certain types of RFI noise, such as HAM radio, make it too difficult to track the interference. In that case, a passive method with predetermined compensator parameters may be a better choice. Also, quite often a passive method is used to support a subsequent active method.

Active methods operating in the analog domain can be further separated depending on how the compensation signal is generated. One possibility is a reference-based method, where the compensation signal is derived from a reference signal that is correlated with the RFI disturber. Such a reference signal could be the common-mode signal which, in most cases, is strongly related to the differential-mode RFI noise.

Similarly, active digital methods can be split into methods that do not have any a priori knowledge about the interference, and methods that assume a pre-defined model for the interference and try to adapt the model parameters with respect to certain error-minimization criteria. Different types of models, either deterministic or stochastic, are possible.

It should be mentioned that it is, of course, possible to further classify suppression strategies. For example, not all suppression methods are equally applicable for a certain DSL modulation type, typically multi-carrier and single-carrier modulation. In fact, most of the methods presented in this chapter refer to multi-carrier modulation. This is a direct consequence of its splitting of the transmission into many narrowband subchannels and the inherent availability of time and frequency domain information, with the result that multi-carrier modulation has good inherent potential for suppressing RFI.

13.3 Analog Suppression Techniques

Analog suppression techniques combat RFI before the receiver's ADC. The ADC, the heart of a modem's analog front-end, is characterized by its resolution, usually expressed in number of effective bits, and the bandwidth over which this resolution is achieved. The purpose of analog suppression is to avoid an overload of the receiver's ADC, which may be caused by strong RFI, typically HAM. In principle, it is possible to move analog suppression efforts into the digital domain by increasing the ADC's resolution so that RFI can be sampled together with the received signal. However, improving the ADC's resolution while maintaining the required bandwidth and keeping the high-volume production cost low is a challenge.

As described earlier, the worst-case RFI levels occur in the neighborhood of radio amateur transmitters. According to [De Clercq 2000], the ingress level at the input of the DSL receiver can be as high as -5 dBm or 207 mV into 135 Ω. This value should be compared to the levels of the DSL signal. For a long VDSL loop,[1] the wideband receive power at the CO side is calculated as -39 dBm or 4 mV into 135 Ω.

Even if perfect digital RFI suppression were possible, it would still require approximately 4–6 extra bits in the analog-to-digital conversion, which would be very difficult and expensive to provide. In these cases, an analog suppression of at least 25–35 dB would be needed. Following the nomenclature defined previously, the suppression techniques can be classified as follows:

- Passive: Common-mode choke, analog filtering
- Active: Reference-based cancellation

13.3.1 Common-Mode Choke

A common-mode choke is an inductor that is intended to filter (or to "choke out") common-mode (CM) signals. By winding the incoming pair of wires around a toroidal magnetic core, the impedance, which the CM signal sees, increases. Consequently, a common-mode choke

---

[1] 1000 m of ETSI VDSL loop 1 (*i.e.*, 0.5 mm BT-dwug); using band plan 997 with mask M2, and implementing upstream power back-off. This yields a bit-rate of approximately 14/10 Mbit/s.

attenuates the CM ingress and, hence, also the differential-mode (DM) ingress. However, because the choke represents an additional longitudinal element in the loop, it has an impact on the differential-mode signal. Another drawback of the common-mode choke is the relatively high cost.

13.3.2 Analog Filtering

A straightforward analog RFI suppression technique is filtering. The simple approach of attenuating entire HAM bands using notch filters, however, causes a degradation of the modem performance that may be unacceptable. The design of analog notch filters that adapt their center frequency to the HAM disturber requires sophisticated signal processing algorithms, the implementation effort of which could exceed the effort of improving the receiver's ADC, thus turning it into an infeasible approach. Apart from the signal processing aspect, the parameters of analog filters are subject to variations of the manufacturing process and temperature drift; hence, tuning of their center frequency may be required. To summarize, although analog filtering is possible in principle, it bears a number of fundamental problems that limit its practical importance.

13.3.3 Active Analog Reference-Based Cancellation

The principle of a cancellation approach, depicted in Figure 13.1, is to appropriately modify a reference signal $c(t)$, which contains information about the RFI, and subtract it from the primary signal $y(t)$ so that the result $a(t)$ contains only residual RFI that does not overload the ADC. The reference signal could be obtained from an antenna, an approach that is used in wireless communications and radar applications. However, this may be overkill for DSL receivers. Instead, the line itself may be used as an antenna that picks up the reference signal, by extracting the CM signal, as indicated in Figure 13.1. This will be explained in more detail in the sequel.

The principle of CM-reference-based RFI cancellation as well as various suggestions regarding its implementation have been published [Cioffi 1996, Sands 1999, Yeap 1999, Magesacher b) 2001, Ödling 2002] and patented [Yeap 2000, Vitenberg 2002, Yeap 2003] at different levels of detail. Generally, all methods perform an orthogonal representation of the reference signal using some basis signals in order to construct the "counter-interferer" as a linear combination of these basis signals using a set of properly adjusted weights. Different

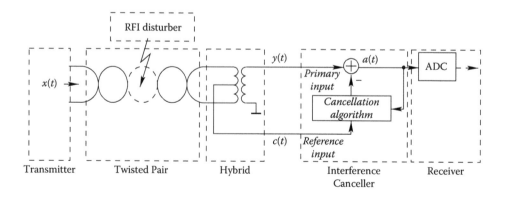

FIGURE 13.1
Principle of reference-based RFI cancellation, where the reference is the common-mode signal available at the center tap of the hybrid.

methods employ different weight-adaptation strategies and consequently exhibit different performance and complexity.

The method reported in [Sands 1999] splits the CM signal into I and Q components, correlates them with the DM signal, and uses the resulting signals to increment or decrement two counters at a fixed clock rate depending on whether the signals are smaller or larger than zero. The counter values correspond to the weights used to construct the counter-interferer from the I and Q components. Because the weights change by a fixed quantization interval every clock cycle, this control loop constitutes a sign-sign-LMS algorithm entirely implemented in the analog domain, which allows an architecture of low complexity. However, the speed of adaptation, often assessed in terms of the convergence time required to reach a certain level of interference suppression starting from a defined state of the canceller, is limited.

The approach described in [Yeap 1999] samples the CM reference at the same rate and with similar resolution as the DM receive signal and uses digital filter banks to construct a noise estimate. In contrast to the previous method, the adaptive algorithm is implemented entirely in the digital domain. The method has a high performance potential, but its complexity is considerable.

A compromise between purely digital and purely analog methods is the mixed-signal approach described and evaluated in [Magesacher b) 2001, Ödling 2002], which is used as an example canceller to demonstrate some principles of analog RFI cancellation in Section 13.3.3.2.

13.3.3.1 Coupling Mechanism and Model

The transmit signal $x(t)$ is applied as a voltage between two wires, causing a differential-mode current. Any radio frequency interferer located closely enough to the wire will cause RFI ingress due to electromagnetic coupling. When talking about RFI ingress, two types should be distinguished, as depicted in Figure 13.2. First, the interference causes an additional DM current in the loop formed by the two wires. The DM signal $y(t)$ measured at the termination impedance Z_d consists of the desired signal $x(t)$, the narrowband disturbance component $s(t)$, and a noise component $v_d(t)$; i.e.,

$$y(t) = x(t) \star h(t) + s(t) + v_d(t), \quad -\infty < t < \infty, \tag{13.18}$$

where $h(t)$ is the channel impulse response and $\star$ denotes convolution. It is assumed that the interference $s(t)$ and the desired signal $x(t)$ are uncorrelated, which holds in practice. The disturbance $s(t)$ interferes additively with the desired signal $x(t)$ and should be kept

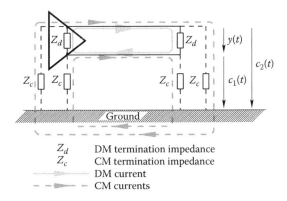

Z_d	DM termination impedance
Z_c	CM termination impedance
	DM current
	CM currents

FIGURE 13.2
Differential-mode (DM) and common-mode (CM) currents and voltages in a wire pair.

as low as possible in order not to saturate the receiver's ADC. Shielded cables would be the preferred choice, but they are rarely installed in the access network. Twisting the two wires lowers the ingress substantially, because the induced currents change their direction from twist to twist and cancel themselves to a certain extent. Secondly, RFI ingress also appears in the loops formed by each of the two wires and ground. These loops are closed by the coupling impedances Z_c. Their values depend on a variety of parameters, for example, the type of the cable, its position relative to ground, the hybrids used for two-wire to four-wire conversion, etc. These CM currents find their return path via ground. The resulting CM signal $c(t) = \frac{c_1(t) + c_2(t)}{2}$, which depends on the system balance discussed previously (see Equation 13.8), consists primarily of the disturbance caused by radio ingress.

There are several ways to obtain the CM signal from a circuit implementation point of view. Using a transformer with a center-tap or employing a CM-choke with an extra winding [Vitenberg 2002] are two practically relevant approaches. The direct implementation of the arithmetic mean relation for $c(t)$ is of minor practical value because it requires two ADCs.

In general, the CM interference is much larger than its corresponding DM component because the CM loops have larger areas and are thus more susceptible to RFI. Due to the unbalance of the wire pair and of other components in the system, $c(t)$ may also contain a small portion of the desired signal. In practice, however, the power of the interference components will be substantially higher than the power of the desired signal component.

Based on the observations at the physical layer, the general coupling model depicted in Figure 13.3 is presented. The coupling impulse responses from DM to CM and vice versa are denoted by $h_{d2c}(t)$ and $h_{c2d}(t)$, respectively. The DM output signal $y(t)$ consists of the desired receive signal $x(t) \star h(t)$, a noise component $v_d^{(c)}(t) + v_c^{(c)}(t) \star h_{c2d}(t)$, which is correlated with the CM output signal $c(t)$ due to the two signal coupling paths, an additional noise component $v_d^{(u)}(t)$, and the RFI component $s(t)$. Analogously, the CM signal $c(t)$ is made up of the narrowband disturber $v_{RFI}(t)$, a noise component $v_c^{(u)}(t)$ uncorrelated with the DM output signal $y(t)$, a correlated part $v_c^{(c)}(t) + v_d^{(c)}(t) \star h_{d2c}(t)$, and the signal component $x(t) \star h(t) \star h_{d2c}(t)$. The coupling between CM and DM and vice versa is linear but frequency dependent. As discussed before, the CM interference $v_{RFI}(t)$ is generally much stronger than the signal $x(t)$. Note that, as the canceller uses the CM $c(t)$ as a reference signal, having a component $x(t) \star h(t) \star h_{d2c}(t)$ of the desired signal at the CM input might, in principle, cause the canceller to, at least partly, cancel the desired DM signal $x(t)$. However, because the purpose of analog RFI suppression is to address the case where $v_{RFI}(t)$ is strong, the component $x(t) \star h(t) \star h_{d2c}(t)$ caused by the DM to CM coupling of the signal has negligible power. Hence, the CM signal is given by $c(t) \approx v_{RFI}(t) + v_c(t)$, where $v_c(t)$

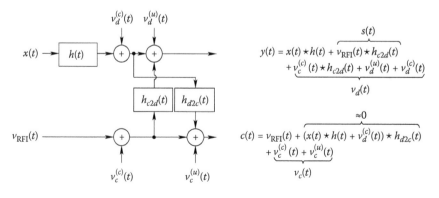

FIGURE 13.3
Signals and coupling model.

represents the total noise at the CM input. Taking into account that the RFI disturber $v_{RFI}(t)$ is of very narrow bandwidth, *i.e.*, essentially sinusoidal, the DM RFI ingress $s(t) = v_{RFI}(t) \star h_{c2d}(t)$ can be written as

$$s(t) = \frac{1}{a_{c2d}} v_{RFI}(t + \tau_{lag}). \tag{13.19}$$

Except for the scaling by $1/a_{c2d}$ and the shift in time by τ_{lag}, the DM interference $s(t)$ is equal to the CM interference $v_{RFI}(t)$. As discussed in Section 13.1.1, depending on the type of wire, the CM to DM coupling can be as high as $a_{c2d,dB} = 30$ dB; *i.e.*, the attenuation from CM to DM can be as low as 30 dB. In the presence of strong HAM signals, the resulting DM RFI levels may be high enough to saturate the ADC, and the desired signal is lost.

VDSL1 occupies the spectrum of the copper channel up to 12 MHz. As Table 13.1 indicates, there are several HAM bands used by radio amateurs between 1 MHz and 12 MHz. The bands themselves are well defined, but within them, the radio amateur transmitter may change its transmit frequency arbitrarily. Although the carrier frequencies are high, the maximum bandwidth of the disturbing signals is only several kHz, as defined by national and international regulations. The canceller needs to track changes in the coupling from the CM signal to the DM signal, *i.e.*, a_{c2d} and τ_{lag} in Equation 13.19 as functions of time. They are both frequency dependent and will change when the RFI disturber changes its frequency. However, their change is virtually zero within the few kHz of bandwidth of a HAM disturber. Also, the disturber does not traverse along the line at any speeds that would cause rapid changes in the coupling. Thus, in practice it can be assumed that a_{c2d} and τ_{lag} are constant for a given RFI disturber.

13.3.3.2 An Example: Mixed-Signal RFI Canceller

In the following, the mixed-signal approach introduced in [Magesacher a) 2001; Ödling 2002] is described in more detail. Figure 13.4 shows a block diagram of the canceller, which is based on the principles described in [Widrow 1975]. The ADCs and the digital-to-analog converters (DACs) operate at the sampling frequency F_{sa}, which corresponds to the algorithm's update rate. This update rate can be chosen to be rather low compared to the

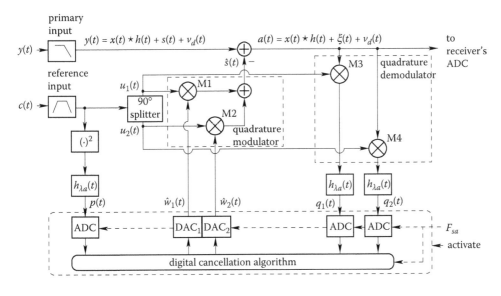

FIGURE 13.4
Canceller block diagram.

frequency of the RFI disturber. The circuit has two inputs: a primary input for the DM signal $y(t)$ and a reference input for the CM signal $c(t)$ (see Figure 13.1). Every $T = 1/F_{sa}$ seconds, the cancellation algorithm calculates a new coefficient vector

$$\hat{\underline{w}}[n] = \begin{bmatrix} \hat{w}_1[n] \\ \hat{w}_2[n] \end{bmatrix}, \qquad n = 0, 1, \ldots, \tag{13.20}$$

which is converted into the time-continuous weight signals $\hat{w}_1(t)$ and $\hat{w}_2(t)$ by the two DACs in Figure 13.4; i.e.,

$$\hat{\underline{w}}(t) = \begin{bmatrix} \hat{w}_1(t) \\ \hat{w}_2(t) \end{bmatrix} = \hat{\underline{w}}[n], \quad nT \le t < (n+1)T. \tag{13.21}$$

The CM signal $c(t)$ serves as a reference and is decomposed into two orthogonal signals, collected in the vector

$$\underline{u}(t) = \begin{bmatrix} u_1(t) \\ u_2(t) \end{bmatrix}, \tag{13.22}$$

by means of a 90°-phase splitter. The elements of $\underline{u}(t)$ are weighted by $\hat{\underline{w}}(t)$ to generate the interference-cancelling signal

$$\hat{s}(t) = \hat{\underline{w}}^T(t)\,\underline{u}(t), \tag{13.23}$$

where $(\cdot)^T$ denotes the transpose. Note that Equation 13.23 is realized by the quadrature modulator in Figure 13.4. With the two parameters $\hat{w}_1(t)$ and $\hat{w}_2(t)$, both amplitude and phase of the sinusoidal interference-cancelling signal $\hat{s}(t)$ can be arbitrarily adjusted. The resulting a priori estimation error is given by

$$\xi(t) = y(t) - \hat{s}(t). \tag{13.24}$$

The quadrature demodulator generates the two-component baseband error signal $\underline{q}(t)$, which is the low-pass filtered product of $\underline{u}(t)$ and the a priori estimation error $\xi(t)$ caused by the current weight vector; i.e.,

$$\underline{q}(t) = \begin{bmatrix} q_1(t) \\ q_2(t) \end{bmatrix} = h_{\lambda a}(t) \star (\underline{u}(t)\xi(t)), \tag{13.25}$$

where $h_{\lambda a}(t)$ is the impulse response of the low-pass filters (see Figure 13.4). The signal

$$p(t) = h_{\lambda a}(t) \star c^2(t) \tag{13.26}$$

provides a measure of the power of the reference signal, which is needed for both the weight-updating and detecting the presence of a disturber, again using a low-pass filter with impulse response $h_{\lambda a}(t)$. The signals $q_1(t)$, $q_2(t)$, and $p(t)$ are sampled at the rate F_{sa} by three ADCs. Because each of these signals is essentially a low-pass filtered product of sinusoids having the same frequency, they can be interpreted as down-converted, DC-like signals. The canceller has to track only these slowly varying levels; thus, the sampling frequency F_{sa} of the converters can be in the range of only a few kHz. The weights are updated according to a mixed-signal version of the recursive least-square (RLS) algorithm [Magesacher a) 2001]. The algorithm is derived by minimizing the cost function

$$\mathcal{E}[n] = \frac{1}{T} \int_0^{nT} \lambda^{(nT-t)/T} e^2(t)\,\mathrm{d}t, \tag{13.27}$$

where $e(t) = y(t) - s(t)$ is the estimation error and the constant $\lambda \leq 1$ is a forgetting factor weighting recent data higher and older data lower. Minimization of Equation 13.27 yields the update rule for the weight vector

$$\underline{\hat{w}}[n] = \underline{\hat{w}}[n-1] + \frac{1}{P[n]}\underline{q}(nT), \qquad (13.28)$$

where the baseband error components are scaled inversely proportional to the CM noise power via

$$P[n] = \lambda P[n-1] + p(nT). \qquad (13.29)$$

Note that the coefficients remain fixed during the observation interval $0 \leq t < nT$. Equations 13.28 and 13.29 constitute the digital part of the update algorithm. The analog part comprises the weighting within one period T, carried out by the three low-pass filters in Equations 13.25 and 13.26. By minimizing Equation 13.27, we also obtain the optimum low-pass filter impulse response

$$h_{\lambda a}(t) = \begin{cases} \dfrac{1}{T}\lambda^{t/T}, & 0 \leq t \leq T \\ 0, & \text{otherwise} \end{cases}, \qquad (13.30)$$

which can be realized efficiently by a split into an analog and a digital part [Magesacher a) 2001].

Performance Evaluation   The performance of an analog canceller can be assessed by three parameters:

- RFI suppression: measured in terms of ΔSIR, which denotes the ratio of the signal-to-RFI power ratio at the output and the signal-to-RFI power ratio at the input of the canceller.
- SNR loss ΔSNR: degradation of the signal-to-noise ratio caused by the canceller.
- Convergence speed: time, or equivalently, number of iterations until the canceller weights reach their steady state.

In the following, the performance of the canceller described above is discussed using exemplary prototype measurement results [Ödling 2002]. In principle, the thoughts are applicable to any analog interference canceller.

Figure 13.5 shows the measured PSD of the signals at the canceller input (left plot) for a scenario with a VDSL-like signal with 3.3 MHz bandwidth and a power of -10 dBm corrupted with a strong RFI disturber represented by a tone whose power is 0 dBm and a noise floor of -125 dBm/Hz. The VDSL signal and the noise were generated using a high-speed digital-to-analog converter unit. The VDSL signal consists of adjacent tones modulated with constant magnitude and random phase. More details about the setup and the methods used to inject the RFI disturber can be found in [Ödling 2002].

According to the PSD of the canceller output signal captured after the canceller's weight update algorithm has converged (right plot), the RFI disturber is suppressed by roughly 25 dB. There is no significant increase in the out-of-band noise floor.

The upper plot in Figure 13.6 shows the measured RFI suppression ΔSIR versus RFI disturber power level for different signal power levels. The RFI suppression rises with increasing RFI power and decreasing signal power. The corresponding SNR loss ΔSNR, depicted in the lower plot of Figure 13.6, rises with RFI power, which is due to the noise generated by the canceller's multipliers.

For illustration, the convergence speed is evaluated by means of the internal canceller signals depicted in Figure 13.7. For time $t < 0$, the canceller is idle. The weights w_1 and w_2 are set to the levels attained during an offset compensation procedure. The RFI disturber

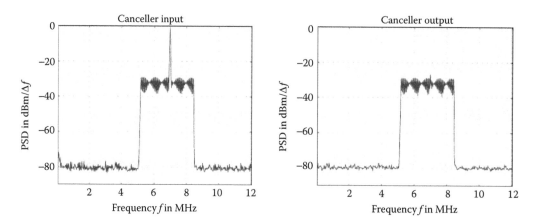

FIGURE 13.5

PSD measured at the input (left) and at the output (right) of the reference-based analog interference canceller in steady state. Note that the PSD is given in dBm/Δf with $\Delta f = 30$ kHz. The background noise PSD of -80 dBm/Δf corresponds to -125 dBm/Hz.

causes the baseband error signals q_1 and q_2 to be different from zero. At time instant $t = 0$, the canceller starts to adapt the weights. After about 20 iterations, which corresponds to 1 ms, the weights have reached their steady state. The baseband error levels q_1 and q_2 tend toward zero.

To summarize, the experimental results confirm the validity of the reference-based cancellation approach but indicate also that the design of an analog canceller must be carried out with extreme care to keep the impact of the canceller on the receive signal as low as possible.

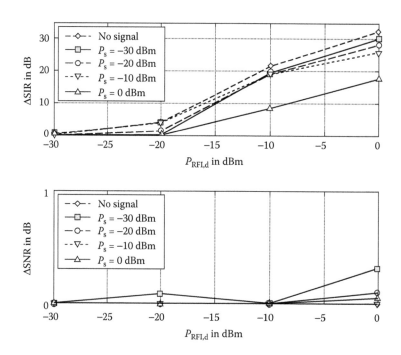

FIGURE 13.6

Measured RFI suppression ΔSIR (top) and measured SNR loss ΔSNR (bottom) versus power of the RFI disturbance.

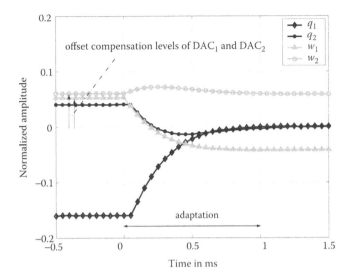

FIGURE 13.7

Measured canceller signals during adaptation: update frequency $F_{sa} = 20\,\text{kHz}$, adaptation starts at time $t = 0\,\text{ms}$.

13.4 Digital Suppression Techniques

Digital-domain processing offers effective and efficient possibilities to suppress embedded RF signals as long as clipping of the received signal by the ADC can be avoided. The risk of clipping by the ADC due to RFI can be reduced by using analog cancellation techniques, such as those described in the previous sections.

Before the ADC, the RF signal $s(t)$ can be modelled as an additive disturbance on the received signal per Equation 13.18:

$$y(t) = x(t) \star h(t) + s(t) + v(t), \quad -\infty < t < \infty, \tag{13.31}$$

where $x(t) \star h(t)$ represents the received DSL-signal after the analog front-end filters, and $v(t)$ represents other background noise such as crosstalk and additive white Gaussian noise (AWGN). For simplicity, the AFE filters are not explicitly modelled. Hence, when no clipping of the received signal $y(t)$ occurs, the corresponding sampled signal after the ADC can be described as

$$y[n] = y(n/f_s) = x[n] \star h[n] + s[n] + v[n], \quad -\infty < n < \infty. \tag{13.32}$$

Before describing different digital RFI suppression techniques that face the scenario described by Equation 13.32, the techniques are classified into two main groups describing how they perform RFI suppression.

- *Passive Suppression Techniques — Filtering, Windowing* (see Section 13.5) The passive techniques do not cancel on-line estimates of the RFI. Instead, they mitigate the RFI by passively suppressing the received signal without active efforts of differentiating between the noise and the signal parts. This is how band-stop (or notch) filters operate, as well as smoother receiver windows than the rectangular window in block processed multi-carrier systems. The definition is not absolute, as the filtering methods also can be designed to be semi-active by adaptively steering the notch center frequency to the location of the RF signal [Pazaitis 1998]. Furthermore, the windowing function for multi-carrier systems can be optimized for suppressing the spectral leakage in certain frequency bands [Redfern 2002]. Still, windowing operates in the same way, independent of the momentary RFI signal variations.

- *Active Suppression Techniques (RFI cancellation)* (see Section 13.6)   Active techniques perform suppression by calculating estimates of the RFI based on on-line measurements. The estimates are then subtracted (cancelled) from the received signal in order to suppress the RFI. In principle, this can be performed in the time or frequency domain. Active methods can be further divided into two subgroups.

    - *Nonmodel-based methods*, which do not use any specific RF signal model to derive the RFI estimation parameters. Instead, the parameters need to be trained using some adaptation method such as least mean-square (LMS) or recursive least square (RLS) algorithms.

    - *Model-based methods*, which use some appropriate RF signal model to describe the RFI. The model may be either deterministic or stochastic, but requires some a priori information about the RFI, such as its maximum expected bandwidth. The a priori information in the RF-model replaces the training of the model parameters needed in the nonmodel-based methods. As long as the RF signal model is robustly selected (typically when the true RF signal falls within the subspace of the RF signal model), good performance can be achieved. Therefore, model-based methods are often more robust than nonmodel-based methods, which may need frequent retraining when the RFI varies in bandwidth or frequency location, as with radio amateurs.

13.5   Passive Digital Suppression Techniques

13.5.1   Digital Receiver Filtering

The most straightforward way to combat RFI in the digital domain is to use a band-stop, or notch, filter $g[n]$. The notch filter operates directly on the received signal and attempts to attenuate the narrow frequency band in which the RF signal resides. The output of the notch filter is:

$$y[n] \star g[n] = x[n] \star h[n] \star g[n] + s[n] \star g[n] + v[n] \star g[n]. \tag{13.33}$$

A notch filter that effectively suppresses the RFI part, $s[n] \star g[n]$, has to be steep and have a deep frequency notch. However, this means that the impulse response of the notch filter has to be long. Hence, besides suppressing the RF signal efficiently, the notch filter will also smear out the channel impulse response, $h[n] \star g[n]$, and as a side effect cause increased inter-symbol interference (ISI). Therefore, with a notch filter, there is a trade-off between effective RFI suppression and amount of increased ISI.

For single-carrier modulation, this trade-off is often acceptable because using a notch filter still improves the final performance in the presence of RFI. Nevertheless, this requires that the channel equalizer be able to remove the larger amount of ISI introduced by the long-tailed filter.

For multi-carrier modulation such as DMT, the long impulse response of $g[n]$ requires an increased cyclic extension[2] (CE) to avoid the ISI. A longer CE reduces the efficiency. For a

---

[2] The length of the cyclic extension is the sum of the cyclic prefix and (if present) the cyclic suffix.

similar reason, a time domain equalizer (TEQ) used before demodulation will encounter a more difficult task in shortening the length of the increased channel impulse response. Furthermore, the combined effect of a notch filter and a TEQ in series is unclear, because they have counteracting missions.

13.5.2 Adaptive Digital Notch Filters

In [Pazaitis 1998], an adaptive digital IIR notch filter is proposed with the z-transform

$$G(z) = \prod_{k=1}^{Q} \frac{1 - 2\cos \omega_k z^{-1} + z^{-2}}{1 - 2\alpha_k \cos \omega_k z^{-1} + \alpha_k^2 z^{-2}}, \tag{13.34}$$

where ω_k are the notch frequencies and α_k are the corresponding pole concentration factors, which relate to the bandwidths of the notches. The filter consists of Q second-order notch filters in series, where each filter aims to suppress a different RFI at angular frequency ω_k. The notches of the filter are obtained by placing the zeros in pairs on the unit circle, $e^{\pm j\omega_k}$, and the pole pair locations, $\alpha_k e^{\pm j\omega_k}$, steer the bandwidth of the notches, which are approximately equal to $\pi(1 - \alpha_k)$ [Pazaitis 1998].

Adaptation of the parameters $\delta_n = -2\cos \omega_n$ for each notch filter in Equation 13.34 is given by

$$\delta_{n+1} = \delta_n + R_n^{-1}\psi_n e_n, \tag{13.35}$$

$$R_n = \rho_n R_{n-1} + \psi_n^2, \tag{13.36}$$

$$e_n = y_n + \delta_n y_{n-1} + y_{n-2} - \alpha_k \delta_n e_{n-1} - \alpha_k^2 e_{n-2}, \tag{13.37}$$

where ρ_n is a forgetting factor and ψ_n is an approximation of the negative prediction error gradient

$$\psi_n = -\frac{\partial e_n}{\partial \delta}\Big|_{\delta=\delta_n} = -y_{n-1} + \alpha_k e_{n-1} - \alpha_k \delta_n \psi_{n-1} - \alpha_k^2 \psi_{n-2}. \tag{13.38}$$

In Equation 13.37, y_n and e_n are the input and output of each second-order notch filter given by Equation 13.34. In the evaluation in Section 13.6.5, the impulse response of each second-order notch filter is analytically derived and some of its behavior evaluated.

13.5.3 Receiver Windowing

DMT systems process blocks of samples, rather than a stream. That is, blocks of (real-valued) samples form time-domain DMT symbols in the transmitter. Before they are transmitted, blocks of samples are cyclically extended to avoid ISI and inter-channel interference (ICI) in the receiver.[3] In this chapter, the samples in the CE are assumed to be distributed both as a cyclic prefix and a cyclic suffix in order to allow windowing. In a similar way, the received samples are processed in blocks in the receiver. This block processing in the receiver affects how the RFI disturbs the frequency-domain subsymbols. In principle, RFI occurs on most subcarriers, even though the spectrum of the RF signal (derived over an infinite time) overlaps only a very small portion (only a few subcarriers) of the spectrum. For example, with the standard rectangular window in a DMT receiver, a single interfering RF-tone on a frequency f_c spreads its power proportional to $\sin^2(f - f_c)/(f - f_c)^2$, before sampling. Note that this is an effect of the block processing rather than the sampling. The latter just

---

[3] For DMT-based VDSL1, the cyclic extension can also be used to avoid nonorthogonal NEXT and near echoes.

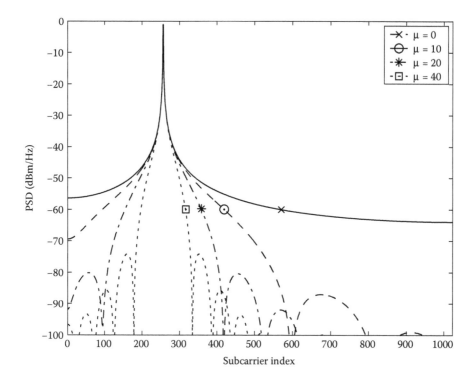

FIGURE 13.8

Receiver windowing in DMT with a raised-cosine shaped window using μ samples. $\overline{N} = 1024$ subcarriers.

folds that spectrum around $\pm f_s/2$. However, by utilizing a smoother receiver window than the standard rectangular (for example, a raised-cosine), effective suppression of the RFI is obtained on subcarriers well away from f_c. However, subcarriers closer to f_c are quite unaffected by windowing. Therefore, it is not sufficient to avoid using a few subcarriers at and around f_c to completely avoid the RFI, whether windowing is used or not.

Figure 13.8 shows an example of the RFI spectral leakage after the FFT receiver block processing. In this figure, $\mu = 0$ corresponds to a rectangular shaped window and $\mu = 10, 20, 40$ corresponds to a raised-cosine shaped receiver window, which utilizes μ additional cyclically extended samples used for the windowing. The extra cyclic extension is needed in order to preserve the subcarrier orthogonality. In this example, the average power of the RFI is -27.6 dBm, and the RF signal has a double-sided bandwidth of 4.3125 kHz, the same as the subcarrier spacing.

Windowing requires few operations per DMT symbol and operates independently of the RF signal (*i.e.*, it is passive). It reduces the overall modem complexity, because fewer subcarriers need active RFI cancellation after windowing, especially those that are far away from f_c. Windowing also reduces complexity required for the active digital cancellation, because fewer model parameters need to be included when the RFI cancellation is combined with windowing.

Next, how to perform receiver windowing while maintaining the DMT subcarrier orthogonality is discussed [Spruyt 1996]. Consider the received signal in Equation 13.32. The nonrectangular windowing is performed after the ADC. The theoretical condition for maintaining the subcarrier orthogonality is that half the number of samples that are used for nonrectangular windowing need to be taken from the cyclic extension. Furthermore, these samples should be free from inter-block interference (ISI from adjacent DMT symbols).

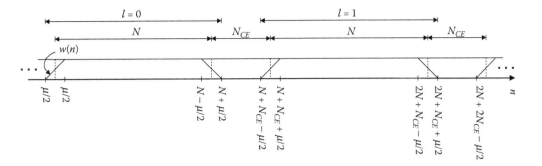

FIGURE 13.9
Framing samples into DMT symbols.

Therefore, an increased symbol shaping decreases the efficiency, because the CE needs to be increased as well, unless the shaping can be performed on already redundant CE samples in the system. However, to obtain a certain suppression performance from windowing (given by μ), the efficiency can be increased by increasing the DMT symbol length (*i.e.,* using more subcarriers with a fixed sampling frequency).

Before the DFT processing in the receiver, the sampled signal is framed into blocks and the CE is removed, as shown in Figure 13.9. Here, N_{CE} denotes the total number of samples of the CE, which is composed of a prefix and a suffix (the latter used in VDSL1). Let a long column vector of $N + \mu$ samples, which represents one DMT symbol, be denoted as

$$\mathbf{y}[l] = [\, y[l\,(N + N_{CE}) - \mu/2] \cdots y[(l+1)N + l \cdot N_{CE} + \mu/2 - 1]\,]^T \qquad (13.39)$$
$$= \mathbf{x}_h[l] + \mathbf{s}[l] + \mathbf{v}[l], \qquad (13.40)$$

where l is a symbol index, with $l = 0$ representing the current DMT symbol being processed by the receiver. Note that the μ samples used for windowing are part of the CE samples ($\mu \leq N_{CE}$).

The DFT of one windowed DMT symbol can be expressed as

$$Y[k] \underset{0 \leq k \leq \overline{N}-1}{=} \sum_{n=-\mu/2}^{N+\mu/2-1} w[n]y[n]e^{-j2\pi nk/N} \qquad (13.41)$$

$$= \sum_{n=0}^{\mu/2-1} (w[n]y[n] + w[n+N]y[n+N])\, e^{-j2\pi nk/N} + \sum_{n-\mu/2}^{N-\mu/2-1} w[n]y[n]e^{-j2\pi nk/N}$$

$$+ \sum_{n=N-\mu/2}^{N-1} (w[n]y[n] + w[n-N]y[n-N])\, e^{-j2\pi nk/N}, \qquad (13.42)$$

where $\overline{N} = N/2$ is the number of DMT subcarriers. The N-point DFT taken over $N + \mu$ samples can be expressed as an N-point DFT taken over N samples. Hence, independent of the window size μ, the often fixed (by hardware) N-size FFT unit in a DMT receiver can be maintained when using a nonrectangular window. Thus, the subcarrier orthogonality will be maintained when using a nonrectangular window if the CE is increased by μ samples[4]

---

[4] $\mu = 0$ corresponds to a rectangular window.

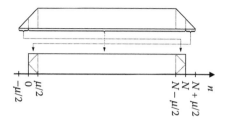

FIGURE 13.10
The windowing operation for DMT.

and if the window function ($w[n]$, $-\mu/2 \leq n \leq N + \mu/2 - 1$) is a pulse that fulfills the Nyquist criterion for ISI free signalling. Then the windowing will only have an impact on the noise parts of the received signal while leaving the DSL signal unaffected.

The windowing operation in Equation 13.42 can be interpreted as a frequency-domain decimation from $N + \mu$ down to N samples, which leads to a time-domain folding, as sketched in Figure 13.10. It can be represented by an $N \times (N + \mu)$ size matrix

$$\mathbf{W} = \begin{bmatrix} \mathbf{0}_{\frac{\mu}{2}} & \mathbf{I}_{\frac{\mu}{2}} & \mathbf{0} & \mathbf{0}_{\frac{\mu}{2}} & \mathbf{I}_{\frac{\mu}{2}} \\ \mathbf{0} & \mathbf{0} & \mathbf{I}_{N-\mu} & \mathbf{0} & \mathbf{0} \\ \mathbf{I}_{\frac{\mu}{2}} & \mathbf{0}_{\frac{\mu}{2}} & \mathbf{0} & \mathbf{I}_{\frac{\mu}{2}} & \mathbf{0}_{\frac{\mu}{2}} \end{bmatrix} \begin{bmatrix} w[-\mu/2] & 0 & 0 \\ 0 & \ddots & 0 \\ 0 & 0 & w[N+\mu/2-1] \end{bmatrix}, \tag{13.43}$$

where $\mathbf{0}_x$ is a square zero matrix and $\mathbf{I}_x$ is an identity matrix, both of size x. With this, a windowed DMT symbol can now be compactly expressed as $\mathbf{W}\mathbf{y}[l]$.

13.6 Active Digital RFI Cancellation

Digital cancellation of RFI can, in principle, be performed both in the time and frequency domains of the DSL signal. However, in contrast to single-carrier modulated DSL, it is possible with DMT-based DSL to have a "clean view" of the RFI on some silent tones after the DFT processor. This is a reason why the most efficient RFI suppression methods are based on frequency-domain signal processing for DMT.

Figure 13.11 shows a DMT receiver with an analog and a digital RFI canceller. The digital canceller operates after the DFT processor in the frequency domain. It relies on the presence of a CE to preserve subcarrier orthogonality. $\overline{N} = N/2$ denotes the number of subcarriers,

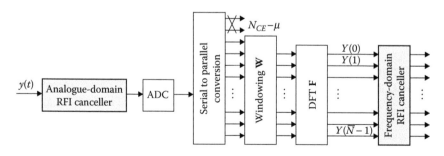

FIGURE 13.11
DMT receiver with analog- and digital-domain RFI cancellation.

where N is the IDFT/DFT size, and μ denotes the samples from the CE that are used for nonrectangular receiver windowing, as discussed in Section 13.5.3.

13.6.1 Model-Based Digital RFI Cancellation

For the model-based RFI cancellers, some type of mathematical model describes, a priori, the RF signal. This model can be of either a deterministic or a stochastic nature. For the deterministic model case described in this chapter, a pre-selected number of deterministic but unknown model parameters need to be estimated continuously (online). In the stochastic RF signal model case, a pre-selected power spectral density serves as a priori information about the RF signal.

The model, either deterministic or stochastic, presumes the RF signal is narrowband, typically not more than the spacing between a few DMT subcarriers, and that it can represent any type of analog modulation, amplitude modulation, double sideband modulation, single sideband modulation, etc. With continuous RFI measurements combined with the RF signal model, accurate estimates of the RFI can then be derived and cancelled online from the received DSL signal.

13.6.2 Deterministic RF Signal Model

This RFI canceller models the RFI as a deterministic, but unknown, signal. To separate between known and unknown signal parts after the DFT processor, Taylor expansion is used to obtain a linear polynomial-based RF signal model. The linearization is a justified approximation because the RF signal is narrowband. With the linear model, the unknown signal parts will be represented by unknown model parameters. These parameters are then continuously estimated and updated during each symbol period by using online measurements of the RFI. With the model parameters, the RFI can then be accurately estimated and cancelled from the received signal.

In line with Equations 13.9 and 13.17, this canceller models the RF signal $s(t)$ as a narrowband signal [Sjöberg 2004]

$$s(t) = b(t)e^{j2\pi f_c t} + b^*(t)e^{-j2\pi f_c t}, \tag{13.44}$$

where $b(t)$ is modelled as a deterministic narrowband baseband equivalent of the RF signal, and the asterisk denotes complex conjugate. Based on this model, the RFI suppression will be performed in the frequency domain, after the DFT processor.

Sampling and demodulation of the received signal $y(t)$ from Equation 13.31 yields

$$Y[k] = \sum_{n=-\mu/2}^{N+\mu/2-1} w[n]y(nT_s)e^{-j2\pi kn/N} \tag{13.45}$$

$$= X[k]H[k] + S[k] + V[k], \quad (0 \le k \le \overline{N} - 1), \tag{13.46}$$

where $T_s = 1/f_s$ is the sampling period, $w[n]$ is the window used in the receiver, $Y[k]$ is the received baseband signal after the DFT, $S[k]$ are the narrowband RF signal's DFT coefficients, and $V[k]$ is the DFT of the other noise.

RFI suppression is now achieved by subtracting frequency-domain estimates $\widehat{S}[k]$ of $S[k]$ from the received signal $Y[k]$:

$$\grave{Y}[k] = Y[k] - \widehat{S}[k] = X[k]H[k] + (S[k] - \widehat{S}[k]) + V[k], \tag{13.47}$$

which is depicted in Figure 13.12.

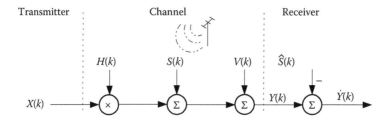

FIGURE 13.12
Frequency-domain RFI cancellation.

An estimate $\widehat{S}[k]$ can be derived by first examining the relation between $S[k]$ and the Fourier transform of a deterministic, but unknown, disturbance $S(f) = \mathcal{F}\{s(t)\}$, namely,

$$S[k] = \sum_{n=-\mu/2}^{N+\mu/2-1} w[n] \left(\int_{-\infty}^{\infty} S(f)e^{j2\pi nff_s}df \right) e^{-j(2\pi n/N)k} \tag{13.48}$$

$$= \int_{-\infty}^{\infty} S(f) \sum_{n=-\mu/2}^{N+\mu/2-1} w[n]e^{j2\pi nf/f_s - j(2\pi n/N)k}df \tag{13.49}$$

$$= \int_{-\infty}^{\infty} S(f)G_k(f)df, \tag{13.50}$$

where $G_k(f)$ can be thought of as a transfer function (Dirichlet kernel [Haykin 1996]) from the unknown RF signal's Fourier transform $S(f)$ to its DFT coefficients $S[k]$. $G_k(f)$ can be parameterized by means of a Taylor series expansion (see Section 13.6.2.1). This leads to the following linear model of the RFI.

$$S[k] \approx \widetilde{S}[k] = \sum_{l=0}^{D_p-1} c_l G_k^{(l)}(+\widehat{f}_c) + \sum_{m=0}^{D_n-1} (-1)^m c_m^* G_k^{(m)}(-\widehat{f}_c), \tag{13.51}$$

where $\widehat{f}_c$ is an estimate of the center frequency, $G_k^{(l)}(\pm\widehat{f}_c)$ is the lth derivative of $G_k(f)$ at $f = \pm\widehat{f}_c$, and $\{c_l\}$ are unknown parameters modelling the RFI. The model in Equation 13.51 can be written in a matrix notation as

$$\underline{S} \approx \underline{\widetilde{S}} \triangleq \mathbf{G}\underline{c}, \tag{13.52}$$

where $\mathbf{G}$ is derived from $G_k(f)$,

$$\mathbf{G} = \begin{pmatrix} G_0(\widehat{f}_c) & \cdots & G_0^{(D_p-1)}(\widehat{f}_c) & G_0(-\widehat{f}_c) & \cdots & G_0^{(D_n-1)}(-\widehat{f}_c) \\ \vdots & \vdots & \vdots & \vdots & \vdots & \vdots \\ G_{\overline{N}-1}(\widehat{f}_c) & \cdots & G_{\overline{N}-1}^{(D_p-1)}(\widehat{f}_c) & G_{\overline{N}-1}(-\widehat{f}_c) & \cdots & G_{\overline{N}-1}^{(D_n-1)}(-\widehat{f}_c) \end{pmatrix}, \tag{13.53}$$

and $\underline{c}$ is a vector containing the unknown RFI-parameters

$$\underline{c} = \begin{pmatrix} c_0 & \cdots & c_{D_p-1} & c_0^* & \cdots & (-1)^{D_n-1} c_{D_n-1}^* \end{pmatrix}^T. \tag{13.54}$$

The conjugated parameters in Equation 13.54 that correspond to contributions from negative frequencies give almost no contribution to $\widetilde{S}$ when windowing is used. This can be seen in Figure 13.8, where the windowed RFI spectrum is nearly symmetric around $f = f_c$ but not the rectangular windowed ($\mu = 0$), which also receives spectral leakage contributions from $f = -f_c$ (i.e., from $b^*(t)e^{-j2\pi f_c t}$) into the positive frequencies. Using windowing, the sizes of the vector $\underline{c}$ and the matrix $\mathbf{G}$ can be reduced to

$$\mathbf{G}_{\text{wind}} = \begin{pmatrix} G_0(\hat{f}_c) & \cdots & G_0^{(D_p-1)}(\hat{f}_c) \\ \vdots & \vdots & \vdots \\ G_{\overline{N}-1}(\hat{f}_c) & \cdots & G_{\overline{N}-1}^{(D_p-1)}(\hat{f}_c) \end{pmatrix} \tag{13.55}$$

and

$$\underline{c}_{\text{wind}} = (c_0 \cdots c_{D_p-1})^T \tag{13.56}$$

with a negligible performance loss.

In the matrix formulation, the model for the received signal on unused (silent) tones ω becomes

$$\underline{Y}_\omega = \underline{S}_\omega + \underline{V}_\omega \approx \tilde{\underline{S}}_\omega + \underline{V}_\omega = \mathbf{G}_\omega \underline{c} + \underline{V}_\omega, \tag{13.57}$$

where $\omega = \{m_1, \ldots, m_M\}$ denotes the index set of the tones on which the RFI is measured. Assuming that the noise $\underline{V}_\omega$ is independent and identically Gaussian distributed, a maximum likelihood (ML) estimate [Scharf 1991], $\hat{\underline{c}}_{ML}$, of the parameters $\underline{c}$ can be obtained when windowing is used:

$$\hat{\underline{c}}_{ML} = \mathbf{G}_\omega^+ \underline{Y}_\omega, \tag{13.58}$$

where $\mathbf{G}_\omega^+$ is the pseudo-inverse of $\mathbf{G}_\omega$. Here it is assumed that $\mathbf{G}_\omega$ has full column rank. If windowing is not used (i.e., the negative frequencies will also contribute), $\mathbf{G}_\omega$, $\underline{c}$, and $\underline{Y}_\omega$ need to be rewritten slightly to separate the real and imaginary parts before the pseudo-inverse is calculated, because the coefficient vector $\underline{c}$ then contains both the parameters, c_l, and their complex conjugates, c_l^*.

By using the definition of $\hat{\underline{c}}_{ML}$ in Equation 13.58 and $\tilde{\underline{S}}$ in Equation 13.52, the estimates of the disturbance on the modulated tones can be obtained with

$$\hat{\underline{S}} = \mathbf{G}\hat{\underline{c}}_{ML} = \overbrace{\mathbf{G}\mathbf{G}_\omega^+ \underline{Y}_\omega}^{K} = \mathbf{K}\underline{Y}_\omega = \tilde{\underline{S}} + \mathbf{K}\underline{V}_\omega, \tag{13.59}$$

where $\mathbf{K}$ denotes the matrix containing all the estimator coefficients for the canceller. The estimates in the vector $\hat{\underline{S}}$ are then finally subtracted from the data, as in Equation 13.47, to suppress the RFI.

13.6.2.1 Appendix: Taylor Parameterization

This appendix describes the Taylor parametrization of the RFI signal model, but first a short background is provided. To describe the RFI in the formulation of the canceller (Equation 13.50 above and in this appendix), an unknown narrowband signal in the frequency domain, $S(f)$, is used as an RFI model. In the autumn of 1997, this signal model was combined with a Taylor parametrization to achieve a linear RFI canceller, which was published the following year [Sjöberg 1998]. Interestingly, in the same time period but independently, a similar RFI canceller was derived by Wiese and Bingham [Wiese 1997]. They approximated the envelope of the time-domain RFI signal (within the duration of one DMT symbol) with a first-order polynomial. With the two cancellers in hand, it shortly became clear that they were in principle the same! That is, a generalization of the canceller proposed by Wiese and Bingham [Wiese 1997], which is achieved by Taylor expansion of the RFI signal's time envelope, results in the same canceller as the one that is based on a Taylor parametrization of the narrowband frequency signal model $S(f)$ [Sjöberg 2004].

Below it is shown how the signal in Equation 13.50 is parameterized into the linear model of Equation 13.51 by Taylor expanding the Dirichlet kernel $G_k(f)$. The expansion is performed in the frequency domain, but as mentioned above and described in more detail below, a corresponding time-domain expansion leads to exactly the same linear model and RFI canceller.

The development starts with the assumption that the disturbance $S(f)$ is narrowband, centered around f_c and with a bandwidth not exceeding $2W$ Hz; i.e., $S(f) = 0$ when $|f \pm f_c| \geq W$. Using this assumption, the DFT coefficients in Equation 13.50 can be written in an equivalent form

$$S[k] = \int_{-W}^{W} S(f + f_c)G_k(f + f_c)\,df + \int_{-W}^{W} S(f - f_c)G_k(f - f_c)\,df. \tag{13.60}$$

Let $B(f)$ denote the Fourier transform of the baseband equivalent $b(t)$ from Equation 13.44. Substituting $S(f) = B(f - f_c) + B^*(-f - f_c)$ in Equation 13.60 yields

$$S[k] = \int_{-W}^{W} B(f)G_k(f + f_c)\,df + \int_{-W}^{W} B^*(-f)G_k(f - f_c)\,df. \tag{13.61}$$

A Taylor expansion of $G_k(\cdot)$ around $\pm\widehat{f}_c$ separates the known parts from the unknown in Equation 13.61:

$$G_k(f + \widehat{f}_c) = \sum_{l=0}^{D_p-1} \frac{f^l G_k^{(l)}(+\widehat{f}_c)}{l!} + O(f^L), \tag{13.62}$$

$$G_k(f - \widehat{f}_c) = \sum_{m=0}^{D_n-1} \frac{f^m G_k^{(m)}(-\widehat{f}_c)}{m!} + O(f^M). \tag{13.63}$$

Truncation of these series and insertion into Equation 13.61 gives a linear approximation $\widetilde{S}[k]$ of $S[k]$:

$$S[k] \approx \widetilde{S}[k] \triangleq \sum_{l=0}^{D_p-1} G_k^{(l)}(-\widehat{f}_c) \int_{-W}^{W} B(f)\left(\frac{f^l}{l!}\right) df \tag{13.64}$$

$$+ \sum_{m=0}^{D_n-1} G_k^{(m)}(-\widehat{f}_c) \int_{-W}^{W} B^*(-f)\left(\frac{f^m}{m!}\right) df$$

$$= \sum_{l=0}^{D_p-1} c_l G_k^{(l)}(-\widehat{f}_c) + \sum_{m=0}^{D_n-1} d_m G_k^{(m)}(-\widehat{f}_c), \tag{13.65}$$

where $S[k] - \widetilde{S}[k]$ represents the model error from truncation of the Taylor series and where

$$c_l = \frac{1}{l!} \int_{-W}^{W} f^l B(f)\,df \tag{13.66}$$

$$d_m = \frac{1}{m!} \int_{-W}^{W} f^m B^*(-f)\,df \tag{13.67}$$

are the unknown parameters modelling the RFI. Comparing Equations 13.66 and 13.67, the relation between c_m and d_m is

$$d_m = (-1)^m c_m^*. \tag{13.68}$$

Thus, the following linear model describes the RFI in a DMT receiver:

$$\widetilde{S}[k] = \sum_{l=0}^{D_p-1} c_l G_k^{(l)}(+\widehat{f}_c) + \sum_{m=0}^{D_n-1} (-1)^m c_m^* G_k^{(m)}(-\widehat{f}_c), \tag{13.69}$$

which means that there are $\max(D_p, D_n)$ unknown model parameters.

The Taylor parameterization in the frequency domain in Equation 13.65 is equivalent to a Taylor parameterization of the time-domain envelope $b(t)$ around $t = 0$,

$$b(t) = \sum_{k=0}^{\infty} \frac{t^k}{k!} b^{(k)}(0), \tag{13.70}$$

where $b^{(k)}(0)$ represents unknown parameters in the model. The continuous-time RF signal in Equation 13.44 can then be written as

$$s(t) \approx \widetilde{s}(t) = \left(\sum_{l=0}^{D_p-1} \frac{b^{(l)}(0)}{l!} t^l \right) e^{j2\pi f_c t} + \left(\sum_{m=0}^{D_n-1} \frac{b^{*(m)}(0)}{m!} t^m \right) e^{-j2\pi f_c t}. \tag{13.71}$$

It can be shown that the approximation in Equation 13.71 is equivalent to the approximation in Equation 13.64 [Sjöberg 2004]. Hence, the model approximation in Equation 13.65 can also be given by the following equivalent expression.

$$S[k] = \sum_{n=-\mu/2}^{N+\mu/2-1} w[n]s(nT_s) e^{-j(2\pi n/N)k} \approx \tag{13.72}$$

$$\widetilde{S}[k] = \sum_{n=-\mu/2}^{N+\mu/2-1} w[n]\widetilde{s}(nT_s) e^{-j(2\pi n/N)k} \tag{13.73}$$

$$= \sum_{l=0}^{D_p-1} b^{(l)}(0) \sum_{n=-\mu/2}^{N+\mu/2-1} \frac{(nT_s)^l}{l!} w[n] e^{j(2\pi n f_c/F_s)-j(2\pi n/N)k}$$

$$+ \sum_{m=0}^{D_n-1} b^{*(m)}(0) \sum_{n=-\mu/2}^{N+\mu/2-1} \frac{(nT_s)^m}{m!} w[n] e^{-j(2\pi n f_c/F_s)-j(2\pi n/N)k}. \tag{13.74}$$

13.6.3 Stochastic RF Signal Model

An RFI canceller that uses a stochastic RF signal model models the interfering signal as a stochastic (narrowband) process. The estimator coefficients for this model are derived using a linear minimum mean square error (LMMSE) criterion. The RF signal's power spectral density serves as a priori information which need only be roughly known in the model. For example, a rough estimate of f_c (for example, by interpolating between the subcarriers with most RFI) and a flat narrowband PSD shape modelling the RF signal spectrum is sufficient to obtain good results.

This canceller can also be designed to be more robust than the previous canceller. Furthermore, it is easier to determine in which situations the RFI cancellation will work and in which situations it will fail. The cancellation complexity can be reduced by using optimal rank reduction. The rank needed can easily be determined by examining the time-bandwidth product of the modelled RF signal.

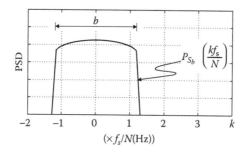

FIGURE 13.13
PSD and relative bandwidth b of RF signal expressed in the DMT subcarrier grid.

13.6.3.1 Narrowband Signal Model

The RF signal model for this canceller is similar to Equation 13.44 and described by

$$s(t) = \text{Re}\left\{ s_b(t)e^{j2\pi f_c t} \right\} \tag{13.75}$$

$$= s_b(t)e^{j2\pi f_c t} + s_b^*(t)e^{-j2\pi f_c t}. \tag{13.76}$$

The difference between Equations 13.76 and 13.44 is that $s_b(t)$ will be treated as a stochastic (wide sense stationary) baseband signal whereas $b(t)$ in Equation 13.44 was regarded as an unknown but deterministic baseband signal. However, both are narrowband compared to the DSL signal. As with $b(t)$, the spectrum of $s_b(t)$ is also defined to be centered around $f = 0$. As before, the frequency location of the RF signal is represented by the center frequency, f_c, with $\widehat{f_c}$ representing its estimate.

Denote the PSD of $s_b(t)$ by $P_{S_b}(f)$ and its autocorrelation function by $r_{s_b}(\tau) \triangleq E\{s_b(t + \tau)s_b^*(t)\}$, where $E\{\cdot\}$ denotes expectation. Figure 13.13 shows a sketch of $P_{S_b}(f)$ whose bandwidth is modelled to be $b \cdot f_s/N$, where f_s/N is the DMT subcarrier spacing, f_s is the sampling frequency of the received signal, and b is the RF signal bandwidth relative to the DMT subcarrier spacing, i.e., the (possibly noninteger) number of subcarriers $P_{S_b}(f)$ spans. The shape and bandwidth of $P_{S_b}(f)$ are design parameters for the LMMSE canceller derived in Section 13.6.3.3. The worst-case narrowband interference PSD for block-processed signals is approximately flat [Naofal 2001], which is a good choice for selecting the a priori PSD for this canceller.

Note that taking the real part in Equation 13.75 doubles the bandwidth of $s(t)$, for $f_c > b/2$, because the frequency content of $s_b(t)$ is translated to both $f = f_c$ and $f = -f_c$. However, because DSL systems can only utilize half of the spectrum for unique data transmission, only one half of the received frequency band (positive or negative) is of unique interest. Still, some power from the RF signal's two components will leak between the negative and the positive frequency halves due to the block processing by the DMT receiver. This was also seen in the derivation of the Taylor polynomial-based deterministic canceller in the previous section. For this reason, it should be recognized that the narrowband signal in Equation 13.76 is not second-order circular (i.e., improper) [Picinbono 1996, Schreier 2003], which needs to be considered when designing the LMMSE canceller.

13.6.3.2 DFT Processing

Continuing from the derivation in Section 13.5.3, after cyclic extension removal and windowing, the time-domain DMT symbols are returned to the frequency-domain by the

DFT processing

$$\underline{Y}[l] = \mathbf{F}\mathbf{W}\underline{y}[l] = \underline{X}_{\mathbf{H}}[l] + \underline{S}[l] + \underline{V}[l] \tag{13.77}$$

$$= [Y_0[l] \cdots Y_{N-1}[l]]^T, \tag{13.78}$$

where the capital letters represent the corresponding frequency-domain signal and $\mathbf{F}$ is the $N \times N$ orthonormal DFT matrix

$$\mathbf{F}_{k,n} = \tfrac{1}{\sqrt{N}} e^{-j2\pi kn/N}, \; 0 \leq k, n \leq N-1. \tag{13.79}$$

13.6.3.3 LMMSE Estimator

This section derives an RFI canceller in the frequency domain of the DMT signal that is based on a linear minimum mean-square error estimator, as shown in Figure 13.11. Similar to the canceller in Section 13.6.2, this canceller uses a few subcarriers located close to the center frequency of the RF signal as measurement tones. This one, however, makes LMMSE estimates of RF signal leakage (RFI) onto all DMT subcarriers [Nilsson 2003]. These estimates, $\widehat{S}_k$, are cancelled from each tone in the same way as in Equation 13.47.

Becuase it is assumed that the interference signal $s(t)$ is narrowband relative to the DMT signal, the correlation function will have support over several DMT symbols; that is, $r_s(\tau) \neq 0$, $|\tau| > (N + N_{CE})/f_s$. Thus, a somewhat improved performance can be achieved if RFI measurements from a few sequential DMT symbols are used by the canceller.[5] Let this be used in the derivation of the estimator. Hence, let

$$\underline{y} = \left[\underline{y}^T[-L_n] \cdots \underline{y}^T[0] \cdots \underline{y}^T[L_p] \right]^T, \tag{13.80}$$

represent an $(L_p+L_n+1)(N+\mu)$ long column-vector, with L_n DMT symbols being processed before and L_p symbols being processed after symbol 0 (the current symbol considered for RFI cancellation).

Let $P = L_p + L_n + 1$ denote the number of symbols from which a few tones are used for RFI measurements. Using $L_p > 0$ introduces an extra buffering delay in the receiver by $L_p(N + N_{CE})/f_s$ seconds, which often, however, cannot be tolerated. By using the weak stationary assumption of $s_b(t)$ in Equation 13.76, the correlation matrix for the RFI component of the received signal becomes

$$\mathbf{R}_{\underline{ss}} = E\{\underline{ss}^H\} = \begin{bmatrix} E\{\underline{s}[-L_n]\underline{s}^H[-L_n]\} & \cdots & E\{\underline{s}[-L_n]\underline{s}^H[L_p]\} \\ \vdots & \ddots & \vdots \\ E\{\underline{s}[L_p]\underline{s}^H[-L_n]\} & \cdots & E\{\underline{s}[L_p]\underline{s}^H[L_p]\} \end{bmatrix} \tag{13.81}$$

$$= \begin{bmatrix} \mathbf{R}_{\underline{ss}}[0] & \cdots & \mathbf{R}_{\underline{ss}}[-L_n - L_p] \\ \vdots & \ddots & \vdots \\ \mathbf{R}_{\underline{ss}}[L_p + L_n] & \cdots & \mathbf{R}_{\underline{ss}}[0] \end{bmatrix}, \tag{13.82}$$

with $\mathbf{R}_{\underline{ss}}[k] = \mathbf{R}_{\underline{ss}}^H[-k]$ symmetry, where the superscript H denotes the Hermitian transpose. The frequency-domain equivalent of $\mathbf{R}_{\underline{ss}}$ is

$$\mathbf{R}_{\underline{SS}} = E\{\underline{SS}^H\} = \begin{bmatrix} \mathbf{R}_{\underline{SS}}[0] & \cdots & \mathbf{R}_{\underline{SS}}[-L_n - L_p] \\ \vdots & \ddots & \vdots \\ \mathbf{R}_{\underline{SS}}[L_p + L_n] & \cdots & \mathbf{R}_{\underline{SS}}[0] \end{bmatrix}, \tag{13.83}$$

---

[5] However, in most practical cases the suppression performance is sufficient by only using one (the current) DMT symbol.

where

$$\mathbf{R_{SS}}[k] = \mathbf{FWR_{ss}}[k]\mathbf{W}^H\mathbf{F}^H = \mathbf{R_{SS}^H}[-k]. \tag{13.84}$$

In the same way as in Equation 13.57, let $\underline{\mathbf{X}}[k] = 0, k \in \omega$, where ω represents the index-set of the silent tones used for measuring the RFI and let

$$Y[k, l] = S[k, l] + V[k, l], \quad \begin{cases} k \in \omega = \{m_1, \dots, m_M\}, \\ l \in \{-L_n, \cdots, L_p\}, \; -L_n \leq 0 \leq L_p \end{cases} \tag{13.85}$$

be the measurements of the RFI on these tones. Furthermore, let the M measurements from each of the $P = L_p + L_n + 1$ symbols be collected in P vectors of length M, $\underline{\mathbf{Y}}_\omega[l] = [Y[m_1, l] \cdots Y[m_M, l]]^T$. These P vectors are then stacked into one column vector $\underline{\mathbf{Y}}_\omega = [\underline{\mathbf{Y}}_\omega[-L_n]^T \dots \underline{\mathbf{Y}}_\omega[L_p]^T]^T$. Finally, let the RFI in the current symbol be represented by the vector $\underline{\mathbf{S}}[0] = [S[0, 0] \cdots S[\overline{N} - 1, 0]]^T$. An LMMSE estimate [Scharf 1991] of $\underline{\mathbf{S}}[0]$ can now be constructed as

$$\hat{\underline{\mathbf{S}}}[0] = \mathbf{K}\underline{\mathbf{Y}}_\omega = \mathbf{R}_{\underline{\mathbf{S}}[0]\underline{\mathbf{Y}}_\omega} \mathbf{R}_{\underline{\mathbf{Y}}_\omega \underline{\mathbf{Y}}_\omega}^{-1} \underline{\mathbf{Y}}_\omega, \tag{13.86}$$

where

$$\mathbf{R}_{\underline{\mathbf{S}}[0]\underline{\mathbf{Y}}_\omega} = E\left\{\underline{\mathbf{S}}[0]\underline{\mathbf{Y}}_\omega^H\right\} = E\left\{\underline{\mathbf{S}}[0]\underline{\mathbf{S}}_\omega^H\right\} = \mathbf{AR_{SS}B}, \tag{13.87}$$

$$\mathbf{R}_{\underline{\mathbf{Y}}_\omega \underline{\mathbf{Y}}_\omega} = E\left\{\underline{\mathbf{Y}}_\omega \underline{\mathbf{Y}}_\omega^H\right\} = E\left\{\underline{\mathbf{S}}_\omega \underline{\mathbf{S}}_\omega^H\right\} + \sigma^2\mathbf{I} = \mathbf{B}^T\mathbf{R_{SS}B} + \sigma^2\mathbf{I}, \tag{13.88}$$

$$\mathbf{A} = [\mathbf{0}_{\overline{N} \times NL_n} \; \mathbf{I}_{\overline{N} \times N} \; \mathbf{0}_{\overline{N} \times NL_p}]_{\overline{N} \times PN}, \; \mathbf{B} = \begin{bmatrix} \mathbf{C}_{N \times M} & \mathbf{0} & \mathbf{0} \\ \mathbf{0} & \ddots & \mathbf{0} \\ \mathbf{0} & \mathbf{0} & \mathbf{C}_{N \times M} \end{bmatrix}_{PN \times PM}, \tag{13.89}$$

and where σ^2 is the variance of the background noise $\underline{\mathbf{V}}[k]$, and $\mathbf{A}$, $\mathbf{B}$, and $\mathbf{C}$ are indicator matrices. The size-$(N \times M)$ matrix $\mathbf{C}$ is all zeros except for M ones positioned on different rows and columns representing how the measurement tones $\omega = \{m_1, \dots, m_M\}$ are located in $[0, \dots, N-1]$. That is, if m_i is a measurement tone, then $\mathbf{C}_{m_i, i} = 1$; otherwise, $\mathbf{C}_{m_i, i} = 0$. Note that $\mathbf{K}$ is reused in Equation 13.86 from the previous, deterministic, canceller in order to simplify notation. When a distinction between the cancellers is needed, it will be emphasized.

A widely linear MMSE estimator [Schreier 2003; Picinbono 1995] of $\underline{\mathbf{S}}[0]$ is easily obtained by also using the complex symmetric pairs of the selected measurement tones, $\underline{\mathbf{Y}}[N - k] = \underline{\mathbf{Y}}^*[k]$, and accounting for them in the formulation of the estimator according to Equation 13.86. This is necessary if receiver windowing is not used, because the RF signal in Equation 13.76 is improper. Then, because Equation 13.76 is real, $\mathbf{R_{\bar{s}}}$ automatically becomes an augmented covariance matrix that contains the complementary (or pseudo) covariance matrix structures, with elements $E\{S[k, l]S[m, n]\}$, in addition to the normal covariance matrix with elements $E\{S[k, l]S^*[m, n]\}$. Hence, all the second-order information about $\underline{\mathbf{S}}[0]$ will then be contained in a widely linear estimator (WLE) matrix $\mathbf{K}$.

With windowing, the spectral leakage between the positive and negative sides of the spectrum becomes effectively suppressed (unless the RF signal center frequency is close to the DSL frequency edges ($f_c \approx 0$ or $f_c \approx f_s/2$)). In practise, this means that the complex symmetric pairs of the measurement tones (at the negative frequencies) do not need to be included in the RFI cancellation (the linear estimator is sufficient). Furthermore, the rank can be reduced one step more than if no receiver windowing is used, as will be discussed in Section 13.6.3.5. This is similar for the deterministic-based RFI canceller in that the model parameters and the measurement tones that correspond to the negative frequencies do not need to be incorporated to achieve a good performance.

The elements in the correlation matrix in Equation 13.83 for constructing $\mathbf{K}$ are of the form

$$E\left\{S[k, p]S^*[l, q]\right\}$$

$$= \frac{1}{N} \sum_{n=0}^{N-1} \sum_{m=0}^{N-1} E\left\{s\left(\frac{n + p(N + N_{CE})}{f_s}\right) s^*\left(\frac{m + q(N + N_{CE})}{f_s}\right)\right\} e^{-j2\pi[nk-ml]/N} \quad (13.90)$$

$$= \frac{1}{N} \sum_{n=0}^{N-1} \sum_{m=0}^{N-1} r_{s_b}\left(\frac{n - m + (p - q)(N + N_{CE})}{f_s}\right)$$

$$\times \cos\left(\frac{2\pi f_c (n - m + (p - q)(N + N_{CE}))}{f_s}\right) e^{-j2\pi[nk-ml]/N} \quad (13.91)$$

$$= \frac{1}{2N} \sum_{n=0}^{N-1} \sum_{m=0}^{N-1} \left(\int_{-\infty}^{\infty} \left(P_{S_b}(f - f_c) + P_{S_b}(f + f_c)\right) e^{j2\pi f[n-m+(p-q)(N+N_{CE})]/f_s} df\right)$$

$$\times e^{-j2\pi[nk-ml]/N}, \quad (13.92)$$

where $P_{S_b}(f)$ and f_c are a priori information about $s(t)$ for this LMMSE estimator, as illustrated in Figure 13.13. In the above expression, it is assumed that the modelled PSD, $P_{S_b}(f)$, is an even function. If not, an additional autocorrelation part, representing the odd part of the PSD, which is modulated on a sine function, also appears in the summation in Equation 13.91.

It suffices to assume a flat narrowband PSD model of $s(t)$ (which is even), with a relative bandwidth b that is at least as wide as the bandwidth of the true RF signal and centered around f_c to obtain a robust canceller. Figure 13.14 illustrates the PSD assumption. This strategy will effectively suppress RF signals that have a bandwidth less than or equal to $b \cdot f_s/N$. A simple estimate of f_c can be obtained by using the squared magnitude of the FFT outputs, as in a periodogram, searching for the subcarriers with the strongest interference and interpolating between those for the location of f_c [Wiese 1997; Sjöberg 1998].

13.6.3.4 Optimal Low-Rank Approximation

To lower the runtime computational complexity, the Wiener estimator $\widehat{\underline{\mathbf{S}}}[0] = \mathbf{K}\underline{\mathbf{Y}}_\omega$ in Equation 13.86 can be simplified with a low-rank approximation $\widehat{\underline{\mathbf{S}}}_r[0] = \mathbf{K}_r\underline{\mathbf{Y}}_\omega$ using singular value decomposition (SVD). For this estimator, the optimal (in terms of mean-squared error)

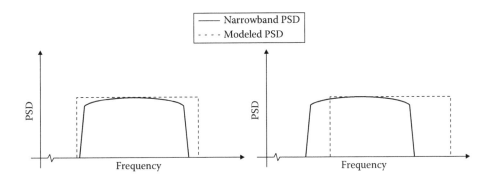

FIGURE 13.14

Narrowband PSD models for the LMMSE canceller. Left, good PSD model. Right, poor PSD model.

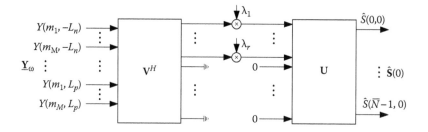

FIGURE 13.15
Low-rank estimator. M tones in $(L_p - L_n + 1)$ symbols are used to estimate the RFI on the N subcarriers in the current symbol (0).

low-rank approximation of $\mathbf{K}$ is [Scharf 1991]

$$\mathbf{K}_r = \mathbf{U}\Sigma_r\mathbf{Q}^H\mathbf{R}_{\underline{Y}_\omega\underline{Y}_\omega}^{-1/2} = \mathbf{U}\Sigma_r\mathbf{V}^H, \tag{13.93}$$

where $\mathbf{U}$ and $\mathbf{Q}$ are unitary matrices from the SVD of $\mathbf{K}\mathbf{R}_{\underline{Y}_\omega\underline{Y}_\omega}^{1/2} = \mathbf{U}\Sigma\mathbf{Q}^H$ and where Σ_r is a size $\overline{N} \times MP$ matrix containing the r most significant singular values (out of MP), $\lambda_1 \geq \cdots \geq \lambda_r$, from Σ along its diagonal

$$\Sigma_r = \mathrm{diag}[\,\lambda_1 \cdots \lambda_r \quad 0 \cdots 0\,]. \tag{13.94}$$

The least significant singular values, $\lambda_{r+1}, \ldots, \lambda_{MP}$, are discarded in the low-rank approximation.

The structure of the rank-reduced estimator is shown in Figure 13.15. Denoting the error of the high-rank estimator as $\underline{e} = \underline{S}[0] - \widehat{\underline{S}}[0]$, the covariance matrix $\mathbf{P}_r$ of the error $\underline{e}_r = \underline{S}[0] - \widehat{\underline{S}}_r[0]$ for the low-rank approximation is given by

$$\mathbf{P}_r = E\{(\underline{S}[0] - \widehat{\underline{S}}_r[0])(\underline{S}[0] - \widehat{\underline{S}}_r[0])^H\} \tag{13.95}$$

$$= E\left\{(\underline{e} + \mathbf{K}\underline{Y}_\omega - \mathbf{K}_r\underline{Y}_\omega)(\underline{e} + \mathbf{K}\underline{Y}_\omega - \mathbf{K}_r\underline{Y}_\omega)^H\right\}. \tag{13.96}$$

In the case of no design mismatch between the modelled and true PSD, the error $\underline{e}$ is orthogonal to the measurement vector $\underline{Y}_\omega$, and

$$\mathbf{P}_r = E\left\{\underline{e}\underline{e}^H + (\mathbf{K} - \mathbf{K}_r)\,\underline{Y}_\omega\underline{Y}_\omega^H\,(\mathbf{K} - \mathbf{K}_r)^H\right\} \tag{13.97}$$

$$= \mathbf{P} + (\mathbf{K} - \mathbf{K}_r)\,\mathbf{R}_{\underline{Y}_\omega\underline{Y}_\omega}\,(\mathbf{K} - \mathbf{K}_r)^H \tag{13.98}$$

$$= \mathbf{P} + \left(\mathbf{K}\mathbf{R}_{\underline{Y}_\omega\underline{Y}_\omega}^{1/2} - \mathbf{K}_r\mathbf{R}_{\underline{Y}_\omega\underline{Y}_\omega}^{1/2}\right)\left(\mathbf{K}\mathbf{R}_{\underline{Y}_\omega\underline{Y}_\omega}^{1/2} - \mathbf{K}_r\mathbf{R}_{\underline{Y}_\omega\underline{Y}_\omega}^{1/2}\right)^H \tag{13.99}$$

$$= \mathbf{P} + \left(\mathbf{U}\Sigma\mathbf{Q}^H - \mathbf{U}\Sigma_r\mathbf{Q}^H\right)\left(\mathbf{U}\Sigma\mathbf{Q}^H - \mathbf{U}\Sigma_r\mathbf{Q}^H\right)^H \tag{13.100}$$

$$= \mathbf{P} + \mathbf{U}\left(\Sigma - \Sigma_r\right)\left(\Sigma - \Sigma_r\right)^H\mathbf{U}^H \tag{13.101}$$

$$= \mathbf{P} + \sum_{k=r+1}^{MP} \lambda_k^2\mathbf{u}_k\mathbf{u}_k^H, \tag{13.102}$$

where $\mathbf{P}$ is the error covariance matrix for the full-rank estimator. The MSE for the rank-r estimator is then

$$\mathrm{MSE}(r) = \mathrm{tr}\mathbf{P}_r = \mathrm{tr}\mathbf{P} + \sum_{k=r+1}^{MP} \lambda_k^2 = \mathrm{MSE}\,(MP) + \sum_{k=r+1}^{MP} \lambda_k^2, \tag{13.103}$$

where MSE(MP) is the MSE of the full-rank LMMSE estimator (13.86), and where the last sum is the extra MSE introduced by the low-rank approximation.

13.6.3.5   *Rank of the RFI*

The minimum rank required for the canceller can be determined by applying the theory of essentially time- and band-limited signals. Landau and Pollak showed that the number of most significant eigenvalues of a time- and band-limited signal equals $2BT + 1$, where B is the one-sided bandwidth of a baseband signal and T is the length of the observation interval [Landau 1962]. Here, B is given by the PSD, $P_{s_b}(f)$, of the RF signal $s_b(t)$.

The bandwidth of the RF signal can be related to the DMT subcarrier spacing as $2B = bf_s/N$, where b is a number, as sketched in Figure 13.13, representing the double-sided bandwidth of $s_b(t)$ relative to the subcarrier spacing. The observation interval, which is considered for RFI cancellation, is one DMT symbol stripped of its cyclic extension, giving $T = N/f_s$. Hence, the effective rank of the RF signal is

$$r_{\text{eff}} = 2BT + 1 = b + 1, \tag{13.104}$$

which indicates that it is possible to reduce the rank of the estimator from MP down to $b+1$ while maintaining most of its performance. However, because DSL typically operates with high SNR and low BER, a higher rank than r_{eff} needs to be used in order not to sacrifice any performance due to RFI. For DSL with receiver windowing, using

$$r_{\text{low}} = r_{\text{eff}} + 2 = b + 3, \tag{13.105}$$

is a pretty good rule of thumb. This results in virtually the same performance for the rank reduced canceller as for the full-rank canceller. Without windowing, this needs to be increased by one. Note that in the above guidelines, the rank also indicates the minimum number of measurement tones needed. To be able to pick up the essential information about the RFI, the measurement, tone placement should be at and around the RFI peak(s).

Figure 13.16 shows the power of the singular values, λ_k, relative to the most significant, λ_1, for some different bandwidths of the RF signal represented by its time-bandwidth product b. This agrees well with Equations 13.104 and 13.105. The solid curves are without receiver windowing, and the corresponding dashed curves are with receiver windowing. The PSD-shape is flat and b subcarriers wide. Note that the subcarrier spacing (here 4.3125 kHz) always corresponds to $b = 1$.

From Figure 13.16 it is also evident that a higher rank needs to be used in case no non-rectangular receiver windowing is used (the solid curves have a "step" around $r = b + 3$). This effect is due to the spectral leakage from negative frequencies into the positive frequencies, which is largely suppressed by using a receiver window. Due to that leakage, the rank needs to be one higher, $r_{\text{low}} = b + 4$, without receiver windowing instead of $r_{\text{low}} = b + 3$ with windowing. This is shown in more detail in Table 13.2. It shows the relative loss of power, $1 - \sum_{k=1}^{r} \lambda_k^2 / \sum_{k=1}^{MP} \lambda_k^2$, for the rank-reduced canceller with and without nonrectangular windowing. It is a measure of how much worse the rank-reduced canceller performs compared to the equivalent full-rank canceller (for example, compare to Equation 13.103).

As the table indicates, the rank-4 canceller without receiver windowing shows the same small relative loss of power as the rank-3 canceller with receiver windowing. The exception for small bandwidths, $b < 1$, is due to the ceiling operation. For $b < 1$, the loss is evident already at $\lceil r = b + 2 \rceil$ (for example, compare with Figure 13.16).

Furthermore, with windowing only, measurement tones around the positive peak, at $f = f_c$, need to be used. Without windowing, the complex conjugated measurement tones around $f = -f_c$ need also to be included. This is in accordance with the Taylor polynomial-based (deterministic) RFI canceller described in the previous section.

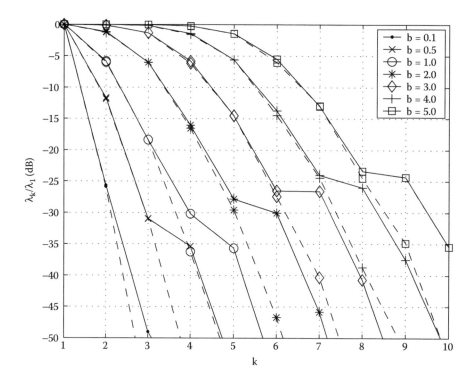

FIGURE 13.16

Relative powers for the singular values, λ_k^2/λ_1^2, for different time-bandwidth products, $b = 2BT$. Solid curves are without windowing, dashed curves with windowing.

13.6.3.6 *Partial RFI Cancellation*

Another way of reducing complexity is to use a lower model order and use fewer measurement tones than the rank of the RFI specifies. This can be accomplished by ignoring suppression of the RFI-peak itself, which is possible within an unused (silent) band, such as a HAM-band. In this case, the few measurement tones should be placed a bit away from the RFI peak, on both sides, and there should be a gap between them where the peak is located [Sjöberg 2004]. Then, provided that the measurement tone placement is correct,[6] good suppression performance can still be obtained on the usable subcarriers outside the silent band, by ignoring the RFI peak in the gap. If, however, the measurement tones are placed too close to the peak, the RFI suppression performance is degraded, because the underdetermined RFI model (either deterministic or stochastic) cannot capture all dimensions of the RFI through the measurement tones in order to make accurate RFI estimates on the used subcarriers. On the other hand, if the measurement tone placement is too far from the RFI peak, they will not pick up sufficient dimensionality needed to derive good RFI estimates.

A price to pay for using a lower model order and fewer measurement tones than the rank specifies is that a strong RFI-peak will remain in the gap between the measurement tones. These tones will therefore become unusable. The desired placement of the measurement tones will also more likely conflict with usable DMT subcarriers (at non-HAM frequencies), because they need to be placed farther away from the RFI peak than if the RFI peak is included in the RFI model. Note that even if the RFI canceller is designed (through the

---

[6] The best distance from the peak depends on the number of subcarriers.

TABLE 13.2

Relative Estimator Power Losses (in dB) for Different Ranks (r), Relative Bandwidths (b), and With or Without Windowing

Relative Power Losses [dB] $1 - \sum_{k=1}^{r}\lambda_k^2 / \sum_{k=1}^{MP}\lambda_k^2$		$b = 0.5$	$b = 1.0$	$b = 2.0$	$b = 3.0$	$b = 4.0$	$b = 5.0$
$r = \lceil b + 3 \rceil$	No windowing	−55.4	−36.7	−33.0	−31.2	−31.7	−31.0
	Windowing	−75.2	−57.1	−49.7	−45.0	−44.5	−41.7
$r = \lceil b + 4 \rceil$	No windowing	−94.5	−57.5	−48.9	−45.4	−43.4	−42.3
	Windowing	−97.2	−70.8	−67.7	−64.8	−57.9	−59.9

selection of an appropriate RFI model) to deal with any DMT subcarrier, estimation and cancellation do not need to be performed on all of them. Subcarriers that correspond to unusable frequency bands, transmission bands, or frequencies where the RFI is much lower than the background noise do not need RFI suppression.

13.6.3.7 Cancellation Complexity

Because the derivation of estimator coefficients can be performed offline, and because they are fixed during receiver operation, the computation burden to calculate them does not need to be incurred by each stand-alone DSL modem. Instead, precalculated RFI coefficients (based on the PSD model) needed for an RFI scenario encountered in a specific cable binder can be downloaded from a server to each DSL receiver, *e.g.*, during their start-up. In this way, the continuous computational complexity required by each receiver in operation can be lowered with the low-rank approximation, as described below.

The rank-r RFI canceller in Equation 13.93 can be formulated as:

$$\widehat{\underline{\mathbf{S}}}_r[0] = \left(\sum_{k=1}^{r} \lambda_k \underline{\mathbf{u}}_k \underline{\mathbf{v}}_k^H \right) \underline{\mathbf{Y}}_\omega = \sum_{k=1}^{r} \mathbf{p}_k \langle \underline{\mathbf{v}}_k, \underline{\mathbf{Y}}_\omega \rangle, \tag{13.106}$$

where $\mathbf{p}_k = \lambda_k \mathbf{u}_k$ and $\mathbf{v}_k$ are vectors of length $\overline{N}$ and MP, respectively. Each inner product, $\langle \mathbf{v}_k, \underline{\mathbf{Y}}_\omega \rangle$, requires MP multiplications, *i.e.*, rMP multiplications in total. The linear combination of r vectors, each of length $\overline{N}$, requires $r\overline{N}$ multiplications. In each symbol, $\overline{N}$ subcarriers are estimated simultaneously, giving

$$\frac{rMP + r\overline{N}}{\overline{N}} = r \left(1 + \frac{MP}{\overline{N}} \right) \tag{13.107}$$

multiplications per subcarrier.

The rank reduction for lowering the complexity is most valuable when more measurement tones are used than the time-bandwidth product specifies. The original estimator in Equation 13.86 requires MP multiplications per subcarrier, resulting in a complexity reduction of $1 - r(1/MP + 1/\overline{N})$ using the low-rank approximation. Table 13.3 shows examples of the complexity reduction using the parameters from the previous subsection and $\overline{N} = N/2 = 2048$ subcarriers.

13.6.4 Nonmodel-Based RFI Cancellation

Nonmodel-based digital RFI cancellation can be achieved in a similar way as the model-based RFI cancellers described for DMT. Figure 13.17 shows the principle of a block-adaptive RFI canceller for DMT working in the frequency domain of the DSL signal. As for the model-based RFI cancellers shown for DMT, this type of nonmodel-based canceller also

TABLE 13.3

Complexity Reductions Using MP Measurement Tones

Reduction of Multiplications (%)		$b = 0.5$	$b = 1.0$	$b = 2.0$	$b = 3.0$	$b = 4.0$	$b = 5.0$
	$MP = 10$	60	60	50	40	30	20
$r = \lceil b + 3 \rceil$	$MP = 20$	80	80	75	70	65	60
	$MP = 40$	90	90	87	85	82	80
	$MP = 10$	50	50	40	30	20	10
$r = \lceil b + 4 \rceil$	$MP = 20$	75	75	70	65	60	55
	$MP = 40$	87	87	85	82	80	77

uses a number of measurement tones $\underline{\mathbf{Y}}_\omega = [Y_{m_1}, \ldots, Y_{m_M}]^T$ to estimate the RFI:

$$\hat{\underline{\mathbf{S}}} = [\hat{S}_0, \ldots, \hat{S}_{\overline{N}-1}]^T = \mathbf{K}\underline{\mathbf{Y}}_\omega. \tag{13.108}$$

An advantage with nonmodel-based cancellers is that little or no a priori information is needed. Instead, they need some adaptation method, such as LMS or RLS, to train the set of filter coefficients for each subcarrier. In Figure 13.17, each set of coefficients is represented by a row in the coefficient filter matrix $\mathbf{K}$ of size $\overline{N} \times M$. During the training period, no data can be transmitted over the line,[7] because instantaneous estimation errors $e_k = \hat{S}_k - (S_k + V_k)$ must be computed on each subcarrier subject to RFI cancellation in order to adapt the coefficient matrix $\mathbf{K}$. The filters will not converge if unknown data is also present on the tones. Note that this is an adaptive block, or a linear combiner, structure in which the input vector does not consist of a tap-delayed sequence of the same signal. Instead, the input is the set of measurement tones, which completely change from symbol to symbol, and the filter coefficients for each cancellation tone are updated independently of the other tones. This means that up to $\overline{N}$ adaptation algorithms are operating in parallel, but all using the same input vector.

13.6.4.1 Convergence Speed

In theory, with an infinite number of iterations, the combiner coefficients of the standard LMS or the RLS algorithm converge to the mean of the optimal Wiener solution, given that the convergence factor for the LMS algorithm, μ, is zero or the forgetting factor, λ, for the RLS algorithm is chosen as one. However, in order to achieve a finite convergence time when measurement noise is present, a non-optimal convergence or forgetting factor must be chosen. Note, however, that it is possible to choose different μ_k (or λ_k) for each tone k subject to RFI cancellation.

With an adaptive combiner, an excess of MSE compared to the minimum MSE occurs [Diniz 1997]. Hence, the cancellation performance is a trade-off between the length of the training period and the degree of optimized performance. The nonmodel-based methods are, therefore, best suited for stationary RF signals (for example, AM-broadcast signals) because a training period is required to suppress the RFI to a level below the background noise. For this type of RFI canceller, the filter coefficients must be trained while the modem is offline and can then be kept fixed during the online modem operation, under the assumption the RF signal properties will not change.

Figure 13.18 shows the learning curves, in terms of residual RFI MSE after cancellation (the errors $|S_k - \hat{S}_k|^2$ averaged over the subcarriers k), for the LMS and the RLS algorithms.

---

[7] It is possible to transmit a known training sequence that is removed in the receiver before the estimation errors $e_0, \ldots, e_{\overline{N}}$ are calculated.

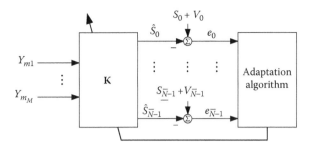

FIGURE 13.17
Adaptive, nonmodel-based, digital RFI canceller for DMT.

Note that one iteration corresponds to one DMT symbol ($N + N_{CE}$ samples). Six measurement tones in a row centered over the RFI peak, plus their complex conjugate twins, and standard rectangular receiver windowing are used. The subcarrier spacing is 4.3125 kHz in this example, and the flat RF signal bandwidth is 5.0 kHz. The average RFI power is 0 dBm/Hz (MSE 0 dB), and the white background noise is set to -40 dBm/Hz. Furthermore, the instantaneous power of the applied RFI is not constant when measured from one DMT symbol to another. Instead, the time-domain baseband RF signal's average amplitude is zero mean, independent (among the DMT symbols), and Gaussian distributed.

This RF signal, which may best represent a radio amateur signal (a single sideband suppressed carrier), would pose quite a challenge to a DMT system that is supposed to operate

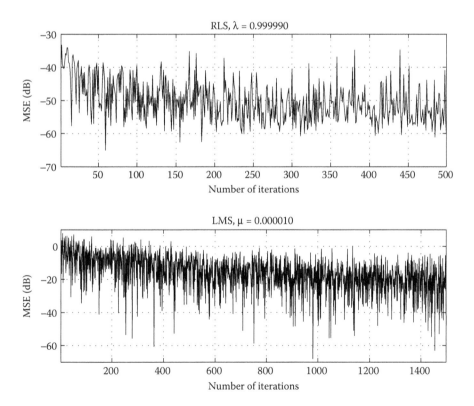

FIGURE 13.18
MSE of cancelled RFI using RLS adaptation (top) with forgetting factor $\lambda = 0.99999$ and LMS adaptation (bottom) with convergence factor $\mu = 0.00001$. The background noise corresponds to an MSE of -40 dB.

virtually error-free in its presence. As seen in Figure 13.18, the LMS algorithm has serious trouble training the estimator coefficients, and the RFI canceller cannot manage to suppress the average MSE below the background noise floor (equivalent to MSE = −40 dB) within 1500 iterations. Selecting a higher convergence factor causes instability problems. However, when a less aggressive RF signal (with less bandwidth and smaller power fluctuations) is applied, the LMS algorithm will be able to more efficiently train the coefficients. The RLS algorithm, on the other hand, has no difficulties training the coefficients, which have converged after fewer than 50 iterations in this example. Note that it performs better from the first iterations, without any tailored algorithm initialization.

Subcarriers located farther from the RF center frequency, f_c, have a slower convergence of their estimator coefficients. This is due to the lower limit, below the background noise, given by the Wiener solution. Therefore, small convergence factors (μ_k) for LMS adaptation, or forgetting factors (λ_k) very close to one for the RLS adaptation are needed on these subcarriers. Nevertheless, subcarriers farther away from f_c will have an increasingly slower convergence also for the case when the same μ or λ is used for all subcarriers. However, the convergence time and steady-state behavior will also largely depend on the bandwidth of the RF signal and its power variations over time. Larger bandwidths and variations in the power result in a slower and more erratic convergence.

13.6.4.2 *Canceller Performance*

Figure 13.19 shows the cancelled RFI PSD averaged over different iteration intervals for the RLS algorithm. This can be compared to the theoretical PSD of the original RFI and the cancelled RFI using the Wiener (LMMSE) solution, the latter resulting in an MSE of −55 dB

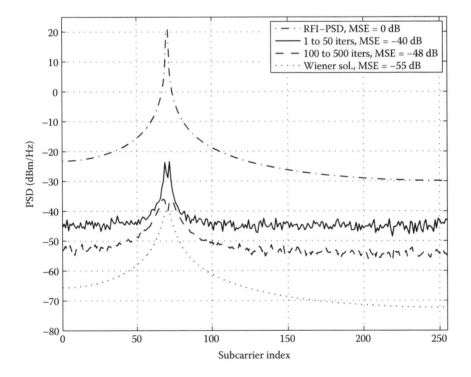

FIGURE 13.19
PSD of cancelled RFI in different iteration intervals using RLS adaptation, forgetting factor $\lambda = 0.99999$. The flat background noise level is at −40 dBm/Hz.

for this example. The PSD of the cancelled RFI for the model-based Taylor canceller in the previous section falls generally between the results of the adaptive RLS and the Wiener solution, but usually closer to the latter.

Although it can take many iterations for the estimation coefficients to converge toward the optimal solution, almost as good performance, in terms of SNR and bit rate, can be reached on many subcarriers with fewer iterations as seen in the example above. This is due to the fact that the other background noise, V_k, dominates on subcarriers farther away from the peak even before the coefficients have converged completely and the RFI is suppressed optimally. However, with finite iterations, a difference (or an SNR loss) between the model-based and adaptive methods can be expected on subcarriers near and around the peak of the RFI where the residual RFI noise of the latter is above, or close, to the other background noise.

13.6.5   Frequency Invariance

Frequency invariance is an interesting complexity-saving property that is shared by all the digital active suppression methods described in this chapter. It follows from their same basic construction with silent measurement tones and is possible due to the properties of the DFT. Frequency invariance applies both to the model-based (in Sections 13.6.1 and 13.6.3) and to the nonmodel-based (in Section 13.6.4) RFI cancellers. It means that if the RF signal changes its frequency location, the RFI estimate coefficients (represented by the matrix **K**), which are derived (either by using a model or by training), do not need to be re-computed. This is possible to achieve provided that a few criteria, which are described below, are met. The technical details showing the frequency invariance for the deterministic and stochastic RFI cancellers can be found in [Sjöberg 2004] and [Nilsson 2003], respectively.

Another complexity advantage with frequency invariance is that RFI from several RF signals can be cancelled independently of each other and by using the same estimator coefficients, **K**. For example, this can be used in a scenario with several amateur radio signals in different HAM bands, or with two sufficiently frequency-separated AM radio stations. To accomplish this, differently placed sets of measurement tones shall be used for differently located RFI peaks. However, the measurement tones within all sets are the same; that is, they use the same number and placing of these tones relative to an RFI peak.

Independent cancellation of different RFIs is possible provided that the frequency locations of the RF signals are sufficiently separated that the spectral leakage from one RFI peak to another is very low (preferably below the other background noise). Then, different sets of measurement tones operating on different RFI peaks do not pick up unmodelled disturbance from other, neighboring RFI peaks. The minimal required frequency distance is highly dependent on the amount of nonrectangular receiver windowing that is applied, the number of subcarriers in the DMT DSL system, and the level of the background noise (compare with Figure 13.8).

The theoretical requirements for obtaining the frequency invariance property are:

1. The RFI signal's frequency shift is an integer number of subcarriers

$$f_{c_1} = f_{c_0} + \frac{k_1}{N} f_s, k_1 \in Z,$$
(13.109)

where f_{c_1} and f_{c_0} represent the new and old RFI center frequencies, respectively. Alternatively, f_{c_0} and f_{c_1} can be interpreted as two (simultaneous) RF signal center frequencies, with f_{c_0} as the center frequency used when deriving **K** (either using an RF model or with training). *Remark:* With a robust RFI canceller design (achieved through a cognizant choice of model parameters), noninteger frequency shifts can also be handled successfully. However, for the nonmodel-based RFI cancellers, which require training, it is a challenge to obtain the robustness needed also to cover noninteger frequency shifts.

2. Sufficient nonrectangular receiver windowing is applied so that RFI spectral leakage from the negative frequency component of the RF signal can be neglected.

   Remark: Specifically, for the deterministic canceller, this means that G_{wind} in Equation 13.55 is used. However, this also means that the frequency invariance property cannot be obtained for RFIs located very close to the lowest (near DC) and highest subcarriers, where spectral leakage occurring from the positive-frequency and negative-frequency RFI peaks inevitably mix at the location of the measurement tones.

3. The measurement tones are circularly shifted to around the new frequency position, f_{c_1}. That is, $\omega_1 = \{(k + k_1) \bmod N \mid k \in \omega_0\}$, where ω_1 and ω_0 are the new and old sets of measurement tone indices, respectively, and mod represents the modulo operator. In other words, the measurement tone placement should always be the same relative to the RFI center frequency.

4. Only measurement tones from one DMT symbol at a time are used.

   Remark: Specifically, for the stochastic model-based canceller in Section 13.6.3, this means that $P = 1$ (with $L_p = L_n = 0$).

With these conditions fulfilled, the estimator-coefficients in $\mathbf{K}$ can remain unchanged. Only a simple circular shift of the RFI estimates is required in order to get them into the right position before they are cancelled from the subcarriers, as in Equation 13.47:

$$\widehat{\mathbf{S}}^1 = \widehat{\mathbf{S}}^0_{((k+k_1)\bmod N)}, \ 0 \le k \le N - 1, \tag{13.110}$$

where the superscripts indicate the corresponding frequency positions for each RFI, located at frequencies f_{c_1} and f_{c_0}, respectively.

To summarize, if a frequency shift of the RF signal is suddenly detected (for example, by monitoring the FFT outputs as a periodogram to locate the RFI peak), the measurement tones should only be shifted to around the new RFI-peak location, as they were at the previous frequency location, and the estimated interference rotated into position as in Equation 13.110 before the cancellation. The same procedure can be used in the case of several simultaneous RFIs (when several separated RFI peaks are detected). No other changes need to be performed in the cancellation procedure.

13.7 Alternative Methods to Suppress RFI

There exist other methods to suppress RFI than direct cancellation. A different method, more similar to a code, spreads the energy for each DMT subsymbol over all subcarriers using orthogonal Hadamard sequences [Gerakoulis 2002] in order to spread the effect of RFI equally over all subcarriers.

Other methods perform joint equalization and RFI suppression [Cuypers 2003, Darsena 2003]. The method in [Cuypers 2003] performs joint so-called per-tone equalization [Van Acker 2001] and windowing. The per-tone equalizer is a type of frequency-domain equalization (FEQ) and maintains the DFT by means of performing a sliding DFT. The method in [Darsena 2003] performs MMSE estimation on the subcarriers by using one DMT symbol, including the cyclic extension. However, this method cannot be implemented by means of the DFT and, hence, its computational complexity is high for systems with many subcarriers.

Another promising method is the so-called generalized DMT (GDMT) (or FEQ-DMT) method, which was developed to perform equalization after the DFT [Trautmann 2002]. This is an alternative to using a TEQ before the DFT to shorten the channel impulse response (see Chapter 11). In many situations, the GDMT method performs better than the combination of a TEQ plus the traditional single-tap FEQ. It also opens the possibility to shorten the CE or even remove it completely. With GDMT, the redundancy of a DMT symbol can be split freely between the time-domain (by using the normal CE) and the frequency-domain (by introducing linear combiners for each tone, which cancel the ISI/ICI). In principle, L CE samples can be exchanged for $L + 1$-tap linear combiners, which equalize each tone separately, where each combiner uses L silent (unused) tones and the tone to be equalized. Due to its similarity to the frequency-domain RFI cancellers, GDMT can achieve excellent RFI suppression. In the case of a CE longer than the channel impulse response, GDMT can be designed to perform essentially the same task as a frequency-domain RFI canceller followed by the common single-tap FEQ. In the case where the CE is shorter than the channel impulse response length, GDMT can jointly perform FEQ and RFI suppression with good results.

13.8 Evaluation of Digital Suppression Methods

This section evaluates the performance of different passive and active RFI suppression methods. The performance metrics are PSD levels of the RFI signals, and bit rates and symbol error rates of DMT DSL systems with and without RFI suppression. Canceller complexities are also compared by examining the number of operations needed to derive the estimator coefficients, *i.e.*, the initial complexity, and the runtime operations needed to derive RFI estimates, *i.e.*, runtime complexity.

13.8.1 Suppression Performance

13.8.1.1 Passive RFI Suppression

Below, the performance of the digital notch filter described in Section 13.5.2 is evaluated. First, its impulse response is derived analytically, and then the time and frequency responses are plotted with different parameters. Also investigated are the effects in the case when the infinite impulse response (IIR) of the filter is truncated to a finite impulse response (FIR).

Each adaptive notch filter, $G_k(z)$, within the multiple notch filter $G(z) = \prod_{k=1}^{Q} G_k(z)$ from Equation 13.34 can be partially expanded to

$$G_k(z) = 1 + \frac{C_k z^{-1}}{1 - \alpha_k e^{-j\omega_k} z^{-1}} + \frac{C_k^* z^{-1}}{1 - \alpha_k e^{j\omega_k} z^{-1}}, \tag{13.111}$$

where C_k is a complex constant (in terms of z) given by

$$C_k = \frac{1 - \alpha_k + \left(\alpha_k^2 - \alpha_k\right) e^{-j2\omega_k}}{\alpha_k (e^{-j\omega_k} - e^{j\omega_k})}. \tag{13.112}$$

The impulse response obtained by the inverse transformation of Equation 13.111 then becomes

$$g_k[n] = \delta[n] + 2\alpha_k^{(n-1)} \text{Re} \left\{ C_k e^{-j\omega_k(n-1)} \right\} u[n-1], \tag{13.113}$$

where $\delta[n]$ is the discrete impulse sequence, $u[n]$ is the unit step sequence, and Re$\{\cdot\}$ denotes the real part. In theory, each notch filter in Equation 13.113 has an IIR, but in practice, the

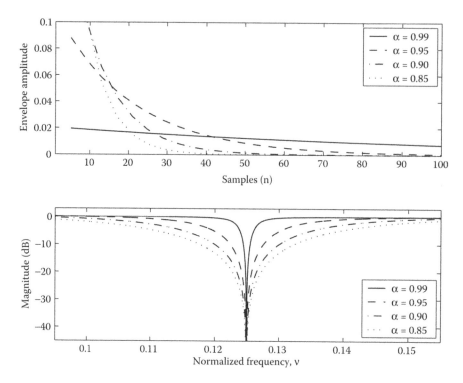

FIGURE 13.20
Time and frequency responses of a notch filter. Top figure: Envelope amplitude of the impulse response. Bottom figure: Normalized frequency response.

parameter α determines the relevant length and characteristics of the impulse response, as well as the bandwidth of the notch. Figure 13.20 shows time and frequency responses of a notch, $g_k[n]$, for different values of α. The upper figure shows the envelope amplitude of the tail of $g_k[n]$ (envelope of $2\alpha_k^{(n-1)}\mathrm{Re}\{C_k e^{-j\omega_k(n-1)}\}u[n-1]$, the part of $g_k[n]$ causing the notch and also ISI) for different values of α. The frequency response, $G_k(e^{j\omega})$, of the IIR notch filter $g_k[n]$ has a zero at $\omega = \omega_k$ (with $\omega = 2\pi v$), as shown in the lower figure in Figure 13.20. The frequency axis is normalized to the sampling frequency $v = 1 \Leftrightarrow f_s$. Note, however, that no ISI is included in this figure, which would be introduced by the IIR of $g_k[n]$. This can be a major concern when using notch filtering for RFI suppression: for a DMT system, it assumes that the cyclic extension is quite long, and for a single-carrier system, it assumes the equalizer is able to equalize the long tail without losing the RFI suppression effect of the notch.

If the filter is truncated into a corresponding FIR filter in order to reduce ISI, the frequency response of the truncated filter limits the depth of the notch. A "natural" truncation may arise due to the use of a TEQ for DMT or a feedforward filter (as part of a decision feedback equalizer) for single-carrier systems to reduce ISI. This is the case particularly when α is close to one, due to the long tail of $g_k[n]$, as seen in Figure 13.20. The frequency response of the truncated impulse response is $\tilde{G}_k(e^{j\omega}) = G_k(e^{j\omega}) \otimes T(e^{j\omega})$, where $\otimes$ denotes the circular convolution over 2π, and where

$$T(e^{j\omega}) = \frac{\sin(\omega(T_w + 1)/2)}{\sin(\omega/2)},$$ (13.114)

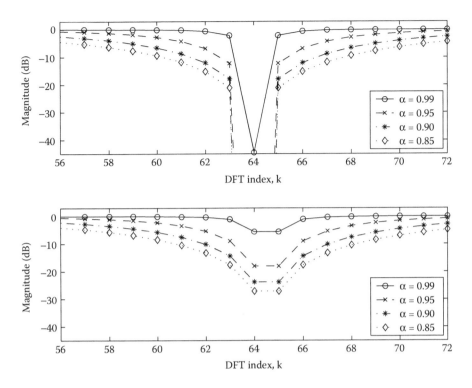

FIGURE 13.21
Truncated frequency responses (of length $T_w = 512$) for a notch filter with notch at ω_k. Top figure: $\omega_k = 2\pi 64/T_w$. Bottom figure: $\omega_k = 2\pi 64.5/T_w$.

is the frequency response of a rectangular truncation "window" $t(n)$

$$\tilde{g}_k(n) = g_k(n)t(n) = \begin{cases} g_k(n), & |n| \leq T_w/2 \\ 0, & |n| > T_w/2, \end{cases} \tag{13.115}$$

where $T_w + 1$ is the length of the window.

The effect of the convolution in Equation 13.114 depends mainly on the zeros of $T(e^{j\omega})$, which are located at $\omega = 2\pi l/T_w$ with $l = 1, \ldots, T_w - 1$. These zeros may or (more likely) may not coincide with the sampled frequency axis by a N-point DFT (the positions of the subcarriers of a DMT system, $\omega = 2\pi k/N$, with $k = 0, \ldots, N - 1$). As a consequence, if the notch frequency, ω_k, coincides with one notch of $T(e^{j\omega})$ and a subcarrier position, $\omega_k = 2\pi l/T_w = 2\pi k/N$, a quite deep notch is maintained, even though the impulse response is truncated, as shown in the top figure of Figure 13.21, which shows the case of $\omega_k = 2\pi 64/N$, using $N = 512$ as for ADSL. Compare this result with Figure 13.20, in which the notch is located at $v_k = 64/512 = 0.125$. However, when the notch frequency ω_k falls between two notches of $T(e^{j\omega})$, much less suppression is achieved. The lower figure in Figure 13.21 shows the effect when $\omega_k = 2\pi 64.5/N$. This effect resembles the spectral leakage in the DFT as a result of truncating the RFI signal due to the block processing by DMT systems.

For DMT DSL systems, the combined (serial) effect of a notch filter for RFI suppression and a TEQ for shortening the overall channel impulse response has not been thoroughly investigated. Placing the notch filter after the TEQ smears out the combined impulse response, thereby losing some effect from the TEQ. On the other hand, placing the notch filter before the TEQ results in a possibly poorer RFI suppression performance, because an optimized TEQ should be designed to make the combined impulse response as short as possible. This is similar to the effect when a TEQ smooths out deep notches of the cable

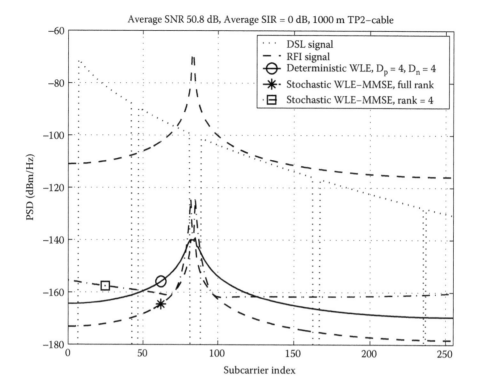

FIGURE 13.22
Performance example when using widely linear estimation (WLE) for RFI cancellation in DMT, without receiver windowing and design mismatch.

impulse response, which may be due to bridge taps, to make the overall channel impulse response shorter.

A similar effect can also occur for single-carrier DSL using a decision feedback equalizer. There, the feedforward filter may have a similar degrading effect on the notch filtering performance as the TEQ for DMT, by shortening the effective length (but also RFI suppression effects) of the notch filter.

13.8.1.2　*Active RFI Cancellation Methods for DMT*

In this section, the two active digital RFI cancellation methods for DMT-based DSL are evaluated. Specifically, characteristics of the RFI PSD before and after cancellation with the different methods are shown, and then the corresponding effect on the bit rate and symbol error rate is considered.

Figures 13.22 and 13.23 show the cancellation performance in terms of RFI PSDs before and after cancellation for the deterministic and stochastic model-based cancellers. The sampling frequency is $f_s = 22\,\mathrm{MHz}$ and the number of subcarriers is $\overline{N} = 256$ and $\overline{N} = 1024$, for Figures 13.22 and 13.23, respectively. The RFI bandwidth is 0.5 subcarriers wide (corresponding to 21.5 kHz and 5.4 kHz, respectively). The RFI center frequency is located at $f_c = 3.6\,\mathrm{MHz}$ and adjusted precisely in between two subcarriers (*i.e.*, it is in the worst-case position). The average RFI power is equal to the average received DSL signal power after the DFT (with a signal-to-interference ratio (SIR) of 0 dB), and the background AWGN floor was set to $-140\,\mathrm{dBm/Hz}$. The model parameters for the cancellers in these two figures are carefully selected to illustrate some "on the edge" characteristics of the cancellation

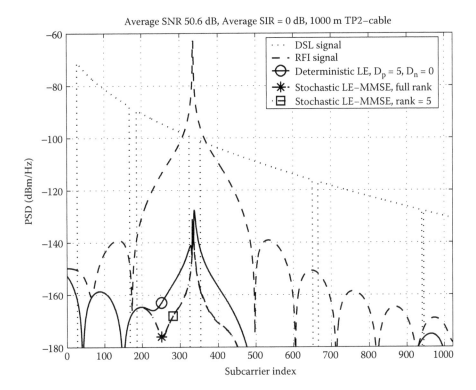

FIGURE 13.23
Performance example when using linear estimation for RFI cancellation in DMT, with receiver windowing and design mismatch.

performance rather than a normal scenario when the cancellers are properly designed to perform flawlessly. In the normal case, however, they all perform so well that their performance can be difficult to distinguish.

In Figure 13.22, there is no model mismatch, and the widely linear estimator (WLE) structure uses in total 14 measurement tones. Half of the measurement tones, Y_k, are placed around the RFI's positive center frequency, $f = f_c$, within the second HAM band (positive frequencies). The other half is selected as their complex conjugates $Y_k^* = Y_{N-k}$ (hence a WLE), which correspond to the negative frequencies around $f = -f_c$. For the deterministic canceller, $L_p = L_n = 4$, and the reduced rank stochastic canceller has rank 4.

For this scenario, the use of rank = 4 is slightly low; the residual peak extends above the AWGN floor on some of the used subcarriers. Reducing the rank further produces worse results, and selecting the rank equal to 5 suppresses the remaining peak to a level under the AWGN floor. In this case, the loss of performance for the "rank = 4" estimator depends on the spectral leakage from the negative frequencies. With nonrectangular receiver windowing, however, the "rank = 4" estimator would suffice to suppress the RFI completely under the AWGN floor similar to a rank-5 canceller without windowing, as shown in Figure 13.16 and Table 13.2.

In Figure 13.23, however, there is a model mismatch of the RFI PSD: a frequency offset $\hat{f}_c - f_c \approx 2.7$ kHz corresponding to 0.25 subcarriers wide (toward the left edge of the HAM band). For the stochastic canceller, the modelled RFI bandwidth is 1.0 subcarriers wide (10.7 kHz). In this example, LE is used with eight measurement tones only on the positive side of the spectrum. The deterministic canceller has $L_p = 5$, $L_n = 0$; *i.e.*, no negative frequencies of the RF signal are included in the model. Windowing is used with $\mu = 20$, corresponding to 1 percent of the DMT symbol length (in samples). Note that the modelled bandwidth for the stochastic canceller is at the limit to completely span the true RFI PSD.

Although the spectral leakage from the negative frequency components is ignored in this example, the RFI suppression still performs quite well. This is due to the windowing, which largely suppresses the leakage from the negative frequency component, $b^*(t)e^{-j2\pi \tilde{f}_c t}$ or $s_b^*(t)e^{-j2\pi \tilde{f}_c t}$ from Equations 13.44 and 13.76, respectively. Without windowing, however, all the LEs would produce much worse results. Here, however, the stochastic "rank = 5" canceller performs as well as the full rank canceller. Reducing the rank to four would cause the canceller to perform slightly worse, which was also the case for the deterministic canceller.

It is possible to achieve excellent suppression results also with very few measurement tones if a narrow residual RFI peak can be accepted. For example, by using only two measuring tones that are placed at a careful distance on each side of the peak, a residual peak remains between the measurement tones. For a system with many subcarriers, such as VDSL, this can be tolerated within a silent HAM band. Figure 13.24 shows the performance for $N = 2048$ subcarriers, with only two measurement tones and using a linear estimator combined with windowing. The windowing uses $\mu = 20$ (extra) cyclic extension samples, which corresponds to less than 0.5 percent of the DMT symbol length. In practise, such a small windowing size would not require any increased cyclic extension. As seen in Figure 13.24, the suppression using each method is virtually identical (all curves are indistinguishable in this scale); there is no SNR degradation on any useful subcarrier. Each method requires only two complex multiplications per subcarrier. Here, "rank = 2" is the same as the full-rank canceller because only two measurement tones are used. For this case,

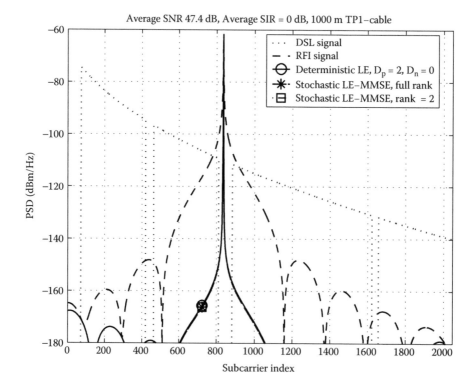

FIGURE 13.24

DMT VDSL system with $\overline{N} = 2048$ subcarriers, subcarrier spacing 4.3125 kHz, 5 kHz RFI bandwidth, and 1 kHz frequency offset.

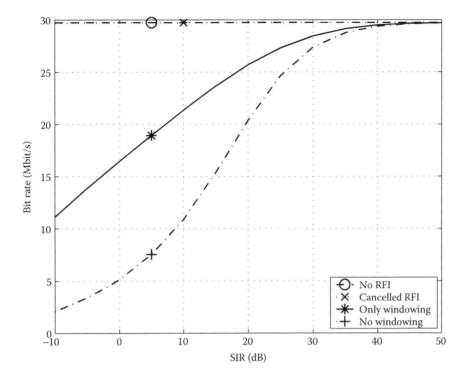

FIGURE 13.25
Bit rates as a function of SIR with and without RFI, windowing, and RFI cancellation.

the stochastic canceller is marginally better than the deterministic canceller, but in practise they are equivalent. Furthermore, due to the windowing, fewer than half of the subcarriers would need RFI cancellation, and it also enables the use of LE instead of using the more complex WLE equivalents.

The bit rates for a DMT VDSL system with properly designed model-based RFI cancellers are shown in Figure 13.25 for different SIR levels. Here, the cancellation was successful in suppressing the RFI to below the background noise. In this evaluation, the bit allocation results in the system maintaining a symbol error rate of less than 10^{-7}. With cancellation, the bit rates are practically identical to the case when no RFI is present.

Figure 13.26 shows the corresponding SERs for the case when the bit allocation is performed without any RFI present. In this mismatched bit allocation scenario, which can occur when an RFI suddenly becomes active or changes characteristics (such as a radio amateur who starts to transmit or change frequency), the effect of windowing is small if no RFI cancellation is performed. Symbol errors are frequent also for quite high SIR, and they are most common on subcarriers close to the RFI center frequency, where windowing has little effect (see Figure 13.8). With cancellation, however, the increased SER can be avoided.

13.8.2 Complexity

This subsection considers the complexity of RFI suppression for the passive and active methods described in earlier sections. An important general difference between the methods described for single-carrier and multi-carrier systems is that RFI suppression methods for single-carrier systems need to be performed at the sampling rate, f_s (typically using time-domain processing), whereas methods for multi-carrier systems often can be performed at the symbol rate, $f_s/(N + N_{CE})$.

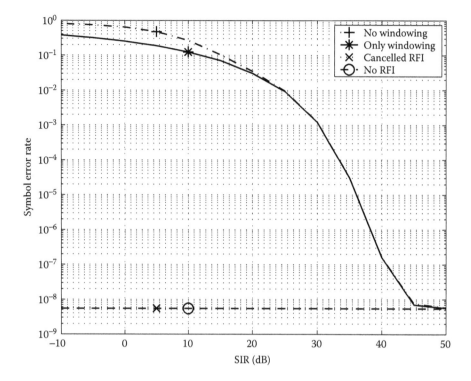

FIGURE 13.26
Symbol error rates when the bit allocation is performed when no RFI is present.

13.8.2.1 Passive methods

Adaptive Notch Filtering Each notch in the filter of $g[n]$, which was described in Section 13.5.2, has two zeros and two poles, which was expressed on the general recursive form in Equation 13.37:

$$e_n = y_n + \delta_n y_{n-1} + y_{n-2} - \alpha_k \delta_n e_{n-1} - \alpha_k^2 e_{n-2}. \tag{13.116}$$

Hence, at each sample interval, $1/f_s$, five real multiplications need to be computed for each notch. In addition, to steer each notch to the right (RFI) location, the adaptation in Equations 13.35, 13.36, and 13.38 requires ten real multiplications per sample interval (by assuming that the computation of the inverse R_n^{-1} takes one multiplication).

Receiver Windowing for DMT A fixed receiver window (for example, with raised cosine shaped tails), of length 2μ samples (using μ extra CE samples) requires only 2μ real multiplications per DMT symbol (during a time interval of $(N + N_{CE})/f_s$ seconds).

13.8.2.2 Active Methods for DMT

The complexity of the active RFI suppression methods is best compared in the two different states: initial complexity for calculating or training the coefficients during an offline period, and runtime complexity for the online cancellation. Still, the initial complexity is of different importance for the model and the nonmodel-based cancellers.

For the model-based cancellers, the initial coefficient computations are best performed offline and remotely (*e.g.*, by a central network coordinator) and then downloaded to the modems. Alternatively, different sets of coefficients can be pre-stored in each modem to represent different RFI scenarios (*e.g.*, the typically RF signal bandwidth). For this reason, the initial complexity is of lower concern than if the coefficients must be computed by each stand-alone modem.

For the nonmodel-based cancellers, the adaption of the coefficients needs to be performed in real time and by each stand-alone modem (at least at the customer side), and it must be performed during an offline period. Hence, the initial adaption complexity for the nonmodel-based cancellers is of higher concern.

Initial Complexity Without going into great detail, this section shows what needs to be calculated in the initialization step in order to derive the RFI canceller coefficients, represented by the matrix $\mathbf{K}$, and considers roughly how many complex multiplications this requires for each method.

Deterministic Model-Based Canceller The initial computations required for this method consist mainly of computing the matrices $\mathbf{G}$ and $\mathbf{G}_\omega^+$ in Equations 13.53 and 13.58, respectively, and then multiplying them together. $\mathbf{G}$ is of size $\overline{N} \times (D_p + D_n)$ where $D_p + D_n$ is the total number of model parameters (terms in the Taylor series). $\mathbf{G}_\omega$ is a subset of $\mathbf{G}$, of size $M \times (D_p + D_n)$ using the rows corresponding to the measurement tones in ω. Hence, the matrix multiplication $\mathbf{K} = \mathbf{G}\mathbf{G}_\omega^+$ requires $\overline{N}M(D_p + D_n)$ multiplications. Because $D_p + D_n \leq M$, this is not more than $O(\overline{N}M^2)$ multiplications, with $M \ll \overline{N}$. To derive $\mathbf{G}$, $(D_p + D_n)$ DFTs, each of length N, need to be computed, which requires $(D_p + D_n)N \log N$ multiplications, where DFT input sequences are of the form $C_l n^l e^{\pm j2\pi n f_c/f_s}$, with a constant C_l. Hence, this is of $O(NM \log N)$ multiplications. Finally, $\mathbf{G}_\omega^+$ is derived by computing a pseudo-inverse, which requires $O(M^3)$ multiplications.

Stochastic Model-Based Canceller For this canceller, the matrix product $\mathbf{K} = \mathbf{R}_{\underline{S}[0]\underline{Y}_\omega} \mathbf{R}_{\underline{Y}_\omega \underline{Y}_\omega}^{-1}$ needs to be computed, which is of $O(\overline{N}M^2)$ multiplications. To derive the inverse $\mathbf{R}_{\underline{Y}_\omega \underline{Y}_\omega}^{-1}$, an additional $O(M^3)$ multiplications are needed. The most straightforward way to find the two matrices that produce $\mathbf{K}$ for the LMMSE canceller is to first compute $\mathbf{R}_{SS}$ in Equation 13.83 and then proceed by selecting the rows and columns that correspond to the measurement tones, ω, to obtain the matrices $\mathbf{R}_{\underline{S}[0]\underline{Y}_\omega}$ and $\mathbf{R}_{\underline{Y}_\omega \underline{Y}_\omega}$. Following this strategy, and by using only one DMT frame ($P = 1$, $L_n = L_p = 0$), the matrix $\mathbf{R}_{SS} = \mathbf{R}_{SS}[0] = \mathbf{F}\mathbf{W}\mathbf{R}_{ss}[0]\mathbf{W}^H\mathbf{F}^H$ from Equation 13.84 must be calculated. Hence, $2N$ DFTs, each of length N, must be computed, resulting in $O(N^2 \log N)$ multiplications. The elements in $\mathbf{R}_{SS}[0]$ can be derived by constructing a Toeplitz matrix from the N-length sequence $r_{S_b}(n/f_s) \cos(2\pi n/f_s)$, where $r_{S_b}(t)$ is the autocorrelation function representing the RFI PSD model $P_{S_b}(f) = \mathcal{F}\{r_{S_b}(t)\}$. In the case of rank reduction, an additional SVD of the matrix $\mathbf{R}_{\underline{S}[0]\underline{Y}_\omega} \mathbf{R}_{\underline{Y}_\omega \underline{Y}_\omega}^{-1/2}$ must be derived, which is of $O(M^3)$ multiplications.

Nonmodel-Based Adaptive Canceller The initialization complexity depends on which adaptation algorithm is used to train $\mathbf{K}$. The LMS algorithm is of $O(M)$ multiplications per subcarrier and iteration, where M represents the number of measurement tones. The RLS algorithm is of $O(M^2)$ multiplications per subcarrier and iteration. Note that the so-called "fast" RLS algorithms, with linear complexity $O(M)$, cannot be applied, because the cancellers linearly combine measurement tones that all change values from one DMT symbol to the next, which is in contrast to a normal filtering scenario where the input is a delayed tap-line sequence.

The conclusion is that among the model-based cancellers, the stochastic LMMSE method is most complex to initialize. Its most demanding computation is of $O(N^2 \log N)$ multiplications, followed by $O(\overline{N}M^2)$, because $M \ll N$ (and M does not typically need to grow as fast as N). For the deterministic Taylor model method, the most demanding computation takes $O(NM \log N)$ multiplications followed by $O(M^3)$. Again, note that the initialization of these model-based cancellers does not have to be performed in real time and not by each modem, because this can be performed by a central network coordinator.

For the nonmodel-based cancellers, the initialization must take place in real time by each modem, but during an offline period, and requires $O(M)$ and $O(M^2)$ multiplications per sample for the LMS and RLS algorithms, respectively.

Runtime Complexity   Once the estimator coefficients have been derived, the runtime cancellation complexity during online operation is similar for all the active digital methods. If $N_c \leq \overline{N}$ tones are subject to active RFI cancellation, MN_c complex multiplications and $2(M-1)N_c$ additions will be needed per DMT symbol (during $(N + N_{CE})/f_s$ seconds) for all the methods. However, if many measurement tones, M, are used and the RF signal bandwidth is low, it is worthwhile to perform rank-reduction of the LMMSE canceller, as described in Section 13.6.3.4, to reduce the runtime complexity to $r(1 + M/N_c)$ multiplications per DMT symbol.

13.9   Summary

This chapter focused on methods for suppressing radio frequency interference (RFI) in DSL. Two major sources of RFI in DSL were identified and characterized: amateur (HAM) radio and AM radio broadcasting. Both of them are relatively narrowband compared to the DSL signal. Compared to the more wideband crosstalk noise, the limited bandwidth of RFI allows effective suppression with low-complexity methods.

Suppression methods for both analog and digital domains were presented and categorized into two classes: active suppression (typically cancellation) and passive suppression (typically time-domain filters). Analog domain RFI suppression is important mainly to avoid saturation of the analog to digital converter (ADC) caused by strong RF ingress. This is crucial because no effective countermeasures can be taken in the digital domain if the ADC has severely clipped the received analog signal. On the other hand, if the received signal passes through the ADC without clipping, powerful suppression techniques can be applied in the digital domain. There it is possible to suppress the RFI to negligible levels, which results in practically no SNR or bit-rate degradation. Although passive suppression methods can achieve good RFI suppression, they have limited performance, and the best results are obtained with active methods specifically designed to cancel the RFI while leaving the information-bearing signal intact.

In the literature available publicly, single-carrier transmission is limited to passive RFI suppression methods, like notch filters. Multi-carrier transmission, on the other hand, can use both passive and active RFI suppression methods and also combine them effectively. An example is passive (nonrectangular) receiver windowing for DMT performed in the digital time domain and combined with frequency-domain RFI cancellation. This combination offers the best RFI suppression performance and the lowest requirements on complexity.

References

[Cioffi 1996] J.M. Cioffi, M. Mallory, and J. Bingham. Analog RF Cancelation with SDMT. Technical Report T1E1.4/96-084, ANSI, 1996.

[Couch 2001] L.W. Couch II. *Digital and Analog Communication Systems*. Prentice-Hall, Upper Saddle River, NJ, 6th edition, 2001.

[Cuypers 2003] G. Cuypers, K. Vanbleu, G. Ysebaert, and M. Moonen. Combining raised cosine windowing and per tone equalization for RFI mitigation in DMT receivers. In *IEEE International Conference on Communications*, 4:2852–2856, May 2003.

[Daecke a) 2000] D. Daecke. The electromagnetic field environment in Europe. Technical Report TM6 TD36 003t36, ETSI, Vienna, Sept. 2000.

[Daecke b) 2000] D. Daecke. Frequency dependence of cable balance. Technical Report TM6 TD37 003t37, ETSI, Vienna, Sept. 2000.

[Daecke c) 2000] D. Daecke. RFI ingress tests for SDSL. Technical Report TM6 TD35 003t35, ETSI, Vienna, Sept. 2000.

[Darsena 2003] D. Darsena, G. Gelli, L. Paura, and F. Verde. Joint equalisation and interference suppression in OFDM systems. *Electronic Letters*, 39(11):873–874, May 2003.

[De Clercq 2000] L. de Clercq, M. Peeters, S. Schelstraete, and T. Pollet. Mitigation of radio interference in xDSL transmission. *IEEE Communications Magazine*, 38(3):168–173, Mar. 2000.

[Diniz 1997] P.S.R. Diniz. *Adaptive Filtering Algorithms and Practical Implementation*. Kluwer Academic, Hingham, MA, 1997.

[Foster 1995] K.T. Foster and J.W. Cook. The radio frequency interference (RFI) environment for high-rate transmission over metallic access wire-pairs. Technical Report TM3 TD29 95xt29, ETSI, Bristol, Apr. 1995.

[Gerakoulis 2002] D. Gerakoulis and P. Salmi. An interference suppressing OFDM system for wireless communications. In *IEEE International Conference on Communications*, volume 1, pages 480–484, 28 April–2 May 2002.

[Haykin 1996] S. Haykin. *Adaptive Filter Theory*. Prentice-Hall, Englewood Cliffs, NJ, 3rd edition, 1996.

[ITU-R 368-7] ITU-R. Ground-wave propagation curves for frequencies between 10 kHz and 30 MHz. *ITU-R Rec. 368-7*, 1992.

[Landau 1962] H.J. Landau and H.O. Pollak. Prolate spheriodal wave functions, Fourier analysis and uncertainty – III: The dimension of the space of essentially time- and band-limited signals. *Bell System Technical Journal*, 41:1295, 1962.

[Magesacher a) 2001] T. Magesacher, S. Haar, R. Zukunft, P. Ödling, T. Nordström, and P.O. Börjesson. Splitting the Recursive Least-Squares Algorithm. In *Proc. of the Int. Symp. on Signal Processing and its Applications*, volume I, pages 319–322, Kuala Lumpur, Malaysia, Aug. 2001.

[Magesacher b) 2001] T. Magesacher, P. Ödling, T. Nordström, T. Lundberg, M. Isaksson, and P.O. Börjesson. An Adaptive Mixed-Signal Narrowband Interference Canceller for Wireline Transmission Systems. In *Proc. of the IEEE International Symposium on Circuits and Systems, ISCAS*, volume IV, pages 450–453, Sydney, Australia, May 2001.

[Naofal 2001] A.-D. Naofal and S.H. Diggavi. Maximum throughput loss of noisy ISI channels due to narrow-band interference. *IEEE Communications Letters*, 5(6):233–235, June 2001.

[Nilsson 2003] R. Nilsson, F. Sjöberg, and J.P. LeBlanc. A rank-reduced LMMSE canceller for narrowband interference suppression in OFDM-based systems. *IEEE Transactions on Communications*, 51(12):2126–2140, Dec. 2003.

[Ödling 2002] P. Ödling, P.O. Börjesson, T. Magesacher, and T. Nordström. An approach to analog mitigation of RFI. *IEEE Journal on Selected Areas in Communications*, 20(5):974–986, June 2002.

[Pazaitis 1998] D.I. Pazaitis, J. Maris, S. Vernalde, M. Engels, and I. Bolsens. Equalisation and Radio Frequency Interference Cancellation in Broadband Twisted Pair Receivers. In *Proc. of the IEEE Global Telecommunications Conference, GLOBECOM*, volume 6, pages 3503–3508, Sydney, Australia, Nov. 1998.

[Picinbono 1996] B. Picinbono. Second-order complex random vectors and normal distributions. *IEEE Transactions on Signal Processing*, 44(10):2637–2640, Oct. 1996.

[Picinbono 1995] B. Picinbono and P. Chevalier. Widely linear estimation with complex data. *IEEE Transactions on Signal Processing*, 43(8):2030–2033, Aug. 1995.

[Redfern 2002] A.J. Redfern. Receiver window design for multicarrier communication systems. *IEEE Journal on Selected Areas in Communications*, 20(5):1029–1036, June 2002.

[Reusens a) 2002] P. Reusens. RFI for DSL: Lists of AM stations, analysis of location and power. Technical Report TM6 TD05 023t05, ETSI, Praha, Sept. 2002.

[Reusens b) 2002] P. Reusens. RFI test for xDSL: Modeling AM carrier precision and modulation. Technical Report TM6 TD10 023t10, ETSI, Praha, Sept. 2002.

[Sands 1999] N.P. Sands, E. Naviasky, W. Evans, M. Mengele, K. Faison, C. Frost, M. Casas, and M. Williams. An integrated analog front-end for VDSL. In *Digest of Technical Papers ISSCC99*, pages 246–247, 1999.

[Scharf 1991] L.L. Scharf. *Statistical Signal Processing: Detection, Estimation, and Time Series Analysis.* Addison-Wesley, Reading, MA, 1991.

[Schreier 2003] P.J. Schreier and L.L. Scharf. Second-order analysis of improper complex random vectors and processes. *IEEE Transactions on Signal Processing*, 51:714–725, Mar. 2003.

[Sjöberg 2004] F. Sjöberg, R. Nilsson, P.O. Börjesson, P. Ödling, B. Wiese, and J.A.C. Bingham. Digital RFI suppression in DMT-based VDSL systems. Accepted for publication IEEE Transactions on Circuits and Systems I, Jan. 2004. In press.

[Sjöberg 1998] F. Sjöberg, R. Nilsson, N. Grip, P.O. Börjesson, S.K. Wilson, and P. Ödling. Digital RFI Suppression in DMT-based VDSL Systems. In *Proc. of the International Conference on Telecommunications, ICT*, volume 2, pages 189–193, Chalkidiki, Greece, June 1998.

[Spruyt 1996] P. Spruyt, P. Reusens, and S. Braet. Performance of improved DMT transceiver for VDSL. Technical Report T1E1.4/96-104, ANSI, Colorado Springs, CO, Apr. 1996.

[Stolle a) 2000] R. Stolle. RFI egress — electric field radiated by a twisted pair cable. Technical Report TM6 TD38 003t38a0, ETSI, Vienna, Sept. 2000.

[Stolle b) 2000] R. Stolle. RFI ingress — coupling of an electromagnetic field into a twisted-pair cable. Technical Report TM6 TD39 003t39a0, ETSI, Vienna, Sept. 2000.

[Trautmann 2002] S. Trautmann and N.J. Fliege. Perfect equalization for DMT systems without guard interval. *IEEE Journal on Selected Areas in Communications*, 20(5):987–996, June 2002.

[Van Acker 2001] K. Van Acker, G. Leus, M. Moonen, O. Van de Wiel, and T. Pollet. Per tone equalization for DMT-based systems. *IEEE Transactions on Communications*, 49(1):109–119, Jan. 2001.

[Vitenberg 2002] R. Vitenberg. Method and Apparatus for RF Common-Mode Noise Rejection in a DSL Receiver. *US patent No. 6459739*, Oct. 1, 2002.

[Widrow 1975] B. Widrow. Adaptive noise cancelling: Principles and applications. *Proceedings of the IEEE*, 63:1692–1716, Dec. 1975.

[Wiese 1997] B. Wiese and J. Bingham. Digital Radio Frequency Cancellation for DMT VDSL. Technical Report T1E1.4/97-460, ANSI, Sacramento, CA, Dec. 1997.

[Yeap 1999] T. Yeap. A Digital Common-Mode Noise Canceller For Twisted-Pair Cable. Technical Report T1E1.4/99-260, ANSI, 1999.

[Yeap 2000] T. Yeap. Adaptive Multiple Sub-Band Common-Mode RFI Suppression. *US patent No. 6052420*, Apr. 18, 2000.

[Yeap 2003] T. Yeap. Suppression of Radio Frequency Interference and Impulse Noise in Communications Channels. *US patent No. 6546057*, Apr. 8, 2003.

Index

μ -law, 27

2B1Q, 128–130

3CXT (Third Circuit Crosstalk), 74, 78, 89

A wire, 35
A-law, 27
ABCD matrix, 115
Activation of DMT modems, 201
Adaptive digital notch filters, 417
Adaptive equalizer, 306, 310, 318, 331,
          332, 335, 336, 349
Adaptive filter, 17
ADSL, 235, 242–245, 253, 256–259,
          265–269, 272, 273, 284, 285, 291,
          294, 295
ADSL technology overview, 131
ADSL2, 191, 234, 246, 259, 265–268, 272
ADSL2, ADSL2plus (overview), 133
ADSL2plus, 191
Advice of Charge, 29
AFE (Receive Analog Front End), 138
AFE (Transmit Analog Front End), 137
AM radio ingress, 402
Amateur ("HAM") radio ingress, 404
ARQ (Automatic Re-Transmission
          Request), 234
ATM, 135
Automatic gain control (AGC), 138, 201,
          334
Automatic Re-Transmission Request
          (ARQ), 234

B wire, 35
Balance about earth, 27
Balance impedance, 16
Band duplexing, 83
Band-edge timing recovery (BETR)
          methods, 359
Barrel effect, 17
Basic-rate ISDN (BRI), 127, 144
Battery, 18

BCH codes, 242, 247–249
BCJR decoding algorithm, 278, 280, 281,
          295, 296
Belief-propagation (BP) algorithm,
          274–278, 281
Bell, Alexander Graham, 2, 34, 35, 37, 38
Bell tap, 24
Berlekamp–Massey algorithm, 248
BGR blind equalization algorithm, 337
Bipartite graph, 273
Bit allocation, 204, 206
Bit distribution, 204
Bit loading, 175
Bit swapping, 133, 207
Blind equalization, 336–338
BPSK, 288
Bridged taps, 39
Burst mode operation, 171
Bus topology, 9

Cable, 37
Cable corrosion, 20
Cable insulation, 37
Cable modem, 123
Cable oxidation, 20
Calling Line Identification, 29
CAP, 128, 136, 138–140, 143, 144,
          159–161, 163, 164, 167, 169, 170,
          172, 322, 327, 338, 353, 355, 360
Capacitive coupling, 90
Capacity (channel capacity), 98, 102–104,
          115, 116
Capacity of additive white Gaussian
          noise channels, 102, 104
Capacity of DMT DSL systems, 115
Capacity of PAM DSL systems, 105, 107,
          108
Capacity of QAM/CAP DSL systems,
          108, 110, 111, 115
Carrier recovery, 139, 157
Carrier serving area (CSA), 129
Causality, 303

Channel analysis performed by DMT
modems, 202
Channel capacity, 98, 102–104, 115, 116
Channel capacity ("Shannon limit"), 104
Channel discovery of DMT modems, 201
Channel eigenfunction, 186
Channel identification performed by
DMT modems, 202
Channel probing, 327, 330
Characteristic impedance, 16
Characteristic impedance (Z_0), 42
Cholesky factorization, 324
CLASS signalling, 29
CLI, 20, 29
Clipping, 197–199
Code weight distribution, 240
Coded modulation, 212
Coding gain, 114, 211, 222, 234, 248,
253–257, 266, 267
Coding methods, 199
Coherent demodulation, 153
Common-mode (CM) signals, 26, 36, 407
Common-mode choke, 407
Concatenated coding, 266
Conditional entropy, 100, 101
Constant modulus algorithm (CMA),
167, 168, 337
Constellation diagram, 146–148, 154,
157, 161–164, 168, 170, 175
Convolutional coding, 212
Convolutional encoder, 223, 224, 226
Convolutional encoding, 278–281, 293
Convolutional interleaver, 259, 262
CRC (Cyclic redundancy check), 234,
238, 241–244, 246–248
CRC polynomials, 241
Crosstalk, 27, 72, 73, 75, 86, 104, 115, 134,
169, 171, 291, 300, 305, 306, 338
Customer premise wiring, 9
Cyclic codes, 238
Cyclic extension (CE), 194, 196, 416
Cyclic prefix, 189, 191–196, 311, 322, 323
Cyclic redundancy check (CRC), 234,
238, 241–244, 246–248
Cyclic suffix, 191, 194–196, 202, 208, 209

Data over Cable Service Interface
Specifications (DOCSIS), 123
Data scrambler, 236
DC signalling, 18
Decision circuits, 155

Decision feedback equalization (DFE),
111, 112, 154, 168, 169, 311–318,
321, 325–327, 339–341, 348, 349
Decision feedback filter (DFF), 313, 315,
321, 327, 329, 340, 341, 343–345
Descrambler, 136
Deterministic RFI canceller, 421
Dialing, 23
Differential encoding, 148, 157, 158
Differential mode, 27
Differential mode signal, 36
Digital duplexing, 193
Digital loop carriers, 3
Digital milliwatt, 14
Digitalization, 4
Digitally duplexed DMT, 194–196
Discrete Fourier transform (DFT), 187,
322, 348
Discrete multi-tone (DMT) modulation,
114, 132, 136, 140, 175, 178, 186,
188, 189, 191–194, 197–202, 204,
205, 207, 209, 272, 286–288, 291,
294, 295, 311, 316, 322, 326, 327,
345, 347–349
Distribution Point, 6
DLC, 3, 144
DMT modem training phase, 201
DMT synchronization, 386
DOCSIS (Data over Cable Service
Interface Specifications), 123
Double-sideband suppressed carrier
(DSB-SC), 405
Dropwires, 6, 7
Dry loops, 20
DSL technology (overview), 125
DSLAM, 126
DTMF, 4, 11, 23, 25
Dual code, 244
Dualtone, 30
Dynamic clip scaling, 199
Dynamic spectrum management (DSM),
135

E1 link, 128
Echo cancellation, 128
Echo canceller, 17, 138, 171
Encoder (convolutional), 218
Encoder (minimal encoder), 218
Encoder (nonsystematic nonrecursive),
218
Encoder (systematic recursive), 218

Encoder (TCM encoder), 223
Entropy (conditional entropy), 100, 101
Entropy (joint entropy), 101
Entropy of an information source, 99, 100, 101
Equalization, 98, 137, 154, 155, 157, 167–169, 171, 182, 184, 186, 191, 193, 195, 202, 207–209, 300–304, 306–316, 318, 319, 321–327, 331, 332, 334–349
Equalization (blind equalization), 167
Equalizer training, 335
Erasures, 265
Error control coding, 235
Error locator polynomial, 248, 250
Error propagation, 238
Error-correction coding (ECC), 284
Etherloop, 144, 171
Ethernet, 135
Ethernet in the First Mile (EFM), 135
ETSI crosstalk model, 76, 77
ETSI NEXT model, 76, 77
Expanded foam cables, 37
Extrinsic information, 280, 282, 283

Far-end Crosstalk (FEXT), 73
Fast Fourier transform (FFT), 186, 322
FEC (Forward error correction), 114, 234, 247, 253
Feedforward filter (FFF), 313, 321, 329, 340, 343–345
FEXT, 73, 75, 77–79, 83, 87, 88, 93–95
FEXT (worst-case FEXT), 76
FEXT transfer functions, 77
Fiber-to-the-home (FTTH), 122
Forney's triangular interleaver, 262
Forward error correction (FEC), 114, 150, 234, 247, 253
Fractionally spaced equalizer, 113, 312, 315, 323, 327, 342–345
Fractionally spaced linear equalizer (FSLE), 154, 168
Frequency-division duplexing (FDD), 133, 134
Frequency-domain equalization (FEQ), 193, 195, 202, 207–209, 321
Frequency-domain equalizer (FEQ), 193
FSAN (Full-service access network), 77, 78, 80, 86–88, 94, 95
FSK modulation, 12, 29, 121

Full-service access network (FSAN), 77, 78, 80, 86–88, 94, 95
Fuses, 28

G.991.2 recommendation, 130, 272
G.992.1 recommendation, 132, 208
G.992.2 recommendation ("G.lite"), 133
G.992.3 recommendation ("ADSL2"), 133, 208, 272
G.994.1 recommendation ("G.hs"), 131
G.shdsl, 130–132, 136, 137
Gallager's decoding algorithm, 277
Galois fields, 235
Gardner timing function, 375
Gatherer/Polley method, 200
Generalized multi-modulus algorithm (GMMA), 168
Generator matrix, 239
Gradient-Descent Timing Recovery, 381
Gray code, 287
Gray mapping, 148–150, 157, 159

HAM radio ingress, 402
Hamming distance, 246
Hamming weight, 240
Handshaking, 127, 170
HDLC, 245
HDSL, 128, 136, 138, 144, 338
HDSL (overview), 128
HDSL2, 131, 136, 137, 144, 338
HDSL2, HDSL4 (overview), 129
HDSL4, 136, 137
Hilbert transformation, 161
HPNA, 8
Hybrid circuit, 13, 137, 171
Hybrid fiber-coax (HFC), 123, 124

ICI (Inter-channel interference), 417
Idle state, 11
Impulse shortening equalizer (ISE), 322, 345, 348, 349
Inductive coupling, 90
Information (measure of information), 98
Information (mutual information), 100
Information metric, 98
Initialization of DMT modems, 200
Insulation (cable), 37
Inter-symbol interference (ISI), 154–156, 161, 163, 169, 184, 189, 191–196, 303–308, 310, 314, 316, 319, 327, 339, 340, 417

Interleaving, 84, 253, 258–262, 267,
        278–280, 286, 287, 292–295
Intermodulation distortion, 197
Intersymbol interference (ISI), 154, 184,
        189, 191–196
Inverse discrete Fourier transform
        (IDFT), 186–189, 191, 192, 199,
        200
ISDN, 8, 27, 127

Jitter, 157
Joint entropy, 101

Key equation, 248

Last mile, 2
Latency, 292
LCL (Longitudinal Conversion Loss), 27,
        401
Least mean-squares algorithm (LMS),
        208, 310, 331–337, 349
LFSR (Linear Feedback Shift Register),
        237, 238
Line interface circuits, 4
Linear Feedback Shift Register (LFSR),
        237, 238
Linear minimum mean square error
        (LMMSE) canceller, 426–429,
        447
Linear minimum mean square error
        (LMMSE) criterion, 425
Load coils, 39
Local exchange battery, 18
Local exchanges, 3
Local loop, 2
Local loop unbundling (LLU), 83
Local multi-point distribution system
        (LMDS), 123
Log-likelihood ratios (LLRs), 275, 276,
        281, 282
Log-MAP algorithm, 281, 283
Longitudinal Conversion Loss (LCL), 27,
        401
Longitudinal mode, 27
Longitudinal signals, 36
Low-density parity-check (LDPC),
        271–273, 275, 279, 280, 285–287,
        289, 292–296

MacWilliams identity, 244
Main Distribution Frames, 5

Matched filter (MF), 304–306, 310, 312,
        327, 344, 345, 348, 349
Max-log-MAP algorithm, 281, 283
Maximum a posteriori (MAP) detection,
        310, 311
Maximum a posteriori (MAP) estimate,
        356
Maximum likelihood (ML), 304, 311, 320,
        321, 348
Maximum likelihood (ML) estimate,
        356
Maximum-likelihood (ML) decoding,
        280
Maximum-likelihood detection,
        156
Maximum-likelihood sequence
        estimation (MLSE), 306, 311,
        319, 321, 327, 339, 348, 349
Maximum-Likelihood Timing Recovery,
        355
MDF, 5, 28
Mean-square error (MSE), 310
Message Waiting Indication, 29
Metallic mode, 27
Metering, 26
Meuller–Müller synchronization
        methods, 376, 379, 381
Minimal encoder, 218
Minimum mean-square error decision
        feedback equalizer
        (MMSE-DFE) , 111, 112, 315,
        316, 321, 327, 328, 340–342
Minimum mean-square error (MMSE),
        304, 306, 308–310, 314, 315, 324,
        326, 327, 329, 340, 346–349
Minimum mean-square error (MMSE)
        criterion, 304
Minimum mean-square linear equalizer
        (MMSE-LE), 312, 340, 342
Minimum squared subset distance
        (MSSD), 214, 215, 217
Morse Samuel, 34
MSSD (Minimum squared subset
        distance), 214, 215, 217
Multi-carrier modulation, 140, 151, 182,
        184–186, 188, 189, 191–195,
        198–202, 205–207, 209
Multi-carrier systems, 182
Multi-modulus algorithm (MMA),
        168
Mutual information, 100

Near-end Crosstalk (NEXT), 73
Near-Shannon-limit coding techniques, 272
Network Interface (NI), 8
Network Interface Device, 8
Network Terminating Equipment, 8
NEXT, 72–79, 83, 88, 89, 93–95, 171, 172
NEXT (simplified NEXT model), 76
NEXT (Unger NEXT model), 76
NEXT (worst-case NEXT), 76
NEXT transfer functions, 77
NI, 8
NID, 8, 28
Noise identification performed by DMT modems, 203
Noise margin, 127, 154, 168–171, 175, 199, 202, 204, 206, 207, 209
Noise predictor (NP), 311, 316–318
Nonlinearly induced spectral line (NISL) methods, 359, 365
Normalized LMS, 334
Notch filter, 417
NTE, 8, 28
Nyquist criterion, 155, 161
Nyquist first criterion, 303

Off-hook state, 11
Offline state, 11
On or off hook detection, 23
On-hook state, 11
Optical network unit (ONU), 134

PAM, 126, 130, 136, 139, 140, 143, 144, 157–159, 162–164, 166, 185, 319, 322, 327, 329–331, 337, 338, 353–355, 360, 361, 370, 381, 385, 387, 393
Parameter exchange between DMT modems, 206
Parity-check, 273, 274, 278, 286, 292
Parity-check polynomials, 220
Passive optical network (PON), 122
PCP, 5
Peak-to-average ratio (PAR), 161–163, 197–200, 209
Phantom circuits, 90
Phase-locked loop (PLL), 352, 353
PIC (Polyolefin Insulated Cables), 37
Pilot tones, 178
Pilot-based timing acquisition and tracking, 391

Plain Old Telephone Service, 2
PLC (Power Line Communications), 125
PMD-frame, 173
Polarity reversal, 20
Polyolefin insulated cables (PIC), 37
Polyvinyl Chloride (PVC), 37
POTS, 2
Power Line Communications (PLC), 125
Power transmission, 46
PRBS (Pseudo-random binary sequence), 236–238
Precoder, 223
Precoding, 137
Primary Connection Point, 5
Primary parameters (of cables), 41
Primitive polynomial, 236
Private metering, 26
Propagation constant, 42
Pseudo-random binary sequence (PRBS), 236–238
Public switched telephone network (PSTN), 2, 3, 120
Pulse dialing, 23
Pupin's technique, 39
PVC (Polyvinyl Chloride), 37

QAM, 123, 136, 138–140, 143–147, 151–155, 157–161, 163–166, 168–172, 185, 186, 188, 198, 204, 284–288, 292–295, 322, 327, 338, 353, 355, 359–362, 387, 393
QAM demodulation, 153
QAM transport capability, 166
QPSK, 124, 198
Quantization, 27
Quiescent state, 11

Radio frequency egress, 169
Radio frequency ingress, 171
Radio-frequency interference (RFI), 36, 93, 197, 400–402, 404, 406–409, 411, 413, 415–417, 420, 421, 425–427, 429, 431–433, 437–439, 441, 442, 445
Receiver windowing, 417
Reduced constellation algorithm (RCA), 167, 168
Reed–Solomon (RS) codes, 233, 234, 242, 246–250, 252–259, 261, 265–269, 272, 294
Reflection coefficient, 46, 53

Resistance of local loop, 19
Return loss, 47
RFI, 93, 133, 197, 400–402, 404, 406–409,
    411, 413, 415–417, 420, 421,
    425–427, 429, 431–433, 437–439,
    441, 442, 445
RFI analog suppression techniques, 407,
    408
RFI cancellation complexity, 433
RFI canceller rank, 431
RFI digital suppression techniques, 406,
    415–417, 420, 421, 425–427, 429,
    431–433, 437–439, 441, 442, 445
RFI ingress calculation, 401
RFI suppression, 406, 415–417, 420, 421,
    425–427, 429, 431–433, 437–439,
    441, 442, 445
Ring, 35
Ring Trip, 22
Ringing, 21
RLCG parameters, 90

Sato blind equalization algorithm, 336,
    337
Scattering parameter s_{21}, 47
Schlaefli lattice, 217
SCM, 143, 144, 146, 148, 161, 164, 170–175
SCPs, 6
Scrambler, 136, 236
SDSL, 128, 130, 338
Secondary connection points, 6
Secondary NEXT, 73
Secondary parameters, 46
SED (Squared Euclidean distance), 213,
    214, 218
Shannon Claude E., 97, 98, 104
Shannon formula, 166
Shannon's communication model, 97
Shannon's equation, 29
SHDSL, 144, 338
Shortening SNR (SSNR), 325, 345–348
Sidetone signal, 16, 17
Single-sideband suppressed carrier
    (SSB-SC), 405
SLIC, 18, 19
Spectral compatibility, 135
Spectrum management (overview), 134
Spectrum shaping filters, 150
Speech, 13
Speech coding, 26
Splices, 20

Splitter, 132
Splitterless ADSL, 133
SPM, 26
Square root raised-cosine filters, 305
Squared Euclidean distance (SED), 213,
    214, 218
Steady state adaptation of DMT
    modems, 206
Steepest descent algorithm, 332, 357
Stochastic RFI canceller, 425
Strowger system, 4
Subchannels, 182, 184–186, 188, 189,
    191–195, 198–202, 205–207, 209
Subscriber Line Interface Circuits, 18
Subscriber private metering, 26
Surge arresters, 28
Symbol-error rate (SER), 291, 292, 294,
    295
Synchronization of DSL Modems, 351,
    386, 391
Syndrome, 248, 249
Systematic block codes, 235
Systematic matrix, 240

T-spaced equalizer, 312, 323, 324, 342–348
T/2-spaced equalizer, 343–348
T1 link, 128
Tanner graph, 273, 274, 280
Taps, 39
TAS, 29
TCM (Trellis-coded modulation),
    211–214, 216–220, 222–225, 230
TCM codes (Optimum TCM codes), 220
TCM encoder, 217, 223
TE, 10
Telegrapher's equations, 42, 90
Telephony system, 2
Tellado's tone injection method, 200
Terminal equipment, 10
Terminal equipment Alerting Signal, 29
Terminating impedance, 16
Testing, 27
Thermal Fusing, 28
Thermal noise, 104
Third Circuit Crosstalk (3CXT), 74, 78, 89
Time-domain equalization (TEQ), 191,
    202, 311, 345–347, 417
Timing advance, 196
Timing recovery, 157
Tinkle, 24
Tip, 35

Toeplitz matrix, 447
Token back frame, 171
Tomlinson–Harashima precoder (THP), 106, 169, 223, 311, 318, 319, 321
Tone dialing, 25
Tong's interleaver, 261
Training sequence generator (TSG), 336
Transmission coefficient, 53
Transmission rate, 101, 102
Transmission standard, 3
Transverse mode, 27
Trellis coding, 114, 130, 133, 136, 139, 150, 193, 211–214, 216–220, 222–225, 230, 272, 273, 285, 295, 318, 321, 348
Trellis coding in ADSL, 225
Trellis decoding, 258
Turbo coding, 271–273, 278–281, 286, 287, 289, 292, 293, 295, 296

Unger NEXT model, 76

V.22 , V.22bis standards, 121
V.32, V.32bis, V.34 standards, 121
V.34bis, 29
V.90, 29
V.90 recommendation, 121

V.92, 29
VDSL, 7, 27, 134, 181, 191, 194, 202, 204, 208, 241, 253, 265, 272
VDSL (Overview), 134
VDSL1, 134, 138–140, 144, 157, 168, 172–175, 188, 191, 208, 241, 263, 265
VDSL2, 241
Viterbi algorithm, 213, 284, 319, 321
Viterbi decoding, 137
Viterbi detector, 258
Voice band modems, 29
Voice over DSL (VoDSL), 135
Voice over IP, 29
Voltage drop, 12, 19

Wei's code, 223
Wetting current, 20, 21
Wiener Estimator, 429
Windowing, 197, 417
Wireless local loop (WLL), 123
Wireless remote access, 123

Zero forcing equalizers (ZFE), 304, 306–308, 310, 315, 327, 348, 349
Zero forcing linear equalizer (ZF-LE), 312, 340
Zipper, 194